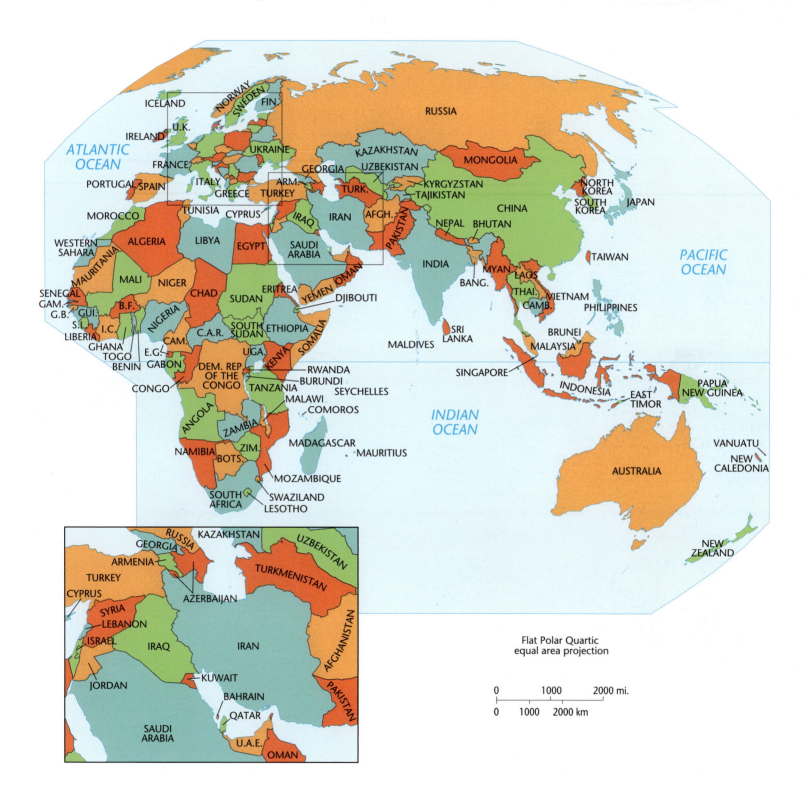

ATLANTIC
OCEAN

PACIFIC
OCEAN

INDIAN
OCEAN

ICELAND
NORWAY
SWEDEN
FIN.
IRELAND
U.K.
UKRAINE
FRANCE
RUSSIA
KAZAKHSTAN
GEORGIA
UZBEKISTAN
MONGOLIA
ARM.
TURKEY
TURK.
KYRGYZSTAN
NORTH
KOREA
JAPAN
PORTUGAL
SPAIN
ITALY
GREECE
IRAQ
IRAN
AFGH.
TAJIKISTAN
SOUTH
KOREA
MOROCCO
TUNISIA
CYPRUS
CHINA
WESTERN
SAHARA
ALGERIA
LIBYA
EGYPT
SAUDI
ARABIA
PAKISTAN
NEPAL
BHUTAN
TAIWAN
MAURITANIA
MALI
NIGER
CHAD
ERITREA
YEMEN
OMAN
INDIA
BANG.
MYAN.
LAOS
SENEGAL
GAM.
G.B.
GUI.
B.F.
SUDAN
DJIBOUTI
THAI.
VIETNAM
PHILIPPINES
S.L.
I.C.
NIGERIA
C.A.R.
SOUTH
SUDAN
ETHIOPIA
SOMALIA
SRI
LANKA
CAMB.
LIBERIA
GHANA
TOGO
E.G.
CAM.
UGA.
KENYA
MALDIVES
BRUNEI
BENIN
GABON
DEM. REP.
OF THE
CONGO
RWANDA
BURUNDI
MALAYSIA
CONGO
TANZANIA
SEYCHELLES
SINGAPORE
INDONESIA
EAST
TIMOR
PAPUA
NEW GUINEA
ANGOLA
MALAWI
COMOROS
ZAMBIA
NAMIBIA
ZIM.
BOTS.
MADAGASCAR
MAURITIUS
VANUATU
NEW
CALEDONIA
SOUTH
AFRICA
SWAZILAND
LESOTHO
MOZAMBIQUE
AUSTRALIA
NEW
ZEALAND

RUSSIA
KAZAKHSTAN
GEORGIA
UZBEKISTAN
ARMENIA
TURKEY
TURKMENISTAN
CYPRUS
AZERBAIJAN
SYRIA
LEBANON
AFGHANISTAN
ISRAEL
IRAQ
IRAN
JORDAN
KUWAIT
PAKISTAN
BAHRAIN
QATAR
SAUDI
ARABIA
U.A.E.
OMAN

Flat Polar Quartic
equal area projection

| 0 | 1000 | 2000 mi. |
| 0 | 1000 | 2000 km |

THE HUMAN MOSAIC

THE HUMAN MOSAIC

A Cultural Approach to Human Geography

Twelfth Edition

Mona Domosh
Dartmouth College

Roderick P. Neumann
Florida International University

Patricia L. Price
Florida International University

Terry G. Jordan-Bychkov
University of Texas at Austin

W. H. Freeman and Company
New York

Executive Editor: Steven Rigolosi
Developmental Editor: Janice Wiggins-Clarke
Director of Marketing: John Britch
Media and Supplements Editor: Kerri Russini
Editorial Assistant: Stephanie Ellis
Project Editor: Jane O'Neill
Text and Cover Designer: Blake Logan
Illustration Coordinator: Janice Donnola
Illustrations: maps.com
Photo Editor: Bianca Moscatelli
Photo Researcher: Christina Micek
Production Coordinator: Julia DeRosa
Composition and Layout: Sheridan Sellers
Manufacturing: RR Donnelley

Library of Congress Control Number: 2011937942
ISBN-13: 978-1-4292-4018-5
ISBN-10: 1-4292-4018-0

W. H. Freeman and Company
41 Madison Ave.
New York, NY 10010
Houndmills, Basingstoke RG21 6XS, England
www.whfreeman.com/geography

CONTENTS IN BRIEF

CONTENTS

Chapter 11

Inside the City

A Cultural Mosaic **385**

Chapter 12

One World or Many?

The Cultural Geography of the Future **431**

PREFACE

Geography is a diverse academic discipline. It concerns place and region and employs diverse methodologies from the social sciences, humanities, and earth sciences. Geographers deal with a wide range of subjects, from spatial patterns of human occupancy to the interaction between people and their environments. The geographer strives for a holistic view of Earth as the home of humankind.

Because the world is in constant flux, geography is an ever-changing discipline. Geographers necessarily consider a wide range of topics and view them from several different perspectives. They continually seek new ways of looking at the inhabited Earth. For example, the rise of feminist perspectives has enabled geographers to see the world anew by pointing out that the spaces we use every day are shaped and used differently because our societies are profoundly structured by gender roles. Another example is the importance of globalization to our world today. This has led geographers to new and incisive engagements with concepts such as transnationalism and postcolonialism. Every revision of an introductory text such as *The Human Mosaic* requires careful attention to such changes and innovations ongoing in the dynamic field that is human geography.

The Five Themes

The Human Mosaic has always been built around **five themes.** The five themes we explore in this book are region, mobility, globalization, nature-culture, and cultural landscape. These five themes are introduced and explained in the first chapter and serve as the framework for the 11 topical chapters that follow. Each theme is applied to a variety of geographical topics: demography, language, ethnicity, politics, religion, agriculture, industry, the city, and types of culture. This thematic organization allows students to relate to the most important aspects of human geography at every point in the text. As instructors, we have found that beginning students learn best when provided with a precise and useful framework, and the five-themes approach provides such a framework for understanding human geography. A small icon accompanies each theme (see icons at right) as a visual reminder to students when these themes recur throughout the book.

In our classroom experience, we have found the thematic framework to be highly successful. Our *region* theme appeals to students' natural curiosity about the differences among places. *Mobility* conveys the dynamic aspect of people and place particularly relevant to this age of incessant and rapid change. Students acquire an appreciation for how

people and cultural traits move (or do not move) from place to place. The topics employed to illustrate the concepts of mobility include many examples to which college students can relate—for instance, reggae and rap music, computer technology and the Internet, and the impact of globalization on consumer goods around the world. *Globalization* permits students to understand the complex processes that link the various economies, cultures, and societies around the world. An understanding of globalizing processes is necessary for explaining how those linkages can create economic and cultural similarities as well as disparities. *Nature-culture* addresses the complicated relationship between culture and

Region

Mobility

Globalization

Nature-Culture

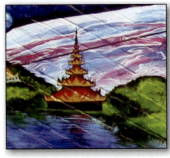

Cultural Landscape

the physical environment. With today's complex and often controversial relationship between the natural environment and our globalizing world, both the tensions and the alliances that arise in regard to this relationship are now at the forefront of this theme. Last, the theme of *cultural landscape* heightens students' awareness of the visible character of places and regions.

New to the Twelfth Edition

In response to instructor and student input from the eleventh edition of *The Human Mosaic*, the new edition includes a set of features designed to make the twelfth edition relevant and accessible to today's students of human geography:

- A **new, visually appealing, modern design** further enhances the themes, pedagogy, and coverage of the book while emphasizing photo and map analysis.

- A new feature—**Author's Notebook**—explores the personal travels and experiences of the authors as they discover and analyze various concepts related to human geography. Locations and topics presented in the Notebooks include:

  Analyzing early-twentieth-century globalization vs. contemporary globalization in Russia

  Conducting research and encountering nature in Tanzania

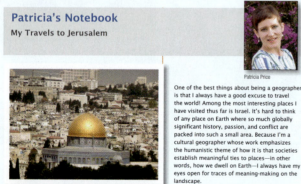

Patricia's Notebook
My Travels to Jerusalem

Patricia Price

One of the best things about being a geographer is that I always have a good excuse to travel the world! Among the most interesting places I have visited thus far is Israel. It's hard to think of any place on Earth where so much globally significant history, passion, and conflict are packed into such a small area. Because I'm a cultural geographer whose work emphasizes the humanistic theme of how it is that societies establish meaningful ties to places—in other words, how we dwell on Earth—I always have my eyes open for traces of meaning-making on the landscape.

In Jerusalem, for instance, you can walk the Via Dolorosa route, visiting all 14 Stations of the Cross walked by Jesus from his condemnation by the Romans to the site of his crucifixion and death. You can see the iconic golden Dome of the Rock on Jerusalem's Temple Mount, which

Panoramic shot of Jerusalem, Israel, site of some of the world's most contested sacred spaces. In the foreground is the iconic Dome of the Rock, sacred to Muslims, which is located on the Temple Mount, site of the destroyed Second Temple and thus also sacred to Jews. *(Courtesy of Patricia L. Price.)*

shelters the rock (hence the name) from which Muhammad is said to have ascended to heaven. Jews also believe the site to be sacred because it is said to be the place where Abraham prepared to sacrifice his son Isaac and is the site of the Western Wall, the remnant of the ancient Jewish temple. Non-Muslims are forbidden from entering the Dome of the Rock for reasons of political conflict and control as well as pollution taboos (which are held by most religions, including Islam). Some Jews voluntarily decline to set foot on the Temple Mount as well, believing it too holy for mere mortal presence.

So much of what we hear and see about this place on our nightly news depicts only the political tensions existing between Israelis and Palestinians. So it is eye-opening to wander the streets of Jerusalem and witness Jews, Christians, and Muslims, orthodox and secular alike, living their daily lives in this city. Yes, there is tension. Mobility and access are often thwarted or at least slowed down, thanks mostly to security concerns. But, as with anyplace else on Earth, most people in Jerusalem are merely going about their business.

I always tell my students that, if given the chance to travel anywhere in the world, they should seriously consider visiting the Middle East. Directly experiencing peoples and places goes a long way toward dispelling the myths and stereotypes about those peoples and places, which are particularly prevalent in this part of the world.

Posted by Patricia Price

Demystifying Sunday crowds in Hong Kong

Exploring places of a genocidal state in Poland

Discovering the globally significant history of Jerusalem

Investigating the importance of place in the global food system as seen in Spain

- The **Subject to Debate** and **Doing Geography** features have been revised to provide more opportunities for critical thinking and further exploration of geographical concepts.

- Several **Seeing Geography** features have been enhanced with new photographs. The authors have added exploration and analysis of these photographs.

- The **Cultural Landscape** sections have been enhanced with additional photographs, and some sections have been completely rewritten and updated. For example, the Cultural Landscape section in Chapter 3, "Population Geography," now includes more images and is more tightly tied to demographic landscapes with new material on landscape adaptations to population declines and increases.

- In addition to completely updated information and data in the text and figures throughout the book, the twelfth edition offers a number of new examples, concepts, and topics, including discussion of the worldwide recession, local indigenous cultures going global as a result of new networking possibilities, religious life in America, the ongoing green revolution, and the latest information from the 2010 Census. All maps reflect the borders of the world's newest nation, South Sudan.

Media and Supplements

The twelfth edition is accompanied by a media and supplements package that facilitates student learning and enhances the teaching experience.

Student Supplements

eBook

The eBook allows instructors and students access to the full textbook online anywhere, anytime. It is also available as a download for use offline. The eBook text is fully searchable and can be annotated with note-taking and highlighting features. Students can copy and paste from the eBook text to augment their own notes, and important sections can be printed. For more information, ask your W. H. Freeman sales representative for an electronic examination copy, or view the ebook on Geography Portal at www.yourgeographyportal.com.

Atlas

Rand McNally's Atlas of World Geography,
ISBN 1-4292-2980-2

The atlas contains 52 physical, political, and thematic maps of the world and continents and 49 regional, physical, political, and thematic maps. It is available packaged with the textbook, with the textbook and *Student Study Guide,* or with the textbook and *Exploring Human Geography with Maps,* second edition.

Mapping Exercise Workbook

Exploring Human Geography with Maps, second edition, by Margaret Pearce, Ohio University, and Owen Dwyer, Indiana University–Purdue University Indianapolis, ISBN 1-4292-2981-0

This full-color workbook introduces students to the diverse world of maps as fundamental tools for exploring and presenting ideas in human geography. It directly addresses the concepts of *The Human Mosaic,* chapter by chapter, and it includes activities accessible through *The Human Mosaic Online* at **www.whfreeman.com/domosh12e**.

The second edition of *Exploring Human Geography with Maps* provides:

- Nine new activities featuring current topics suggested by geography instructors

- Web and text updates for all other activities

- An instructor's guide to help integrate *Exploring Human Geography with Maps* into the curriculum

- Assessment questions for instructors that draw from both *The Human Mosaic* and *Exploring Human Geography with Maps,* further incorporating map-reading skills into the classroom

- PowerPoint slides with maps from the text and information from the instructor's guide to help in lectures

Study Guide

Student Study Guide, by Michael Kukral, Rose-Hulman Institute of Technology, ISBN 1-4292-5352-5

The new and updated *Student Study Guide* provides a significant learning advantage for students using *The Human Mosaic.* This best-selling supplement contains updated practice tests, chapter learning objectives, key terms, and sections on map reading and interpretation. A highly integrated manual, the *Student Study Guide* supports and enhances the material covered in *The Human Mosaic* and guides the student to a clear understanding of human geography.

Companion Web Site

The Human Mosaic Online:
www.whfreeman.com/domosh12e

The companion web site serves as an online study guide. The core of the site includes a range of features that encourage critical thinking and assist in study and review:

- Chapter quizzes to test student understanding and retention of key concepts

- Electronic flashcards for all key terms

- Web activities from *Exploring Human Geography with Maps,* second edition

- Web links to important and informative geographical sites

- Blank outline maps

- MapBuilder instructions and exercises. MapBuilder is a custom mapping tool that allows students to create, display, and print custom thematic maps.

- An audio pronunciation guide for world place names

- More than 200 video clips, each 2 to 6 minutes long with an accompanying five-question quiz. (Note: Videos are a premium resource that must be accessed via the Premium portion of the companion web site or via Geography Portal.)

Instructor Supplements

Videos and Presentation

W. H. Freeman Human/ Cultural Geography 2-DVD Set, ISBN 1-4292-7392-5

This DVD set, available free to adopters of the twelfth edition, contains 25 projection-quality video clips from 3 to 7 minutes in length, each keyed to a specific chapter in *The Human Mosaic.* **An instructor's video guide** is also included on the DVD.

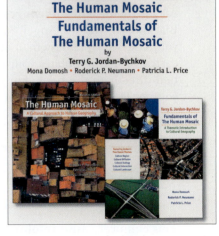

Instructor's Resource CD-ROM and Web Site

- Instructor Resources available at **www.whfreeman.com/domosh12e** (username/password required)

- CD-ROM, ISBN 1-4292-9139-7

Both resources contain the following:

- All the text images available in .JPEG format and as Microsoft PowerPoint slides for use in classroom presentations. Images have been optimized for high-quality projection in large classrooms.

- A set of additional images, organized by theme, for lecture preparation.

- PowerPoint lecture slides, for instructors who prefer a pre-made presentation summarizing each chapter's key concepts, terms, and images.

- A test bank file with approximately 2,000 multiple-choice and true/false questions. Microsoft Word files allow instructors to easily modify questions.

- Clicker questions, one set for each chapter, for use with any personal response system, such as i>clicker.

- The W. H. Freeman Human/Cultural Geography 2-DVD set of 25 projection-quality videos, along with an instructor's video guide.

Assessment: Computerized Test Bank

Expanded and based on the original work of Ray Sumner, Long Beach City College; Jose Javier Lopez, Minnesota State University; Roxane Fridirici, California State University–Sacramento; and Douglas Munski, University of North Dakota, ISBN 1-4292-9137-0.

The computerized test bank, powered by Brownstone, contains approximately 150 to 200 multiple-choice and true/false questions for each chapter in the text. To request the computerized test bank, please contact your W. H. Freeman sales representative.

Course Management

All instructor and student resources are also available via Blackboard, WebCT, Angel, Sakai, and Moodle to enhance your course. W. H. Freeman and Company offers a course cartridge that populates your course web site with content tied directly to the book. For more details, please contact your W. H. Freeman sales representative.

geographyportal

www.yourgeographyportal.com

This comprehensive resource is designed to offer a complete solution for today's classroom. Geography Portal offers all the instructor and student resources available on the book companion web site, as well as premium resources available only on the Portal:

- An eBook of *The Human Mosaic*, twelfth edition, complete and customizable; allows students to quickly search the text and personalize it just as they would the printed version, with highlighting, bookmarking, and note-taking features

- **A Guide to Using Google Earth** for the novice, plus step-by-step **Google Earth** exercises for each chapter

- The full student **Study Guide**, in .pdf format

- Selected articles from *Focus on Geography* magazine (one for each chapter in the textbook) and accompanying quizzes for each article

- More than 200 **videos clips**, each 2 to 6 minutes long, with accompanying five-question quizzes

- A wide variety of **mapping activities**, including drag-and-drop matching activities and MapBuilder (thematic map-building) activities

- Online **news feeds** for highly respected magazines such as *Scientific American* and *The Economist*

- **Gallup WorldWind** polling data links (across world regions)

- Full gradebook functionality and integration

- Professional links and listservs

For more information or to schedule a demo, please contact your W. H. Freeman sales representative.

Acknowledgments

No textbook is ever written single-handedly or even "double-handedly." An introductory text covering a wide range of topics must draw heavily on the research and help of others. In various chapters, we have not hesitated to mention a great many geographers on whose work we have relied. We apologize for any misinterpretations or oversimplifications of their findings that may have resulted because of our own error or the limited space available.

Many geographers contributed advice, comments, ideas, and assistance as this book moved from outline through draft to publication from the first edition through the eleventh. We would like to thank those colleagues who contributed their helpful opinions during the revision of the twelfth edition:

Mathias Le Bossé, Kutztown University of Pennsylvania
Larry E. Brown, University of Missouri, Columbia
Anthony J. Dzik, Shawnee State University
Kari B. Jensen, Hofstra University
Olaf Kuhlke, University of Minnesota, Duluth
Debra J. Taylor Matthews, Boise State University
Blake L. Mayberry, University of Kansas
Vincent Mazeika, Ohio University
Norman Moline, Augustana College
Thomas Orf, Las Positas College
Paul E. Phillips, Ft. Hays (KS) State University
Mary Kimsey Tacy, James Madison University

We would also like to thank those colleagues who offered helpful comments during the preparation of earlier editions:

Jennifer Adams, Pennsylvania State University
Paul C. Adams, University of Texas, Austin
W. Frank Ainsley, University of North Carolina, Wilmington
Christopher Airriess, Ball State University
Nigel Allan, University of California, Davis
Thomas D. Anderson, Bowling Green State University
Timothy G. Anderson, Ohio Wesleyan University
Patrick Ashwood, Hawkeye Community College
Nancy Bain, Ohio University
Timothy Bawden, University of Wisconsin, Eau Claire
Brad A. Bays, Oklahoma State University
A. Steele Becker, University of Nebraska, Kearney
Sarah Bednarz, Texas A&M University
Gigi Berardi, Western Washington University
Daniel Borough, California State University, Los Angeles
Patricia Boudinot, George Mason University
Wayne Brew, Montgomery Community College
Michael J. Broadway, Northern Michigan University
Sarah Osgood Brooks, Central Michigan University
Scott S. Brown, Francis Marion University
Craig S. Campbell, Youngstown State University
Kimberlee J. Chambers, Sonoma State University
Wing H. Cheung, Palomar Community College
Carolyn A. Coulter, Atlantic Cape Community College
Merel J. Cox, Pennsylvania State University, Altoona
Marcelo Cruz, University of Wisconsin, Green Bay
Christina Dando, University of Nebraska, Omaha
Robin E. Datel, California State University, Sacramento
James A. Davis, Brigham Young University
Richard Deal, Western Kentucky University
Jeff R. DeGrave, University of Wisconsin, Eau Claire
Lorraine Dowler, Pennsylvania State University
Christine Drake, Old Dominion University
Owen Dwyer, Indiana University–Purdue University Indianapolis
Matthew Ebiner, El Camino College
James D. Ewing, Florida Community College, Jackson
Maria Grace Fadiman, Florida Atlantic University
David Albert Farmer, Wilmington College
Kim Feigenbaum, Santa Fe Community College
D. J. P. Forth, West Hills Community College
Carolyn Gallaher, American University
Charles R. Gildersleeve, University of Nebraska, Omaha
M. A. Goodman, Grossmont College
Jeffrey J. Gordon, Bowling Green State University
Richard J. Grant, University of Miami
Charles F. Gritzner, South Dakota State University
Sally Gros, University of Oklahoma, Norman
Qian Guo, San Francisco State University
Joshua Hagen, Marshall University
Daniel J. Hammel, University of Toledo
Ellen R. Hansen, Emporia State University
Jennifer Helzer, California State University, Stanislaus
Andy Herod, University of Georgia
Elliot P. Hertzenberg, Wilmington College
Deryck Holdsworth, Pennsylvania State University
Cecelia Hudleson, Foothill College

Ronald Isaac, Ohio University
Gregory Jean, Samford University
Brad Jokisch, Ohio University
Edward L. Kinman, Longwood University
James R. Keese, California Polytechnic State University
Artimus Keiffer, Wittenberg University
Edward L. Kinman, Longwood University
Marti L. Klein, Saddleback College
Vandara Kohli, California State University, Bakersfield
Jennifer Kopf, West Texas A&M
John C. Kostelnick, Illinois State University
Debra Kreitzer, Western Kentucky University
Olaf Kuhlke, University of Minnesota, Duluth
Michael Kukral, Ohio Wesleyan University
Hsiang-te Kung, Memphis University
William G. Laatsch, University of Wisconsin, Green Bay
Paul R. Larson, Southern Utah University
Ann Legreid, Central Missouri State University
Peter Li, Tennessee Technological University
Ronald Lockmann, California State University, Dominiguez Hills
Jose Javier Lopez, Minnesota State University
Michael Madsen, Brigham Young University, Idaho
Jesse O. McKee, University of Southern Mississippi
Wayne McKim, Towson State University
Douglas Meyer, Eastern Illinois University
Klaus Meyer-Arendt, Mississippi State University
John Milbauer, Northeastern State University
Cynthia A. Miller, Syracuse University
Glenn R. Miller, Bridgewater State University
Don Mitchell, Syracuse University
Edris Montalvo, Texas State University, San Marcos
Karen M. Morin, Bucknell University
James Mulvihill, California State University, San Bernardino
Douglas Munski, University of North Dakota
Gareth A. Myers, University of Kansas
David J. Nemeth, University of Toledo
James W. Newton, University of North Carolina, Chapel Hill
Michael G. Noll, Valdosta State University
Ann M. Oberhauser, West Virginia University
Stephen M. O'Connell, Oklahoma State University
Thomas Orf, Prestonburg Community College
Brian Osborne, Queen's University
Kenji Oshiro, Wright State University
Bimal K. Paul, Kansas State University
Eric Prout, Texas A&M University
Darren Purcell, University of Oklahoma
Virginia M. Ragan, Maple Woods Community College
Jeffrey P. Richetto, University of Alabama
Henry O. Robertson, Louisiana State University, Alexandria
Robert Rundstrom, University of Oklahoma
Norman H. Runge, University of Delaware
Stephen Sandlin, California State University, Pomona
Lydia Savage, University of Southern Maine
Steven Schnell, Kutztown University of Pennsylvania
Andrew Schoolmaster III, University of North Texas
Cynthia S. Simmons, Michigan State University
Emily Skop, University of Colorado, Colorado Springs
Christa Smith, Clemson University
Anne K. Soper, Indiana University

Roger W. Stump, State University of New York, Albany
Ray Sumner, Long Beach City College
David A. Tait, Rogers State University
Jonathan Taylor, California State University, Fullerton
Thomas A. Terich, Western Washington University
Thomas M. Tharp, Purdue University
Benjamin F. Timms, California Polytechnic State University
Ralph Triplette, Western Carolina University
Daniel E. Turbeville III, Eastern Washington University
Elisabeth S. Vidon, Indiana University
Ingolf Vogeler, University of Wisconsin
Timothy M. Vowles, Colorado State University
Philip Wagner, Simon Fraser State University
Barney Warf, Florida State University
Henry Way, University of Kansas
Barbara Weightman, California State University, Fullerton
John Western, Syracuse University
W. Michael Wheeler, Southwestern Oklahoma State University
David Wilkins, University of Utah
Douglas Wilms, East Carolina State University
Donald Zeigler, Old Dominion University

Our thanks also go to various staff members of W. H. Freeman and Company, whose encouragement, skills, and suggestions have created a special working environment and to whom we express our deepest gratitude. In particular, we thank Steven Rigolosi, executive editor for geography; Janice Wiggins-Clarke, developmental editor par excellence; John Britch, director of marketing; Kerri Russini, media and supplements editor; Jane O'Neill, project editor; Blake Logan, interior and cover designer; Sheridan Sellers, page makeup designer; Janice Donnola, illustration coordinator; Bianca Moscatelli, photo editor; Julia DeRosa, production manager; Philip McCaffrey, managing editor; and Ellen Cash, vice president of production. In addition, we'd like to thank several others whose fine work helped ensure the overall quality of the book: Christina Micek, photo researcher; Karen Osborne, copyeditor; Tina Hastings, proofreader; and Ellen Brennan, indexer. The beneficial influence of all these people can be detected throughout the book.

ABOUT THE AUTHORS

Mona Domosh is professor of geography at Dartmouth College. She earned her PhD at Clark University. Her research has examined the links between gender ideologies and the cultural formation of large American cities in the nineteenth century, particularly in regard to such critical but vexing distinctions as consumption/production, public/private, masculine/feminine. She is currently engaged in research that takes the ideological association of women, femininity, and space in a more postcolonial direction by asking what roles nineteenth-century ideas of femininity, masculinity, consumption, and "whiteness" played in the crucial shift from American nation-building to empire-building. Domosh is the author of *American Commodities in an Age of Empire* (2006); *Invented Cities: The Creation of Landscape in 19th-Century New York and Boston* (1996); the coauthor, with Joni Seager, of *Putting Women in Place: Feminist Geographers Make Sense of the World* (2001); and the coeditor of *Handbook of Cultural Geography* (2002).

Roderick P. Neumann is professor of geography and chair of the Department of Global and Sociocultural Studies in the School of International and Public Affairs at Florida International University. He earned his PhD at the University of California, Berkeley. He studies the complex interactions of culture and nature through a specific focus on national parks and natural resources. In his research, he combines the analytical tools of cultural and political ecology with landscape studies. He has pursued these investigations through historical and ethnographic research mostly in East Africa, with comparative work in North America and England. His current research explores interwoven narratives of nature, landscape, and identity in the European Union, with a particular emphasis on Spain. His scholarly books include *Imposing Wilderness: Struggles over Livelihoods and Nature Preservation in Africa* (1998), *Making Political Ecology* (2005), and *The Commercialization of Non-Timber Forest Products* (2000), the latter coauthored with Eric Hirsch.

Patricia L. Price is associate professor of geography at Florida International University. She earned her PhD at the University of Washington. Connecting the long-standing theme of humanistic scholarship in geography to more recent critical approaches best describes her ongoing intellectual project. From her initial field research in urban Mexico, she has extended her focus to the border between Mexico and the United States and, most recently, to south Florida as a borderland of sorts. Her most recent field research is on comparative ethnic neighborhoods, conducted with colleagues and graduate students in Phoenix, Chicago, and Miami, and funded by the National Science Foundation. Price is the author of *Dry Place: Landscapes of Belonging and Exclusion* (2004) and coeditor (with Tim Oakes) of *The Cultural Geography Reader* (2008).

Terry G. Jordan-Bychkov was the Walter Prescott Webb Professor in the Department of Geography at the University of Texas, Austin. He earned his PhD at the University of Wisconsin, Madison. A specialist in the cultural and historical geography of the United States, Jordan-Bychkov was particularly interested in the diffusion of Old World culture in North America that helped produce the vivid geographical mosaic evident today. He served as president of the Association of American Geographers in 1987 and 1988 and received an Honors Award from that organization. He wrote on a wide range of American cultural topics, including forest colonization, cattle ranching, folk architecture, and ethnicity. His scholarly books include *The European Culture Area: A Systematic Geography*, fourth edition (with Bella Bychkova Jordan, 2002), *Anglo-Celtic Australia: Colonial Immigration and Cultural Regionalism* (with Alyson L. Grenier, 2002), *Siberian Village: Land and Life in the Sakha Republic* (with Bella Bychkova Jordan, 2001), *The Mountain West: Interpreting the Folk Landscape* (with Jon Kilpinen and Charles Gritzner, 1997), *North American Cattle Ranching Frontiers* (1993), *The American Backwoods Frontier* (with Matt Kaups, 1989), *American Log Building* (1985), *Texas Graveyards* (1982), *Trails to Texas: Southern Roots of Western Cattle Ranching* (1981), and *German Seed in Texas Soil* (1966).

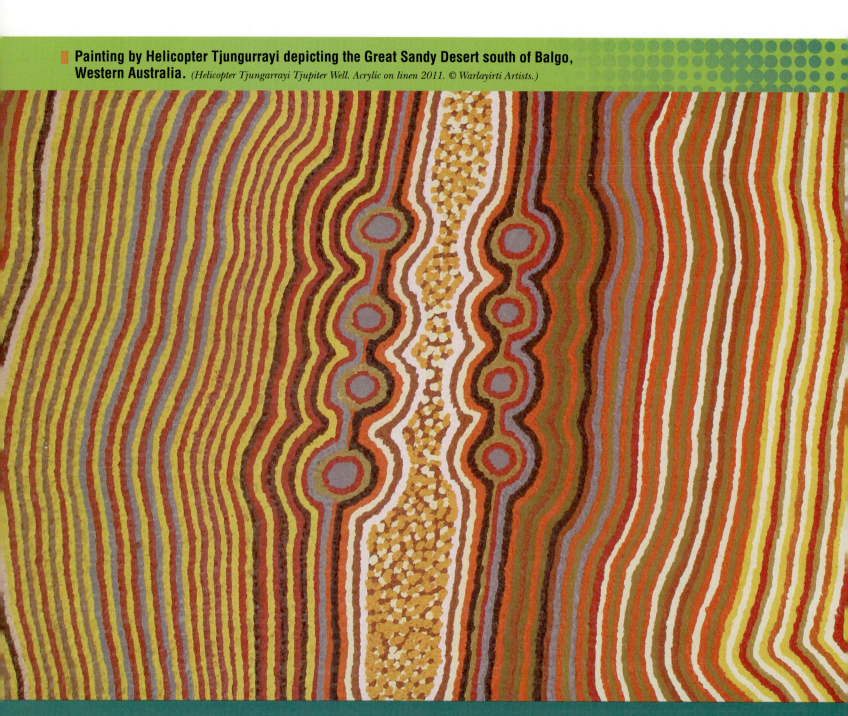

Why is it difficult for most of us to interpret this image as a map?

Go to "Seeing Geography" on page 28 to learn more about this image.

HUMAN GEOGRAPHY
A Cultural Approach

Most of us are born geographers. We are curious about the distinctive character of places and peoples. W vink in terms of territory and space. Take a look outside your window right now. The houses and commercial buildings, streets and highways, gardens and lawns all tell us something interesting and profound about who we are as a culture. If you travel down the road, or on a jet to another region or country, that view outside your window will change, sometimes subtly, sometimes drastically. Our geographical imaginations will push us to look and think and begin to make sense of what is going on in these different places, environments, and landscapes. It is this curiosity about the world—about how and why it is structured the way it is, what it means, and how we have changed it and continue to change it— that is at the heart of human geography. You are already geographers; we hope that our book will make you better ones.

If all places on Earth were identical, we would not need geography, but each is unique. Every place, however, does share characteristics with other places.

region
A grouping of like places or the functional union of places to form a spatial unit.

Geographers define the concept of **region** as a grouping of similar places or of places with similar characteristics. The existence of different regions endows the Earth's surface with a mosaiclike quality. Geography as an academic discipline is an outgrowth of both our curiosity about lands and peoples other than our own and our need to come to grips with the place-centered element within our souls. When professional, academic geographers consider the differences and similarities among places, they want to understand what they see. They first find out exactly what variations exist among regions and places by describing them as precisely as possible. Then they try to decide what forces made these areas different or alike. Geographers ask *what? where? why?* and *how?*

Our natural geographical curiosity and intrinsic need for identity were long ago reinforced by pragmatism, the practical motives of traders and empire builders who wanted information about the world for the purposes of commerce and conquest. This concern for the practical aspects of geography first arose thousands of years ago among the ancient Greeks, Romans, Mesopotamians, and Phoenicians, the greatest traders and empire builders of their times. They cataloged factual information about locations, places, and products. Indeed, **geography** is a Greek word meaning literally "to describe the Earth." Not content merely to chart

geography
The study of spatial patterns and of differences and similarities from one place to another in environment and culture.

and describe the world, these ancient geographers soon began to ask questions about why cultures and environments differ from place to place, initiating the study of what today we call geography.

What Is a Cultural Approach to Human Geography?

Human geography forms one part of the discipline of geography, complementing physical geography (which deals with the natural environment). Human geography examines the relationships between people and the places and spaces in which they live using a variety of scales ranging from the local to the global. Human geographers explore how these relationships create the diverse spatial arrangements that we see around us, arrangements that include homes, neighborhoods, cities, nations, and regions. A cultural approach to the study of human geography implies an emphasis on the meanings, values, attitudes, and beliefs that different groups of people around the world lend to and derive from places and spaces. To understand the scope of a cultural approach to human geography, we must first discuss the various meanings of **culture.**

There are many definitions of culture, some broad and some narrow. For the purposes of this book, we define *culture* as learned, collective human behavior, as opposed to innate, or inborn, behavior. Learned similarities in speech, behavior, ideology, livelihood, technology, value systems, and society form a way of life common to a group of people. *Culture,* defined in this way, involves a means of communicating these learned beliefs, memories, perceptions, traditions, and attitudes that serves to shape behavior. As geographers, we tend to be interested in how these various aspects of culture take shape in particular places, environments, and landscapes.

A particular culture is not a static, fixed phenomenon, and it does not always govern its members. Rather, as geographers Kay Anderson and Fay Gale put it, "culture is a *process* in which people are actively engaged." Individual members can and do change a culture, which means that ways of life constantly change and that tensions between opposing views are usually present. Cultures are never internally homogeneous because individual humans never think or behave in exactly the same manner.

A cultural approach to human geography, then, studies the relationships among space, place, environment, and culture. It examines the ways in which culture is expressed and symbolized in the landscapes we see around us, including homes, commercial buildings, roads, agricultural patterns, gardens, and parks. It analyzes the ways in which language, religion, the economy, government, and other cultural phenomena vary or remain constant from one place to another and provides a perspective for understanding how people function spatially and identify with place and region (**Figure 1.1**).

In seeking explanations for cultural diversity and place identity, geographers consider a wide array of factors that cause this diversity. Some of these involve the **physical environment:** terrain, climate, natural vegetation, wildlife, variations in soil,

human geography
The study of the relationships between people and the places and spaces in which they live.

culture
A total way of life held in common by a group of people, including such learned features as speech, ideology, behavior, livelihood, technology, and government; or the local, customary way of doing things—a way of life; an ever-changing process in which a group is actively engaged; a dynamic mix of symbols, beliefs, speech, and practices.

physical environment
All aspects of the natural physical surroundings, such as climate, terrain, soils, vegetation, and wildlife.

FIGURE 1.1 Two traditional houses of worship. Geographers seek to learn how and why cultures differ, or are similar, from one place to another. Often the differences and similarities have a visual expression. *In what ways are these two structures—one a Catholic church in Honduras and the other a Buddhist temple in Laos—alike and different?* (Left: Rob Crandall/Stock Connection/Alamy; Right: Peter Adams/Alamy.)

and the pattern of land and water. Because we cannot understand a culture removed from its physical setting, human geography offers not only a spatial perspective but also an ecological one.

Many complex forces are at work on cultural phenomena, and all of them are interconnected in very complicated ways. The complexity of the forces that affect culture can be illustrated by an example drawn from agricultural geography: the distribution of wheat cultivation in the world. If you look at **Figure 1.2**, you can see important areas of wheat cultivation in Australia but not in Africa, in the United States but not in Chile, in China but not in Southeast Asia. Why does this spatial pattern exist? Partly it results from environmental factors such as climate, terrain, and soils. Some regions have always been too dry for wheat cultivation. The land in others is too steep or infertile. Indeed, there is a strong correlation between wheat cultivation and midlatitude climates, level terrain, and good soil.

Still, we should not place exclusive importance on such physical factors. People can modify the effects of climate through irrigation; the use of hothouses; or the development of new, specialized strains of wheat. They can conquer slopes by terracing, and they can make poor soils

productive with fertilization. For example, farmers in mountainous parts of Greece traditionally wrested an annual harvest of wheat from tiny terraced plots where soil had been trapped behind hand-built stone retaining walls. Even in the United States, environmental factors alone cannot explain the curious fact that major wheat cultivation is concentrated in the semiarid Great Plains, some distance from states such as Ohio and Illinois, where the climate for growing wheat is better. The human geographer knows that wheat has to survive in a cultural environment as well as a physical one.

Ultimately, agricultural patterns cannot be explained by the characteristics of the land and climate alone. Many factors complicate the distribution of wheat, including people's tastes and traditions. Food preferences and taboos, often backed by religious beliefs, strongly influence the choice of crops to plant. Where wheat bread is preferred, people are willing to put great efforts into overcoming physical surroundings hostile to growing wheat. They have even created new strains of wheat, thereby decreasing the environment's influence on the distribution of wheat cultivation. Other factors, such as public policies, can also encourage or discourage wheat cultivation. For example,

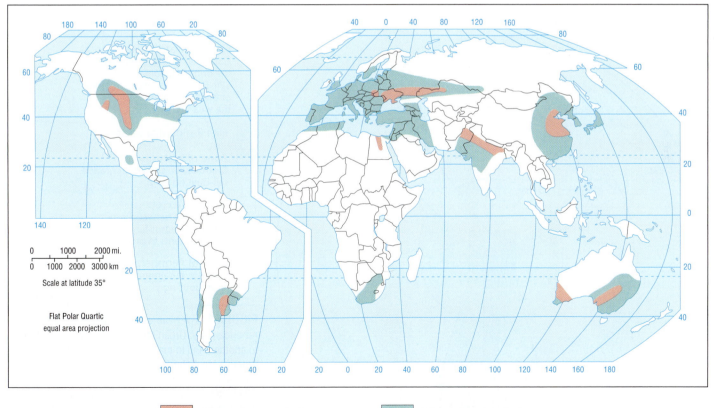

Major wheat-producing areas   Minor wheat-producing areas

FIGURE 1.2 Areas of wheat production in the world today. These regions are based on a single trait: the importance of wheat in the agricultural system. This map tells us what and where. It raises the question of why. *What causal forces might be at work to produce this geographical distribution of wheat farming?*

tariffs protect the wheat farmers of France and other European countries from competition with more efficient American and Canadian producers.

This is by no means a complete list of the forces that affect the geographical distribution of wheat cultivation. The distribution of all cultural elements is a result of the constant interplay of diverse factors. Human geography is the discipline that seeks such explanations and understandings.

How to Understand Human Geography

Generally speaking, geographers have taken three different perspectives in studying and understanding the complexity of the human mosaic. Each of these perspectives brings a different emphasis to studying the diversity of human patterns on the Earth.

Spatial Models Some geographers seek patterns and regularities amid the complexity and apply the scientific method to the study of people. Emulating physicists and chemists, they devise theories and seek regularities or universal spatial principles that apply across cultural lines, explaining all of humankind. These principles ideally become the basis for laws of human spatial behavior. **Space**—a term that refers to an abstract location on a map—is the word that perhaps best

connotes this approach to cultural geography (see Doing Geography, pages 26–27).

Social scientists face a difficult problem because, unlike physical scientists, they cannot limit the effects of diverse factors by running experiments in controlled laboratories. One solution to this problem is the technique known as **model** building. Aware that many causal forces are involved in the real world, they set up artificial situations to focus on one or more potential factors. Torsten Hägerstrand's diagrams of different ways in which ideas and people move from one place to another are examples of spatial models. Some model-building geographers devise culture-specific models to describe and explain certain facets of spatial behavior within specific cultures. They still seek regularities and spatial principles but within the bounds of individual cultures. For example, several geographers proposed a model for Latin American cities in an effort to stress similarities among them and to understand why cities are formed the way they are (**Figure 1.3**). Obviously, no actual city in Latin America

space
A term used to connote the objective, quantitative, theoretical, model-based, economics-oriented type of geography that seeks to understand spatial systems and networks through application of the principles of social science.

model
An abstraction, an imaginary situation, proposed by geographers to simulate laboratory conditions so that they can isolate certain causal forces for detailed study.

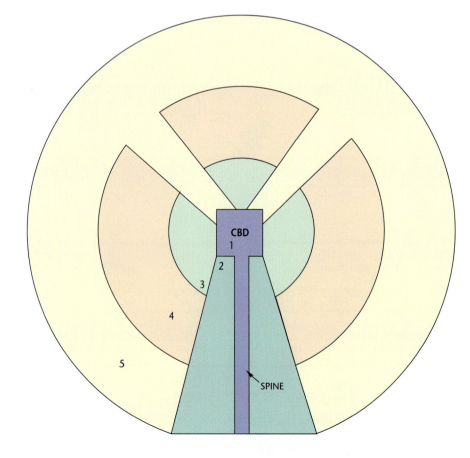

Commercial/Industrial Areas
CBD = Central business district, the original colonial city
SPINE = High-quality expansion of the CBD, catering to the wealthy

Elite Residential Sector

Zone of Maturity
Gradually improved, upgraded, self-built housing

Zone of Accretion
Transitional between zones 3 and 5, modest housing, improvements in progress

Zone of Peripheral
Squatter settlements
Slum housing

FIGURE 1.3 A generalized model of the Latin American city. Urban structure differs from one culture to another, and in many ways the cities of Latin America are distinctive, sharing much in common with one another. Geographers Ernst Griffin and Larry Ford developed the model diagrammed here to help describe and explain the processes at work shaping the cities of Latin America. *In what ways would this model not be applicable to cities in the United States and Canada?* (After Griffin and Ford, 1980: 406.)

conforms precisely to their uncomplicated geometric plan. Instead, they deliberately generalized and simplified so that an urban type could be recognized and studied. The model will look strange to a person living in a city in the United States or Canada, for it describes a very different kind of urban environment, based in another culture.

Sense of Place   Other geographers seek to understand the uniqueness of each region and place. Just as *space* identifies the perspective of the model-building geographer, **place** is the key word connoting this more humanistic view of geography. The geographer Yi-Fu Tuan coined the word *topophilia*, literally "love of place," to describe the characteristic of people who exhibit a strong sense of place and the geographers who are attracted to the study of such places and peoples. Geographer Edward Relph tells us that "to be human is to have and know your place" in the geographical sense. This perspective on cultural geography values subjective experience over objective scientific observation.

It focuses on understanding the complexity of different cultures and how those cultures give meaning to and derive meaning from particular places. For example, many geographers are interested in understanding how and why certain places continue to evoke strong emotions in people, even though those people may have little direct connection with those places. Denis Cosgrove (see Practicing Geography below) has studied why Venice continues to stir people's imaginations, people as diverse as tourists from Japan and farmers from Iowa, despite the facts that the city hasn't held any political or economic power in hundreds of years and that the cultures out of which it was formed have long since ceased to exist. However, some geographers are interested in the opposite kind of places—ordinary places—and ask how and why people become attached to and derive

> **place**
> A term used to connote the subjective, idiographic, humanistic, culturally oriented type of geography that seeks to understand the unique character of individual regions and places, rejecting the principles of science as flawed and unknowingly biased.

PRACTICING GEOGRAPHY

When cultural geographer Denis Cosgrove rode the bus through West Los Angeles on his way to work at UCLA, or when he took his Sunday walks through the greenspaces of London, as he did often in the summer months on his visits to his former hometown, he was "practicing" geography. As he said, "the world/landscape around me is a primary source of questions . . . life is a field course."

Cosgrove, who was the Humboldt Chair of Geography at UCLA and one of the most prominent cultural geographers in the English-speaking world, explored, through a series of scholarly articles and books (for some of these, see Ten Recommended Books on a Cultural Approach to Human Geography at the end of this chapter), the relationships between landscape and culture in Renaissance Italy, nineteenth-century England, and twentieth-century America. His fascination with these places and times, and his enthusiasm for the study of geography in general, began early, when he was a child in Liverpool, England. His walks were important then, too. "Being raised in a great port city where our Sunday walks were often along the docks, seeing great cargo ships with words like Montevideo and Cape Town and Lagos on their sterns . . . and being given a globe at the age of eight and seeing dot-

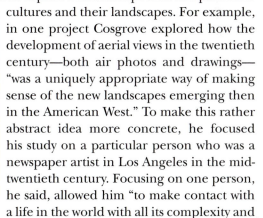

(Courtesy of Denis E. Cosgrove.)

**Denis Cosgrove
(1948–2008)**

ted lines crossing the oceans to these same places with 'Distance to Liverpool' printed on these steamship routes . . . made me realize that I lived in a place that mattered on the globe."

As the years went by, Cosgrove found himself working as much inside as out, often in archives, looking at historical documents and closely reading maps and other images that express relationships between particular cultures and their landscapes. For example, in one project Cosgrove explored how the development of aerial views in the twentieth century—both air photos and drawings—"was a uniquely appropriate way of making sense of the new landscapes emerging then in the American West." To make this rather abstract idea more concrete, he focused his study on a particular person who was a newspaper artist in Los Angeles in the mid-twentieth century. Focusing on one person, he said, allowed him "to make contact with a life in the world with all its complexity and use it to make more general points about how geographies come about." His approach was based more in the humanities than in the social sciences, interpretative instead of explanatory. And, above all, it involved "a great respect for the role of imagination (my own and others') in the ways that we shape the world."

meaning from their local neighborhoods or communities and how those meanings can often come into conflict with each other. Many of the debates that you see in newspapers and hear about on the evening news—debates about the construction of a new high-rise building or the location of a highway, for example—can only be understood by examining the meanings and values different groups of people give to and derive from particular places.

Power and Ideology Cultures are rarely, if ever, homogeneous. Often certain groups of people have more power in society, and their beliefs and ways of life dominate and are considered the norm, whereas other groups of people with less power may participate in alternative cultures. These divisions are often based on gender, economic class, racial categories, ethnicity, or sexual orientation. The social hierarchies that result are maintained, reinforced, and challenged through many means. Those means can include such things as physical violence, but often social hierarchies are maintained in ways far more subtle. For example, some geographers study ideology—a set of dominant ideas and beliefs—in relationship to place, environment, and landscape in order to understand how power works culturally. For example, most nations maintain a set of powerful beliefs about their relationships to the land, some holding to the idea that there is a deep and natural connection between a particular territory and the people who have inhabited it. These ideas often form part of a national identity and are expressed so routinely in poems, music, laws, and rituals that people accept these ideas as truths. Many American patriotic songs, for example, express the idea that the country naturally spreads from "sea to sea." Yet immigrants to that culture and country, and people who have been marginalized by that culture, may hold very different ideas of identity with the land. Native Americans have claims to land that far predate those of the U.S. government and would argue against an American national identity that includes a so-called natural connection to all the land between the Atlantic and the Pacific. Uncovering and analyzing the connections between ideology and power, then, are often integral to the geographer's task of understanding the diversity within a culture.

These different approaches to thinking about human geography are both necessary and healthy. These groups ask different questions about place and space; not surprisingly, they often obtain different answers. The model-builders tend to minimize diversity through their search for universal causal forces; the humanists examine diversity *among* cultures and strive to understand the unique; those who look to power and ideology focus on diversity and contestation *within* cultures. All lines of inquiry yield valuable findings. We present all of these perspectives throughout *The Human Mosaic.*

Themes in Human Geography

Our study of the human mosaic is organized around five geographical concepts or themes: region, mobility, globalization, nature-culture, and cultural landscape. We use these themes to organize the diversity of issues that confront human geography and have selected them because they represent the major concepts that human geographers discuss. Each of them stresses one particular aspect of the discipline, and even though we have separated them for purposes of clarity, it is important to remember that the concepts are related to each other. When discussing the theme of mobility, for example, we will inevitably bring up issues related to globalization, and vice versa. These themes give a common structure to each chapter and are stressed throughout the book.

Region

Phrased as a question, the theme of region could be "How are people and their traits grouped or arranged geographically?" Places and regions provide the essence of geography. How and why are places alike or different? How do they mesh together into functioning spatial networks? How do their inhabitants perceive them and identify with them? These are central geographical questions. A region, then, is a geographical unit based on characteristics and functions of culture. Geographers recognize three types of regions: formal, functional, and vernacular.

Formal Regions

A **formal region** is an area inhabited by people who have one or more traits in common, such as language, religion, or a system of livelihood. It is an area, therefore, that is relatively homogeneous with regard to one or more cultural traits. Geographers use this concept to map spatial differences throughout the world. For example, an Arabic-language formal region can be drawn on a map of languages and would include the areas where Arabic is spoken, rather than, say, English or Hindi or Mandarin. Similarly, a wheat-farming formal region would include the parts of the world where wheat is a major crop (look again at Figure 1.2).

The examples of Arabic speech and of wheat cultivation represent the concept of formal region at its simplest

> **formal region**
> A cultural region inhabited by people who have one or more cultural traits in common.

FIGURE 1.4 **An Inuit hunter with his dogsled team.** Various facets of a multitrait formal region can be seen here, including the clothing, the use of dogsleds as transportation, and hunting as a livelihood system. *(Bryan and Cherry Alexander Photography/Alamy.)*

level. Each is based on a single cultural trait. More commonly, formal regions depend on multiple related traits (**Figure 1.4**). Thus, an Inuit (Eskimo) culture region might be based on language, religion, economy, social organization, and type of dwellings. The region would reflect the spatial distribution of these five Inuit cultural traits. Districts in which all five of these traits are present would be part of the culture region.

Formal regions are the geographer's somewhat arbitrary creations. No two cultural traits have the same distribution, and the territorial extent of a culture region depends on what and how many defining traits are used. Why *five* Inuit traits, not four or six? Why not *foods* instead of (or in addition to) dwelling types? Consider, for example, Greeks and Turks, who differ in language and religion. Formal regions defined on the basis of speech and religious faith would separate these two groups. However, Greeks and Turks hold many other cultural traits in common. Both groups are monotheistic, worshipping a single god. In both groups, male supremacy and patriarchal families are the rule. Both enjoy certain folk foods, such as shish kebab. Whether Greeks and Turks are placed in the same formal region or in different ones depends entirely on how the geographer chooses to define the region. That choice in turn depends on the specific purpose of research or teaching that the region is designed to serve. Thus, an infinite number of formal regions can be created. It is unlikely that any two geographers would use exactly the same distinguishing criteria or place cultural boundaries in precisely the same location.

The geographer who identifies a formal region must locate borders. Because cultures overlap and mix, such boundaries are rarely sharp, even if only a single cultural trait is mapped. For this reason, we find **border zones** rather than lines. These zones broaden with each additional trait that is considered because no two traits have the same spatial distribution. As a result, instead of having clear borders, formal regions reveal a center or core where the defining traits are all present. Moving away from the central core, the characteristics weaken and disappear. Thus, many formal regions display a **core-periphery** pattern. This refers to a situation where a region can be divided into two sections, one near the center where the particular attributes that define the region (in this case, language and religion) are strong, and other portions of the region farther away from the core, called the periphery, where those attributes are weaker.

In a real sense, then, the human world is chaotic. No matter how closely related two elements of culture seem to be, careful investigation always shows that they do not cover exactly the same area. This is true regardless of the degree of detail involved. What does this chaos mean to the human geographer? First, it tells us that every cultural trait is spatially unique and that the explanation for each spatial variation differs in some degree from all others. Second, it means that culture changes continually throughout an area and that every inhabited place on Earth has a unique combination of cultural features. No place is exactly like another.

Does this cultural uniqueness of each place prevent geographers from seeking explanatory theories? Does it doom them to explaining each locale separately? The answer must be no. The fact that no two hills or rocks, no two planets or stars, no two trees or flowers are identical has not prevented geologists, astronomers, and botanists from formulating theories and explanations based on generalizations.

border zones
The areas where different regions meet and sometimes overlap.

core-periphery
A concept based on the tendency of both formal and functional culture regions to consist of a core or node, in which defining traits are purest or functions are headquartered, and a periphery that is tributary and displays fewer of the defining traits.

Functional Regions

The hallmark of a formal region is cultural homogeneity. Moreover, it is abstract rather than concrete. By contrast, a

functional region
A cultural area that functions as a unit politically, socially, or economically.

node
A central point in a functional culture region where functions are coordinated and directed.

functional region need not be culturally homogeneous; instead, it is an area that has been organized to function politically, socially, or economically as one unit. A city, an independent state, a precinct, a church diocese or parish, a trade area, a farm, and a Federal Reserve Bank district are all examples of functional regions. Functional regions have **nodes,** or central points where the functions are coordinated and directed. Examples of such nodes are city halls, national capitals, precinct voting places, parish churches, factories, and banks. In this sense, functional regions also possess a core-periphery configuration, in common with formal regions.

Many functional regions have clearly defined borders. A metropolitan area is a functional region that includes all the land under the jurisdiction of a particular urban government (**Figure 1.5**). The borders of this functional region may not be so apparent from a car window, but they will be clearly delineated on a regional map by a line distinguishing one jurisdiction from another. Similarly, each state in the United States and each Canadian province is a functional region, coordinated and directed from a capital, with government control extended over a fixed area with clearly defined borders.

Not all functional regions have fixed, precise borders, however. A good example is a daily newspaper's circulation area. The node for the paper would be the plant where it is produced. Every morning, trucks move out of the plant to distribute the paper throughout the city. The newspaper may have a sales area extending into the city's suburbs, local bedroom communities, nearby towns, and rural areas. There its sales area overlaps with the sales territories of competing newspapers published in other cities. It would be futile to try to define exclusive borders for such an area. How would you draw a sales area boundary for the *New York Times?* Its Sunday edition is sold in some quantity even in California, thousands of miles from its node, and it is published simultaneously in different cities.

Functional regions generally do not coincide spatially with formal regions, and this disjuncture often creates problems for the functional region. Germany provides an example (**Figure 1.6**). As an independent state, Germany forms a functional region. Language provides a substantial basis for political unity. However, the formal region of the German language extends beyond the political borders of Germany and includes part or all of eight other independent states. More important, numerous formal regions have borders cutting through German territory. Some of these have endured for millennia, causing differences among northern, southern, eastern, and western Germans. These contrasts make the functioning of the German state more difficult and help explain why Germany has been politically fragmented more often than unified.

Vernacular Regions

vernacular region
A culture region perceived to exist by its inhabitants, based in the collective spatial perception of the population at large and bearing a generally accepted name or nickname (such as "Dixie").

A **vernacular region** is one that is *perceived* to exist by its inhabitants, as evidenced by the widespread acceptance and use of a special regional name. Some vernacular regions are based on physical environmental features. For example, there are many regions called simply "the valley." Wikipedia lists more than 30 different regions in the United States and Canada that are referred to as such, places as varied as the Sudbury Basin in Ontario and the Lehigh Valley in Pennsylvania. In the 1980s, "the valley" became synonymous with the San Fernando Valley in Southern California (**Figure 1.7**) and became associated with a type of landscape (suburban), person (a white, teenage girl, called the "valley girl"), and

FIGURE 1.5 Aerial view of Denver. This image clearly illustrates the node of a functional region—here, the dense cluster of commercial buildings—that coordinates activities throughout the area that surrounds it. *Can you identify the border of this functional region? Why or why not?* (Jim Wark/Airphoto.)

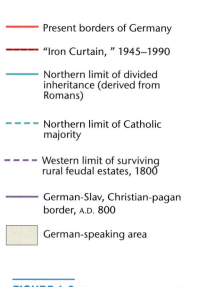

Present borders of Germany

"Iron Curtain," 1945–1990

Northern limit of divided inheritance (derived from Romans)

Northern limit of Catholic majority

Western limit of surviving rural feudal estates, 1800

German-Slav, Christian-pagan border, A.D. 800

German-speaking area

FIGURE 1.6 **East versus west and north versus south in Germany.** As a political unit and functional culture region, Germany must overcome the disruptions caused by numerous formal regions that tend to make the sections of Germany culturally different. Formal and functional regions rarely coincide spatially. ***How might these sectional contrasts cause problems for modern Germany?***

way of speaking ("valspeak"). Other vernacular regions find their basis in economic, political, or historical characteristics. Vernacular regions, like most regions, generally lack sharp borders, and the inhabitants of any given area may claim residence in more than one such region. They vary in scale from city neighborhoods to sizable parts of continents.

At a basic level, a vernacular region grows out of people's sense of belonging to and identification with a particular region. By contrast, many formal or functional regions lack this attribute and, as a result, are often far less meaningful for people. You're more likely to hear people say "we're fighting to preserve 'the valley' from further urban development" than to see people rally under the banner of "wheat-growing areas of the world"! Self-conscious regional identity can have major political and social ramifications.

Reflecting on Geography

What examples can you think of that show how identification with a vernacular region is a powerful political force?

Vernacular regions often lack the organization necessary for functional regions, although they may be centered around a single urban node, and they frequently do not display the cultural homogeneity that characterizes formal regions.

FIGURE 1.7 **The San Fernando Valley.** Referred to locally as "the valley," the San Fernando Valley is located in the northwestern area of Los Angeles. Despite its reputation as a white, suburban area, portions of the valley are densely populated, and the area is home to a wide range of peoples from many different backgrounds. ***Can you think of other examples of regions that are locally known as "the valley"?*** *(Robert Landau/drr.net.)*

Mobility

The concept of regions helps us see that similar or related sets of elements are often grouped together in space. Equally important in geography is understanding how and why these different cultural elements move through space and locate in particular settings. Regions themselves, as we have seen, are not stable but are constantly changing as people, ideas, practices, and technologies move around in space. Are there some patterns to these movements? These questions define our second theme, **mobility.**

One important way to study mobility is through the concept of diffusion. **Diffusion** can be defined as the movement of people, ideas, or things from one location outward toward other locations where these items are not initially found. Through the study of diffusion, the human geographer can begin to understand how spatial patterns in culture emerged and evolved. After all, any culture is the product of almost countless innovations that spread from their points of origin to cover a wider area. Some innovations occur only once, and geographers can sometimes trace a cultural element back to a single place of origin. In other cases, **independent invention** occurs: the same or a very similar innovation is separately developed at different places by different peoples.

mobility
The relative ability of people, ideas, or things to move freely through space.

diffusion
The movement of people, ideas, or things from one location outward toward other locations.

independent invention
A cultural innovation that is developed in two or more locations by individuals or groups working independently.

Types of Diffusion

Geographers, drawing heavily on the research of Hägerstrand, recognize several different kinds of diffusion (**Figure 1.8**). **Relocation diffusion** occurs when individuals or groups with a particular idea or practice migrate from one location to another, thereby bringing the idea or practice to their new homeland. Religions frequently spread this way. An example is the migration of Christianity with European settlers who came to America. In **expansion diffusion,** ideas or practices spread throughout a population, from area

relocation diffusion
The spread of an innovation or other element of culture that occurs with the bodily relocation (migration) of the individual or group responsible for the innovation.

expansion diffusion
The spread of innovations within an area in a snowballing process, so that the total number of knowers or users becomes greater and the area of occurrence grows.

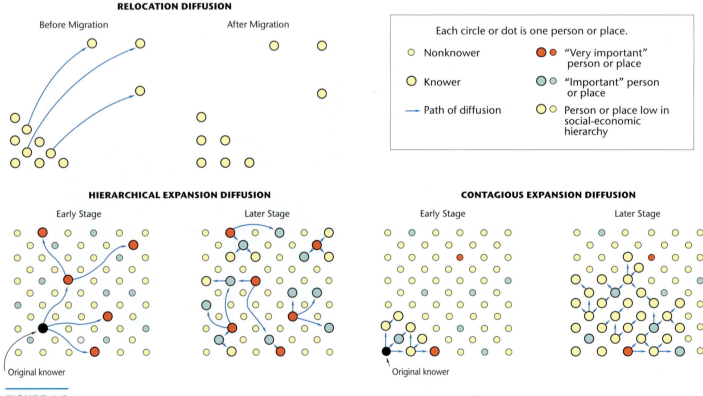

FIGURE 1.8 Types of cultural diffusion. These diagrams are merely suggestive; in reality, spatial diffusion is far more complex. In hierarchical diffusion, different scales can be used, so that, for example, the category "very important person" could be replaced by "large city."

hierarchical diffusion
A type of expansion diffusion in which innovations spread from one important person to another or from one urban center to another, temporarily bypassing other persons or rural areas.

contagious diffusion
A type of expansion diffusion in which cultural innovation spreads by person-to-person contact, moving wavelike through an area and population without regard to social status.

to area, in a snowballing process, so that the total number of knowers or users and the areas of occurrence increase.

Expansion diffusion can be further divided into three subtypes. In **hierarchical diffusion,** ideas leapfrog from one important person to another or from one urban center to another, temporarily bypassing other persons or rural territories. We can see hierarchical diffusion at work in everyday life by observing the acceptance of new modes of dress or foods. For example, sushi restaurants originally diffused from Japan in the 1970s very slowly because many people were reluctant to eat raw fish. In the United States, the first

sushi restaurants appeared in the major cities of Los Angeles and New York. Only gradually throughout the 1980s and 1990s did sushi eating become more common in the less urbanized parts of the country. By contrast, **contagious diffusion** involves the wavelike spread of ideas in the manner of a contagious disease, moving throughout space without regard to hierarchies. Hierarchical and contagious diffusion often work together. The worldwide spread of HIV/AIDS provides a sobering example of how these two types of diffusion can reinforce each other. As you can see from **Figure 1.9**, HIV/AIDS diffused first to urban areas (hierarchical diffusion) and from

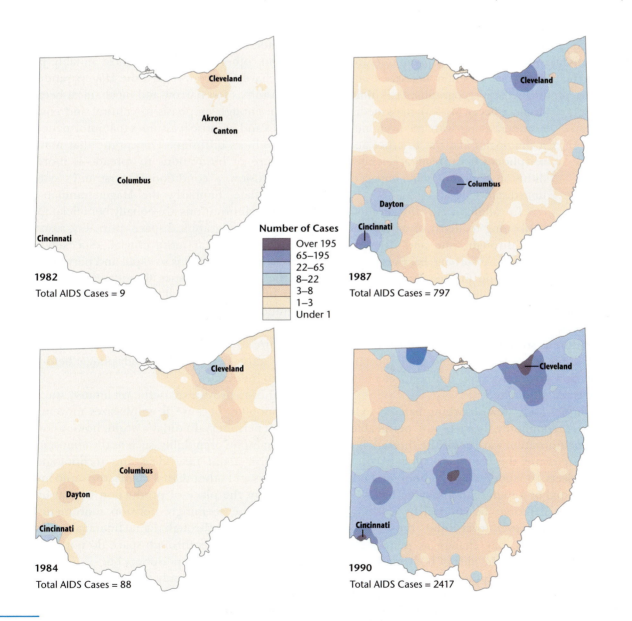

FIGURE 1.9 Diffusion of HIV/AIDS in Ohio. As you can see from this map, HIV/AIDS spread through both hierarchical and contagious diffusion processes. *Do you think a similar pattern is evident at the national scale? At the global scale?* (Source: Gould, 1993.)

stimulus diffusion
A type of expansion diffusion in which a specific trait fails to spread but the underlying idea or concept is accepted.

there spread outward (contagious diffusion). Sometimes a specific trait is rejected but the underlying idea is accepted, resulting in **stimulus diffusion.** For example, early Siberian peoples domesticated reindeer only after exposure to the domesticated cattle raised by cultures to their south. The Siberians had no use for cattle, but the idea of domesticated herds appealed to them, and they began domesticating reindeer, an animal they had long hunted.

If you throw a rock into a pond and watch the spreading ripples, you can see them become gradually weaker as they move away from the point of impact. In the same way, diffusion becomes weaker as a cultural innovation moves away from its point of origin. That is, diffusion decreases with distance. An innovation will usually be accepted most thoroughly in the areas closest to where it originates. Because innovations take increasing time to spread outward, time is also a factor. Acceptance generally decreases with distance and time, producing what geographers call **time-**

time-distance decay
The decrease in acceptance of a cultural innovation with increasing time and distance from its origin.

absorbing barrier
A barrier that completely halts diffusion of innovations and blocks the spread of cultural elements.

distance decay. Modern mass media, however, along with relatively new technologies like the Internet and cell phones, have greatly accelerated diffusion, diminishing the impact of time-distance decay.

In addition to the gradual weakening or decay of an innovation through time and distance, barriers can retard its spread. **Absorbing barriers** completely halt diffusion, allowing no further progress. For example, in 1998 the fundamentalist Islamic Taliban government of Afghanistan decided to abolish television, videocassette recorders, and videotapes, viewing them as "causes of corruption in society." As a result, the cultural diffusion of television sets was reversed, and the important role of television as a communication device to facilitate the spread of ideas was eliminated.

Extreme examples aside, few absorbing barriers exist in the world. More commonly, barriers are permeable, allowing part of the innovation wave to diffuse through but acting to weaken or retard the continued spread. When a school board objects to students with tattoos or body piercings, the principal of a high school may set limits by mandating that these markings be covered by clothing.

permeable barrier
A barrier that permits some aspects of an innovation to diffuse through it but weakens and retards continued spread; an innovation can be modified in passing through a permeable barrier.

However, over time, those mandates may change as people get used to the idea of body markings. More likely than not, though, some mandates will remain in place. In this way, the principal and school board act as a **permeable barrier** to cultural innovations.

Reflecting on Geography

The Internet has certainly made the diffusion of many forms of cultural change much more rapid. Some scholars have argued that, in fact, the Internet has eliminated barriers to diffusion. Can you think of examples where this is true? Untrue?

Although all places and communities hypothetically have equal potential to adopt a new idea or practice, diffusion typically produces a core-periphery spatial arrangement. Hägerstrand offered an explanation of how diffusion produces such a regional configuration. The distribution of innovations can be random, but the overlap of new ideas and traits as they diffuse through space and time is greatest toward the center of the region and least at the peripheries. As a result of this overlap, more innovations are adopted in the center, or core, of the region.

Some other geographers, most notably James Blaut and Richard Ormrod, regard the Hägerstrandian concept of diffusion as too narrow and mechanical because it does not give enough emphasis to cultural and environmental variables and because it assumes that information automatically produces diffusion. They argue that nondiffusion—the failure of innovations to spread—is more prevalent than diffusion, a condition Hägerstrand's system cannot accommodate. Similarly, the Hägerstrandian system assumes that innovations are equally beneficial to all people throughout geographical space. In reality, susceptibility to an innovation is far more crucial, especially in a world where communication is so rapid and pervasive that it renders the friction of distance almost meaningless. The inhabitants of two regions will not respond identically to an innovation, and the geographer must seek to understand this spatial variation in receptiveness to explain diffusion or the failure to diffuse. Within the context of their culture, people must perceive some advantage before they will adopt an innovation.

Diffusion provides a useful, yet limited, way of thinking about mobility because it emphasizes movement from a core to a periphery. In today's world, however, we see many other examples of mobility, such as the almost instant communication about new ideas and technologies through computers and other digital media, the rapid movement of goods from the place of production to that of consumption, and the seemingly nonstop movement of money around the globe through digital financial networks. These types of movements through space do not necessarily follow the pattern of core-periphery but instead create new and different types of patterns. The term **circulation** might better fit many of these forms of mobility because this term implies an ongoing set of movements

circulation
A term that implies an ongoing set of movements of people, ideas, or things that have no particular center or periphery.

migrations
The large-scale movements of people between different regions of the world.

transnational migrations
The movements of groups of people who maintain ties to their homelands after they have migrated.

with no particular center or periphery. Other types of mobilities, such as large-scale movements of people between different regions, can be best thought of as **migrations** from one region or country to another through particular routes. In today's globalizing world, with better and faster communication and transportation technologies, many migrants more easily maintain ties to their homelands even after they have migrated, and some may move back and forth between their home countries and those to which they have migrated. Scholars refer to these groups of people as **transnational migrants** (**Figure 1.10**). We will discuss much more about these different patterns of mobility—diffusion, circulation, and migration—in the rest of the book.

FIGURE 1.10 Demonstrations in Hamburg, Germany, against the deportation of immigrants. Many migrant-worker groups are fighting to maintain rights of transnational migrants who would like to be able to move freely between their home countries and those that currently provide work for them. *(Vario Images GmbH & Co. KG/Alamy.)*

Globalization

How and why are different cultures, economies, and societies linked around the world? And given all these new linkages, why do so many differences exist between different groups of people in the world? The modern technological age, in which improved worldwide transport and communications allow the instantaneous diffusion of ideas and innovations, has accelerated the phenomenon called **globalization.** This term refers to an increasingly linked world in which international borders are diminished in importance and a worldwide marketplace is created. This interconnected world has been created from a set of factors: faster and more reliable transportation, particularly the jet plane; the almost-instantaneous communication that computers, phones, faxes, and so on have allowed; and the creation of digital sources of information and media, such as the Internet. Thus, globalization in this sense is a rather recent phenomenon, dating from the late twentieth century. Yet we know that long before that time different countries and different parts of the world were linked. For example, in early medieval times overland trade routes connected China with other parts of Asia, the British East India Company maintained maritime trading routes between England and large portions of South Asia as early as the seventeenth century, and religious and political wars in Europe and the Middle East brought different peoples into direct contact with each other. Some geographers refer to such moments as early global encounters and suggest that they set the background for contemporary globalization. Beginning in the early twentieth century, but strengthening in the 1970s with new and advanced communication technologies, encounters between different cultures began to take place not face-to-face but rather mediated through technologies such as telephones, film, computers, and the Internet. These new media forms allowed those encounters to happen at any time, in many different places, and all at the same time. This new sense of interlinked and spontaneous communication between different peoples around the world is what most people mean by globalization.

These increasingly linked economic, political, and cultural networks around the world might lead many to believe that different groups of people around the globe are becoming more and more alike. In some ways this is true, but what these new global encounters have enabled is an increasing recognition of the differences between groups of people, and some of those differences have been caused by

globalization
The binding together of all the lands and peoples of the world into an integrated system driven by capitalistic free markets, in which cultural diffusion is rapid, independent states are weakened, and cultural homogenization is encouraged.

globalization itself. Some groups of people have access to advanced technologies, more thorough health care, and education, whereas others do not. Even within our own neighborhoods and cities, we know that there are people who are less able to afford these things. If we mapped certain indicators of human well-being on a global scale, such as life expectancy, literacy, and standard of living, we would find quite an uneven distribution. **Figure 1.11** shows us that

different cultures around the world have different access to these types of resources. These differences are what scholars mean when they refer to stages of development. In Figure 1.11, you can see that there are regions of the world that have a fairly high Human Development Index (HDI) and those that have a relatively low HDI. Scholars often refer to these two types of regions as developed (relatively high HDI) and developing (relatively low HDI). This inequitable

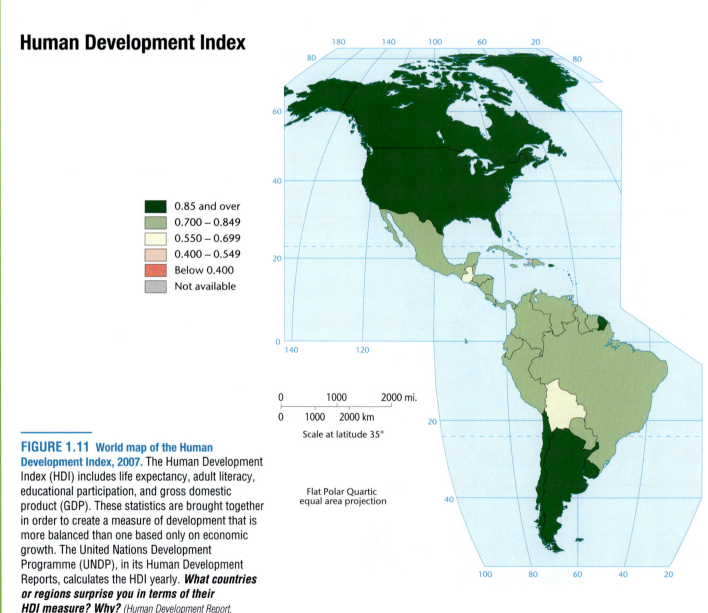

Human Development Index

0.85 and over
0.700 – 0.849
0.550 – 0.699
0.400 – 0.549
Below 0.400
Not available

0   1000   2000 mi.
0   1000   2000 km
Scale at latitude 35°

Flat Polar Quartic
equal area projection

FIGURE 1.11 World map of the Human Development Index, 2007. The Human Development Index (HDI) includes life expectancy, adult literacy, educational participation, and gross domestic product (GDP). These statistics are brought together in order to create a measure of development that is more balanced than one based only on economic growth. The United Nations Development Programme (UNDP), in its Human Development Reports, calculates the HDI yearly. *What countries or regions surprise you in terms of their HDI measure? Why?* (Human Development Report, 2007/2008. Accessed at http://hdr.undp.org.)

uneven development
The tendency for industry to develop in a core-periphery pattern, enriching the industrialized countries of the core and impoverishing the less industrialized periphery. This term is also used to describe urban patterns in which suburban areas are enriched while the inner city is impoverished.

distribution of resources is referred to as **uneven development.** We will be discussing much more about development in Chapters 3 and 9.

Globalization helps make us aware of these uneven developments and contributes to some of them. How does this happen? Globalization

can be thought of as both a set of processes that are economic, political, and cultural in nature, and as the effects of those processes. For example, economic globalization refers to the interlinked networks of money, production, transportation, labor, and consumption that allow, say, the parts of an automobile to be manufactured in several countries, assembled in yet another, and then sold throughout large portions of the world. These economic networks and

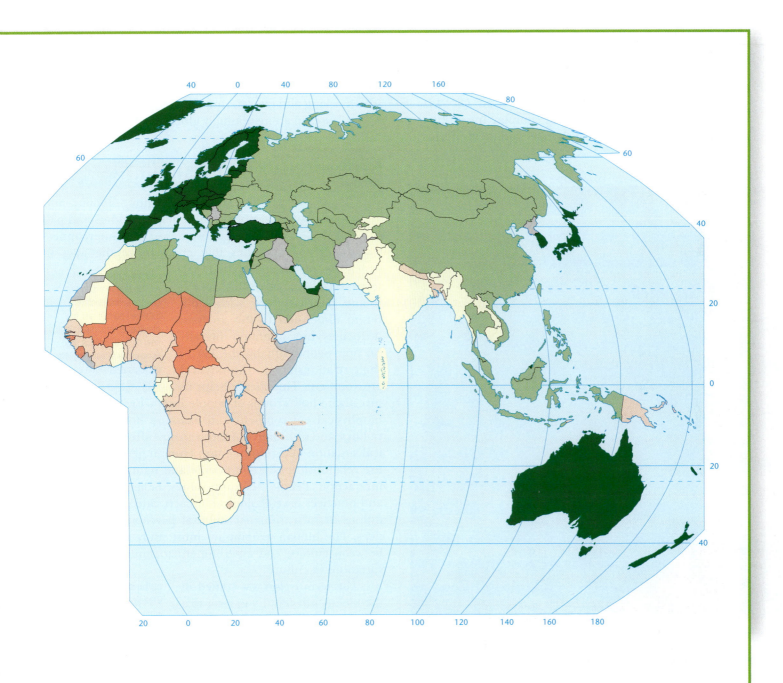

processes in turn have significant and often uneven effects on the economies of different countries and regions. Some regions gain employment, whereas others lose jobs; some consumers are able to afford these cars because they are less expensive, whereas other potential consumers who have become unemployed cannot afford to purchase these cars. These global economic processes and effects are in turn linked to politics and culture. In other words, globalization entails not only certain processes and effects but also the relationships among these things. For example, those countries chosen by the automobile manufacturer as sites of production might see the standard of living of their populations improve, leading to larger consumer markets, better communications and media, and often changing political sensibilities. And these changing political ideas in turn will shape economic decisions and so on. Globalization, therefore, involves looking at complex interconnections between a set of related processes and their effects.

Culture, of course, is a key variable in these interactions and interconnections. In fact, as we have just suggested, globalization is occurring through cultural media, for example, in films, on television, and on the Internet (**Figure 1.12** and Mona's Notebook). In addition, if we consider culture as a way of life, then globalization is a key shaper of culture and in turn is shaped by it. Some scholars have suggested that globalizing processes and an increase in mobility will work to homogenize different peoples, breaking down culture regions and eventually producing a single global culture. Other scholars see a different picture, one where new forms of media and communication will allow local cultures to maintain their distinct identities, reinforcing the diversity of cultures around the world. Throughout *The Human Mosaic*, we will return to these issues, asking and considering the complex role of culture and cultures in an increasingly global world.

Nature-Culture

The themes of region, mobility, and globalization help us understand patterns, movements, and interconnections that characterize the ways in which people create spatial patterns on the Earth. Our fourth theme, **nature-culture,** adds a different dimension to this analysis; it focuses our attention literally on how people inhabit the Earth, their relationships to the physical environment. This theme helps us investigate how groups of people interact with the Earth's biophysical environment and examine how the culture, politics, and economies of those groups affect their ecological situation and resource use. Human geographers view the relationship between people and nature as a two-way interaction. People's cultural values, beliefs, perceptions, and practices have ecological impacts, and ecological conditions in turn influence cultural perceptions and practices. The human geographer must study the interaction between culture and environment to understand spatial variations in culture.

> **nature-culture**
> A term that refers to the complex relationships between people and the physical environment, including how culture, politics, and economies affect people's ecological situation and resource use.

The term *ecology* was coined in the nineteenth century to refer to a new biological science concerned with studying the complex relationships among living organisms and their physical environments. Geographers borrowed this term in the mid-twentieth century and joined it with the term *culture* in order to

> **cultural ecology**
> Broadly defined, the study of the relationships between the physical environment and culture; narrowly (and more commonly) defined, the study of culture as an adaptive system that facilitates human adaptation to nature and environmental change.

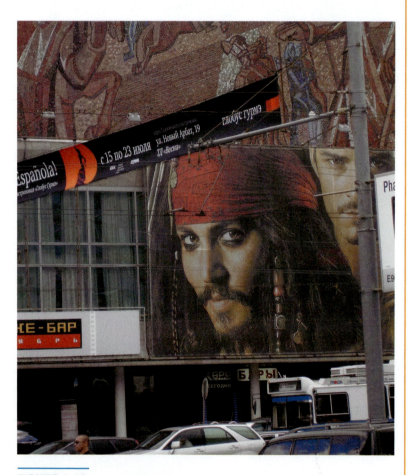

FIGURE 1.12 Global culture on the streets of Moscow. A karaoke bar (a much globalized Japanese cultural form) occupies the storefront of a Soviet-era building on the Arbat (a main street) in downtown Moscow; the building is now almost hidden behind an advertisement for the *Pirates of the Caribbean* movie. Meanwhile, an advertisement for a Spanish festival drapes across the street. *(Courtesy of Mona Domosh.)*

Mona's Notebook

Early-Twentieth-Century Globalization vs. Contemporary Globalization in Russia

Mona Domosh

Nineteenth-century commercial buildings being renovated in downtown Moscow. *What other signs of globalization can you see in this image?* (Courtesy of Mona Domosh.)

I took the photo at left during my second research trip to Moscow in 2006, when I hoped to spend my time in archives hunting down information about how and in what ways U.S.-based corporations had expanded into Russia in the early twentieth century. Russia was America's largest foreign market at the time, so I knew it held one of the keys to figuring out my primary research question: how and why early globalization was different from and similar to contemporary globalization. The archives were indeed filled with information (though extremely difficult to access), but I soon understood that I could learn almost as much from walking the streets as I could from sitting, squinting, and typing in closed-off archival rooms. What I saw was a city in the midst of a huge consumer revolution, funded with money from Russian natural gas and oil, and capitalized by global corporations. Construction cranes were everywhere because large, globally based development companies saw profitable opportunities in the skyrocketing demand for housing, entertainment, and shopping spaces.

One hundred years earlier, as I found out in the archives, Moscow was in a similar situation, though the scale and pace of globalization was much smaller and slower. In addition to German, French, and British companies, U.S.-based companies sought out Russia as a market because of its large population and relatively underdeveloped industrial sector. Parts of downtown Moscow in 1915 began to resemble Wall Street and Broadway: a landscape filled with corporate headquarters, bank and insurance buildings, department stores, and even a stock exchange. Of course, as we all know from our history classes, all of this came to an abrupt halt in 1917, when those not benefiting from globalization (or the czar's government!) changed the entire system. Today the outcome seems markedly different. If what I learned from walking the streets of Moscow today is indicative, twenty-first-century globalization is there to stay.

Posted by Mona Domosh

---

delineate a field of study—**cultural ecology**—that dealt with the interaction between culture and physical environments. Later, another concept, the ecosystem, was introduced to describe a territorially bounded system consisting of inter-

acting organic and inorganic components. Plant and animal species were said to be adapted to specific conditions in the ecosystem and to function so as to keep the system stable over time.

It soon became clear, however, that human cultural interactions with the environment were far too complex to be analyzed with concepts borrowed from biology. Furthermore, the idea that groups of people interacted with their ecosystems in isolation from larger-scale political, economic, and social forces was difficult to defend. We can readily observe, for example, that trade goods come into communities, agricultural commodities flow out, money circulates, taxes are collected, and people migrate in and out for work. So geographers now use the term *nature-culture* to refer to the complex interactions among all these variables and to reflect the fact that studies of local human-environment relations need to include political, economic, and social forces operating on national, and even global, scales.

The theme of nature-culture, the meeting ground of cultural and physical geographers, has traditionally provided a focal point for the academic discipline of geography. In fact, some geographers have proposed that *the* theme of geography is nature-culture, that the study of the intricate relationships between people and their physical environments unites cultural and physical geography to form the entire academic discipline. Although few accept this narrow definition of geography, most will agree that an appreciation of the complex people-environment relationship is necessary for concerned citizens of the twenty-first century.

Through the years, human geographers have developed various perspectives on the interaction between humans and the land. Four schools of thought have developed: environmental determinism, possibilism, environmental perception, and humans as modifiers of the Earth.

Environmental Determinism

environmental determinism
The belief that cultures are directly or indirectly shaped by the physical environment.

During the late nineteenth and early twentieth centuries, many geographers accepted **environmental determinism:** the belief that the physical environment is the dominant force shaping cultures and that humankind is essentially a passive product of its physical surroundings. Humans are clay to be molded by nature. Similar physical environments produce similar cultures.

For example, environmental determinists believed that peoples of the mountains were predestined by the rugged terrain to be simple, backward, conservative, unimaginative, and freedom loving. Desert dwellers were likely to believe in one God but to live under the rule of tyrants. Temperate climates produced inventiveness, industriousness, and democracy. Coastlands pitted with fjords produced great navigators and fishers. Environmental determinism had serious consequences, particularly during the time of European colonization in the late nineteenth century. For example, many Europeans saw Latin American native inhabitants as lazy, childlike, and prone to vices such as alcoholism because of the tropical climates that cover much of this region. Living in a tropical climate supposedly ensured that people didn't have to work very hard for their food. Europeans were able to rationalize their colonization of large portions of the world in part along climatic lines. Because the natives were "naturally" lazy and slow, the European reasoning went, they would benefit from the presence of the "naturally" stronger, smarter, and more industrious Europeans who came from more temperate lands.

Determinists overemphasize the role of environment in human affairs. This does not mean that environmental influences are inconsequential or that geographers should not study such influences. Rather, the physical environment is only one of many forces affecting human culture and is never the sole determinant of behavior and beliefs.

Possibilism

Since the 1920s, **possibilism** has been the favored view among geographers. Possibilists claim that any physical environment offers a number of possible ways for a culture to develop. In this way, the local environment helps shape its resident culture. However, a culture's way of life ultimately depends on the choices people make among the possibilities that are offered by the environment. These choices are guided by cultural heritage and are shaped by a particular political and economic system. Possibilists see the physical environment as offering opportunities and limitations; people make choices among these to satisfy their needs. **Figure 1.13** provides an interesting example: the cities of San Francisco and Chongqing both were built on similar physical terrains that dictated an overall form, but differing cultures lead to very different street patterns, architecture, and land use. In short, local traits of culture and economy are the products of culturally based decisions made within the limits of possibilities offered by the environment.

possibilism
A school of thought based on the belief that humans, rather than the physical environment, are the primary active force; that any environment offers a number of different possible ways for a culture to develop; and that the choices among these possibilities are guided by cultural heritage.

Most possibilists think that the higher the technological level of a culture, the greater the number of possibilities and the weaker the influences of the physical environment. Technologically advanced cultures, in this view, have achieved some mastery over their physical surroundings. Geographers Jim Norwine and Thomas Anderson, however, warn that even in these advanced societies "the quantity and quality of human life are still strongly influenced by the

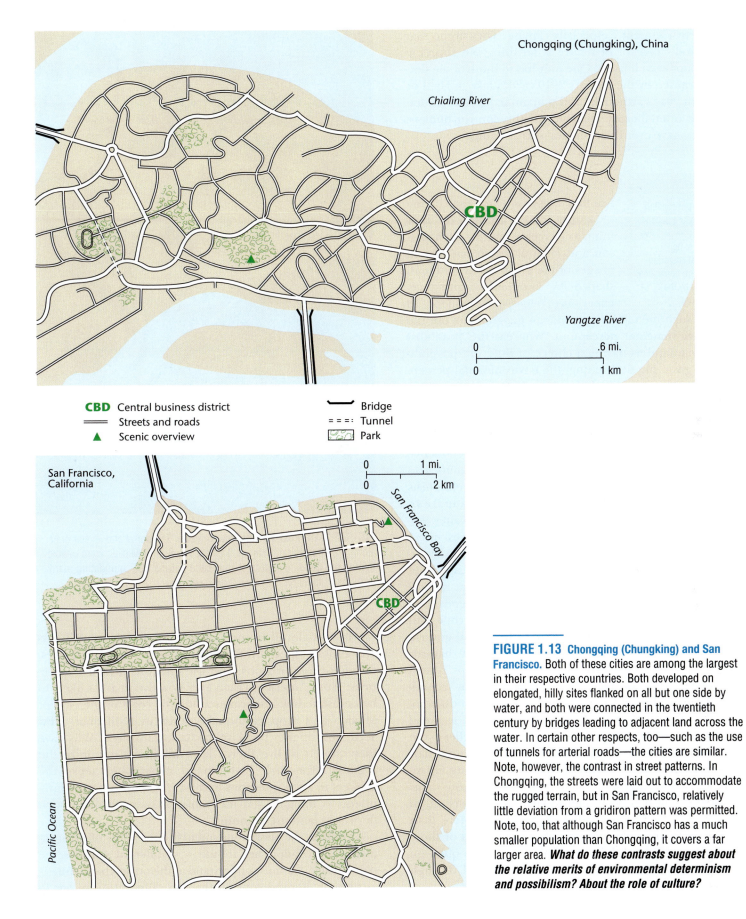

CBD Central business district ⌒ Bridge
≡ Streets and roads = = =: Tunnel
▲ Scenic overview Park

FIGURE 1.13 Chongqing (Chungking) and San Francisco. Both of these cities are among the largest in their respective countries. Both developed on elongated, hilly sites flanked on all but one side by water, and both were connected in the twentieth century by bridges leading to adjacent land across the water. In certain other respects, too—such as the use of tunnels for arterial roads—the cities are similar. Note, however, the contrast in street patterns. In Chongqing, the streets were laid out to accommodate the rugged terrain, but in San Francisco, relatively little deviation from a gridiron pattern was permitted. Note, too, that although San Francisco has a much smaller population than Chongqing, it covers a far larger area. *What do these contrasts suggest about the relative merits of environmental determinism and possibilism? About the role of culture?*

natural environment," especially climate. They argue that humankind's control of nature is anything but supreme and perhaps even illusory. One only has to think of the devastation caused by the January 2010 earthquake in Haiti or the March 2011 earthquake and tsunami in Japan to underscore the often-illusory character of the control humans think they have over their physical surroundings.

Environmental Perception

environmental perception
The belief that culture depends more on what people perceive the environment to be than on the actual character of the environment; perception, in turn, is colored by the teachings of culture.

Another approach to the theme of nature-culture focuses on how humans perceive nature. Each person and cultural group has mental images of the physical environment, shaped by knowledge, ignorance, experience, values, and emotions. To describe such mental images, human geographers use the term **environmental perception.** Whereas the possibilist sees humankind as having a choice of different possibilities in a given physical setting, the environmental perceptionist declares that the choices people make will depend more on what they perceive the environment to be than on the actual character of the land. Perception, in turn, is colored by the teachings of culture.

natural hazard
An inherent danger present in a given habitat, such as floods, hurricanes, volcanic eruptions, or earthquakes; often perceived differently by different peoples.

Some of the most productive research done by geographers who are environmental perceptionists has been on the topic of **natural hazards,** such as floods, hurricanes, volcanic eruptions, earthquakes, insect infestations, and droughts. All cultures react to such hazards and catastrophes, but the reactions vary greatly from one group to another. Some people reason that natural disasters and risks are unavoidable acts of the gods, perhaps even divine retribution. Often they seek to cope with the hazards by placating their gods. Others hold government responsible for taking care of them when hazards yield disasters. In Western culture, many groups regard natural hazards as problems that can be solved by technological means. In the United States, one of the most common manifestations of this belief has been the widespread construction of dams to prevent flooding. This belief in the use of technology to mitigate the impacts of natural hazards was sorely tested in the United States during Hurricane Katrina in August of 2005, when the numerous dams and levees that had been built to keep floodwaters out of the city of New Orleans failed (**Figure 1.14**). More than 1800 people lost their lives as a result, and tens of thousands of people were left homeless, calling into question the belief that natural disasters can be avoided through technology.

FIGURE 1.14   Post-Katrina New Orleans. The dams and levees built along the Mississippi River as it enters the Gulf of Mexico were unable to stop the flooding waters of Hurricane Katrina from inundating many portions of the city. *(U.S. Coast Guard/Getty Images.)*

In virtually all cultures, people knowingly inhabit hazard zones, especially floodplains, exposed coastal sites, drought-prone regions, and the environs of active volcanoes. More Americans than ever now live in areas likely to be devastated by hurricanes along the coast of the Gulf of Mexico and atop earthquake faults in California. How accurately do they perceive the hazard involved? Why have they chosen to live there? How might we minimize the eventual disasters? The human geographer seeks the answers to such questions and aspires, with other geographers, to mitigate the inevitable disasters through such devices as land-use planning.

Perhaps the most fundamental expression of environmental perception lies in the way different cultures see nature itself. We must understand at the outset that nature is a culturally derived concept that has different meanings to

organic view of nature
The view that humans are part of, not separate from, nature and that the habitat possesses a soul and is filled with nature-spirits.

mechanistic view of nature
The view that humans are separate from nature and hold dominion over it and that the habitat is an integrated mechanism governed by external forces that the human mind can understand and manipulate.

different peoples. In the **organic view,** held by many traditional groups, people are part of nature. The habitat possesses a soul, is filled with nature-spirits, and must not be offended. By contrast, most Western peoples believe in the **mechanistic view of nature.** Humans are separate from and hold dominion over nature. They see the habitat as an integrated system of mechanisms governed by external forces that can be rendered into natural laws and understood by the human mind.

Humans as Modifiers of the Earth

Many human geographers, observing the environmental changes people have wrought, emphasize humans as modifiers of the habitat. This presents yet another facet of the theme of nature-culture. In a sense, the human-as-modifier school of thought is the opposite of environmental determinism. Whereas the determinists proclaim that nature molds humankind, and possibilists believe that nature presents possibilities for people, those geographers who emphasize the human impact on the land assert that humans mold nature.

In addition to deliberate modifications of the Earth through such activities as mining, logging, and irrigation, we now know that even seemingly innocuous behavior, repeated for millennia, for centuries, or in some cases for mere decades, can have catastrophic effects on the environment. Plowing fields and grazing livestock can eventually denude regions (**Figure 1.15**). The use of certain types of air conditioners or spray cans apparently has the potential to destroy the planet's very ability to support life. And the increasing release of fossil-fuel emissions from vehicles

and factories—what are known as greenhouse gases—is arguably leading to global warming, with potentially devastating effects on the Earth's environment (see Subject to Debate, page 22). Clearly, access to energy and technology is the key variable that controls the magnitude and speed of environmental alteration. Geographers seek to understand and explain the processes of environmental alteration as they vary from one culture to another and, through applied geography, to propose alternative, less destructive modes of behavior.

Human geographers began to concentrate on the human role in changing the face of the Earth long before the present level of ecological consciousness developed. They learned early on the fact that different cultural groups have widely different outlooks on humankind's role in changing the Earth. Some, such as those rooted in the mechanistic tradition, tend to regard environmental modification as divinely approved, viewing humans as God's helpers in completing the task of creation. Other groups, organic in their view of nature, are much more cautious, taking care not to offend the forces of nature. They see humans as part of nature, meant to be in harmony with their environment (for more on this topic, see Chapter 7).

ecofeminism
A doctrine proposing that women are inherently better environmental preservationists than men because the traditional roles of women involved creating and nurturing life, whereas the traditional roles of men too often necessitated death and destruction.

Gender differences can also play a role in the human modification of the Earth. **Ecofeminism,** a term derived from a book by Karen Warren, maintains that because of socialization, women have been better ecologists and environmentalists than men. (We should not forget that the modern environmental preservation movement grew in no small part out of Rachel Carson's 1962 book *Silent Spring.*) Traditionally, women—as childbearers, gardeners, and nurturers of the family and home—dealt with the daily chores of gaining food from the Earth, whereas men—as hunters, fishers, warriors, and forest clearers—were involved with activities that were more associated with destruction. Regardless of whether we agree with this rather deterministic and essentializing viewpoint (understanding gender differences as biologically determined rather than culturally constructed), we can see that in many situations through time, and around the globe, women and men have had different relationships to the natural world.

FIGURE 1.15 **Human modification of the Earth includes severe soil erosion.** This erosion could have been caused by road building or poor farming methods. The scene is in the Amazon Basin of Brazil. *How can we adopt less destructive ways of modifying the land?* (Michael Nichols/ National Geographic.)

Subject to Debate

Human Activities and Global Climate Change

One of the most important and vexing scientific and political issues of the early twenty-first century is understanding the causes of recent global climate change and deciding what policies to pursue to mitigate its effects. It's also an issue that human geographers, with their emphasis on understanding nature-culture relationships, are well prepared to discuss.

According to a report by the National Academies of the United States (a joint body comprised of the National Academy of Science, the National Academy of Engineering, the Institute of Medicine, and the National Research Council), the Earth's temperatures are rising. Since the early twentieth century, the surface temperature of the Earth has risen 1.4°F, with the greatest amount of increase occurring since 1978. Global climate, of course, is always changing. What is critical today, however, is the degree to which scientists have been able to correlate global warming trends with the rise in the levels of carbon dioxide in the atmosphere. Carbon dioxide is one of several greenhouse gases that keep radiative energy (and therefore warmth) trapped in the Earth's atmosphere. It occurs naturally in the atmosphere but is also released when fossil fuels such as coal, oil, and natural gas are burned. Changes in the levels of carbon dioxide in the atmosphere, therefore, lead to changes in the Earth's temperature. Changes in temperature, in turn, lead to other changes in the environment, such as the melting of glacial caps and the rising of sea levels.

To what degree are our activities—particularly our energy demands that lead to the burning of fossil fuels—responsible for this climate change? This is where the real debate starts. Some scientists are wary of pointing the finger at carbon dioxide emissions as the primary culprit, arguing along several lines that it is far from certain whether human activities have had such impacts on the Earth's climate; some are critical of the data itself that show increases in surface temperature; some believe that the recent fluctuation in climate is far more a natural occurrence than one induced by humans; and some believe that the Earth's atmospheric and climatic systems are simply so complex that it is premature to isolate one factor. Yet there is growing worldwide consensus that human activities—particularly our reliance on the burning of fossil fuels—are the primary factors responsible for the recent global warming. The Intergovernmental Panel on Climate Change (IPCC), a group of scientists from many different countries, concluded that because of the increase in greenhouse gases in the atmosphere, by 2100 average temperatures on the surface of the Earth are likely to rise between 2.5°F and 10.4°F above 1990 levels. The question for this group of scientists is not what is causing these changes but what to do about it. The first step, these scientists argue, is to find ways to decrease levels of carbon dioxide released by looking to new technologies and alternative energy sources. But because the changes in climate are already occurring, the second step is finding ways to deal with the effects of global warming.

Geography is a discipline well suited to dealing with this debate because, as we have seen, one of the primary sets of issues it deals with is understanding nature-culture relationships. Figuring out to what degree, how, and why human activities interact with our physical environment is clearly the heart of what is being disputed here.

Continuing the Debate

As geographers, we know that various cultures interact with the environment differently and have varied beliefs and ideas about the role of science in explaining physical phenomena. Keeping this all in mind, consider these questions:

- How might these cultural differences affect people's conclusions about the causes and effects of global climate change?

- How are your ideas about global climate change impacted by your position in the world?

Polar bear on breaking ice floe, North Pole.
One of the many consequences of global climate change is the disappearance of habitats for many species, including the polar bear. *Are there any endangered habitats in your local region?* (© Yi Lu/Corbis.)

CULTURAL LANDSCAPE

What are the visible expressions of culture? How are peoples' interactions with nature materially expressed? What do regions actually look like? These questions provide the basis of our fifth and final theme, **cultural landscape.** The human or cultural landscape is comprised of all the built forms that cultural groups create in inhabiting the Earth—roads, agricultural fields, cities, houses, parks, gardens, commercial buildings, and so on. Every inhabited area has a cultural landscape, fashioned from the natural landscape, and each uniquely reflects the culture or cultures that created it (**Figure 1.16**). Landscape mirrors a culture's needs, values, and attitudes toward the Earth, and the human geographer can learn much about a group of people by carefully observing and studying the landscape. Indeed, so important is this visual record of cultures that some geographers regard landscape study as geography's central interest.

Why is such importance attached to the human landscape? Perhaps part of the answer is that it visually reflects the most basic strivings of humankind: shelter, food, and clothing. In addition, the cultural landscape reveals people's different attitudes toward the modification of the Earth. It also contains valuable evidence about the origin, spread, and development of cultures because it usually preserves various types of archaic forms. Dominant and alternative cultures use, alter, and manipulate landscapes to express their diverse identities (**Figure 1.17**)

Aside from containing archaic forms, landscapes also convey revealing messages about the present-day inhabitants and cultures. According to geographer Pierce Lewis, "the cultural landscape is our collective and revealing autobiography, reflecting our tastes, values, aspirations, and fears in tangible forms." Cultural landscapes offer "texts" that geographers read to discover dominant ideas and prevailing practices within a culture as well as less dominant and alternative forms within that culture. This "reading," however, is often a very difficult task, given the complexity of cultures, cultural change, and recent globalizing trends that can obscure local histories.

Reflecting on Geography

As we will learn throughout this book, landscapes are often created from more than one set of cultural values and beliefs, oftentimes in conflict with each other. How then can we "read" conflict into the landscape, when it appears so natural and unified? What sort of information would we need?

Geographers have pushed the idea of "reading" the landscape further in order to focus on the symbolic and ideological qualities of landscape. In fact, as geographer Denis Cosgrove has suggested (see Practicing Geography on page 5), the very idea of landscape itself was ideological, in that its development in the Renaissance served the interests of the new elite class for whom agricultural land was valued not for its productivity but for its use as a visual subject. Land, in other words, was important to look at as a scene, and the actual activities that were necessary for agriculture were thus

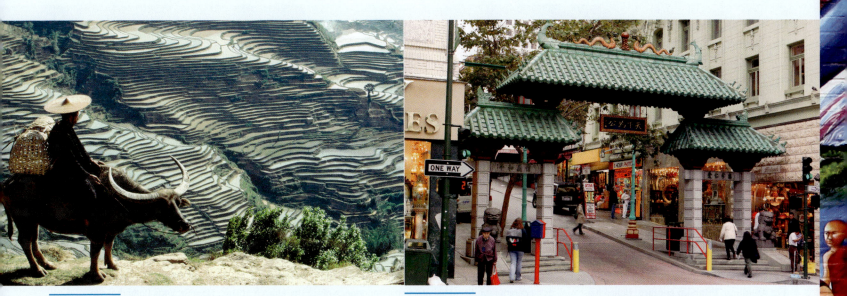

FIGURE 1.16 **Terraced cultural landscape of an irrigated rice district in Yunnan Province, China.** In such areas, the artificial landscape made by people overwhelms nature and forms a human mosaic on the land. *Why is rice cultivated in such hilly areas in Asia, whereas in the United States rice farming is confined to flat plains?* (Stone.)

FIGURE 1.17 **Chinatown, San Francisco.** Gates are symbolically important in denoting the entry into sacred space in traditional Chinese culture; they are often used in American cities like San Francisco to mark a similar transition into a neighborhood dominated by immigrant Chinese. Most of our landscapes are dotted with such cultural symbols. *Can you identify some in your neighborhood?* (© Ron Niebrugge/Alamy.)

hidden from these views. If you go to an art museum, for example, it will be difficult, if not impossible, to find in the Italian Renaissance room any landscape paintings that depict agricultural laborers (**Figure 1.18**).

Closer to home, we need only to look outside our windows to see other symbolic and ideological landscapes. One of the most familiar and obvious **symbolic landscapes** is the urban skyline. Composed of tall buildings that normally house financial service industries, it represents the power and dominance of finance and economics within that culture (**Figure 1.19**). However, other cities are dominated by tall structures that have little to do with economics but more with religion. In medieval Europe, for example, cathedrals and churches rose high above other buildings, symbolizing the centrality and dominance of Catholicism in this culture (**Figure 1.20**).

Even the most mundane landscape element can be interpreted as symbolic and ideological. The typical, middle-class, suburban, American or Canadian home, for example, can be interpreted as an expression of a dominant set of ideas about culture and family structure (**Figure 1.21**). These homes are often comprised of a living room, dining room, kitchen, and bedrooms, all separated by walls. The cultural assumptions built into this division of space include the assumed value of individual privacy (everyone has his or her own bedroom); the idea that certain functions should be spatially separate from others (cooking, eating, socializing, sleeping); and the notion that a family

> **symbolic landscapes**
> Landscapes that express the values, beliefs, and meanings of a particular culture.

is composed of a mother, father, and children (indicated by the "master" bedroom and smaller "children's" bedrooms). Thus, even the most common of landscapes can be seen as symbolic of a particular culture and built from ideological assumptions.

As we have seen, the physical content of the **cultural landscape** is both varied and complex. To better study and understand these complexities, geographical studies focus on three principal aspects of landscape: settlement forms, land-division patterns, and architectural styles. In the study of **settlement forms,** human geographers describe and explain the spatial arrangement of buildings, roads, and other features that people construct while inhabiting an area. One of the most basic ways in which geographers categorize settlement forms is to examine their degree of **nucleation,** a term that refers to the relative density of landscape elements. Urban centers are of course very nucleated, whereas rural farming areas tend to be much less nucleated, what geographers call **dispersed.** Another common way to think about settlement forms is the degree to which they appear standardized and planned, such as the grid form of much of the American West (**Figure 1.22**), versus the degree to which the forms appear to be organic, that is, to have

> **cultural landscape**
> The visible human imprint on the land.

> **settlement forms**
> The spatial arrangement of buildings, roads, towns, and other features that people construct while inhabiting an area.

> **nucleation**
> A relatively dense settlement form.

> **dispersed**
> A type of settlement form in which people live relatively distant from each other.

FIGURE 1.18 Landscape triptych panel by Fra Angelico, fifteenth century. Notice the depiction of the beautiful and orderly agricultural landscape outside the city walls but the absence of people actually doing the work to maintain that order. **Why aren't the laborers depicted?** (Archivo Iconografico, S.A./Corbis.)

FIGURE 1.19 Yokohama at dusk. This skyline is a powerful symbol of the economic importance of the world's largest city, Tokyo-Yokohama. **What landscape form best symbolizes your town/city?** (Jose Fuste Raga/Corbis.)

FIGURE 1.20 Prague's skyline is dominated by church spires. Here St. Vitus Cathedral sits majestically overlooking the Vltava River and the Small Town. (Courtesy of Mona Domosh.)

been built without any apparent geometric plan, such as the central areas of most European cities. Thinking about settlement forms in terms of these two basic categories helps geographers begin their analysis of the relationships between cultures and the landscapes they produce.

land-division patterns A term that refers to the spatial patterns of different land uses.

Land-division patterns indicate the uses of particular parcels of land and as such reveal the way people have divided the land for economic, social, and political uses. Within a particular nucleated settlement form—a city, for example—you can see different patterns of land use. Some areas are devoted to economic uses, others to residential, political (city hall, for example), social, and cultural uses. Each of these areas can be further divided. Economic uses can include offices for financial services, retail stores, warehouses, and factories. Residential areas are often divided into middle-class, upper-class, and working-class districts and/or are grouped by ethnicity and race (see Chapter 11). Such patterns of course vary a great deal from place to place and culture to culture, as we will see throughout this book. One of the best ways to glimpse settlement and land-division patterns is through an airplane window. Looking down, you can see the multicolored abstract patterns of planted fields, as vivid as any modern painting, and the regular checkerboard or chaotic tangle of urban streets.

Perhaps no other aspect of the human landscape is as readily visible from ground level as the architectural style of a culture. Geographers look at the exterior materials and decoration, as well as the layout and design of the interiors. Styles tend to vary both through time, as cultures change, and across space, in the sense that different cultures adopt and invent their own stylistic detailing according to their own particular needs, aesthetics, and desires. Thus, examining architectural style is often useful when trying to date a particular landscape element or when trying to understand the particular values and beliefs that cultures may hold. In North American culture, different building styles catch the eye: modest white New England churches and giant urban cathedrals, hand-hewn barns and geodesic domes, wooden one-room schoolhouses and the new windowless school buildings of urban areas, shopping malls and glass office buildings. Each tells us something about the people who designed, built, or inhabit these spaces. Thus, architecture provides a vivid record of the resident culture (Figure 1.23). For this reason, cultural geographers have traditionally devoted considerable attention to examining architecture and style in the cultural landscape.

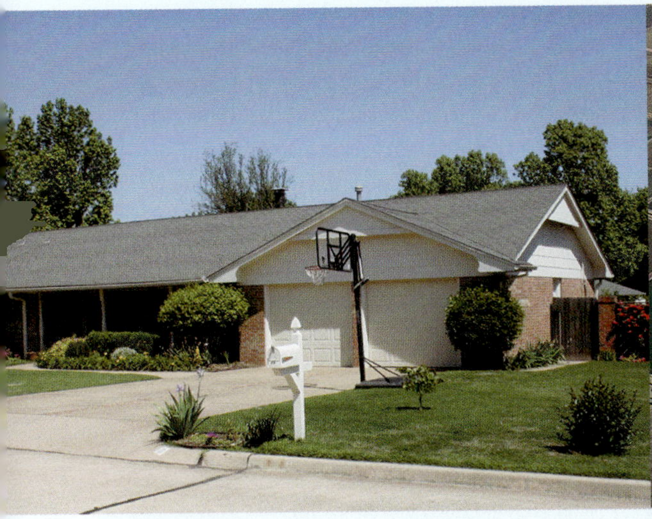

FIGURE 1.21 American ranch house. The horizontally expansive ranch is a common house form in the United States and Canada. ***Can you think of other countries where ranch houses are also common?*** *(Courtesy of Brad Bays.)*

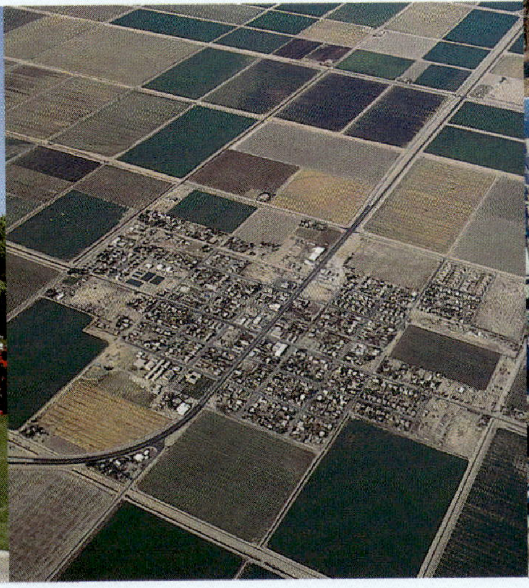

FIGURE 1.22 The town of Westmoreland in the Imperial Valley of California. It's difficult to find a more regularized, geometric land pattern than this. ***Why do you think much of the American West was divided into these rectangles?*** *(Jim Wark/Airphoto.)*

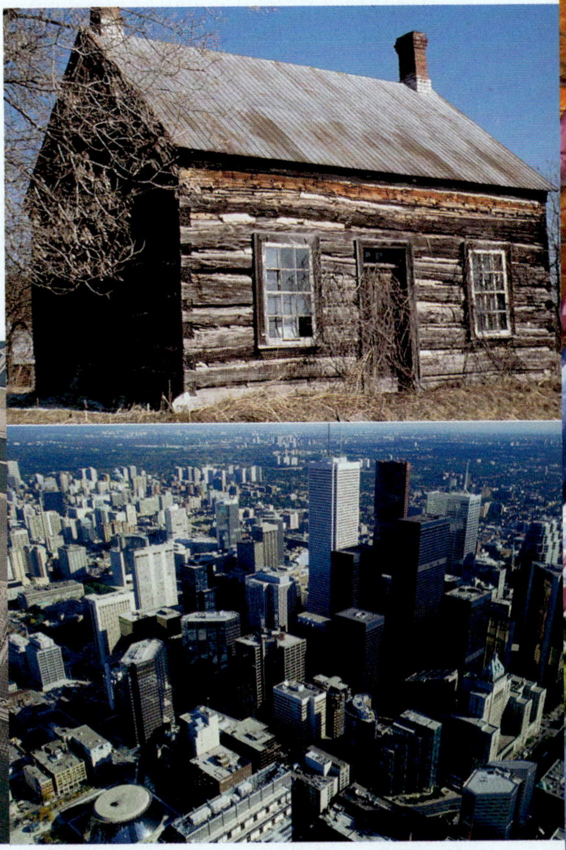

FIGURE 1.23 Architecture as a reflection of culture. This log house (*top*), near Ottawa in Canada, is a folk dwelling and stands in sharp contrast to the professional architecture of the Toronto skyline (*bottom*). ***What conclusion might a perceptive person from another culture reach (considering the "virtues" of height, durability, and centrality) about the ideology of the culture that produced the Toronto landscape?*** *(Top: Courtesy of Terry G. Jordan-Bychkov; Bottom: Photodisc.)*

CONCLUSION

As we have seen and will continue to see, the interests of human geographers are diverse. It might seem to you, confronted by the various themes, subject matter, viewpoints, and methodologies described in this chapter, that geographers run off in all directions, lacking unity of purpose. What does a geographer who studies architecture have in common with a colleague who studies the political and cultural causes of environmental degradation? What interests do an environmental perceptionist and a student of diffusion share? Why do scholars with such apparently different interests belong in the same academic discipline? Why are they all geographers?

The answer is that regardless of the particular topic the human geographer studies, she or he necessarily touches on several or all of the five themes we have discussed. The themes are closely related segments of a whole. Spatial patterns in culture, as revealed by maps of regions, are reflected in and expressed through the cultural landscape, require an ecological interpretation, are the result of mobility, and are inextricably linked with globalization.

As an example of how the various themes of human geography overlap and intertwine, let's look at one element of architecture that most North Americans will be familiar with: the ranch-style, single-family house (see Figure 1.21). This house type is defined by its one-story height and its linear form. These houses are found throughout much of the United States and to a lesser extent in Canada, though they are rare in other countries. They are obviously part of the cultural landscape, and their spatial distribution constitutes a formal region that can be mapped.

Geographers who study such houses also need to employ the other themes of human geography to gain a complete understanding. They can use the concept of diffusion to learn when and by what routes this building style emerged and diffused and what barriers hindered its diffusion. In this particular case, geographers would be led back to the early years of the twentieth century, when the first suburban houses were built outside of urban centers. In this case, land was relatively inexpensive, allowing for a house type that occupied a wide expanse of space. What's more, they would learn that a dominant design motif in the United States in the early twentieth century was based on the notion that buildings should fit in with their natural surroundings instead of dominate them. As a result, low-slung housing styles like the ranch house were particularly popular. Further, the geographer would need an ecological interpretation of the ranch house. What materials were required to build such a house? Did the style vary in the different climatic and ecological regions in which such houses were built? Finally, the human geographer would want to know how the use of ranch houses was related to

globalization. Did economic changes in the world raise the standard of living, leading people to accept ranch houses? Did changes in technology lead to more elaborate houses? Why did it become the quintessential house type in post–World War II America, featured in many of its popular TV shows? Do these humble structures possess a symbolism related to traditional American values and virtues? Thus, the geographer interested in housing is firmly bound by the total fabric of human geography, unable to segregate a particular topic such as ranch houses from the geographical whole. Region, cultural landscape, nature-culture relationships, mobility, and globalization are interwoven.

In this manner, the human geographer passes from one theme to another, demonstrating the holistic nature of the discipline. In no small measure, it is this holism—this broad, multithematic approach—that distinguishes the human geographer from other students of culture. We believe that, by the end of the course, you will have gained a new perspective on the Earth as the home of humankind

DOING GEOGRAPHY

Space, Place, and Knowing Your Way Around

We started this chapter by saying that most of us are born geographers, with a sense of curiosity about the places and spaces around us. Think, for example, of the place you call home. You are probably familiar enough with its streets and buildings and greenspaces, and with the people who inhabit these spaces, to make connections among them—you know how to "read" the place. There are many other places, however, where this is not the case. Most of you have had the experience of going somewhere new, where it is difficult to find your way around, literally and metaphorically. Many of you, for example, are attending a college or university far from home, whereas others have experienced the disorienting feeling of moving from one home to another, whether across town or across continents. How did you find your way? How did the "strange" space become a familiar place?

This activity requires you to draw on your own experiences to understand two fundamental concepts in human geography: space and place. Geographers tend to use the term *space* in a much more abstract way than *place*—as a term that describes a two-dimensional location on a map. *Place,* however, is a less dry term, one that is used to describe a location that has meaning. Your college campus, for example, may have been simply an abstract space located on a map when you applied to the school, yet now it is a place because you have filled it with your own meanings. Here are the steps for your activity.

Steps to Understanding Space, Place, and Knowing Your Way Around

Step 1: Draw on your own experiences by picking one particular space that has become a place for you.

Step 2: Identify what you knew about this space beforehand and how you knew this (for example, perhaps you looked up the place in an atlas or saw it in a movie).

Step 3: In narrative form, describe the process whereby that space became a place for you.

Step 4: Identify the ways in which you learned how to "read" this place. In other words, how did you learn to see this particular location as a three-dimensional, meaningful place that is different from what you knew of it as a "space"?

Use your experiences to answer the following questions:

- Do you have to live in a place to really "know" it?
- Do you have to experience a space for it to become a place?

Father and son in front of college dorm. For many young people, the transition to college is the first time they have had to find their own way and create homes for themselves. *(© Corbis Super RF/Alamy.)*

Key Terms

Geography on the Internet

You can learn more about the discipline of geography and the subdiscipline of human geography on the Internet at the following web sites:

American Geographical Society
http://www.amergeog.org/
America's oldest geographical organization, with a long and distinguished record; publisher of the *Geographical Review.*

Association of American Geographers
http://www.aag.org/
The leading organization of professional geographers in the United States. This site contains information about the discipline, the association, and its activities, including annual and regional meetings.

National Geographic Society
http://www.nationalgeographic.com/
An organization that has, for more than a century, served to popularize geography with active programs of publishing and television presentations prepared for the public.

SEEING GEOGRAPHY
Aboriginal Topographical Painting of the Great Sandy Desert
Why is it difficult for most of us to interpret this image as a map?

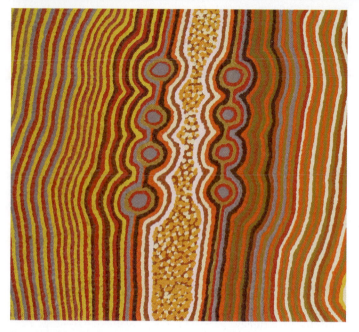

Painting by Helicopter Tjungurrayi depicting the Great Sandy Desert south of Balgo, Western Australia.

As the title indicates, this is an Aboriginal painting that depicts a landscape in the Great Sandy Desert, which is located in the northwest of Western Australia. Like all maps, it is a two-dimensional rendering of three-dimensional space. In other words, it is an attempt to represent land and location on a flat surface. All cultures devise certain symbols that allow for these representations to be understood. On United States Geological Survey (USGS) maps, for example, a standardized set of symbols represent such things as roads, rivers, and cities. Similarly, this map is filled with symbols that represent such features as watering holes (the circular figures) and sandhills (the wavy lines).

As we've learned throughout this chapter, different cultures create and experience landscapes differently. Here we can see that different cultures also represent their landscapes differently. This image is part of a long tradition of Australian Aboriginal topographic representations that reflect their particular culture and their views of the lands they occupy. In general, these representations are religious in nature, depicting myths about the sites and travels of ancestors. Most of these myths concern a spiritual identification between people and their lands. Hence, these maplike representations are not meant as objective measurements of land, as in degrees of longitude and latitude or miles and kilometers. Instead, they communicate meanings about the sacred relationships between people and their physical environments.

To most Western eyes, then, these images do not look like maps because we neither recognize the symbols that Aboriginal peoples use to translate three-dimensional spaces into two dimensions nor think of maps as meaningful in and of themselves. Most likely, Western maps would look very odd to Aboriginal eyes. One of the goals of studying human geography, as we will see throughout this book, is to appreciate this diversity—this mosaic—of relationships between peoples and the places they inhabit, shape, and represent.

Royal Geographic Society/Institute of British Geographers

http://www.rgs.org/

Explore the activities of these allied British organizations, whose collective history goes back to the Age of Exploration and Discovery in the 1800s . . . and don't forget to visit The Human Mosaic Online at http://www.whfreeman.com/domosh12e/

Sources

Anderson, Kay, and Fay Gale. 1999. *Cultural Geographies*. Melbourne: Pearson Education.

Blaut, James M. 1977. "Two Views of Diffusion." *Annals of the Association of American Geographers* 67: 343–349.

Carson, Rachel. 1962. *Silent Spring*. Boston: Houghton Mifflin.

Gould, Peter. 1993. *The Slow Plague: A Geography of the AIDS Pandemic*. Cambridge: Blackwell.

Griffin, Ernst, and Larry Ford. 1980. "A Model of Latin American City Structure." *Geographical Review* 70: 397–422.

Hägerstrand, Torsten. 1967. *Innovation Diffusion as a Spatial Process*. Allan Pred (trans.). Chicago: University of Chicago Press.

Lewis, Peirce. 1983. "Learning from Looking: Geographic and Other Writing About the American Cultural Landscape." *American Quarterly* 35: 242–261.

Norwine, Jim, and Thomas D. Anderson. 1980. *Geography as Human Ecology?* Lanham, Md.: University Press of America.

Ormrod, Richard K. 1990. "Local Context and Innovation Diffusion in a Well-Connected World." *Economic Geography* 66: 109–122.

Relph, Edward. 1981. *Rational Landscapes and Humanistic Geography.* New York: Barnes & Noble.

Staudt, Amanda, Nancy Huddleston, and Sandi Rudenstein. 2006. *Understanding and Responding to Climate Change, a Report Prepared by the National Research Council based on National Academies' Reports.* Washington, D.C.: National Academy of Science.

Tuan, Yi-Fu. 1974. *Topophilia: A Study of Environmental Perception, Attitudes, and Values.* Englewood Cliffs, N.J.: Prentice-Hall.

Warren, Karen J. (ed.). 1997. *Ecofeminism: Women, Culture, Nature.* Bloomington: Indiana University Press.

Ten Recommended Books on a Cultural Approach to Human Geography

(For additional suggested readings, see *The Human Mosaic* web site: www.whfreeman.com/domosh12e)

Anderson, Kay, Mona Domosh, Steve Pile, and Nigel Thrift (eds.). 2003. *Handbook of Cultural Geography.* London: Sage Publications. An edited collection of essays that push the boundaries of cultural geography into such subdisciplines as economic, social, and political geography.

Blunt, Alison, Pyrs Gruffudd, Jon May, Miles Ogborn, and David Pinder (eds.). 2003. *Cultural Geography in Practice.* London: Arnold Publishing. An edited collection of essays that take a very practical view of what it means to actually conduct research in the field of cultural geography.

Cosgrove, Denis. 1998. *Social Formation and Symbolic Landscape.* Madison: University of Wisconsin Press. The landmark study that outlines the relationships between the idea of landscape and social and class formation in such places as Italy, England, and the United States.

Cosgrove, Denis, and Stephen Daniels (eds.). 1990. *The Iconography of Landscape: Essays on the Symbolic Representation, Design and Use of Past Environments.* Cambridge: Cambridge University Press. An important collection of essays that foreground an ideological reading of landscape.

Duncan, James, Nuala Johnson, and Richard Schein (eds.). 2007. *A Companion to Cultural Geography.* New York: Blackwell. A set of essays that represent contemporary thinking about the state of cultural geography in the English-speaking world.

Foote, Kenneth E., Peter J. Hugill, Kent Mathewson, and Jonathan M. Smith (eds.). 1994. *Re-Reading Cultural Geography.* Austin: University of Texas Press. A beautifully compiled representative collection of some of the best works in American cultural geography at the end of the twentieth century, and a useful companion to the book edited by Wagner and Mikesell.

Mitchell, Don. 1999. *Cultural Geography: A Critical Introduction.* New York: Blackwell. An introductory text on cultural geography that emphasizes the material and political elements of the discipline.

Radcliffe, Sarah (ed.). 2006. *Culture and Development in a Globalizing World: Geographies, Actors, and Paradigms.* London: Routledge. A series of essays that provide case studies from around the world showing the various ways that culture and economic development are integrally related.

Tuan, Yi-Fu. 1974. *Topophilia: A Study of Environmental Perception, Attitudes, and Values.* Englewood Cliffs, N.J.: Prentice-Hall. A Chinese-born geographer's innovative and imaginative look at people's attachment to place, a central concern of the cultural approach to human geography.

Wagner, Philip L., and Marvin W. Mikesell (eds.). 1962. *Readings in Cultural Geography.* Chicago: University of Chicago Press. A classic collection, edited by two distinguished Berkeley-trained cultural geographers, presenting the subdiscipline as it was in the mid-twentieth century and developing the device of five themes.

Journals in Human Geography

Annals of the Association of American Geographers. Volume 1 was published in 1911. The leading scholarly journal of American geographers.

Cultural Geographies (formerly known as *Ecumene*). Volume I was published in 1994.

Geographical Review. Published by the American Geographical Society. Volume 1 was published in 1916.

Journal of Cultural Geography. Published semiannually by the Department of Geography, Oklahoma State University, Stillwater, Oklahoma. Volume 1 was published in 1980.

Progress in Human Geography. A quarterly journal providing critical appraisal of developments and trends in the discipline. Volume 1 was published in 1977.

Social and Cultural Geography. Volume 1 was published in 2000 by Routledge, Taylor, & Francis Ltd. in Great Britain.

Enjoying nature in a national park campground.
(Courtesy of Roderick Neumann.)

What can this scene tell us about culture-nature relations in North American popular culture?

Go to "Seeing Geography" on page 66 to learn more about this image.

MANY WORLDS
Geographies of Cultural Difference

No matter where we live, if you look carefully, you will be reminded constantly of how important the expression of cultural identity is to people's daily lives. The geography of cultural difference is evident everywhere—not only in the geographic distribution of different cultures but also in the way that difference is created or reinforced by geography. For example, in the United States, the history of legally enforced spatial segregation of "whites" and "blacks" has been important in establishing and maintaining cultural differences between these groups.

In Chapter 1 we noted that human geographers are interested in studying the geographic expression of difference both among and within cultures. For example, using the concept of formal region, we can identify and map differences among cultures. This sort of analysis is usually done on a very large geographic scale, such as a continent or even the entire world. But geographers are also interested in analyses at smaller scales. When we look closer at a formal culture region, we begin to see that differences appear along racial, ethnic, gender, and other lines of distinction. Sometimes groups within a dominant culture become distinctive enough that we label them **subcultures.** These can be the result of resistance to the dominant culture or of a distinct religious, ethnic, or national group forming an enclave community within a larger culture.

subcultures
Groups of people with norms, values, and material practices that differentiate them from the dominant culture to which they belong.

In this chapter, we are going to explore the geographies of cultural difference using three broad cat-egories of classification: folk, popular, and indigenous cultures. Popular culture, as we will see, is synonymous with mass culture and so, by definition, is the dominant form of cultural expression. Folk and indigenous cultures are, to a large degree, distinguished in relation to popular culture. The term *difference* implies a relationship and a set of criteria for comparison and assessment; that is, cultures are defined relationally.

But what does it mean to speak of "geographies" in the plural? Isn't there only one "geography"? The plural form emphasizes that there is no single way of seeing the land and the landscape. Recall from Chapter 1 our discussion on the concept of subjective experience in the sense of place, which emphasizes the multiplicity of meanings versus a single, universally shared meaning. Cultural geography studies have shown, for example, that women and men often experience the same places in different ways. A certain street corner or tavern might be a comfortable and familiar hangout for men but a threatening or uncomfortable zone that women avoid. To speak of geographies, then, is to go beyond the idea of a single, objectively observable world and raise new questions about the different meanings that people give to places and landscapes; how these relate to their sense of self and belonging; and how, in multicultural societies, we deal with these different meanings politically and socially. As we explore folk, indigenous, and popular cultures in this chapter, we need to keep in mind the multiple subjectivities that operate both within and among cultures.

Many Cultures

Cultures are classified using many different criteria. The concept of culture includes both material and nonmaterial elements. **Material culture** includes all objects or "things" made and used by members of a cultural group: buildings, furniture, clothing, artwork, musical instruments, and other physical objects. The elements of material culture are visible. **Nonmaterial culture** includes the wide range of beliefs, values, myths, and symbolic meanings that are transmitted across generations of a given society. Cultures may be categorized and geographically located using criteria based on either or both of these features.

Let's explore how these criteria are used to identify, categorize, and graphically delineate cultures. According to literary critic and cultural theorist Raymond Williams, *culture* is one of the two or three most complicated words in the English language. In the late eighteenth and early nineteenth centuries, people began to speak of "cultures" in the plural form. Specifically, they began thinking about "European culture" in relation to other cultures around the world. As Europe industrialized and urbanized in the nineteenth century, a new term was invented, **folk culture,** to distinguish traditional ways of life in rural spaces from those in the new urban and industrial ones. Thus, folk culture was defined and made sense only in relation to an urban, industrialized culture. Urban dwellers began to think—in increasingly romantic and nostalgic terms—of rural spaces as inhabited by distinct folk cultures.

The word **folk** describes a rural people who live in an old-fashioned way—a people holding onto a lifestyle less influenced by modern technology. Folk cultures are rural, cohesive, largely self-sufficient groups that are homogeneous in custom and ethnicity. In terms of nonmaterial culture, folk cultures typically have strong family or clan structures and highly localized rituals. Order is maintained through sanctions based in

material culture
All physical, tangible objects made and used by members of a cultural group, such as clothing, buildings, tools and utensils, instruments, furniture, and artwork; the visible aspect of culture.

nonmaterial culture
The wide range of tales, songs, lore, beliefs, values, and customs that pass from generation to generation as part of an oral or written tradition.

folk culture
A small, cohesive, stable, isolated, nearly self-sufficient group that is homogeneous in custom and race; characterized by a strong family or clan structure, order maintained through sanctions based in the religion or family, little division of labor other than that between the sexes, frequent and strong interpersonal relationships, and a material culture consisting mainly of handmade goods.

folk
Traditional, rural; the opposite of "popular."

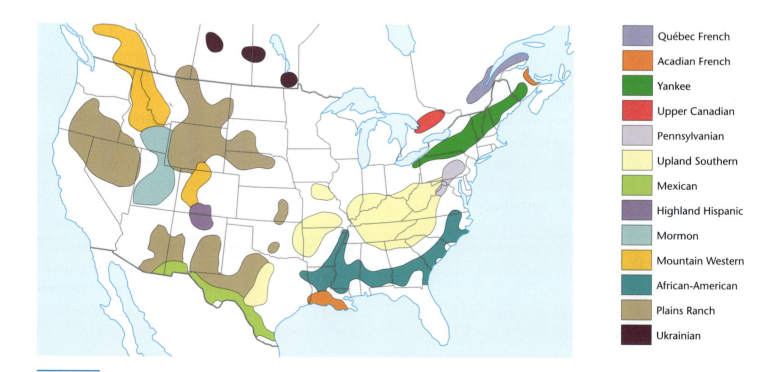

FIGURE 2.1 Folk cultural survival regions of the United States and southern Canada. All are now in decay and retreat, and no true folk cultures survive in North America.

religion or the family, and interpersonal relationships are strong. In terms of material culture, most goods are handmade, and a subsistence economy prevails. Individualism is generally weakly developed in folk cultures, as are social classes.

In the poorer countries of the underdeveloped world, some aspects of folk culture still exist, though few if any peoples have been left untouched by the forces of globalization. In industrialized countries, such as the United States and Canada, unaltered folk cultures no longer exist, though many remnants can be found (**Figure 2.1**). **Folk geography,** a term coined by Eugene Wilhelm, may be defined as the study of the spatial patterns and ecology of these traditional groups.

folk geography
The study of the spatial patterns and ecology of traditional groups; a branch of cultural geography.

Popular culture, by contrast, is generated from and concentrated mainly in urban areas (**Figure 2.2**). Popular material goods are mass-produced by machines in factories, and a cash economy, rather than barter or subsistence, dominates. Relationships among individuals are more numerous but less personal than in folk cultures, and the family structure is weaker. Mass media such as film, print, television, radio, and, increasingly, the Internet are more influential in shaping popular culture. People are more mobile, less attached to place and environment. Secular institutions of authority—such as the police, army, and courts—take the place of family and church in maintaining order. Individualism is strongly developed.

Another major category is **indigenous culture.** A simple definition of

popular culture
A dynamic culture based in large, heterogeneous societies permitting considerable individualism, innovation, and change; having a money-based economy, division of labor into professions, secular institutions of control, and weak interpersonal ties; producing and consuming machine-made goods.

indigenous culture
A culture group that constitutes the original inhabitants of a territory, distinct from the dominant national culture, which is often derived from colonial occupation.

FIGURE 2.2 Popular culture is reflected in every aspect of life, from the clothes we wear to the recreational activities that occupy our leisure time. *(Left: Scott Olson/Getty Images; Right: Mikhael Subotzky/Corbis.)*

indigenous is "native" or "of native origin." In the modern world of sovereign nation-states, the word has acquired much greater cultural and political meanings. In fact, the International Labour Organization's (ILO) Indigenous and Tribal Peoples Convention 169 (Article 1.1) presents a legal definition that recognizes indigenous peoples as comprising a distinct culture. According to the ILO, indigenous peoples are self-identified tribal peoples whose social, cultural, and economic conditions distinguish them from the national society of their host state. Indigenous peoples are regarded as descending from peoples present in the state territory at the time of conquest or colonization. Although they may share some of the material and nonmaterial characteristics of folk cultures, their histories (and geographies) are quite distinct. Indigenous cultures are, in effect, those peoples who were colonized—mostly, but not exclusively—by European cultures and are now minorities in their homelands.

This definition is applied globally, suggesting that indigenous cultures worldwide share common traits and face similar perils and opportunities. The United Nations helped focus global attention on indigenous cultures when it declared 1995–2004 to be the International Decade of the World's Indigenous People. Then, in 2007, the UN General Assembly adopted the United Nations Declaration on the Rights of Indigenous Peoples. Though not binding in international law, the declaration prohibits discrimination against indigenous peoples and defends their right to maintain their cultures, identities, traditions, and institutions. Only the United States and Canada, two countries with a long history of conflict between indigenous peoples and European settlers, declined to sign the declaration.

In most cases, folk and indigenous cultures can be thought of as subcultures in relation to a dominant popular culture. In reality, none of these categories are homogeneous. We can use our five themes—region, mobility, globalization, nature-culture, and cultural landscape—to study geographies of cultural difference.

Region

How do cultures vary geographically? Some cultures exhibit major material and nonmaterial variations from place to place with minor variations over time. Others display less difference from region to region but change rapidly over time. For this reason, the theme of culture region is particularly well suited to the study of cultural difference. Formal culture regions can be delineated on the basis of both material and nonmaterial elements.

Material Folk Culture Regions

Although folk culture has largely vanished from the United States and Canada, vestiges remain in various areas of both countries. Figure 2.1 (page 32) shows culture regions in which the material artifacts of 13 different North American folk cultures survive in some abundance, but even these artifacts are disappearing. Each region possesses many distinctive relics of material culture.

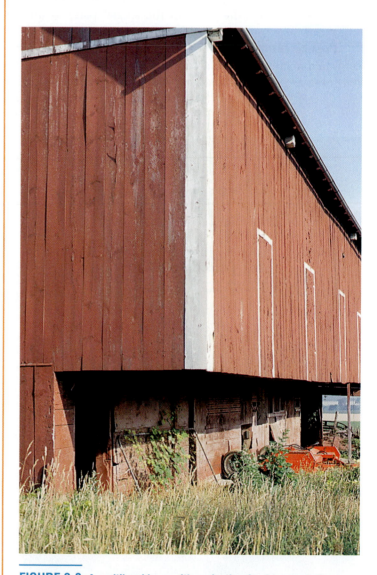

FIGURE 2.3 A multilevel barn with projecting forebay, central Pennsylvania. Every folk culture region possesses distinctive forms of traditional architecture. Of Swiss origin, the forebay barn is one of the main identifying material traits of the Pennsylvanian folk culture region. *(Courtesy of Terry G. Jordan-Bychkov.)*

FIGURE 2.4 A scraped-earth folk graveyard in East Texas. The laborious removal of all grass from such cemeteries is an African-derived custom. Long ago, this practice diffused from the African-American folk culture region to European-Americans in the southern coastal plain of the United States to become simply a "southern" custom. *(Courtesy of Terry G. Jordan-Bychkov.)*

FIGURE 2.5 Beef wheel in the ranching country of the Harney Basin in central Oregon. This windlass device hoists the carcass of a slaughtered animal to facilitate butchering. Derived, as was much of the local ranching culture, from Hispanic Californians, the beef wheel represents the folk material culture of ranching. *(Courtesy of Terry G. Jordan-Bychkov.)*

For example, the strongly Germanic Pennsylvanian folk culture region features an unusual Swiss-German type of barn, distinguished by an overhanging upper-level "forebay" on one side (**Figure 2.3**). In contrast, barns are usually attached to the rear of houses in the Yankee folk region, which is also distinguished by an elaborate traditional gravestone art, featuring "winged death heads." The Upland South is noted in part for the abundance of a variety of distinctive house types built using notched-log construction. The African-American folk region displays such features as the "scraped-earth" cemetery, from which all grass is laboriously removed to expose the bare ground (**Figure 2.4**); the banjo, an instrument of African origin; and head kerchiefs worn by women. Grist windmills with sturdy stone towers and pétanque (a bowling game played with small metal balls), among other traits, characterize the Québec French folk region. The Mormon folk culture is identifiable by distinctive hay derricks and clustered farm villages conforming to a checkerboard street pattern. The western plains ranch folk culture produced such material items as the "beef wheel," a windlass used during butchering (**Figure 2.5**). These examples of material artifacts are only a few of the many that survive from various folk regions.

Is Popular Culture Placeless?

Superficially at least, popular culture varies less from place to place than does folk culture. In fact, Canadian geographer Edward Relph goes so far as to propose that popular culture produces a profound **placelessness,** a spatial standardization that diminishes regional variety and demeans the human spirit. Others observe that one place seems pretty much like another, each robbed of its unique character by the pervasive influence of a continental or even worldwide popular culture (**Figure 2.6**, page 36). When compared with regions and places produced by folk culture, rich in their uniqueness (**Figure 2.7**, page 36), the geographic face of popular culture often seems expressionless. The greater mobility of people in popular culture weakens attachments to place and compounds the problem of placelessness. Moreover, the spread of McDonald's, Levi's, CNN, shopping malls, and much else further adds to the sense of placelessness.

placelessness
A spatial standardization that diminishes regional variety; may result from the spread of popular culture, which can diminish or destroy the uniqueness of place through cultural standardization on a national or even worldwide scale.

FIGURE 2.6 Placelessness exemplified: scenes almost anywhere, developed world. *Guess where these pictures were taken.* The answers are provided on page 69. *(See also Curtis, 1982. Photo at left: Donald Dietz/Stock Boston, Inc.; Top right: David Frazier/Photo Researchers; Bottom right: Courtesy of Terry G. Jordan-Bychkov.)*

But is popular culture truly regionless and placeless? Many cultural geographers are more cautious about making such a sweeping generalization. The geographer Michael Weiss, for example, argues in his book *The Clustering of America* that "American society has become increasingly fragmented" and identifies 40 "lifestyle clusters" based on postal zip codes. "Those five digits can indicate the kinds of magazines you read, the meals you serve at dinner," and what political party you support. "Tell me someone's zip code and I can predict what they eat, drink, drive—even think." The lifestyle clusters, each of which is a formal culture region, bear Weiss's colorful names—such as "Gray Power" (upper-middle-class retirement areas), "Old Yankee Rows" (older

FIGURE 2.7 Retaining a sense of place: a hill town in Cappadocia Province, Turkey. This town, produced by a folk culture, exhibits striking individuality. *How can you tell that this is not a popular culture landscape?* *(Courtesy of Terry G. Jordan-Bychkov.)*

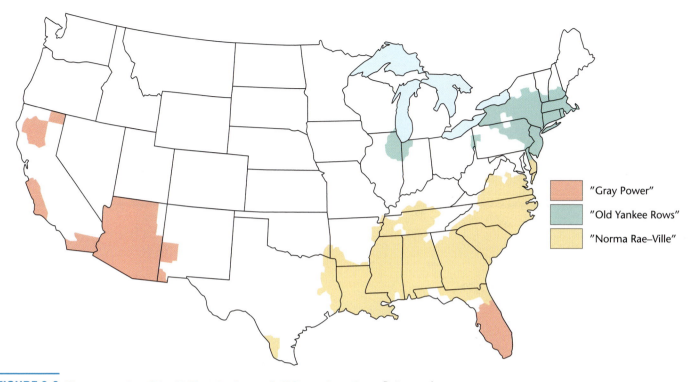

FIGURE 2.8 **Three examples of the 40 lifestyle clusters in U.S. popular culture.** Patterns of consumption within popular culture shift regionally, resulting in "lifestyle clusters." For a description of each lifestyle, see the text. ***Are these regions accurately described? What would you change?***

ethnic neighborhoods of the Northeast), and "Norma Rae–Ville" (lower- and middle-class southern mill towns, named for the Sally Field movie about the tribulations of a union organizer in a textile-manufacturing town) (**Figure 2.8**). Old Yankee Rowers, for example, typically have high school educations, enjoy bowling and ice hockey, and are three times as likely as the average American to live in row houses or duplexes. Residents of Norma Rae–Ville are mostly non-union factory workers, have trouble earning a living, and consume twice as much canned stew as the national average. In short, a whole panoply of popular subcultures exist in America and the world at large, each possessing its own belief system, spokespeople, dress code, and lifestyle.

Reflecting on Geography

Do you live in a "placeless" place, in "nowhere U.S.A."? If not, how is a distinctive regional form of popular culture reflected in your region?

Popular Food and Drink

A persistent formal regionalization of popular culture is vividly revealed by what foods and beverages are con-

sumed, which varies markedly from one part of a country to another and throughout different parts of the world. The highest per capita levels of U.S. beer consumption occur in the West, with the notable exception of Mormon Utah. Whiskey made from corn, manufactured both legally and illegally, has been a traditional southern alcoholic beverage, whereas wine is more common in California.

Foods consumed by members of the North American popular culture also vary from place to place. In the South, grits, barbecued pork and beef, fried chicken, and hamburgers are far more popular than elsewhere in the United States, whereas more pizza and submarine sandwiches are consumed in the North, the destination for many Italian immigrants.

The spread of global brands such as Coca-Cola and Kentucky Fried Chicken would seem to indicate increasing homogenization of food and beverage consumption. Yet many studies show that such brands have different meanings in different places around the world. Coca-Cola may represent modernization and progress in one place and foreign domination in another. Sometimes multinational corporations have to change their foods and beverages to suit local cultural preferences. For example, in Mumbai,

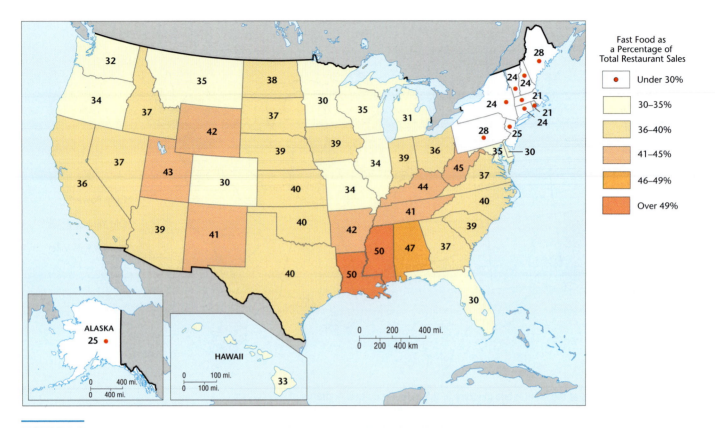

FIGURE 2.9 **Fast-food sales as a share of total restaurant sales, by state, 2007. *What does this illustration suggest about the claim that regional cultures in America are collapsing into a national culture?*** *(Source: U.S. Census Bureau, 2007 Economic Census.)*

India, McDonald's has had to add local-style sauces to its menus. Rather than a simple one-way process of "Americanizing" Indian food preferences, local consumers are "Indianizing" McDonald's.

Fast food might seem to epitomize popular culture, yet its importance varies greatly even within the United States (**Figure 2.9**). The stronghold of the fast-food industry is the American South; the Northeast has the fewest fast-food restaurants. Such differences undermine the geographic uniformity or placelessness supposedly created by popular culture. Music provides another example.

Popular Music

Popular culture has spawned many different styles of music, all of which reveal geographic patterns in levels of acceptance. Elvis Presley epitomized both popular music and the associated cult of personality. Even today, more than a generation after his death, he retains an important place in American popular culture.

Elvis also illustrates the vivid geography of that culture. In the sale of Presley memorabilia, the nation reveals a split personality. The main hotbeds of Elvis worship lie in the eastern states, whereas the King of Rock and Roll is largely forgotten out west. Although it raises more questions than it answers, **Figure 2.10** leaves no doubt that popular culture varies regionally.

Indigenous Culture Regions

Concentrations of indigenous peoples are generally in areas with few roads or modern communications systems, such as mountainous areas, vast arid and semiarid regions, or large expanses of forest or wetlands. These concentrations constitute indigenous culture regions. Worldwide, large concentrations of indigenous populations exist outside of the strong influence of national cultures and the effective control of governments located in faraway capital cities. National control by a central government is often weakened by minimal infrastructure, rough topography, or harsh environmental conditions. In many cases, indigenous peoples either fled or were forcibly removed by central governments to environmentally marginal regions, such as arid lands.

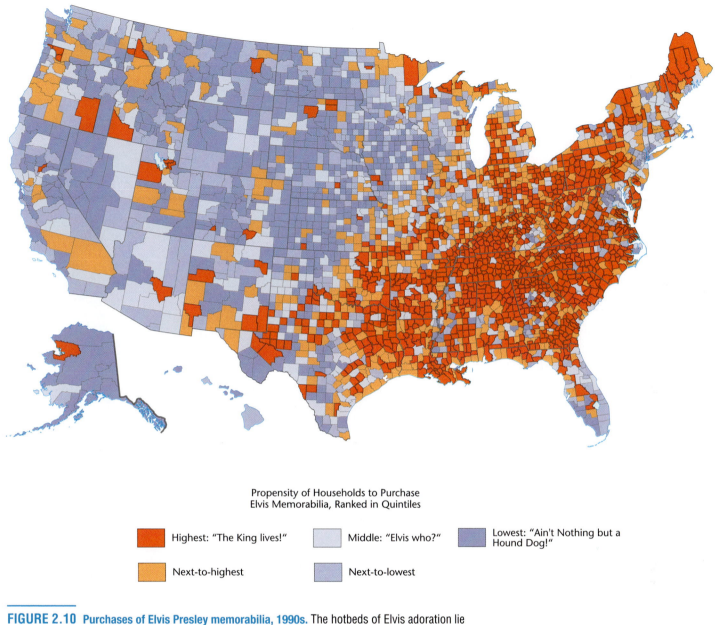

Propensity of Households to Purchase
Elvis Memorabilia, Ranked in Quintiles

■ Highest: "The King lives!"	■ Middle: "Elvis who?"	■ Lowest: "Ain't Nothing but a Hound Dog!"
■ Next-to-highest	■ Next-to-lowest	

FIGURE 2.10 Purchases of Elvis Presley memorabilia, 1990s. The hotbeds of Elvis adoration lie mainly in the eastern United States, while most westerners can take him or leave him. ***What cultural factors might underlie this "fault line" in the geography of popular culture?*** (Redrawn, based on data collected by Bob Lunn of DICI, Bellaire, Texas, and published by Edmonson and Jacobson, 1993.)

In the United States and Canada, this geographic pattern of marginalization is particularly evident. In the United States, the central government has had a complex and often contradictory relationship with indigenous peoples, sometimes treating them as sovereign nations and at other times as second-class citizens. Much of the current pattern of indigenous population distribution reflects both the history of the east-to-west movement of European set-

tlers and nineteenth-century government policies (**Figure 2.11**, page 40). For example, the Indian Removal Act of 1830 was intended to make way for European settlers by relocating eastern American Indian tribes west of the Mississippi River, many to Oklahoma. Western American Indian tribes were also eventually forced onto government-created reservations, generally on the most unproductive and arid lands. Some of the larger reservation complexes in

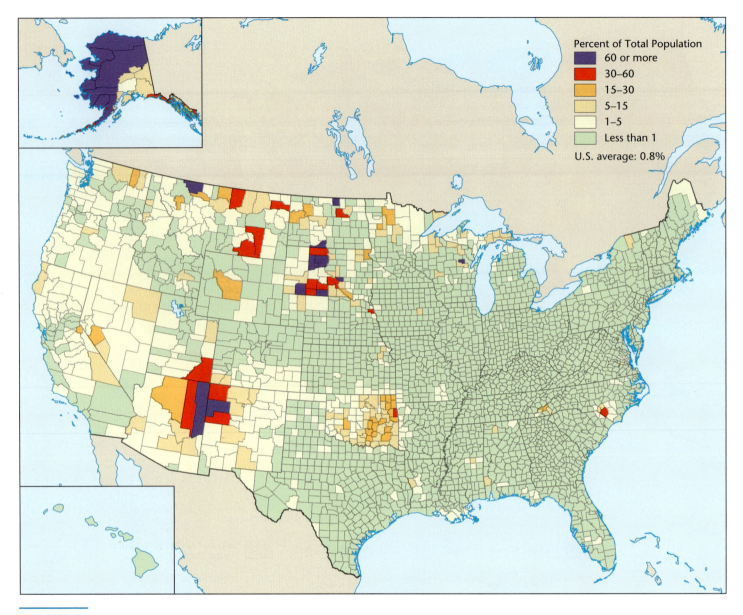

FIGURE 2.11 Indigenous American Indian population distribution in the United States.
(Source: U.S. Census Bureau.)

the arid West constitute an indigenous culture region (**Figure 2.12**).

The so-called Hill Tribes of South Asia are another good example. Mountain ranges, including the Chittagong Hills, the Assam Hills, and the Himalayas, surround the fertile valleys and deltas around which the ancient South Asian Hindu and Islamic civilizations were centered. Various indigenous peoples occupy these highland regions, which are remote from the lowland centers of authority and culturally distinct from them (**Figure 2.13**). A series of indigenous culture regions ringing the valleys of South

Asia thus exists, occupied by what the British colonial authorities referred to as Hill Tribes. Most of these peoples practice some version of swidden agriculture, which involves multiyear cycles of forest clearing, planting, and fallowing (see Chapter 8). Most hold Christian or animist beliefs and speak languages distinct from those spoken in the lowlands. A similar pattern of highland indigenous culture regions can also be identified in the countries of Southeast Asia, such as Myanmar and Thailand, where indigenous peoples such as the Shan and Karen have populations in the millions.

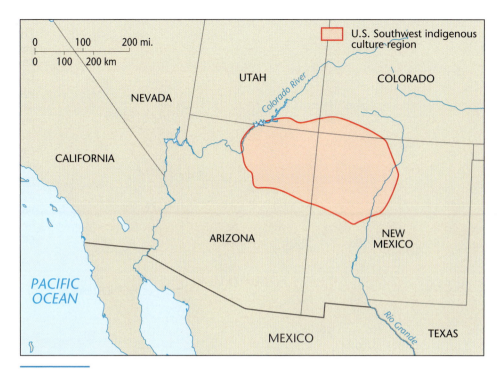

FIGURE 2.12 An indigenous culture region in the United States. *(U.S. Geological Survey.)*

FIGURE 2.13 A Murong tribesman with his children in the indigenous culture region of Bangladesh's Chittagong Hill Tracts. *(Shehzad Noorani/Peter Arnold.)*

Indigenous culture regions also persist in Central and South America. There is a distinct Mayan culture region that encompasses parts of Mexico, Belize, Guatemala, and Honduras (**Figure 2.14**, page 42). Concentrations of Mayan speakers are especially common in rugged highlands and tropical forests. In South America, another concentration of indigenous peoples exists in sections of the Andes Mountains. This area constituted the geographic core of the Inca civilization, which thrived between A.D. 1300 and 1533 and incorporated several major linguistic groups under its rule. Today, up to 55 percent of the national populations of Andean countries, such as Bolivia, Peru, and Ecuador, are indigenous. On the slopes and in the high valleys of the Andes, Quechua and Aymara speakers constitute an overwhelming majority, signifying an indigenous culture region (**Figure 2.15**, page 42).

Vernacular Culture Regions

A **vernacular culture region** is the product of the spatial perception of the population at large—a composite of the mental maps of the people. Such regions vary greatly in size, from small districts covering only part of a city or

vernacular culture region
A culture region perceived to exist by its inhabitants, based in the collective spatial perception of the population at large, and bearing a generally accepted name or nickname (such as "Dixie").

FIGURE 2.14 The Mayan culture region in Middle America. The ancient Mayan Empire collapsed centuries ago, but its Mayan-speaking descendents continue to occupy the region today. In many cases Mayan communities, after centuries of political and economic marginalization, are today actively struggling to have their land rights recognized by their respective governments.

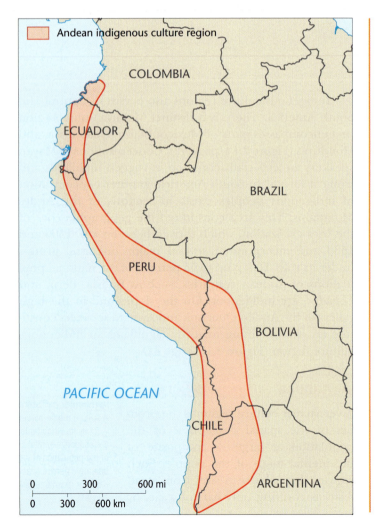

town to huge, multistate areas. Like most other geographical regions, they often overlap and usually have poorly defined borders.

Almost every part of the industrialized Western world offers examples of vernacular regions based in the popular culture. **Figure 2.16** shows some sizable vernacular regions in North America. Geographer Wilbur Zelinsky compiled these regions by determining the most common name for businesses appearing in the white pages of urban telephone directories. One curious feature of the map is the sizable, populous district—in New York, Ontario, eastern Ohio, and western Pennsylvania—where no regional affiliation is perceived. Using a different source of information, geographer Joseph Brownell sought to delimit the popular "Midwest" in 1960 (**Figure 2.17**). He sent out questionnaires to postal employees in the midsection of the United States, from the Appalachians to the Rockies, asking each whether, in his or her opinion, the community lay in the "Midwest." The results identified a vernacular region in

FIGURE 2.15 The indigenous culture region of the Andes, including Quechua- and Aymara-speaking peoples. This is the core of the ancient Inca Empire, where today many of the indigenous people speak their mother language rather than Spanish. *(Adapted from de Blij and Muller, 2004.)*

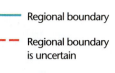

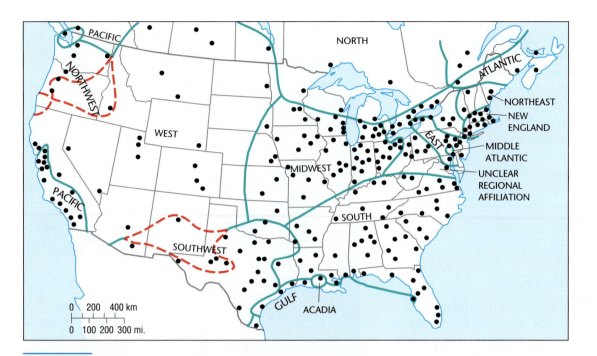

FIGURE 2.16 Some vernacular regions in North America. Cultural geographer Wilbur Zelinsky mapped these regions on the basis of business names in the white pages of metropolitan telephone directories. ***Why are names containing "West" more widespread than those containing "East"? What might account for the areas where no region name is perceived?*** *(Adapted from Zelinsky, 1980a: 14.)*

which the residents considered themselves midwesterners. A similar survey done 20 years later, using student respondents, revealed a core-periphery pattern for the Midwest

(see Figure 2.17). As befits an element of popular culture, the vernacular region is often perpetuated by the mass media, especially radio and television.

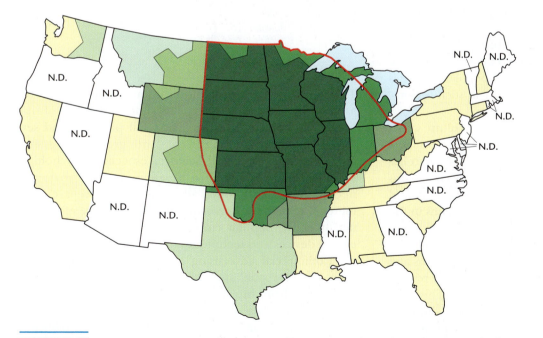

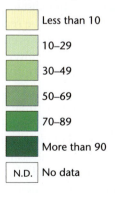

Percentage of University Student Respondents in Different States and Areas Claiming to Live in the "Middle West," 1980

	Less than 10
	10–29
	30–49
	50–69
	70–89
	More than 90
N.D.	No data

—— Border of area in which postmasters claimed to live in the "Middle West," 1960

FIGURE 2.17 The vernacular Middle West or Midwest Two surveys, taken a generation apart and using two different groups of respondents, yielded similar results. *(Sources: Brownell, 1960: 83; Shortridge, 1989.)*

Mobility

Do elements of folk culture spread through geographical space differently from those of popular culture? Whereas folk culture spreads by the same models and processes of diffusion as popular culture, diffusion operates more slowly within a folk setting. The relative conservatism of such cultures produces a resistance to change. The Amish, for example, as one of the few surviving folk cultures, are distinctive today simply because they reject innovations that they believe to be inappropriate for their way of life and values.

Diffusion in Popular Culture

Before the advent of modern transportation and mass communications, innovations usually required thousands of years to complete their areal spread, and even as recently as the early nineteenth century the time span was still measured in decades. In regard to popular culture, modern transportation and communications networks now permit cultural diffusion to occur within weeks or even days. The propensity for change makes diffusion extremely important in popular culture. The availability of devices permitting rapid diffusion enhances the chance for change in popular culture.

Hierarchical diffusion often plays a greater role in popular culture than in folk culture or indigenous culture because popular society, unlike folk culture, is highly stratified by socioeconomic class. For example, the spread of McDonald's restaurants—beginning in 1955 in the United States and, later, internationally—occurred hierarchically for the most part, revealing a bias in favor of larger urban markets (**Figure 2.18**). Further facilitating the diffusion of popular culture is the fact that time-distance decay is weaker in such regions, largely because of the reach of mass media.

Sometimes, however, diffusion in popular culture works differently, as a study of Walmart revealed. Geographers Thomas Graff and Dub Ashton concluded that Walmart initially diffused from its Arkansas base in a largely contagious pattern, reaching first into other parts of Arkansas and neighboring states. Simultaneously, as often happens in the spatial spread of culture, another pattern of diffusion was at work, one Graff and Ashton called reverse hierarchical diffusion. Walmart initially located its stores in smaller towns and markets, only later spreading into cities—the precise reverse of the way hierarchical diffusion normally works. This combination of contagious and reverse hierarchical diffusion led Walmart to become the nation's largest retailer in only 30 years.

FIGURE 2.18 Another McDonald's opens in Moscow. McDonald's, which first spread to Moscow around 1987, has always preferred hierarchical diffusion. Of all McDonald's outlets worldwide today, about 45 percent are located in foreign countries, almost always in large cities. *(Courtesy of Terry G. Jordan-Bychkov.)*

Advertising

The most effective device for diffusion in popular culture, as Zelinsky suggests, confronts us almost every day of our lives. Commercial advertising of retail products and services bombards our eyes and ears, with great effect. Using the techniques of social science, especially psychology, advertisers have learned how to sell us products we do not need. The skill with which advertising firms prepare commercials often determines the success or failure of a product. In short, popular culture is equipped with the most potent devices and techniques of diffusion ever devised.

Commercial advertising is limited in its capacity to overcome all spatial and cultural barriers to homogenization. Cases from international advertising are illustrative. When England-based Cadbury decided to market its line of chocolates and confectionaries in China, the company was forced to change its advertising strategy. Unlike much of

Europe and North America, China had no culture of impulse buying and no tradition of self-service. Cadbury had to change the names of its products, eschew mass marketing, focus on a small group of high-end consumers, and even change product content.

Place of product origin is also extremely important in trying to advertise and market internationally. Sometimes place helps sell a product—think of New Zealand wool or Italian olive oil—and sometimes it is a hindrance to sales. There are many examples of products having negative associations among consumers because the country of origin has a tarnished international reputation. A good example is South Africa's advertising efforts during the era of apartheid, when consumer boycotts of the country's products were common. South African industries had to suppress references to country of origin in their advertising in order to market their product internationally.

Communications Barriers

Although the communications media create the potential for almost instant diffusion over very large areas, this can be greatly retarded if access to the media is denied or limited. *Billboard,* a magazine devoted largely to popular music, described one such barrier. A record company executive complained that radio stations and disk jockeys refused to play "punk rock" records, thereby denying the style an equal opportunity for exposure. He claimed that punk devotees were concentrated in New York City, Los Angeles, Boston, and London, where many young people had found the style reflective of their feelings and frustrations. Without access to radio stations, punk rock could diffuse from these centers only through live concerts and the record sales they generated. The publishers of *Billboard* noted that "punk rock is but one of a number of musical forms which initially had problems breaking through nationally out of regional footholds," for Pachanga, ska, pop/gospel, "women's music," reggae, and "gangsta rap" experienced similar difficulties. Similarly, Time Warner, a major distributor of gangsta rap music, endured scathing criticism from the U.S. Congress in 1995 because of potentially offensive or deleterious aspects of this genre. This eventually led the company to sell the subsidiary label that recorded this form of rap. To control the programming of radio and television, or media distribution generally, is to control much of the diffusionary apparatus in popular culture. The diffusion of innovations ultimately depends on the flow of information.

Government censorship, as opposed to mere criticism, also creates barriers to diffusion, though with varying degrees of effectiveness. In 1995, the Islamic fundamentalist regime in Iran, opposed to what it perceived as the corrupting influences of Western popular culture, outlawed television satellite dishes in an attempt to prevent citizens from watching programs broadcast in foreign countries. The Taliban government of Afghanistan went even further in the 1990s, banning all television sets. Control of the media can greatly control people's tastes in, preferences in, and ideas about popular culture. Even so, repressive regimes must cope with a proliferation of communication methods, including fax machines and the Internet. So pervasive has cultural diffusion become that the insular, isolated status of nations is probably no longer attainable for very long, even under totalitarian conditions.

Although newspapers are potent agents of diffusion in popular culture, they also act as selective barriers, often reinforcing the effect of political boundaries. For example, between 20 and 50 percent of all news published in Canadian newspapers is about foreign countries, whereas only about 12 percent of all news appearing in U.S. papers reports on foreign areas. This pattern suggests that newspaper readers living within U.S. borders are less exposed to world events than those within Canada's.

Reflecting on Geography

> Because Canadian newspapers devote so much more coverage to international stories than U.S. newspapers, are Americans more provincial than Canadians as a result?

Diffusion of the Rodeo

Barriers of one kind or another usually weaken the diffusion of elements of popular culture before they become ubiquitous. The rodeo provides an example. Rooted in the ranching folk culture of the American West, it has never completely escaped that setting (**Figure 2.19**, page 46).

Like so many elements of popular culture, the modern rodeo had its origins in folk tradition. Taking their name from the Spanish *rodear,* "to round up," rodeos began simply as roundups of cattle in the Spanish livestock ranching system in northern Mexico and the American Southwest. Anglo-Americans adopted Mexican cowboy skills in the nineteenth century, and cowboys from adjacent ranches began to hold contests at roundup time. Eventually, some cowboy contests on the Great Plains became formalized, with prizes awarded.

The transition to commercial rodeo, with admission tickets and grandstands, came quickly as an outgrowth of the formal cowboy contests. One such affair, at North Platte, Nebraska, in 1882, led to the inclusion of some rodeo events in a Wild West show in Omaha in 1883. These shows, which moved by railroad from town to town in the manner of circuses, were probably the most potent agent

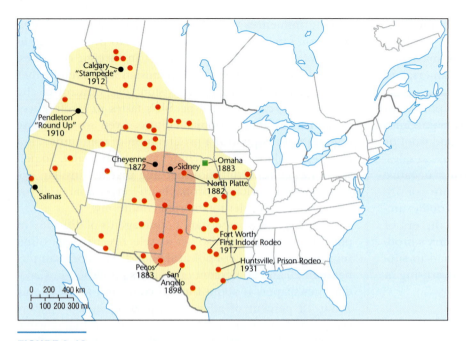

FIGURE 2.19 Origin and diffusion of the American commercial rodeo. Derived originally from folk culture, rodeos evolved through formal cowboy contests and Wild West shows to emerge, in the late 1880s and 1890s, in their present popular culture form. The border between the United States and Canada proved no barrier to the diffusion, although Canadian rodeo, like Canadian football, differs in some respects from the U.S. type. *What barriers might the diffusion have encountered?* *(Sources: Frederickson, 1984; Pillsbury, 1990b.)*

of early rodeo diffusion. Within a decade of the Omaha event, commercial rodeos were being held independently of Wild West shows in several towns, such as Prescott, Arizona. Spreading rapidly, commercial rodeos had appeared throughout much of the West and parts of Canada by the early 1900s. The famous Frontier Days rodeo was first held in 1897 in Cheyenne, Wyoming. By the time of World War I, the rodeo had also become an institution in the provinces of western Canada, where the Calgary Stampede began in 1912.

Today, rodeos are held in 36 states and 3 Canadian provinces. The state of Oklahoma's annual calendar of events lists no fewer than 98 scheduled rodeos. Rodeos have received the greatest acceptance in the popular culture found west of the Mississippi and Missouri rivers (see Figure 2.19). Absorbing and permeable barriers to the diffusion of commercial rodeo were encountered at the U.S. border with Mexico, south of which bullfighting occupies a dominant position, and in the Mormon culture region centered in Utah.

Blowguns: Diffusion or Independent Invention?

Often the path of the past diffusion of an item of material culture is not clearly known or understood, presenting geographers with a problem of interpretation. The blowgun is a good example. A hunting tool of many indigenous peoples, it is a long, hollow tube through which a projectile is blown by the force of one's breath. Geographer Stephen Jett mapped the distribution of blowguns, which he discovered were used in societies in both the Eastern and Western Hemispheres, all the way from the island of Madagascar, off the east coast of Africa, to the Amazon rain forests of South America (**Figure 2.20**).

Indonesian peoples, probably on the island of Borneo, appear to have first invented the blowgun. It became their principal hunting weapon and diffused through much of the equatorial island belt of the Eastern Hemisphere. How, then, do we account for its presence among Native American groups in the Western Hemisphere? Was it independently invented by Native Americans? Was it brought to the Americas by relocation diffusion in pre-Columbian times? Or did it spread to the New World only after the European discovery of America? We do not know the answers to these questions, but the problem presented is one common to cultural geography, especially when studying the traditions of nonliterate cultures, which precludes the use of written records that might reveal such diffusion. Certain rules of thumb can be employed in any given situation to help resolve the issue. For example, if one or more nonfunctional features of blowguns, such as a decorative motif or

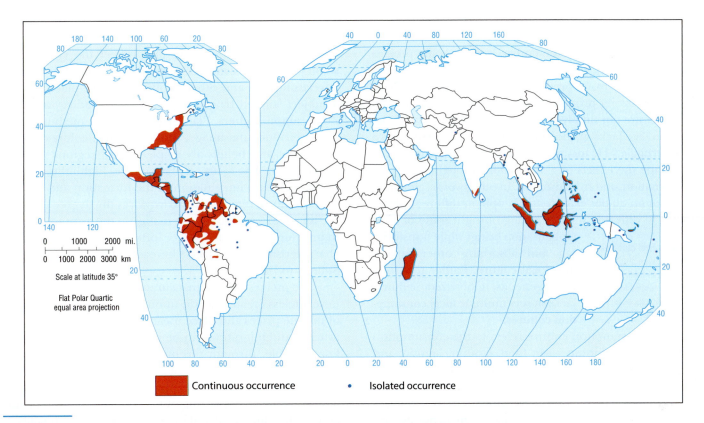

FIGURE 2.20   Former distribution of the blowgun among Native Americans, South Asians, Africans, and Pacific Islanders. The blowgun occurred among folk cultures in two widely separated areas of the world. *Was this the result of independent invention or cultural diffusion? What kinds of data might one seek to answer this question?* Compare and contrast the occurrence in the Indian and Pacific ocean lands to the distribution of the Austronesian languages (see Chapter 4). *(Source: Jett, 1991: 92–93.)*

specific terminology, occurred in both South America and Indonesia, then the logical conclusion would be that cultural diffusion explained the distribution of blowguns.

Globalization

Does globalization homogenize cultural difference? This is the crux of the issue for geographers and other social scientists. As we look for evidence in real world examples, we find that the answer is less than straightforward.

From Difference to Convergence

Globalization is most directly and visibly at work in popular culture. Increased leisure time, instant communications, greater affluence for many people, heightened mobility, and weakened attachment to family and place—all attributes of popular culture—have the potential, through interaction,

to cause massive spatial restructuring. Most social scientists long assumed that the result of such globalizing forces and trends, especially mobility and the electronic media, would be the homogenization of culture, wherein the differences among places are reduced or eliminated. This assumption is called the **convergence hypothesis;** that is, cultures are converging, or becoming more alike. In the geographical sense, this would yield placelessness, a concept discussed earlier in the chapter.

> **convergence hypothesis**
> A hypothesis holding that cultural differences among places are being reduced by improved transportation and communications systems, leading to a homogenization of popular culture.

Impressive geographical evidence can be marshaled to support the convergence hypothesis. Wilbur Zelinsky, for example, compared the given names of people in various parts of the United States for the years 1790 and 1968 and found that a more pronounced regionalization existed in the eighteenth century than in the mid-twentieth century. The personal names that the present generation of parents bestows on children vary less from place to place than did those of our ancestors two centuries ago.

Difference Revitalized

Globalization, we should remember, is an ongoing process or, more accurately, set of processes. It is incomplete and its outcome far from predetermined. Geographer Peter Jackson is a strong proponent of the position that cultural differences are not simply obliterated under the wave of globalization. For Jackson, globalization is best understood as a "site of struggle." He means that cultural practices rooted in place shape the effects of globalization through resistance, transformation, and hybridization. In other words, globalization is not an all-powerful force. People in different places respond in different ways, rejecting outright some of what globalization brings while transforming and absorbing other aspects into local culture. Rather than one homogeneous globalized culture, Jackson sees multiple **local consumption cultures.**

local consumption cultures
Distinct consumption practices and preferences in food, clothing, music, and so forth formed in specific places and historical moments.

Local consumption culture refers to the consumption practices and preferences—in food, clothing, music, and so on—formed in specific places and historical moments. These local consumption cultures often shape globalization and its effects. In some ways, globalization revitalizes local difference. That is, people reject or incorporate into their cultural practices the ideas and artifacts of globalization and in the process reassert place-based identities. The cases of international advertising discussed in the previous section will help illustrate this idea.

Jackson suggests that the introduction of Cadbury's chocolate into China is more than simply another sign of globalization. He argues that the case "demonstrates the resilience of local consumption cultures to which transnational corporations must adapt." In cases where companies' products are negatively associated with their place of origin, such as exports from apartheid South Africa, the global ambitions of multinational companies can be thwarted. "Local" circumstances thus can make a difference to the outcomes of globalization.

consumer nationalism
A situation in which local consumers favor nationally produced goods over imported goods as part of a nationalist political agenda.

Local resistance to globalization often takes the form of **consumer nationalism.** This occurs when local consumers avoid imported products and favor locally produced alternatives. India and China, in particular, have a long history of resisting outside domination through boycotts of imported goods. Jackson discusses a recent case in China in which Chinese entrepreneurs invented a local alternative to Kentucky Fried Chicken called Ronhua Fried Chicken Company. The company uses what it claims are traditional Chinese herbs in its recipe, delivering a product more suitable to Chinese cultural tastes.

Place Images

The same media that serve and reflect the rise of personal preference—movies, television, photography, music, advertising, art, and others—often produce place images, a subject studied by geographers Brian Godfrey and Leo Zonn, among others. Place, portrayer, and medium interact to produce the image, which, in turn, colors our perception of and beliefs about places and regions we have never visited. The focus on place images highlights the role of the collective imagination in the formation and dissolution of culture regions. It also explores the degree to which the image of a region fits the reality on the ground. That is, in imagining a region or place, oftentimes certain regional characteristics are stressed whereas others are ignored (see Subject to Debate).

The images may be inaccurate or misleading, but they nevertheless create a world in our minds that has an array of unique places and place meanings. Our decisions about tourism and migration can be influenced by these images. For example, through the media, Hawaii has become in the American mind a sort of earthly paradise peopled by scantily clad, eternally happy, invariably good-looking natives who live in a setting of unparalleled natural beauty and idyllic climate. People have always formed images of faraway places. Through the interworkings of popular culture, these images proliferate and become more vivid, if not more accurate.

Local Indigenous Cultures Go Global

The world's indigenous peoples often interact with globalization in interesting ways. On the one hand, new global communications systems, institutions of global governance, and international nongovernmental organizations (NGOs) are providing indigenous peoples with extraordinary networking possibilities. Local indigenous peoples around the world are now linked in global networks that allow them to share strategies, rally international support for local causes, and create a united front to defend cultural survival. On the other hand, globalization brings the world to formerly isolated cultures. Global mass communications introduce new values, and multinational corporations' search for new markets and new sources of gas, oil, genetic, forest, and other resources can threaten local economies and environments.

Both aspects of indigenous peoples' interactions with globalization were evident at the World Trade Organization's (WTO) Ministerial Conference in Cancún, Mexico in 2003. Indigenous peoples' organizations from around the world gathered for the conference, hosted by the Mayan community in nearby Quintana Roo. Though not

Subject to Debate

Mobile Identities: Questions of Culture and Citizenship

One of the most dynamic features of cultural difference in the world today is mobility. So-called diaspora communities have sprung up around the globe. Some of these communities have arisen quite recently and are historically unprecedented, such as the movement of Southeast Asians into U.S. cities in the late twentieth century. Some are artifacts of European colonialism, such as the large populations of South Asians in England or West and North Africans in France. Others are deeply historical and express a centuries-old interregional linkage, such as the contemporary movement of North Africans into southern Spain.

The movement and settlement of large populations of migrants have raised questions of belonging and exclusion. How do transplanted populations become English, French, or Spanish, not only in terms of citizenship but also in terms of belonging to that culture? Some observers have argued that diaspora populations find ways to blend symbols from their cultures of origin with those of their host cultures. Thus, people find a sense of belonging through a process of cultural hybridization. Other observers point to long-standing situations of cultural exclusion. In 2005 in France, for example, riots broke out in more than a dozen cities in suburban enclaves of West African and North African populations. The reasons for the riots are complex. However, many observers pointed out that underprivileged youth of African decent feel excluded from mainstream French culture, even though many are second- and third-generation French citizens.

The debate over how to address questions of citizenship and cultural belonging is played out in many venues. In terms of policy, the French, for example, emphasize cultural integration; for example, the French government approved a law in 2010 banning full-face veils (*burqas*) in public, whereas Britain and the United States promote multiculturalism. But many questions remain about the effectiveness of state policies toward diaspora cultures.

Continuing the Debate

Based on the discussion presented above, consider these questions:

- Do the geographic enclaves of Asian and African diaspora populations in the former colonial capitals of Europe reflect an effort by migrants to retain a distinct cultural identity? Or do they reflect persisting racial and ethnic prejudices and efforts to segregate "foreigners"? Or is it a combination of factors?

- How long does it take an Asian immigrant community to become "English" or a West African immigrant community to become "French"? One generation? Two? Never?

- Does the presence of Asians and Africans make the landscape of England and France appear less "English" or less "French"? Why or why not?

Riots and protests, such as this memorial march for two dead teenagers in Clichy sous Bois, spread across several cities in France in 2005. Protesters' complaints centered on the discriminatory treatment of French citizens of African descent. (© Jean-Michel Turpin/Corbis.)

FIGURE 2.21 **A group of indigenous Filipinos participate in the opening of the Forum for Indigenous People** at the Casa de la Cultura in Cancun, Quintana Roo, Mexico, during the World Trade Organization ministerial meetings. Indigenous peoples' groups from around the world organized the forum as a counterpoint to the WTO talks. *(Jack Kurtz/The Image Works.)*

officially part of the conference, they came together there to strategize ways to forward their collective cause of cultural survival and self-determination, gain worldwide publicity, and protest the WTO's vision of globalization (**Figure 2.21**). One outcome of this meeting was the International Cancún Declaration of Indigenous Peoples (ICDIP), a document that is highly critical of current trends in globalization.

According to the ICDIP, the situation of indigenous peoples globally "has turned from bad to worse" since the establishment of the WTO. Indigenous rights organizations claim that "our territories and resources, our indigenous knowledge, cultures and identities are grossly violated" by international trade and investment rules. The document urges governments worldwide to make no further agreements under the WTO and to reconsider previous agreements. The control of plant genetic resources is a particularly important concern. Many indigenous peoples argue that generations of their labor and cumulative knowledge have gone into producing the genetic resources that transnational corporations are trying to privatize for their own profit. The ICDIP asks that future international agreements ensure "that we, Indigenous Peoples, retain our rights to have control over our seeds, medicinal plants and indigenous knowledge."

Globalization is clearly a critical issue for indigenous cultures. Some argue that globalization, because it facili-

tates the creation of global networks that provide strength in numbers, may ultimately improve indigenous peoples' efforts to control their own destinies. The future of indigenous cultural survival ultimately will depend on how globalization is structured and for whose benefit.

Nature-Culture

How is nature related to cultural difference? Do different cultures and subcultures differ in their interactions with the physical environment? Are some cultures closer to nature than others? People who depend on the land for their livelihoods—farmers, hunters, and ranchers, for example—tend to have a different view of nature from those who work in commerce and manufacturing in the city. Indeed, one of the main distinctions between folk culture and popular culture is their differing relationships with nature.

Indigenous Ecology

Many observers believe that indigenous peoples possess a very close relationship with and a great deal of knowledge about their physical environment. In many cases, indigenous cultures have developed sustainable land-use practices over generations of experimentation in a particular environmental setting. As a consequence, academics, journalists, and even corporate advertisers often portray indigenous peoples as defenders of endangered environments, such as tropical rain forests. It was not always this way. Especially during the height of European **colonialism,** indigenous populations (then considered colonial subjects) were often accused of destroying the environment. The then-common belief that indigenous land-use practices were destructive helped Europeans justify colonialism by claiming they were saving colonial subjects from themselves. In hindsight, it is easy to see that this belief was related to now-discredited European ideas about the racial inferiority of colonized peoples.

colonialism
The forceful appropriation of a territory by a distant state, often involving the displacement of indigenous populations to make way for colonial settlers.

Debate continues today, with some observing that, although indigenous cultures may once have lived sustainably, globalization is making their knowledge and practices less useful. That is, globalization introduces new markets, new types of crops, and new technologies that displace existing land-use practices. Others note that it is impossible to generalize about sustainability in indigenous cultures because the way indigenous peoples use their environments varies from place to place and the indigenous societies are internally heterogeneous. A key discussion centers on the role of indigenous peoples in conserving global biodiversity. In part, this reflects the reality that indigenous peoples often occupy territories that Western scientists view as critical to global biodiversity conservation (**Figure 2.22**). For example, 85 percent of national parks and other protected areas in Central America and 80 percent in South America have resident indigenous populations. There is also a close geographic correspondence between indigenous territories and tropical rain forests not only in Latin America but also in Africa and Southeast Asia. Tropical rain forests, although they cover only 6 percent of the Earth's surface, are estimated to contain 60 percent of the world's biodiversity. With the rise of genetic engineering, conservationists and corporations alike view tropical forests as in situ gene banks. As multinational biotechnology companies look to the tropics for genetic resources for use in developing new medicines or crop seeds, indigenous peoples are increasingly vocal about their proprietary rights over the biodiversity of their homelands.

Faced with these issues, cultural geographers generally emphasize the continued importance of indigenous knowledge for environmental management and of indigenous land-use practices for sustainable development. Initially, geographers focused on the adaptive strategies of indigenous

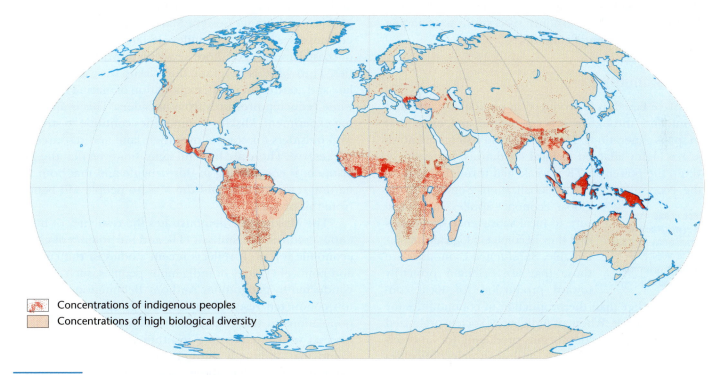

Concentrations of indigenous peoples
Concentrations of high biological diversity

FIGURE 2.22 The global congruence of cultural and biological diversity. Conservationists are aware that many of the world's most biologically diverse regions are occupied by indigenous cultures. Many suggest, therefore, that cultural preservation and biological preservation should go hand in hand. *(Adapted from IDRC, 2004.)*

cultures in relation to ecological conditions. For example, some studied the social norms and land-use practices that helped some cultures adapt to periodic drought. Later, they came to realize that externally generated political and economic forces were just as important in shaping nature-culture relationships. A look at the work of a few key cultural geographers will illustrate the implications of this idea.

Local Knowledge

indigenous technical knowledge (ITK)
Highly localized knowledge about environmental conditions and sustainable land-use practices.

Much of the interest in indigenous perceptions and practices falls under the rubric of **indigenous technical knowledge (ITK).** This is a concept that anthropologists and geographers developed to describe the detailed local knowledge about the environment and land use that is part of many indigenous cultures. Geographer Paul Richards, for example, suggested that ITK is, in many cases, superior to Western scientific knowledge and, therefore, should be considered in environmental management and agricultural planning. For example, in his study of West African cultures, *Indigenous Agricultural Revolution,* Richards documented the subtle and extensive knowledge about local soils, climate, and plant life. This local-scale knowledge provides the foundation for people to experiment with new crops and agricultural techniques while also allowing them to adjust successfully to changing social and environmental conditions.

Global Economy

ITK is place based. It is produced in particular places and environments through a process of trial and error that often spans generations. Thus, these local systems of knowledge are highly adapted to local conditions. Increasingly, however, the power of ITK is weakened through exposure to the economic forces of globalization. Sometimes the global economy applies such pressure to local **subsistence economies** that they become ecologically unsustainable. Subsistence economies are those that are oriented primarily toward production for local consumption, rather than the production of commodities for sale on the market. When an indigenous society organized for subsistence production begins producing for an external market, social, ecological, and economic difficulties often ensue.

subsistence economies
Economies in which people seek to consume only what they produce and to produce only for local consumption rather than for exchange or export.

Geographer Bernard Nietschmann's classic study of the indigenous Miskito communities living along the Caribbean coast of Nicaragua showed how external markets can undermine local subsistence economies. Miskito communities had developed a subsistence economy founded on land-based gardening and the harvesting of marine resources, including green turtles. Marine resources were harvested in seasons in which agriculture was less demanding. The value of green turtles increased dramatically when companies moved in to process and export turtle products (meat, shells, leather). They paid cash and extended credit so that the Miskito could harvest turtles year-round instead of seasonally. Subsistence production in other areas suffered as labor was directed to harvesting turtles. Turtles became scarce, so more labor time was required to hunt them in a desperate effort to pay debts and buy food. Ultimately, the turtle population was decimated and the subsistence production system collapsed.

This study might sound like yet another tragic story of "disappearing peoples" or a "vanished way of life," but it didn't end there. Nietschmann continued his research with Miskito communities into the 1990s (he died in 2000), discovering, among other things, that the Miskito people did not disappear but continued to defend their cultural autonomy in the face of great external pressures from globalization. They responded to the collapse of their resource base in 1991 by creating, in cooperation with the Nicaraguan government, a protected area as part of a local environmental management plan. They were supported in this endeavor by academics, international conservation NGOs, and the Nicaraguan government. Known as the Cayos Miskitos and Franja Costera Marine Biological Reserve, it encompasses 5019 square miles (13,000 square kilometers) of coastal area and offshore keys with 38 Miskito communities. Through this program, the Miskito were able to regulate and control their own exploitation of marine resources while reducing pressures from outsiders. This is an ongoing experiment. Miskito communities continue to struggle with outsiders for control over their land and resources in the reserve. There are hopeful signs, however, that the government is cooperating in this struggle and that both the natural resource base and the Miskito people will benefit from this project.

The Miskito case demonstrates the resiliency of indigenous cultures, the limits of ITK, and the potency of global economic forces. It reflects recent studies of the cultural and political ecology of indigenous peoples, such as those conducted by geographer Anthony Bebbington. Bebbington conducted research among the indigenous Quichua populations in the Ecuadorian Andes to assess how they interact with modernizing institutions and practices. He found that, although the Quichua people often possess extensive knowledge about local farming and resource management, ITK alone is not sufficient to allow them to prosper in a global economy. For example, there is little indigenous knowledge about the way international markets work and thus little understanding of how to price and market their own produce. As a consequence, they have sought the support and knowledge of government agencies, the Catholic Church, and NGOs. He further found that indigenous Quichua communities use outside ideas and technologies to promote their own cultural survival, attempting, in essence, to negotiate their interactions with globalization on their own terms.

Reflecting on Geography

> Contrary to their current popular image, indigenous cultures do cause environmental damage. Can you think of an example?

Folk Ecology

As with indigenous cultures, ideas persist about the particular abilities of folk cultures to sustainably manage the environment. Although the attention to conservation varies from culture to culture, folk cultures' close ties to the land and local environment enhance the environmental perception of folk groups. This becomes particularly evident when they migrate. Typically, they seek new lands similar to the one left behind. A good example can be seen in the migrations of Upland Southerners from the mountains of Appalachia between 1830 and 1930. As the Appalachians

became increasingly populous, many Upland Southerners began looking elsewhere for similar areas to settle. Initially, they found an environmental twin of the Appalachians in the Ozark-Ouachita Mountains of Missouri and Arkansas. Somewhat later, others sought out the hollows, coves, and gaps of the central Texas Hill Country. The final migration of Appalachian hill people brought some 15,000 members of this folk culture to the Cascade and Coast mountain ranges of Washington State between 1880 and 1930 (**Figure 2.23**).

Gendered Nature

We stressed in Chapter 1 and earlier in this chapter that cultures are heterogeneous and that gender is one of the principal areas of difference within places and regions. Geographers and other social scientists have documented significant differences between men's and women's relationships with the environment. This observation holds for popular, folk, and indigenous cultures. Ecofeminism is one way of thinking about how gender influences our interactions with nature. This concept, however, might seem to suggest that there is something inherent or essential about men and women that makes them think about and behave toward the environment in particular and different ways. Although cultural geographers would argue against such

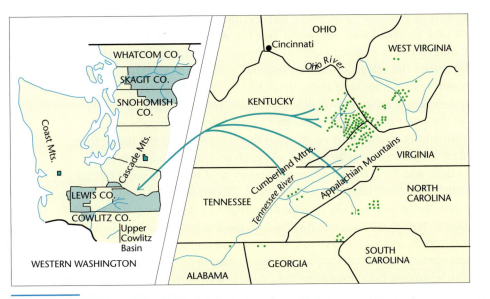

- • Appalachian place of origin of families or individuals migrating to the Upper Cowlitz Basin

- ▭ Settlement areas of Appalachian hill folk in Washington State

FIGURE 2.23 **The relocation diffusion of Upland Southern hill folk from Appalachia to western Washington.** Each dot represents the former home of an individual or family that migrated to the Upper Cowlitz River basin in the Cascade Mountains of Washington State between 1884 and 1937. Some 3000 descendants of these migrants lived in the Cowlitz area by 1940. *What does the high degree of clustering of the sources of the migrants and subsequent clustering in Washington suggest about the processes of folk migrations? How should we interpret their choices of familiar terrain and vegetation for a new home? Why might members of a folk society who migrate choose a new land similar to the old one?* (After Clevinger, 1938: 120; Clevinger, 1942: 4.)

essentialism, few would disagree that gender is an important variable in nature-culture relations.

Diane Rocheleau's work on women's roles in the management of **agroforestry** systems is illustrative. Agroforestry systems are farming systems that combine the growing of trees with the cultivation of agricultural crops. Agroforestry is practiced by folk and indigenous cultures across the tropical world and has been shown to be a highly

agroforestry
A cultivation system that features the interplanting of trees with field crops.

productive and ecologically sustainable practice. It is common in these production systems for men and women to have very distinct roles. Generally, for example, women are involved in seeding, weeding, and harvesting, whereas men take responsibility for clearing and cultivation. Gender differences also often exist in the types of crops men and women control and in the marketing of produce.

After conducting studies for many years—first in East Africa and later in the Dominican Republic—Rocheleau was

Rod's Notebook

Encountering Nature

Rod Neumann

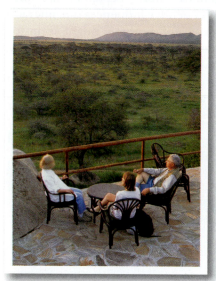

The situation at Tarangire Safari Lodge in Tarangire National Park, Tanzania, where tourists are able to view wildlife from the comfort of a patio bar, is common in many African parks. *(Christina Micek.)*

In popular culture, fewer and fewer of us have direct contact with what we might call wild nature. We don't make our livings by hunting, clearing fields, or pasturing livestock in the mountains. Rather, mass media, particularly television and film documentaries, shape our understanding of nature and wild places. Wild nature has been transformed into a spectacle in popular culture, something for our amusement and entertainment.

I was conducting research in the historical records of Tanzania's Serengeti National Park when these observations were driven home to me in a very powerful way. Serengeti, like the Amazon, is one of those iconic wild places in popular culture. Who hasn't seen a Serengeti cheetah blazing across the television screen in pursuit of a panicked wildebeest?

I spent my days looking through archived files at the park's research headquarters and my evenings at the Seronera Wildlife Lodge, a luxury hotel in the heart of the African "wilds." I would watch vanloads of American and European (but no African) tourists return to the lodge from their daily wildlife safaris. We would all perch on the balcony with beers or cocktails, watching the elephants, gazelles, and baboons gather at the artificial watering hole just meters away. Drawn to water in the arid landscape, these beasts daily provided us with an entertaining spectacle of wild African nature without our having to leave the bar!

After dark, people would move inside and gather around a television in the lounge and watch—wait for it!—television documentaries about wildlife in Serengeti. I was witnessing an almost surreal feedback loop in which urbanized tourists, drawn to East Africa by the television documentaries they'd viewed in their living rooms in the United States, now sat in front of a television set in the middle of Serengeti watching the wildlife they'd seen that day or hoped to see tomorrow. It seemed as if the television in the lounge was needed to reinforce the reality of the actual experience of viewing African wildlife firsthand. Such is the power of mass media in shaping human encounters with wild nature in popular culture.

Posted by Rod Neumann

able to derive general themes regarding the way human-environment relations are gendered not only in agroforestry systems but also in many rural and urban environments. Together with two colleagues, she identified three themes: gendered knowledge, gendered environmental rights, and gendered environmental politics. First, because women and men often have different tasks and move in different spaces, they possess different and even distinct sets of knowledge about the environment. Second, men and women have different rights, especially with regard to the ownership and control of land and resources. Third, for reasons having to do with their responsibilities in their families and communities, women are often the main leaders and activists in political movements concerned with environmental issues. Taken together, these themes suggest that environmental planning or resource management schemes that do not address issues of gendered ecology are likely to have unintended consequences, some of them negative for both women and environmental quality.

Nature in Popular Culture

Popular culture is less directly tied to the physical environment than are folk and indigenous cultures, which is not to say that it does not have an enormous impact on the environment. Urban dwellers generally do not draw their livelihoods from the land. They have no direct experience with farming, mining, or logging activities, though they could not live without the commodities produced from those activities. Gone is the intimate association between people and land known by our folk ancestors. Gone, too, is our direct vulnerability to many environmental forces, although this security is more apparent than real. Because popular culture is so tied to mass consumption, it can have enormous environmental impacts, such as the production of air and water pollution and massive amounts of solid waste. Also, because popular culture fosters limited contact with and knowledge of the physical world, usually through recreational activities, our environmental perceptions can become quite distorted. (See Rod's Notebook.)

Popular culture makes heavy demands on ecosystems. This is true even in the seemingly benign realm of recreation. Recreational activities have increased greatly in the world's economically affluent regions. Many of these activities require machines, such as snowmobiles, off-road vehicles, and jet skis, that are powered by internal combustion engines and have numerous adverse ecological impacts ranging from air pollution to soil erosion. In national parks and protected areas worldwide, affluent tourists in search of nature have overtaxed protected environments and wildlife and produced levels of congestion approaching those of urban areas (**Figure 2.24**).

Such a massive presence of people in our recreational areas inevitably results in damage to the physical environment. A study by geographer Jeanne Kay and her students in Utah revealed substantial environmental damage done by off-road recreational vehicles, including "soil loss and long-term soil deterioration." One of the paradoxes of the modern age and popular culture seems to be that the more we cluster in cities and suburbs, the greater our impact on open areas; we carry our popular culture with us when we vacation in such regions.

FIGURE 2.24 Traffic jam in Yosemite National Park at the height of the summer tourist season. (Tom Meyers Photography.)

CULTURAL LANDSCAPE

Do folk, indigenous, and popular cultures look different? Do different cultures have distinctive cultural landscapes? The theme of cultural landscape reveals the important differences within and between cultures.

Folk Architecture

folk architecture
Structures built by members of a folk society or culture in a traditional manner and style, without the assistance of professional architects or blueprints, using locally available raw materials.

Every folk culture produces a highly distinctive landscape. One of the most visible aspects of these landscapes is **folk architecture.** These traditional buildings illustrate the theme of cultural landscape in folk geography.

Folk architecture springs not from the drafting tables of professional architects but from the collective memory of groups of traditional people (**Figure 2.25**). These buildings—whether dwellings, barns, churches, mills, or inns—are based not on blueprints but on mental images that change little from one generation to the next. Folk architecture is marked not by refined artistic genius or spectacular, revolutionary design but rather by traditional, conservative, and functional structures. Material composition, floor plan, and layout are important ingredients of folk architecture, but numerous other characteristics help classify farmsteads and dwellings. The form or shape of the roof, the placement of the chimney, and even

such details as the number and location of doors and windows can be important classifying criteria. E. Estyn Evans, a noted expert on Irish folk geography, considered roof form and chimney placement, among other traits, in devising an informal classification of Irish houses.

The house, or dwelling, is the most basic structure that people erect, regardless of culture. For most people in nearly all folk cultures, a house is the single most important thing they ever build. Folk cultures as a rule are rural and agricultural. For these reasons, it seems appropriate to focus on the folk house.

Folk Housing in North America

In the United States and Canada, folk architecture today is a relict form preserved in the cultural landscape. For the most part, popular culture, with its mass-produced, commercially built houses, has so overwhelmed folk traditions that few folk houses are built today, yet many survive in the refuge regions of American and Canadian folk culture (see Figures 2.1, page 32, and 2.25).

Yankee folk houses are of wooden frame construction, and shingle siding often covers the exterior walls. They are built with a variety of floor plans, including the New England "large" house, a huge two-and-a-half-story house built around a central chimney and two rooms deep. As the Yankee folk migrated westward, they developed the upright-and-wing dwelling. These particular Yankee houses are often massive, in part because the cold winters of the region forced most work to be done indoors. By contrast, Upland Southern folk houses are smaller and built of notched logs.

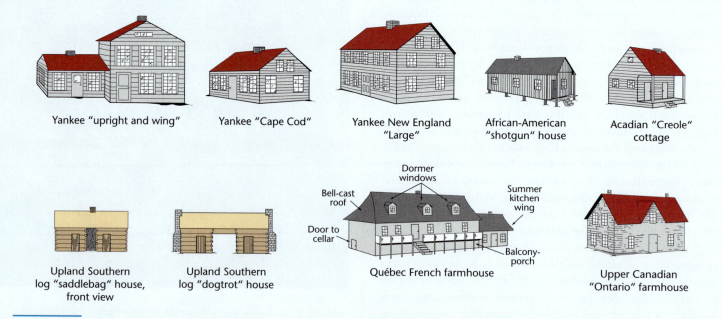

FIGURE 2.25 Selected folk houses. Six of the 15 folk culture regions of North America are represented (see Figure 2.1). *(After Glassie, 1968; Kniffen, 1965.)*

FIGURE 2.26 A dogtrot house, typical of the Upland Southern folk region. The distinguishing feature is the open-air passageway, or dogtrot, between the two main rooms. This house is located in central Texas. *(Courtesy of Terry G. Jordan-Bychkov.)*

Many houses in this folk tradition consist of two log rooms, with either a double fireplace between, forming the saddle-bag house, or an open, roofed breezeway separating the two rooms, a plan known as the dogtrot house (**Figure 2.26**). An example of an African-American folk dwelling is the shotgun house, a narrow structure only one room in width but two, three, or even four rooms in depth. Acadiana, a French-derived folk region in Louisiana, is characterized by the half-timbered Creole cottage, which has a central chimney and built-in porch. Scores of other folk house types survive in the American landscape, although most such dwellings now stand abandoned and derelict.

Reflecting on Geography

The relicts of folk cultural landscapes can be found in most parts of North America. Should we strive to preserve those relicts, such as folk houses? Why or why not?

Canada also offers a variety of traditional folk houses (see Figure 2.25). In French-speaking Québec, one of the common types consists of a main story atop a cellar, with attic rooms beneath a curved, bell-shaped (or bell-cast) roof. A balcony-porch with railing extends across the front, sheltered by the overhanging eaves. Attached to one side of this type of French-Canadian folk house is a summer kitchen that is sealed off during the long, cold winter. Often the folk houses of Québec are built of stone. To the west, in the Upper Canadian folk region, one type of folk house occurs so frequently that it is known as the Ontario farmhouse. One-and-a-half stories in height, the Ontario farmhouse is usually built of brick and has a distinctive gabled front dormer window.

The interpretation of folk architecture is by no means a simple process (**Figure 2.27**, page 58). Folk geographers often work for years trying to "read" such structures, seeking clues to diffusion and traditional adaptive strategies. The old problem of independent invention versus diffusion is raised repeatedly in the folk landscape, as **Figure 2.28**, page 58, illustrates. Precisely because interpretation is often difficult, however, geographers find these old structures challenging and well worth studying. Folk cultures rarely leave behind much in the way of written records, making their landscape artifacts all the more important in seeking explanations.

Folk Housing in Sub-Saharan Africa

Throughout East Africa and southern Africa, rural family homesteads take a common form. Most consist of a compound of buildings called a *kraal*, a term related to the English word *corral* and used across the region. The compound typically includes a main house (or houses in polygamous cultures), a detached building in the rear for cooking, and smaller buildings or enclosures for livestock.

All construction is done with local materials. Small, flexible sticks are woven in between poles that have been driven into the ground to serve as the frame. Then a mixture of clay and animal dung is plastered against the woven sticks, layer by layer, until it is entirely covered. Dried tall grasses are tied together in bundles to form the roof. In a recent innovation, rural people who can afford the expense have replaced the grass with corrugated iron sheets. In some societies the dwellings are round, in others square or rectangular. You can often tell when you have entered a different culture region by the change in house types.

One of the most distinctive house types is found in the Ndebele culture region of southern Africa, which stretches from South Africa north into southern Zimbabwe. In the rural parts of the Ndebele region, people are farmers and

FIGURE 2.27 Four folk houses in North America. *Using the sketches in Figure 2.25, page 56, and the related section of the text, determine the regional affiliation and type of each.* The answers are provided on page 69. *(Courtesy of Terry G. Jordan-Bychkov.)*

FIGURE 2.28 Two polygonal folk houses. (*Left*) A Buriat Mongol yurt in southern Siberia, near Lake Baikal. (*Right*) A Navajo hogan in New Mexico. The two dwellings, almost identical and each built of notched logs, lie on opposite sides of the world, among unrelated folk groups who never had contact with each other. Such houses do not occur anywhere in between. *Is cultural diffusion or independent invention responsible? How might a folk geographer go about finding the answer?* (Left, Courtesy of Terry G. Jordan-Bychkov; Right, Courtesy of Stephen C. Jett.)

FIGURE 2.29 **Ndebele village in South Africa.** While the origins and meaning of Ndebele house painting is debated, there is no question that it creates a visually distinct cultural landscape. *(Ariadne Van Zandbergen/ Alamy.)*

livestock keepers and live in traditional kraals. What makes these houses distinctive is the Ndebele custom of painting brightly colored designs on the exterior house walls and sometimes on the walls and gates surrounding the kraal (**Figure 2.29**). The precise origins of this custom are unclear, but it seems to date to the mid-nineteenth century. Some suggest that it was an assertion of cultural identity in response to their displacement and domination at the hands of white settlers. Others point to a religious or sacred role.

What is clear is that the custom has always been the purview of women, a skill and practice passed down from mother to daughter. Many of the symbols and patterns are associated with particular families or clans. Initially, women used natural pigments from clay, charcoal, and local plants, which restricted their palette to earth tones of brown, red, and black. Today many women use commercial paints—expensive, but longer lasting—to apply a range of bright colors, limited only by the imagination. Another new development is in the types of designs and symbols used. People are incorporating modern machines such as automobiles, televisions, and airplanes into their designs. In many cases, traditional paints and symbols are blended with the modern to produce a synthetic design of old and new. Ndebele house painting is developing and evolving in new directions, all the while continuing to signal a persistent cultural identity to all who pass through the region.

Landscapes of Popular Culture

Popular culture permeates the landscape of countries such as the United States, Canada, and Australia, including everything from mass-produced suburban houses to golf courses and neon-lit strips. So overwhelming is the presence of popular culture in most American settlement landscapes that an observer must often search diligently to find visual fragments of the older folk cultures. The popular landscape is in continual flux, for change is a hallmark of popular culture.

Few aspects of the popular landscape are more visually striking than the ubiquitous commercial malls and strips on urban arterial streets, which geographer Robert Sack calls *landscapes of consumption* (see Figure 2.6, page 36). In an Illinois college town, two other cultural geographers, John Jakle and Richard Mattson, made a study of the evolution of one such strip. During a 60-year span, the street under study changed from a single-family residential area to a commercial district (**Figure 2.30**, page 60). The researchers suggested a five-stage model of strip evolution, beginning with the single-family residential period, moving through stages of increasing commercialization, which drives owner-residents out, and culminating in stage 5, where the residential function of the street disappears and a totally commercial landscape prevails. Business properties expand so that off-street parking can be provided. Public outcries over the ugliness of such strips are common. Even landscapes such as these are subject to interpretation, however, for the people who create them perceive them differently. For example, geographer Yi-Fu Tuan suggests that a commercial strip of stores, fast-food restaurants, filling stations, and used-car lots may appear as visual blight to an outsider, but the owners or operators of the businesses are very proud of them and of their role in the community. Hard work and high hopes color their perceptions of the popular landscape.

Perhaps no landscape of consumption is more reflective of popular culture than the indoor shopping mall,

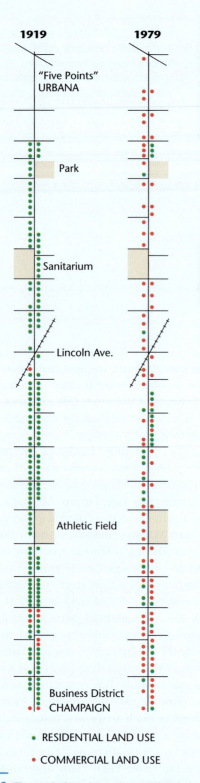

FIGURE 2.30 The evolution of a commercial strip in Champaign-Urbana, Illinois, 1919–1979. Popular culture reshaped a landscape. *Is the older or newer landscape "better"? Why?* (Adapted from Jakle and Mattson, 1981: 14, 20.)

numerous examples of which now dot both urban and suburban landscapes. Of these, the largest is West Edmonton Mall in the Canadian province of Alberta (**Figure 2.31**).

Enclosing some 5.3 million square feet (493,000 square meters) and opened in 1986, West Edmonton Mall employs 23,500 people in more than 800 stores and services, accounts for nearly one-fourth of the total retail space in greater Edmonton, earned 42 percent of the dollars spent in local shopping centers, and experienced 2800 crimes in its first nine months of operation. Beyond its sheer size, West Edmonton Mall also boasts a water park, a sea aquarium, an ice-skating rink, a miniature golf course, a roller coaster, 21 movie theaters, and a 360-room hotel. Its "streets" feature motifs from such distant places as New Orleans, represented by a Bourbon Street complete with fiberglass ladies of the evening. Jeffrey Hopkins, a geographer who studied this mall, refers to this as a "landscape of myth and elsewhereness," a "simulated landscape" that reveals the "growing intrusion of spectacle, fantasy, and escapism into the urban landscape."

Leisure Landscapes

Another common feature of popular culture is what geographer Karl Raitz labeled **leisure landscapes.** Leisure landscapes are designed to entertain people on weekends and vacations; often they are included as part of a larger tourist experience. Golf courses and theme parks such as Disney World are good examples of such landscapes. **Amenity landscapes** are a related landscape form. These are regions with attractive natural features such as forests, scenic mountains, or lakes and rivers that have become desirable locations for retirement or vacation homes. One such landscape is in the Minnesota North Woods lake country, where, in a sampling of home ownership, geographer Richard Hecock found that fully 40 percent of all dwellings were not permanent residences but instead weekend cottages or vacation homes. These are often purposefully made rustic or even humble in appearance.

The past, reflected in relict buildings, has also been incorporated into the leisure landscape. Most often, collections of old structures are relocated to form "historylands," often enclosed by imposing chain-link fences and open only during certain seasons or hours. If the desired bit of visual history has perished, Americans and Canadians do not hesitate to rebuild it from scratch, undisturbed by the lack of authenticity—as, for example, at Jamestown, Virginia, or Louisbourg on Cape Breton Island, Nova Scotia. Normally, the history parks are put in out-of-the-way places and sanitized to the extent that people no longer live in them. Role-playing actors sometimes prowl these parks,

leisure landscapes
Landscapes that are planned and designed primarily for entertainment purposes, such as ski and beach resorts.

amenity landscapes
Landscapes that are prized for their natural and cultural aesthetic qualities by the tourism and real estate industries and their customers.

FIGURE 2.31 The enormous West Edmonton Mall represents the growing infusion of spectacle and fantasy into the urban landscape. In addition to retail outlets, the mall includes a sea aquarium, an ice-skating rink, a miniature golf course, and the Submarine Ride shown here. (© James Marshall/Corbis.)

Daniel Gade, a cultural geographer, coined the term *elitist space* to describe such landscapes, using the French Riviera as an example (**Figure 2.32**). In that district of southern France, famous for its stunning natural beauty and idyllic climate, the French elite applied "refined taste to create an aesthetically pleasing cultural landscape" characterized by the preservation of old buildings and town cores, a sense of proportion, and respect for scale. Building codes and height restrictions, for instance, are rigorously enforced. Land values, in response, have risen, making the Riviera ever more elitist, far removed from the folk culture and poverty that prevailed there before 1850. Farmers and fishers have almost disappeared from the region, though one need drive but a short distance, to Toulon, to find a working seaport. It seems, then, that the different social classes generated within popular culture become geographically segregated, each producing a distinctive cultural landscape (**Figure 2.33**, page 62).

America, too, offers elitist landscapes. An excellent example is the gentleman farm, an agricultural unit operated for pleasure rather than profit (**Figure 2.34**, page 62). Typically, affluent city people own gentleman farms as an avocation, and such farms help to create or maintain a high social standing for those who own them. Some rural landscapes in America now contain many such gentleman farms; perhaps most notable among these places are the inner Bluegrass Basin of north-central Kentucky, the Virginia Piedmont west of Washington, D.C., eastern Long Island in New York, and parts of southeastern Pennsylvania.

pretending to live in some past era, adding "elsewhenness" to "elsewhereness."

Elitist Landscapes

A characteristic of popular culture is the development of social classes. A small elite group—consisting of persons of wealth, education, and expensive tastes—occupies the top economic position in popular cultures. The important geographical fact about such people is that because of their wealth, desire to be around similar people, and affluent lifestyles, they can and do create distinctive cultural landscapes, often over fairly large areas.

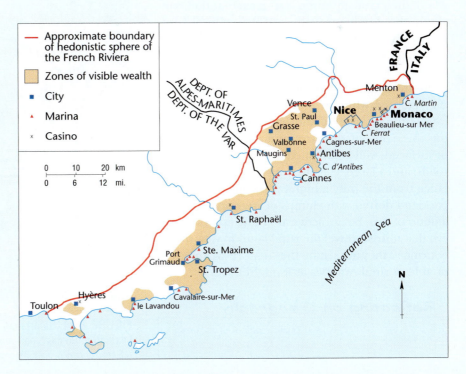

FIGURE 2.32 The distribution of an elitist or hedonistic cultural landscape on the French Riviera. *What forces in the popular culture generate such landscapes?* (Adapted from Gade, 1982: 22.)

FIGURE 2.33 **The cultural landscape of the affluent in Port of Fontvieille, Monaco.**
The huge yachts and Mediterranean residences are out of reach for all but the world's
wealthiest. *(Sergio Pitamitz/Age Fotostock.)*

Gentleman farmers engage in such activities as breeding fine cattle, raising racing horses, or hunting foxes.

Geographer Karl Raitz conducted a study of gentleman farms in the Kentucky Bluegrass Basin, where their concentration is so great that they constitute the dominant feature of the cultural landscape. The result revealed an idyllic scene, a rural landscape created more for appearance than for function. Raitz provided a list of visual indicators of Kentucky gentleman farms: wooden fences, either painted white or creosoted black; an elaborate entrance gate; a fine hand-painted sign giving the name of the farm and owner; a network of surfaced, well-maintained driveways and pasture roads; and a large, elegant house, visible in the distance from the public highway through a lawnlike parkland dotted with clumps of trees and perhaps a pond or two. So attractive are these estates to the eye that tourists travel the rural lands to view them, convinced they are seeing the "real" rural America, or at least rural America as it ought to be.

Reflecting on Geography

Can you think of other types of popular culture landscapes in addition to consumption, leisure, and elitist landscapes?

FIGURE 2.34 **Gentleman farm in the Kentucky Bluegrass region near Lexington.** Here is "real" rural America as it should be (but never was).
(Courtesy of Terry G. Jordan-Bychkov.)

FIGURE 2.35 Oil derrick on the Oklahoma state capitol grounds. The landscape of American popular culture is characterized by such functionality. The public and private sectors of the economy are increasingly linked in the popular culture. *Is criticism of such a landscape elitist and snobbish?* (See also Robertson, 1996. Photo courtesy of Terry G. Jordan-Bychkov.)

towering redwoods of California, and the Rocky Mountains.

Americans, argues Lowenthal, tend to regard their cultural landscape as unfinished. As a result, they are "predisposed to accept present structures that are makeshift, flimsy, and transient," resembling "throwaway stage sets." Similarly, the hardships of pioneer life perhaps preconditioned Americans to value function more highly than beauty and form. The state capitol grounds in Oklahoma City are adorned with little more than oil derricks, standing above busy pumps drawing oil and the wealth that comes with it from the Sooner soil—an extreme but revealing view of the American landscape (**Figure 2.35**).

In summary, American popular culture seems to have produced a built landscape that stresses bigness, utilitarianism, and transience. Sometimes these are opposing trends, as in the case of many of the massive structures previously cited, which are clearly built to last. Sometimes the trends mesh, as in the recent explosion of "big box" retail chains (**Figure 2.36**). These giant retail buildings are no more than oversized metal sheds that one can easily imagine being razed overnight to be replaced by the next big thing.

The American Popular Landscape

In an article entitled "The American Scene," geographer David Lowenthal attempted to analyze the cumulative visible impact of popular culture on the American countryside. Lowenthal identified the main characteristics of popular landscape in the United States, including the "cult of bigness"; the tolerance of present ugliness to achieve a supposedly glorious future; an emphasis on individual features at the expense of aggregates, producing a "casual chaos"; and the preeminence of function over form.

The American fondness for massive structures is reflected in edifices such as the Empire State Building, the Pentagon, the San Francisco–Oakland Bay Bridge, and Salt Lake City's Mormon Temple. Americans have dotted their cultural landscape with the world's largest of this or that, perhaps in an effort to match the grand scale of the physical environment, which includes such landmarks as the Grand Canyon, the

FIGURE 2.36 Walmart is one example among many of the "big box" stores now ubiquitous in suburban landscapes. Such buildings incorporate the ideals of bigness, transience, and utilitarianism common in the architecture of popular culture. *(Courtesy of Roderick Neumann.)*

CONCLUSION

In this chapter we have just scratched the surface of the complex nature of the geography of multiculturalism. Using the cases of folk, popular, and indigenous cultures, we learned that there are many ways of perceiving and being in the landscape. We have also seen how each of these cultural categories is in turn internally heterogeneous, with significant differences occurring among gender, class, and ethnic groups. Religion, often a defining element of cultural difference, is also vitally important. Chapter 7 is devoted to this major cultural trait.

DOING GEOGRAPHY

Self-Representation of Indigenous Culture

Indigenous cultures are reasserting themselves after 500 years of marginalization. For centuries, the roar of dominant national cultures drowned out indigenous peoples' voices. Members of dominant national cultures—academics, missionaries, government officials, and journalists—have been largely responsible for writing about the histories and cultures of indigenous peoples. This is changing as some indigenous groups have prospered economically and as the indigenous rights movement gains increasing support worldwide. Today, more and more indigenous peoples are taking charge of representing themselves and their cultures to the outside world by building museums, producing films, hosting conferences, and creating web sites.

This exercise requires you to carefully study one of these platforms for cultural expression: *self-produced* indigenous peoples' web sites. To do so, follow the steps below.

Steps to Understanding Indigenous Culture

Step 1: Do some background research on the names and locations of major indigenous cultures. You can use the web sites listed at the end of this chapter to help you get started. It is important to verify that the web sites you are studying are self-produced. Confirm that the site is produced by a tribal or indigenous organization, not by an external NGO, national government, corporation, or university.

Step 2: Think about how you are going to analyze the content of the web site in order to draw conclusions about the self-representation of indigenous cultures. Here are a few suggestions and possibilities: focus on questions of geography such as territorial claims, rights over natural resources, culturally significant relations with nature, and homeland self-rule.

Many Native American tribes, such as the Cherokee Nation, create web sites to control their public image, promote business interests, and highlight political agendas. *(Cherokee-nc.com.)*

Based on your research and analysis, systematically analyze how indigenous populations represent their cultural identities. Consider these questions:

- How do indigenous groups speak about their relationship to the land and the environment?
- What do they say about territorial claims and homelands? What roles do maps play on the web sites?
- What are their ideas on biodiversity conservation and bioprospecting (the search for genetic resources and other biological resources)?

In addition, look for discussions of conflict, cooperation, or disagreement with national governments or multinational corporations:

- Is there a project or policy (e.g., disposal of radioactive material) that is disputed?
- What position is taken on the web site? How is the position framed in relation to indigenous rights and culture?
- What major issues and challenges does the site highlight and how do these relate to globalization?

Finally, think about possibilities for comparison:

- Are there regional (on either the U.S. national or global scale) differences in terms of the quantity and content of web sites?

- Do you find common themes across or within regions?
- Are there indigenous cultures that produce contrasting or competing representations?
- Do some indigenous cultures have more than one self-generated web site and do those sites present different ideas?

Key Terms

Cultures on the Internet

You can learn more about the main categories of culture discussed in the chapter on the Internet at the following web sites:

American Memory, Library of Congress
http://memory.loc.gov
A project of the Library of Congress that presents a history of American popular culture, complete with documentation and maps.

Center for Folklife and Cultural Heritage
http://www.folklife.si.edu/center/about_us.html
A research and educational unit of the Smithsonian Institution promoting the understanding and continuity of diverse, contemporary grassroots cultures in the United States and around the world. The center produces exhibitions, films and videos, and educational materials.

Cultural Survival
http://www.cs.org/
The interactive web site of Cultural Survival, an organization that promotes the human rights and goals of indigenous peoples. Many timely indigenous cultural issues and important links can be found here.

First Nations Seeker
http://www.firstnationsseeker.ca
This site is a directory of North American Indian portal web sites. Tribes of Canada and the United States are ordered linguistically.

First Peoples Worldwide
http://www.firstpeoplesworldwide.org/
This group promotes an indigenous-controlled international organization that advocates for indigenous self-governance and culturally appropriate economic development.

Manchester Institute for Popular Culture, Manchester, U.K.
http://www.mmu.ac.uk/h-ss/mipc/
This site is dedicated to the academic study of popular culture, based at the Manchester Metropolitan University.

Native Lands
http://www.nativelands.org/
Native Lands deals with biological and cultural diversity in Latin America. It is very involved in mapping projects to help secure indigenous territorial claims and protect tropical forests.

Popular Culture Association
http://www2.h-net.msu.edu/~pcaaca/
A multidisciplinary organization dedicated to the academic discussion of popular culture, where activities of the association are discussed.

Sources

Abler, Ronald F. 1973. "Monoculture or Miniculture? The Impact of Communication Media on Culture in Space," in David A. Lanegran and Risa Palm (eds.), *An Invitation to Geography*. New York: McGraw-Hill, 186–195.

Alcorn, Janice. 1994. "Noble Savage or Noble State? Northern Myths and Southern Realities in Biodiversity Conservation." *Ethnoecologica* 2(3): 7–19.

Bebbington, Anthony. 1996. "Movements, Modernizations, and Markets: Indigenous Organizations and Agrarian Strategies in Ecuador," in R. Peet and M. Watts (eds.), *Liberation Ecologies: Environment, Development, Social Movements*. London: Routledge, 86–109.

Bebbington, Anthony. 2000. "Reencountering Development: Livelihood Transitions and Place Transformations in the Andes." *Annals of the Association of American Geographers* 90(3): 495–520.

Brownell, Joseph W. 1960. "The Cultural Midwest." *Journal of Geography* 59: 81–85.

SEEING GEOGRAPHY Camping in the "Great Outdoors"

What can this scene tell us about nature-culture relations in North American popular culture?

Enjoying nature in a national park campground.

How we think about and interact with nature reveals a great deal about how we understand our place in the world. For example, within contemporary popular culture we often think of our relationship with nature as something outside the routine of daily life. We seek out nature on weekends, during vacations, or in retirement. Nature is equated with pretty scenery, which in popular culture typically means forests, mountains, and wide-open spaces. This scene from Everglades National Park in Florida reveals a great deal about one of the main ways within popular culture that we express this understanding of nature: camping in the "Great Outdoors."

Of course, there are many ways to camp. This particular form, recreational vehicle (RV) or motor home camping, is extremely energy intensive. Unseen in this photo is the supporting industrial manufacturing complex organized to produce a fleet of vehicles, trailers, and equipment, all of which are intended to be situated on an asphalt slab and connected to an electrical power grid. A lot of natural resources have to be consumed to experience nature in this way. Such an experience is shaped less by direct physical interaction with nature than by interaction with mass-produced commodities. Nature serves mainly as background scenery.

Thinking of nature as background scenery suggests a stage set for actors to interact on. Look carefully at the photo and you will see that this campground is a stage for both extremely private and highly public interactions. On one hand, every set of "campers" has its own private home completely sealed from the outside (e.g., note the satellite dish and rooftop air conditioners). Each RV is self-sufficient, requiring interactions neither with nature nor with human neighbors. On the other hand, the RVs are extremely closely spaced. Interactions among campers are almost forced, for merely stepping out of the RV puts one in public view. Conversations with strangers—even sharing drinks and meals—become a cultural norm in such a setting. Indeed, meeting new people is one of the reasons campers give when explaining why they enjoy camping. Could it be that getting in touch with nature really means getting in touch with one another in ways that would be difficult in the daily routines of popular culture?

Clevinger, Woodrow R. 1938. "The Appalachian Mountaineers in the Upper Cowlitz Basin." *Pacific Northwest Quarterly* 29: 115–134.

Clevinger, Woodrow R. 1942. "Southern Appalachian Highlanders in Western Washington." *Pacific Northwest Quarterly* 33: 3–25.

Curtis, James R. 1982. "McDonald's Abroad: Outposts of American Culture." *Journal of Geography* 81: 14–20.

de Blij, Harm, and Peter Muller. 2004. *Geography: Realms, Regions and Concepts,* 11th ed. New York: Wiley.

Edmonson, Brad, and Linda Jacobsen. 1993. "Elvis Presley Memorabilia." *American Demographics* 15(8): 64.

Evans, E. Estyn. 1957. *Irish Folk Ways.* London: Routledge & Kegan Paul.

Francaviglia, Richard V. 1973. "Diffusion and Popular Culture: Comments on the Spatial Aspects of Rock Music," in David A. Lanegran and Risa Palm (eds.), *An Invitation to Geography.* New York: McGraw–Hill, 87–96.

Frederickson, Kristine. 1984. *American Rodeo from Buffalo Bill to Big Business.* College Station: Texas A&M University Press.

Gade, Daniel W. 1982. "The French Riviera as Elitist Space." *Journal of Cultural Geography* 3: 19–28.

Glassie, Henry. 1968. *Pattern in the Material Folk Culture of the Eastern United States.* Philadelphia: University of Pennsylvania Press.

Godfrey, Brian J. 1993. "Regional Depiction in Contemporary Film." *Geographical Review* 83: 428–440.

Goss, Jon. 1999. "Once-upon-a-Time in the Commodity World: An Unofficial Guide to the Mall of America." *Annals of the Association of American Geographers* 89: 45–75.

Graff, Thomas O., and Dub Ashton. 1994. "Spatial Diffusion of Walmart: Contagious and Reverse Hierarchical Elements." *Professional Geographer* 46: 19–29.

Hecock, Richard D. 1987. "Changes in the Amenity Landscape: The Case of Some Northern Minnesota Townships." *North American Culture* 3(1): 53–66.

Holder, Arnold. 1993. "Dublin's Expanding Golfscape." *Irish Geography* 26: 151–157.

Hopkins, Jeffrey. 1990. "West Edmonton Mall: Landscape of Myths and Elsewhereness." *Canadian Geographer* 34: 2–17.

IDRC. 2004. Web site of the International Development Research Centre, available online at http://web.idrc.ca/en/ev-1248-201-1-DO_TOPIC.html.

International Cancún Declaration of Indigenous Peoples. 2003. Prepared during the Fifth WTO Ministerial Conference, Cancún, Mexico. Available online at http://www.ifg.org/programs/indig/CancunDec.html.

International Labour Organization. 1989. Indigenous and Tribal Peoples Convention. Available online at http://members.tripod.com/PPLP/ILOC169.html.

Jackson, Peter. 2004. "Local Consumption Cultures in a Globalizing World." *Transactions of the Institute of British Geographers* NS 29: 165–178.

Jakle, John A., and Richard L. Mattson. 1981. "The Evolution of a Commercial Strip." *Journal of Cultural Geography* 1(2): 12–25.

Jett, Stephen C. 1991. "Further Information on the Geography of the Blowgun and Its Implications for Early Transoceanic Contacts." *Annals of the Association of American Geographers* 81: 89–102.

Kay, Jeanne, et al. 1981. "Evaluating Environmental Impacts of Off-Road Vehicles." *Journal of Geography* 80: 10–18.

Kimber, Clarissa T. 1973. "Plants in the Folk Medicine of the Texas-Mexico Borderlands." *Proceedings, Association of American Geographers* 5: 130–133.

Kimmel, James P. 1997. "Santa Fe: Thoughts on Contrivance, Commoditization, and Authenticity." *Southwestern Geographer* 1: 44–61.

Kniffen, Fred B. 1965. "Folk-Housing: Key to Diffusion." *Annals of the Association of American Geographers* 55: 549–577.

Kunstler, James H. 1993. *The Geography of Nowhere: The Rise and Decline of America's Man-Made Landscape.* New York: Simon & Schuster.

Lowenthal, David. 1968. "The American Scene." *Geographical Review* 58: 61–88.

McDowell, Linda. 1994. "The Transformation of Cultural Geography," in Derek Gregory, Ron Martin, and Graham Smith (eds.), *Human Geography: Society, Space, and Social Science.* Minneapolis: University of Minnesota Press, 146–173.

Miller, E. Joan Wilson. 1968. "The Ozark Culture Region as Revealed by Traditional Materials." *Annals of the Association of American Geographers* 58: 51–77.

Mings, Robert C., and Kevin E. McHugh. 1989. "The RV Resort Landscape." *Journal of Cultural Geography* 10: 35–49.

Nietschmann, Bernard. 1973. *Between Land and Water: The Subsistence Ecology of the Miskito Indians, Eastern Nicaragua.* New York: Seminar Press.

Nietschmann, Bernard. 1979. "Ecological Change, Inflation, and Migration in the Far West Caribbean." *Geographical Review* 69: 1–24.

Pillsbury, Richard. 1990a. "Striking to Success." *Sport Place* 4(1): 16, 35.

Pillsbury, Richard. 1990b. "Ride 'm Cowboy." *Sport Place* 4: 26–32.

Pillsbury, Richard. 1998. *No Foreign Food: The American Diet in Time and Place.* Boulder, Colo.: Westview Press.

Price, Edward T. 1960. "Root Digging in the Appalachians: The Geography of Botanical Drugs." *Geographical Review* 50: 1–20.

Raitz, Karl B. 1975. "Gentleman Farms in Kentucky's Inner Bluegrass." *Southeastern Geographer* 15: 33–46.

Raitz, Karl B. 1987. "Place, Space and Environment in America's Leisure Landscapes." *Journal of Cultural Geography* 8(1): 49–62.

Relph, Edward. 1976. *Place and Placelessness.* London: Pion.

Richards, Paul. 1975. "'Alternative' Strategies for the African Environment. 'Folk Ecology' as a Basis for Community Oriented Agricultural Development," in Paul Richards (ed.), *African Environment, Problems and Perspectives.* London: International African Institute, 102–117.

Richards, Paul. 1985. *Indigenous Agricultural Revolution.* London: Hutchinson.

Roark, Michael. 1985. "Fast Foods: American Food Regions." *North American Culture* 2(1): 24–36.

Robertson, David S. 1996. "Oil Derricks and Corinthian Columns: The Industrial Transformation of the Oklahoma State Capitol Grounds." *Journal of Cultural Geography* 16: 17–44.

Rocheleau, Diane, Barbara Thomas-Slayter, and Esther Wangari (eds.). 1996. *Feminist Political Ecology: Global Issues and Local Experiences.* New York: Routledge.

Rooney, John F., Jr. 1969. "Up from the Mines and Out from the Prairies: Some Geographical Implications of Football in the United States." *Geographical Review* 59: 471–492.

Rooney, John F., Jr. 1988. "Where They Come From." *Sport Place* 2(3): 17.

Rooney, John F., Jr., and Paul L. Butt. 1978. "Beer, Bourbon, and Boone's Farm: A Geographical Examination of Alcoholic Drink in the United States." *Journal of Popular Culture* 11: 832–856.

Rooney, John F., Jr., and Richard Pillsbury. 1992. *Atlas of American Sport.* New York: Macmillan.

Sack, Robert D. 1992. *Place, Modernity, and the Consumer's World: A Rational Framework for Geographical Analysis*. Baltimore: Johns Hopkins University Press.

Scott, Damon. In progress. "The Gay Geography of San Francisco." PhD dissertation. University of Texas at Austin.

Shortridge, Barbara G. 1987. *Atlas of American Women*. New York: Macmillan.

Shortridge, James R. 1989. *The Middle West: Its Meaning in American Culture*. Lawrence: University Press of Kansas.

Shuhua, Chang. 1977. "The Gentle Yamis of Orchid Island." *National Geographic* 151(1): 98–109.

Tuan, Yi-Fu. *Topophilia*. Englewood Cliffs, N.J.: Prentice Hall, 1974.

Weiss, Michael J. 1988. *The Clustering of America*. New York: Harper & Row.

Wilhelm, Eugene J., Jr. 1968. "Field Work in Folklife: Meeting Ground of Geography and Folklore." *Keystone Folklore Quarterly* 13: 241–247.

Williams, Raymond. 1976. *Keywords: A Vocabulary of Culture and Society*. New York: Oxford University Press.

Zelinsky, Wilbur. 1974. "Cultural Variation in Personal Name Patterns in the Eastern United States." *Annals of the Association of American Geographers* 60: 743–769.

Zelinsky, Wilbur. 1980a. "North America's Vernacular Regions." *Annals of the Association of American Geographers* 70: 1–16.

Zelinsky, Wilbur. 1980b. "Selfward Bound? Personal Preference Patterns and the Changing Map of American Society." *Economic Geography* 50: 144–179.

Zonn, Leo (ed.). 1990. *Place Images in Media: Portrayal, Experience, and Meaning*. Savage, Md.: Rowman & Littlefield.

Ten Recommended Books on Geographies of Cultural Difference

(For additional suggested readings, see *The Human Mosaic* web site: www.whfreeman.com/domosh12e)

Burgess, Jacquelin A., and John R. Gold (eds.). 1985. *Geography, the Media, and Popular Culture*. New York: St. Martin's Press. The geography of popular culture is linked in diverse ways to the communications media, and this collection of essays explores facets of that relationship.

Carney, George O. (ed.). 1998. *Baseball, Barns and Bluegrass: A Geography of American Folklife*. Boulder, Colo.: Rowman & Littlefield. A wonderful collection of readings that, contrary to the title, span the gap between folk and popular culture.

Ensminger, Robert F. 1992. *The Pennsylvania Barn: Its Origin, Evolution, and Distribution in North America*. Baltimore: Johns Hopkins University Press. A common American folk barn, part of the rural cultural landscape, provides geographer Ensminger with visual clues to its origin and diffusion; a fascinating detective story showing how geographers "read" cultural landscapes and what they learn in the process.

Glassie, Henry. 1968. *Pattern in the Material Folk Culture of the Eastern United States*. Philadelphia: University of Pennsylvania Press. Glassie, a student of folk geographer Fred Kniffen, considers the geographical distribution of a wide array of folk culture items in this classic overview.

Jackson, Peter, and Jan Penrose (eds.). 1993. *Constructions of Race, Place, and Nation*. Minneapolis: University of Minnesota Press. An edited collection that examines the way in which the ideas of racial and national identity vary from place to place; rich in empirical research.

Jordan, Terry G., Jon T. Kilpinen, and Charles F. Gritzner. 1997. *The Mountain West: Interpreting the Folk Landscape*. Baltimore: Johns Hopkins University Press. Reading the folk landscapes of the American West, three geographers reach conclusions about the regional culture and how it evolved.

Price, Patricia. 2004. *Dry Place: Landscapes of Belonging and Exclusion*. Minneapolis: University of Minnesota Press. Price explores the narratives that have sought to establish claims to the dry lands along the U.S.–Mexico border, demonstrating how stories can become vehicles for reshaping places and cultural identities.

Skelton, Tracey, and Gill Valentine (eds.). 1998. *Cool Places: Geographies of Youth Cultures*. London: Routledge. The engaging essays in *Cool Places* explore the dichotomy of youthful lives by addressing the issues of representation and resistance in youth culture today. Using first-person vignettes to illustrate the wide-ranging experiences of youth, the authors consider how the media have imagined young people as a particular community with shared interests and how young people resist these stereotypes, instead creating their own independent representations of their lives.

Weiss, Michael J. 1994. *Latitudes and Attitudes: An Atlas of American Tastes, Trends, Politics, and Passions*. New York: Little, Brown. Using marketing data organized by postal zip codes, Weiss reveals the geographical diversity of American popular culture.

Zelinsky, Wilbur. 1992. *The Cultural Geography of the United States*, 2nd ed. Englewood Cliffs, N.J.: Prentice Hall. This revised edition of a sprightly, classic book, originally published in 1973, reveals the cultural sectionalism in modern America in the era of popular culture, with attention also to folk roots.

Journals in Geographies of Cultural Difference

Indigenous Affairs. A quarterly journal published by the International Working Group for Indigenous Affairs thematically focused on issues of indigenous cultures. Volume 1 was published in 1976.

Journal of Popular Culture. Published by the Popular Culture Association since 1967, this journal focuses on the role of popular culture in the making of contemporary society. See in particular Volume 11, No. 4, 1978, a special issue on cultural geography and popular culture.

Material Culture: Journal of the Pioneer America Society. Published twice annually, this leading periodical specializes in the subject of the American rural material culture of the past. Volume 1 was published in 1969, and prior to 1984 the journal was called *Pioneer America.*

Answers

Figure 2.6 The scenes were taken in the following "placeless" places: McDonald's in Tokyo, Wendy's in Idaho, and Pampas Grill in Finland.

Figure 2.27 (a) French-Canadian farmhouse, Port Joli, Québec; (b) New England "large" house, New Hampshire; (c) Yankee upright-and-wing house, Massachusetts; (d) shotgun house, Alleyton, Texas.

Would you feel comfortable walking here? If not, why not?

Go to "Seeing Geography" on page 112 to learn more about this image.

3 POPULATION GEOGRAPHY
Shaping the Human Mosaic

The nineteenth-century French philosopher Auguste Comte famously asserted that "demography is destiny." Though this might be a bit of an exaggeration, it is true that one of the most important aspects of the world's human population is its demographic characteristics, such as age, gender, health, mortality, density, and mobility. In fact, many cultural geographers argue that familiarity with the spatial dimensions of demography provides a baseline for the discipline. Thus, **population geography** provides an ideal topic to launch our substantive discussion of the human mosaic.

The most essential demographic fact is that more than 6.8 billion people inhabit the Earth today. But numbers alone tell only part of the story of the delicate balance between human populations and the resources on which we depend for our survival, comfort, and enjoyment. Think of the sort of lifestyle you may now enjoy or aspire to in the future. Does it involve driving a car? Eating meat regularly? Owning a spacious house with central heating and air conditioning? If so, you are not alone. In fact, these "Western" consumption habits have become so widespread that they may now be more properly regarded as universal. Satisfying these lifestyle demands requires using a wide range of nonrenewable resources—fossil fuels, extensive farmlands, and fresh water among them— whose consumption ultimately limits the number

of people the Earth can support. Indeed, it has been argued that if Western lifestyles are adopted by a significant number of the globe's inhabitants, then our current population of 6.8 billion is already excessive and will soon deplete or contaminate the Earth's life-support systems: the air, soil, and water we depend on for our very survival. Although we may think of our geodemographic choices, such as how many children we will have or what to eat for dinner, as highly individual ones, when aggregated across whole groups they can have truly global repercussions.

As we do throughout this book, we approach our study using the five themes of cultural geography. The demographics of human populations, their size, age, gender compositions, and spatial distribution, are discussed under the regional theme. Population mobility and the related movement of diseases that affect human populations are discussed next. Population debates are considered under the theme of globalization. Although population is experienced locally and policies affecting population are typically set at the national level, the size and impact of human populations involve a debate that is most commonly pitched at the global scale. Next, the theme of nature-culture reveals that the ways in which we interpret the natural world and adapt it to our needs are deeply entwined with

population geography
The study of the spatial and ecological aspects of population, including distribution, density per unit of land area, fertility, gender, health, age, mortality, and migration.

demographic practices. We close with a consideration of the demographic cultural landscape, which highlights the surprisingly diverse ways that places across the world respond to changing population dynamics.

Region

In what ways do demographic traits vary regionally? How is the theme of culture region expressed in terms of population characteristics? The principal

characteristics of human populations—their densities, spatial distributions, age and gender structures, the ways they increase and decrease, and how rapidly population numbers change—vary enormously from place to place. Understanding the demographic characteristics of populations, and why and how they change over time, gives us important clues to their cultural characteristics.

Population Distribution and Density

If the 6,800,000,000 inhabitants of the Earth were evenly distributed across the land area, the **population**

> **population density**
> A measurement of population per unit area (for example, per square mile).

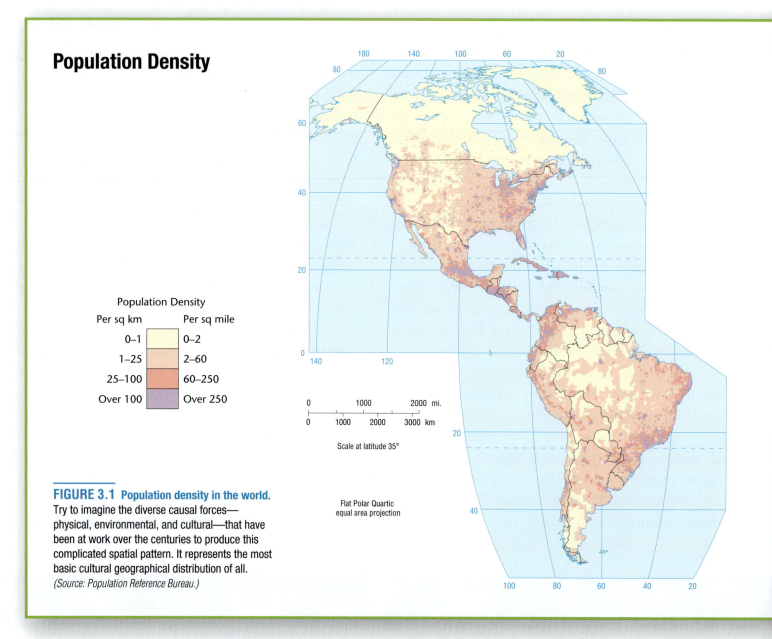

Population Density

Population Density

Per sq km	Per sq mile
0–1	0–2
1–25	2–60
25–100	60–250
Over 100	Over 250

Scale at latitude 35°

Flat Polar Quartic
equal area projection

FIGURE 3.1 Population density in the world.
Try to imagine the diverse causal forces—physical, environmental, and cultural—that have been at work over the centuries to produce this complicated spatial pattern. It represents the most basic cultural geographical distribution of all.
(Source: Population Reference Bureau.)

density would be about 115 persons per square mile (44 per square kilometer). However, people are very unevenly distributed, creating huge disparities in density. Mongolia, for example, has 4.3 persons per square mile (1.7 per square kilometer), whereas Bangladesh has 2300 persons per square mile (890 per square kilometer) (**Figure 3.1**).

If we consider the distribution of people by continents, we find that 72.7 percent of the human race lives in Eurasia —Europe and Asia. The continent of North America is home to only 7.9 percent of all people, Africa to 13.2 percent, South America to 5.7 percent, and Australia and the Pacific islands to 0.5 percent. When we consider population distribution by country, we find that 20 percent of all humans

reside in China, 17 percent in India, and only 4.6 percent in the third-largest nation in the world, the United States (**Table 3.1**, page 74). In fact, one out of every 50 humans lives in only one valley of one province of China: the Sichuan Basin.

For analyzing data, it is convenient to divide population density into categories. For example, in Figure 3.1, one end of the spectrum contains thickly settled areas having 250 or more persons per square mile (100 or more per square kilometer); on the other end, largely unpopulated areas have fewer than 2 persons per square mile (less than 1 per square kilometer). Moderately settled areas, with 60 to 250 persons per square mile (25 to 100 per square kilometer), and thinly settled areas, inhabited by 2 to 60 persons per square mile

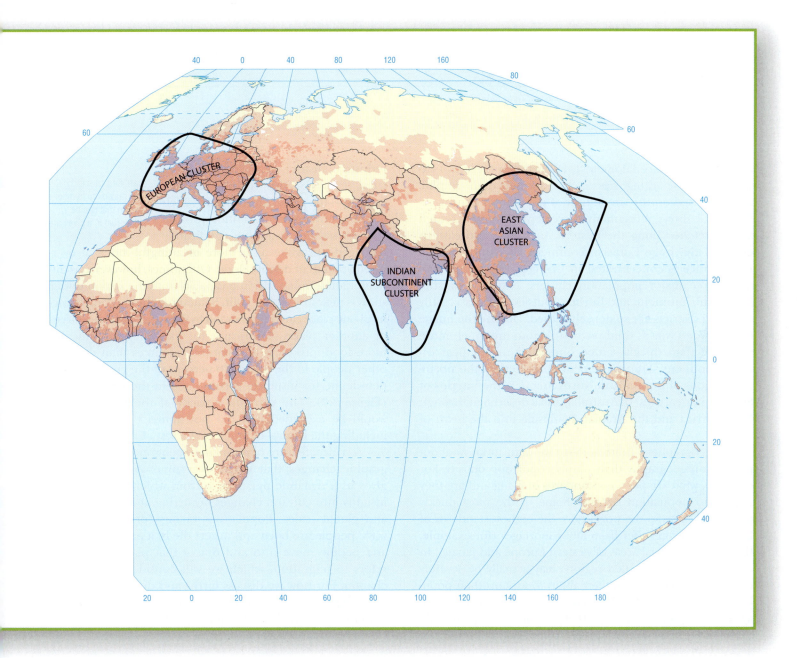

TABLE 3.1 The World's 10 Most Populous Countries, 2010 and 2050

Largest Countries in 2010	Population in 2010 (in millions)	Largest Countries in 2050	Population in 2050 (estimated, in millions)
China	1,330	India	1,657
India	1,173	China	1,304
United States	310	United States	439
Indonesia	243	Indonesia	313
Brazil	201	Pakistan	291
Pakistan	184	Ethiopia	278
Bangladesh	156	Nigeria	264
Nigeria	152	Brazil	261
Russia	139	Bangladesh	250
Japan	127	Democratic Republic of Congo	189

(Source: U.S. Census Bureau International Data Base, 2010.)

(1 to 25 per square kilometer), fall between these two extremes. These categories create formal demographic regions based on the single trait of population density. As Figure 3.1 shows, a fragmented crescent of densely settled areas stretches along the western, southern, and eastern edges of the huge Eurasian continent. Two-thirds of the human race is concentrated in this crescent, which contains three major population clusters: eastern Asia, the Indian subcontinent, and Europe. Outside of Eurasia, only scattered districts are so densely settled. Despite the image of a crowded world, thinly settled regions are much more extensive than thickly settled ones, and they appear on every continent. Thin settlement characterizes the northern sections of Eurasia and North America, the interior of South America, most of Australia, and a desert belt through North Africa and the Arabian Peninsula into the heart of Eurasia.

As geographers, there is more we want to know about population geography than simply population density. For example, what are people's standards of living and are they related to population density? Some of the most thickly populated areas in the world have the highest standards of living—and even suffer from labor shortages (for example, the major industrial areas of western Europe and Japan). In other cases, thinly settled regions may actually be severely overpopulated relative to their ability to support their populations, a situation that is usually associated with marginal agricultural lands. Although 1000 persons per square mile (400 per square kilometer) is a "sparse" population for an industrial district, it is "dense" for a rural area. For this reason, **carrying capacity**—the population beyond which a given environment cannot provide support without becoming significantly damaged—provides a far more meaningful index of overpopulation than density alone. Oftentimes, however, it is difficult to determine carrying capacity until the region under study is near or over the limit. Sometimes the carrying capacity of one place can be expanded by drawing on the resources of another place. Americans, for example, consume far more food, products, and natural resources than do most other people in the world: 26 percent of the entire world's petroleum, for instance, is consumed in the United States (**Figure 3.2**). The carrying capacity of the United States would be exceeded if it did not annex the resources—including the labor—of much of the rest of the world.

A critical feature of population geography is the demographic changes that occur over time. Analyzing these gives us a dynamic perspective from which we can glean insights into cultural changes occurring at local, regional, and global scales. Populations change primarily in two ways: people are born and others die in a particular place, and people move into and out of that place. The latter refers to migration, which we will consider later in this chapter. For now we discuss births and deaths, which can be thought of as additions to and subtractions from a population. They provide what demographers refer to as natural increases and natural decreases.

carrying capacity
The maximum number of people that can be supported in a given area.

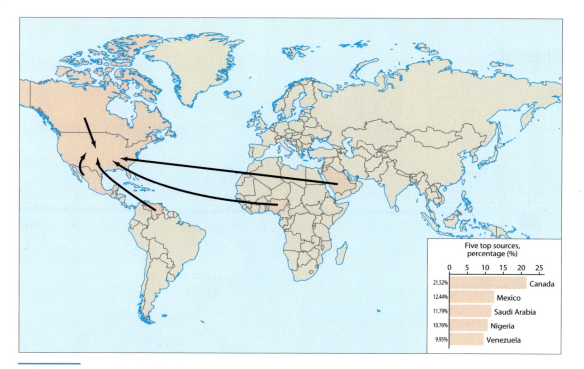

Five top sources, percentage (%)

0   5   10   15   20   25

21.52%	Canada
12.44%	Mexico
11.79%	Saudi Arabia
10.76%	Nigeria
9.95%	Venezuela

FIGURE 3.2 From where does U.S. oil come? The United States has domestic oil resources, but they cannot fully supply the domestic demand, so oil is imported from other areas of the world. The United States consumes 26 percent of all the oil produced in the world. *(Source: U.S. Department of Energy, 2010.)*

Patterns of Natality

Births can be measured by several methods. The older way was simply to calculate the birthrate: the number of births per year per thousand people (**Figure 3.3**, pages 76–77).

total fertility rate (TFR)
The number of children the average woman will bear during her reproductive lifetime (15–44 or 15–49 years of age). A TFR of less than 2.1, if maintained, will cause a natural decline of population.

More revealing is the **total fertility rate (TFR),** which is measured as the average number of children born per woman during her reproductive lifetime, which is considered to be from 15 to 49 years of age (some countries have a lower upper-age limit of 44 years of age). The TFR is a more useful measure than the birthrate because it focuses on the female segment of the population, reveals average family size, and gives an indication of future changes in the population structure. A TFR of 2.1 is needed to produce a stabilized population over time, one that does not increase or decrease. Once achieved, this condition is called **zero population growth.**

zero population growth
A stabilized population created when an average of only two children per couple survive to adulthood, so that, eventually, the number of deaths equals the number of births.

The TFR varies markedly from one part of the world to another, revealing a vivid geographical pattern (**Figure 3.4**, pages 78–79). In southern and eastern Europe, the average TFR is only 1.3. Every country with a TFR of 2.0 or lower will eventually experience population decline. Bulgaria, for example, has a TFR of 1.2 and is expected to lose 38 percent of its population by 2050. Interestingly, the Chinese Special Administrative Regions of Macau and Hong Kong report TFRs of only 0.91 and 1.04, respectively, the lowest in the world as of 2010.

By contrast, sub-Saharan Africa has the highest TFR of any sizable part of the world, led by Niger with 7.68 and Uganda with 7.63 in 2010. Elsewhere in the world, only Afghanistan and Yemen, both in Southwest Asia, can rival the sub-Saharan African rates. However, according to the World Bank, during the past two decades TFRs have fallen in all sub-Saharan African nations.

The Geography of Mortality

Another way to assess demographic change is to analyze death rates: the number of deaths per year per 1000 people (**Figure 3.5**, pages 80–81). Of course, death is a natural part of the life cycle, and there is no way to achieve a death rate of zero. But geographically speaking, death comes in different forms. In the developed world, most people die of age-induced degenerative conditions, such as heart disease, or from maladies caused by industrial pollution of

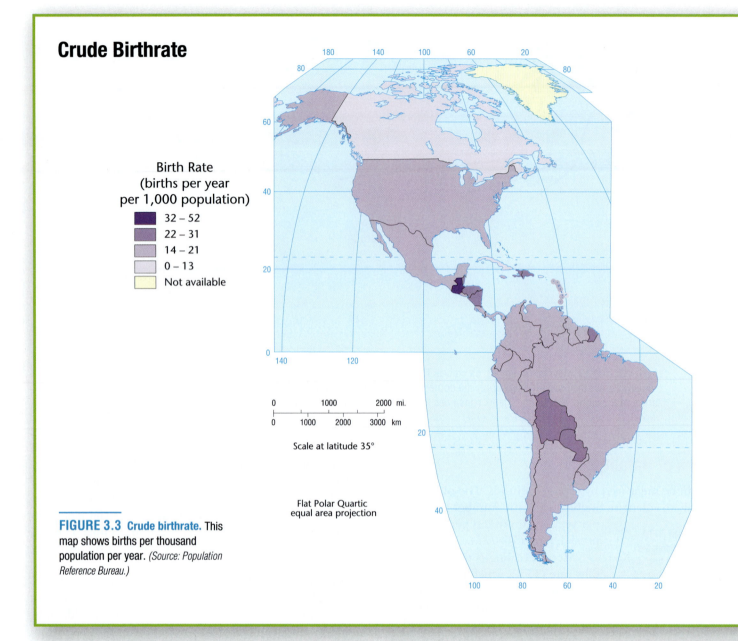

Crude Birthrate

Birth Rate
(births per year
per 1,000 population)

- 32 – 52
- 22 – 31
- 14 – 21
- 0 – 13
- Not available

0   1000   2000 mi.

0   1000   2000   3000 km

Scale at latitude 35°

Flat Polar Quartic
equal area projection

FIGURE 3.3 Crude birthrate. This
map shows births per thousand
population per year. *(Source: Population
Reference Bureau.)*

the environment. Many types of cancer fall into the latter category. By contrast, contagious diseases such as malaria, HIV/AIDS, and diarrheal diseases are a leading cause of death in poorer countries. Civil warfare, inadequate health services, and the age structure of a country's population will also affect its death rate.

The highest death rates occur in sub-Saharan Africa, the poorest world region and most afflicted by life-threatening diseases and civil strife (**Figure 3.6**, pages 82–83, illustrates the geography of HIV/AIDS). In general, death rates of more than 25 per 1000 people are uncommon today. The world's highest death rate as of 2010—just under 24 per 1000 people—was found in Angola, in southwestern Africa,

and is the result of high infant mortality rates coupled with the destruction of Angola's infrastructure and economy after thirty years of civil warfare. High death rates are also found in eastern European nations—Russia, for instance, ranked seventh in the world in 2010, with a death rate of 16 per 1000, thanks to a collapsing public health care system in the post-Soviet era, environmental contamination and increased cancer incidence, poor health choices including smoking and alcohol consumption, and very high rates of diseases such as tuberculosis and HIV/AIDS. By contrast, the American tropics generally have rather low death rates, as does the desert belt across North Africa, the Middle East, and central Asia. In these regions, the

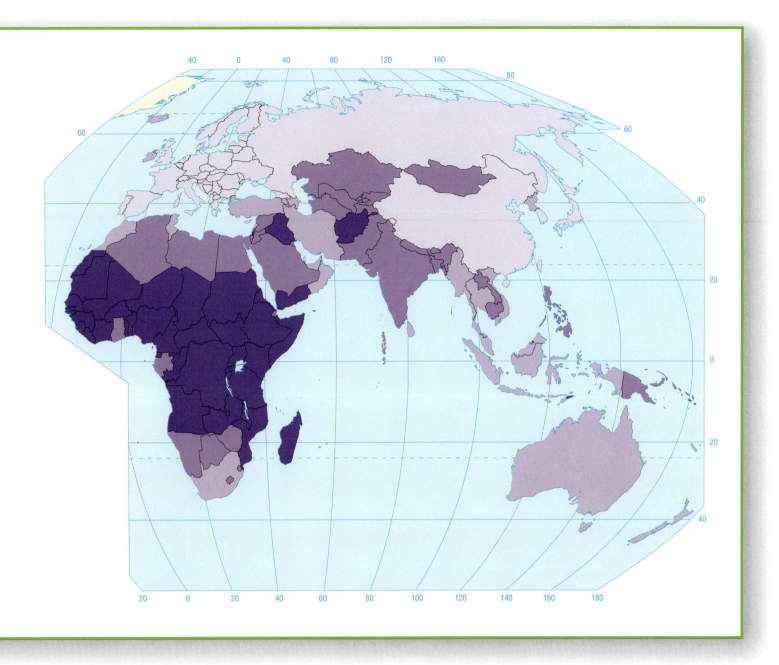

predominantly young population depresses the death rate. Compared to Angola, Nicaragua's death rate of only 4 per 1000 seems quite low. Because of its older population structure, the average death rate in the European Union is 10 per 1000. Australia, Canada, and the United States, which continue to attract young immigrants, have lower death rates than most of Europe. Canada's death rate, for instance, is slightly less than 8 people per 1000.

The Demographic Transition

All industrialized, technologically advanced countries have low fertility rates and stabilized or declining populations, having passed through what is called the **demographic transition** (**Figure 3.7**, page 82). In preindustrial societies, birth and death rates were both high, resulting in almost no population growth. Because these were agrarian societies that depended on family labor, many children meant larger workforces, thus the high birthrates. But low levels of public health and limited access to health care, particularly for the very young, also meant high death rates. With the coming of the industrial era, medical advances and improvements in diet set the stage for a drop in death rates. Human life expectancy in industrialized countries soared from an

demographic transition
A term used to describe the movement from high birth and death rates to low birth and death rates.

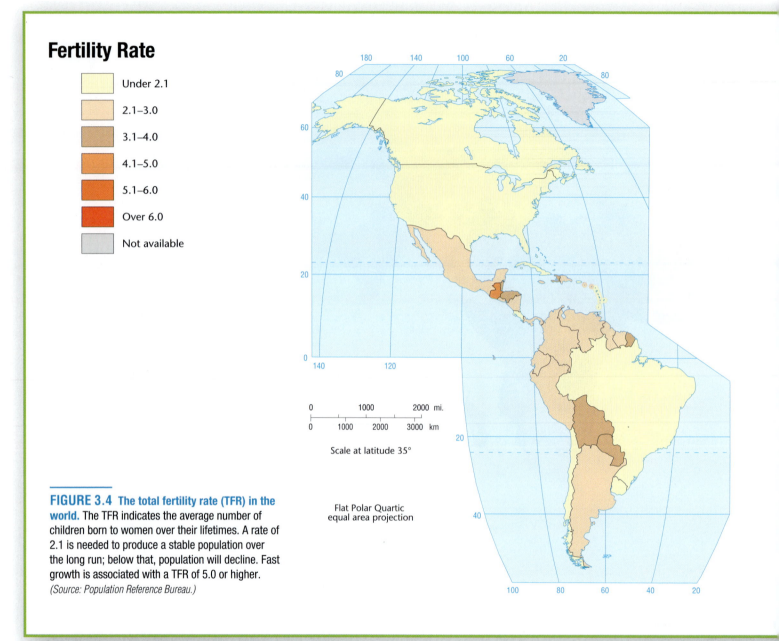

Fertility Rate

- Under 2.1
- 2.1–3.0
- 3.1–4.0
- 4.1–5.0
- 5.1–6.0
- Over 6.0
- Not available

0   1000   2000 mi.
0 1000 2000 3000 km

Scale at latitude 35°

Flat Polar Quartic
equal area projection

FIGURE 3.4 The total fertility rate (TFR) in the world. The TFR indicates the average number of children born to women over their lifetimes. A rate of 2.1 is needed to produce a stable population over the long run; below that, population will decline. Fast growth is associated with a TFR of 5.0 or higher. *(Source: Population Reference Bureau.)*

average of 35 years in the eighteenth century to 75 years or more at present. Yet birthrates did not fall so quickly, leading to a population explosion as fertility outpaced mortality. In Figure 3.7, this is shown in late stage 2 and early stage 3 of the model. Eventually, a decline in the birthrate followed the decline in the death rate, slowing population growth. An important reason leading to lower fertility levels involves the high cost of children in industrial societies, particularly because childhood itself becomes a prolonged period of economic dependence on parents. Finally, in

the postindustrial period, the demographic transition produced zero population growth or actual population decline (Figure 3.7, page 82, and **Figure 3.8**, pages 84–85).

Achieving lower death rates is relatively cost effective, historically requiring little more than the provision of safe drinking water and vaccinations against common infectious disease. Lowering death rates tends to be uncontroversial and quickly achieved, demographically speaking. Getting birthrates to fall, however, can be far more difficult, especially for a government official who wants to be

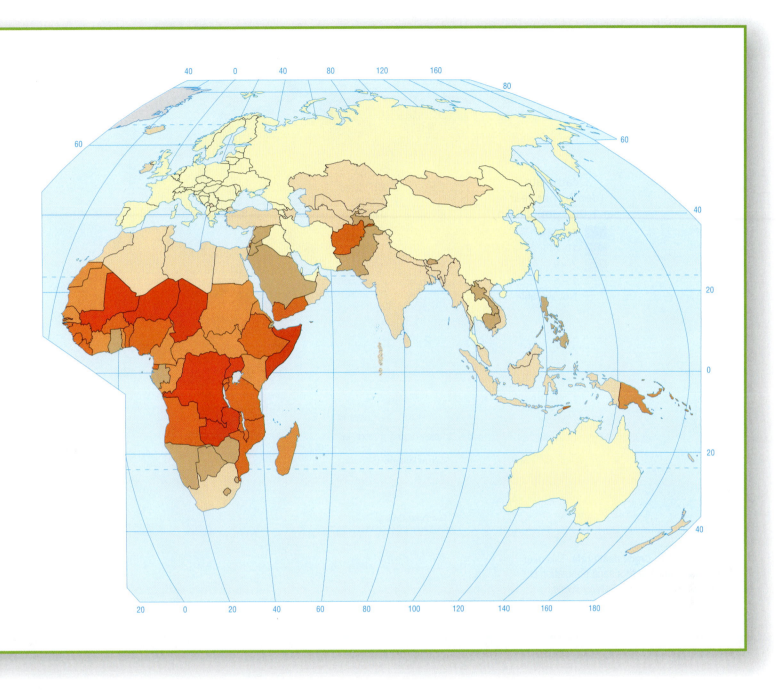

reelected. Birth control, abortion, and challenging long-held beliefs about family size can prove quite controversial, and political leaders may be reluctant to legislate for them. Indeed, the Chinese implementation of its one-child-per-couple policy (see Subject to Debate, page 87) probably would not have been possible in a country with a democratically elected government. In addition, because it involves changing a cultural norm, the idea of smaller families can take three or four generations to take hold. Increased educational levels for women are closely associated with falling fertility levels, as is access to various contraceptive devices (**Figure 3.9**, page 86).

The demographic transition is a model that predicts trends in birthrates, death rates, and overall population levels in the abstract. Yet, like many models used by demographers, it is based on the historical experience of western Europe: it is **Eurocentric.** It does a good job of describing population patterns over time in Europe, as well as in other wealthy regions. However, it has several

Eurocentric
Using the historical experience of Europe as the benchmark for all cases.

Crude Death Rate

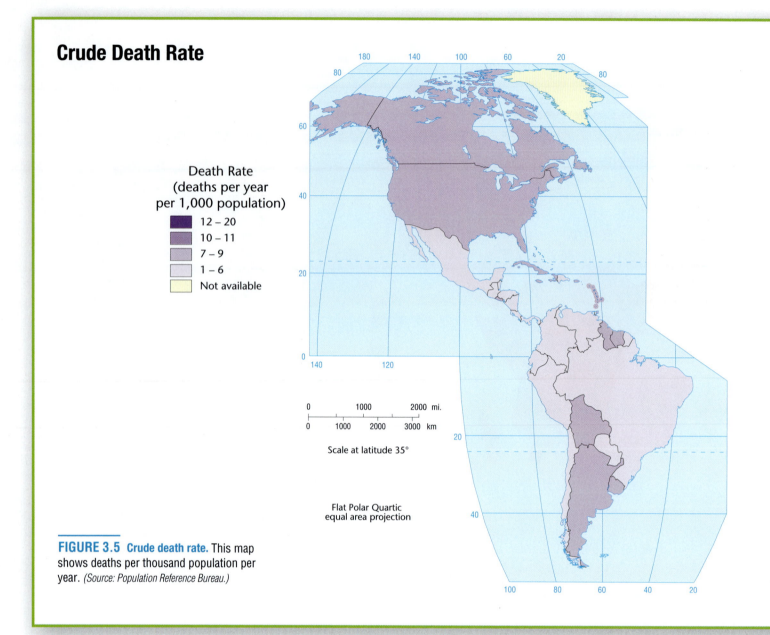

Death Rate
(deaths per year
per 1,000 population)

- 12 – 20
- 10 – 11
- 7 – 9
- 1 – 6
- Not available

0   1000   2000 mi.

0   1000   2000   3000   km

Scale at latitude 35°

Flat Polar Quartic
equal area projection

FIGURE 3.5 Crude death rate. This map shows deaths per thousand population per year. (*Source: Population Reference Bureau.*)

shortcomings. First is the inexorable stage-by-stage progression implicit in the model. Have countries or regions ever skipped a stage or regressed? Certainly. The case of China shows how policy, in this case government-imposed restrictions on births, can fast-forward an entire nation to stage 4 (see Subject to Debate). War, too, can occasion a return to an earlier stage in the model by increasing death rates. For instance, Angola and Afghanistan are two countries with recent histories of conflict and with some of the highest death rates in the world: 24 per 1000 and 18 per 1000, respectively. In other cases, wealth has not led to declining fertility. Thanks to oil exports, residents of Saudi

Arabia enjoy relatively high average incomes; but fertility, too, remains relatively high at nearly 4 children per woman in 2010. Indeed, the Population Reference Bureau has pointed to a "demographic divide" between countries where the demographic transition model applies well and others—mostly poorer countries or those experiencing widespread conflict or disease—where birth and death rates do not necessarily follow the model's predictions. Even in Europe, there are countries where fertility has dropped precipitously, while at the same time death rates have escalated. In countries such as Russia, which, according to Peter Coclanis, has "somehow managed to reverse

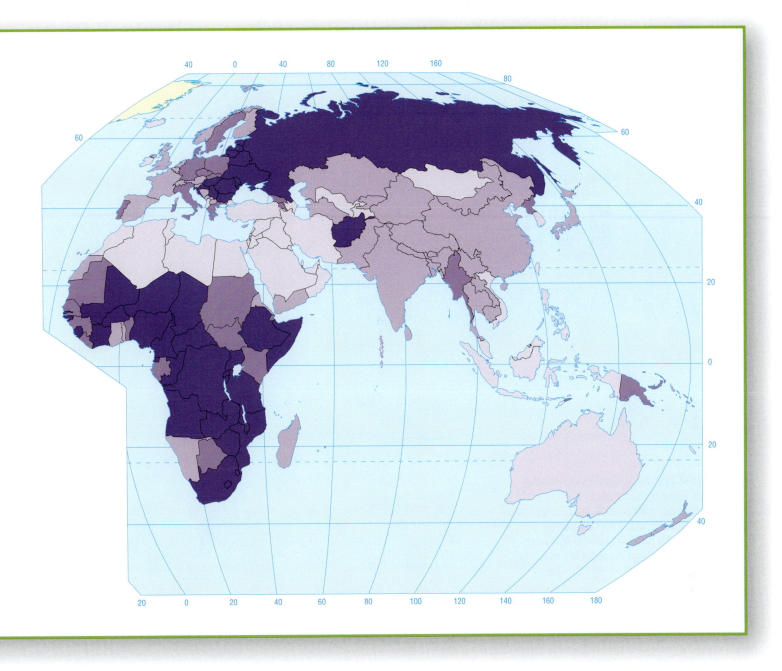

Age Distributions

the so-called demographic transition," the model itself must be questioned.

Some countries have overwhelmingly young populations. In a majority of countries in Africa, as well as some countries in Latin America and tropical Asia, close to half the population is younger than 15 years of age (**Figure 3.10**, pages 88–89). In Uganda, for example, 51 percent of the population is younger than 15 years of age. In sub-Saharan Africa, 44 percent of all people are younger than 15. Other countries, generally those that industrialized early, have a preponderance of middle-aged people in the over 15– under 65 age bracket. A growing number of affluent countries have remarkably aged populations. In Germany, for example, fully 19 percent of the people have now passed the traditional retirement age of 65. Many other European countries are not far behind. A sharp contrast emerges when Europe is compared with Africa, Latin America, or parts of Asia, where the average person never even lives to age 65. In Mauritania, Niger, Afghanistan, Guatemala, and many other countries, only 2 to 3 percent of the people have reached that age.

Adult HIV/AIDS Cases

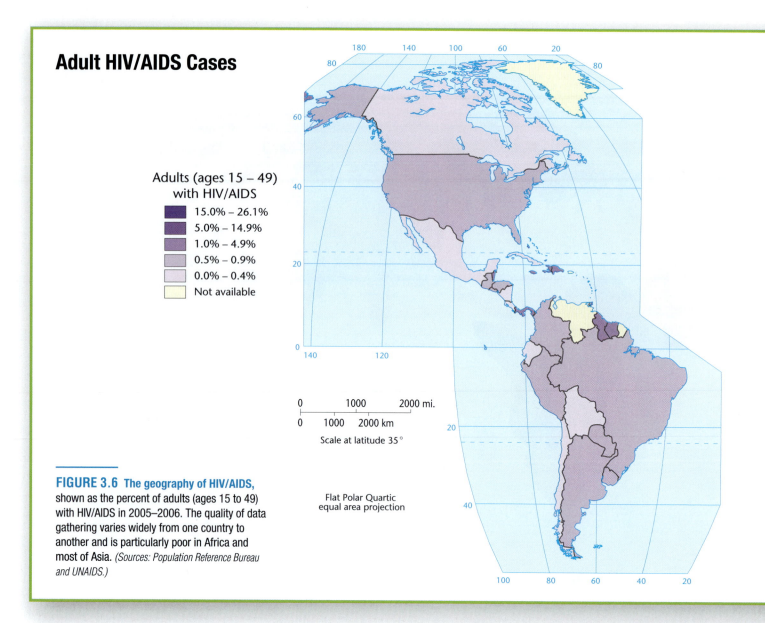

Adults (ages 15 – 49)
with HIV/AIDS

- 15.0% – 26.1%
- 5.0% – 14.9%
- 1.0% – 4.9%
- 0.5% – 0.9%
- 0.0% – 0.4%
- Not available

0   1000   2000 mi.

0   1000   2000 km

Scale at latitude 35°

Flat Polar Quartic
equal area projection

FIGURE 3.6 **The geography of HIV/AIDS,**
shown as the percent of adults (ages 15 to 49)
with HIV/AIDS in 2005–2006. The quality of data
gathering varies widely from one country to
another and is particularly poor in Africa and
most of Asia. *(Sources: Population Reference Bureau
and UNAIDS.)*

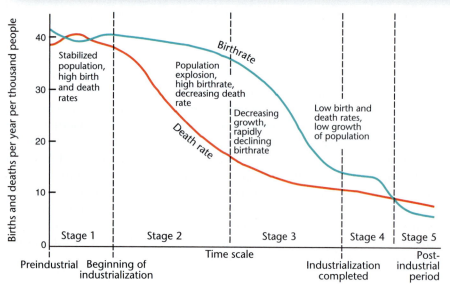

FIGURE 3.7 **The demographic transition
as a graph.** The "transition" occurs in several
stages as the industrialization of a country
progresses. In stage 2, the death rate
declines rapidly, causing a population
explosion as the gap between the number of
births and deaths widens. Then, in stage 3,
the birthrate begins a sharp decline. The
transition ends when, in stage 4, both birth
and death rates have reached low levels, by
which time the total population is many times
greater than at the beginning of the
transformation. In the postindustrial stage,
population decline eventually begins.

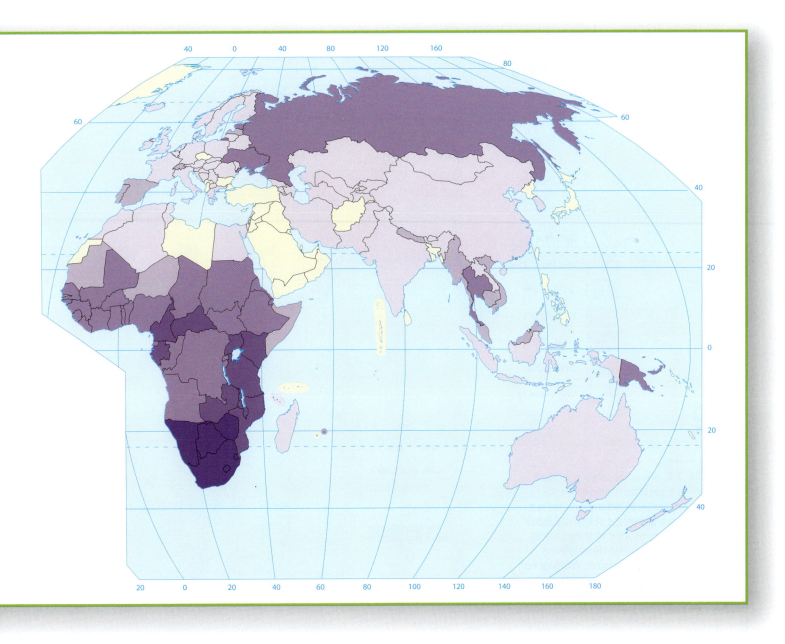

Countries with disproportionate numbers of old or young people often address these imbalances in innovative ways. Italy, for example, has one of the lowest birthrates in the world, with a TFR of only 1.3. In addition, Italy's population is one of the oldest in the world: 20.3 percent of its population is age 65 or older. As a result of both its TFR and age distribution, Italy's population is projected to shrink by 10 percent between 2010 and 2050. Of course, immigrants to Italy from other countries can and doubt-lessly will counteract this trend, but this will entail conten-tious political and cultural debates.

Given that the Italian culture does not embrace the institutionalization of the growing ranks of its elderly, and faced with the reality that more Italian women than ever work outside the home and thus few adult women are will-ing or able to stay at home full time to care for their elders, Italians have gotten creative. Elderly Italians can apply for adoption by families in need of grandfathers or grand-mothers. Frances D'Emilio recounts the experience of one such man, Giorgio Angelozzi, who moved in with the Rivas, a Roman family with two teenagers. Angelozzi said that Marlena Riva's voice reminded him of his deceased wife, Lucia, and this is what convinced him to choose the Riva family. Dagmara Riva, the family's teenage daughter, said that Mr. Angelozzi has helped her with Latin studies and that "Grandpa is a person of great experience, an

Annual Population Change

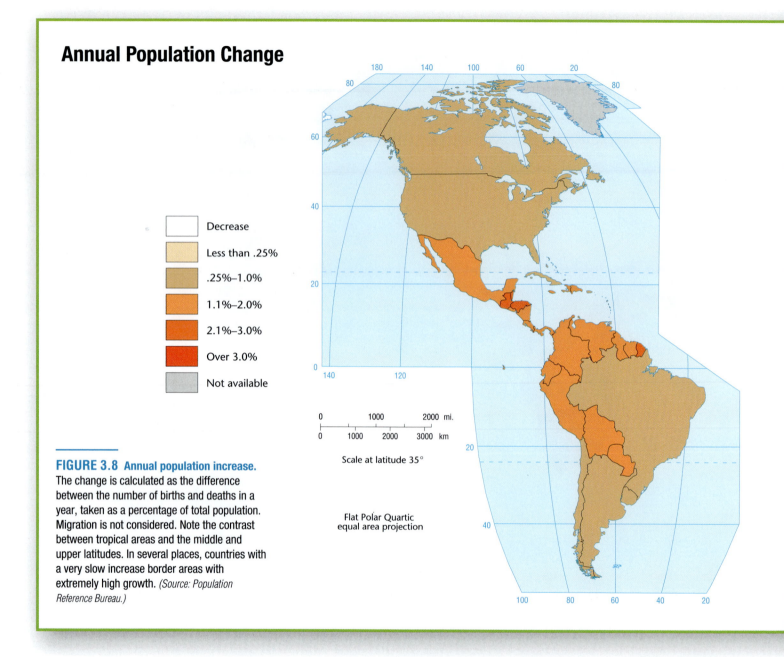

Decrease

Less than .25%

.25%–1.0%

1.1%–2.0%

2.1%–3.0%

Over 3.0%

Not available

0   1000   2000 mi.
0   1000   2000   3000 km

Scale at latitude 35°

Flat Polar Quartic
equal area projection

FIGURE 3.8 Annual population increase.
The change is calculated as the difference
between the number of births and deaths in a
year, taken as a percentage of total population.
Migration is not considered. Note the contrast
between tropical areas and the middle and
upper latitudes. In several places, countries with
a very slow increase border areas with
extremely high growth. (*Source: Population
Reference Bureau.*)

affectionate person. We're very happy we invited him to
live with us."

Age structure also differs spatially within individual
countries. For example, rural populations in the United
States and many other countries are usually older than
those in urban areas. The flight of young people to the cities has left some rural counties in the midsection of the
United States with populations whose median age is 45 or
older. Some warm areas of the United States have become
retirement havens for the elderly; parts of Arizona and
Florida, for example, have populations far above the average age. Communities such as Sun City near Phoenix,

Arizona, legally restrict residence to the elderly (**Figure
3.11**, page 89). In Great Britain, coastal districts have a
much higher proportion of elderly than does the interior,
suggesting that the aged often migrate to seaside locations
when they retire.

A very useful graphic device for
comparing age characteristics is the
population pyramid (**Figure 3.12**,
page 90). Careful study of such pyra-
mids not only reveals the past progress of birth control but
also allows geographers to predict future population
trends. Youth-weighted pyramids, those that are broad at

population pyramid
A graph used to show the
age and sex composition of
a population.

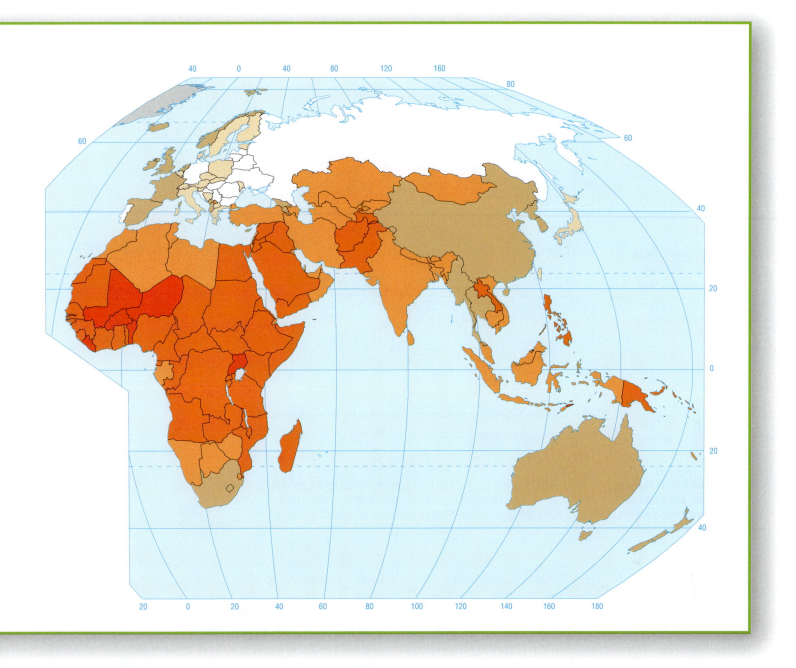

the base, suggest the rapid growth typical of the population explosion. What's more, broad-based population pyramids not only reflect past births but also show that future growth will rely on the momentum of all of those young people growing into their reproductive years and having their own families, regardless of how small those families may be in contrast to earlier generations. Those population pyramids with more of a cylindrical shape represent countries approaching population stability or those in demographic decline. A quick look at a country's population pyramid can tell volumes about its past as well as its future. How many dependent people—the very old and the very young—live there? Has the country suffered the demographic effects of genocide or a massive disease epidemic (**Figure 3.13**, page 92)? Are significantly more boys than girls being born? These questions and more can be explored at a glance using a population pyramid.

The Geography of Gender

Although the human race is divided almost evenly between females and males, geographical differences do occur in the **sex ratio:** the ratio between men and women in a

sex ratio
The numerical ratio of males to females in a population.

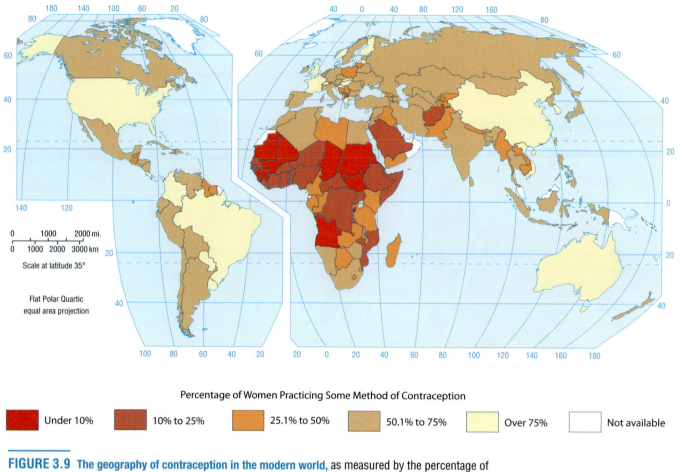

Percentage of Women Practicing Some Method of Contraception

Under 10%	10% to 25%	25.1% to 50%	50.1% to 75%	Over 75%	Not available

FIGURE 3.9 The geography of contraception in the modern world, as measured by the percentage of women using devices of any sort. Contraception is much more widely practiced than abortion, but cultures differ greatly in their level of acceptance. Several different devices are included. *(Source: World Population Data Sheet.)*

population (**Figure 3.14**, pages 92–93). Slightly more boys than girls are born, but infant boys have slightly higher mortality rates than do infant girls. Recently settled areas typically have more males than females, as is evident in parts of Alaska, northern Canada, and tropical Australia. According to the U.S. Census Bureau, in 2009 males constituted 52 percent of Alaska's inhabitants. By contrast, Mississippi's population was 52 percent female, reflecting in part the emigration of young males in search of better economic opportunities elsewhere. Some poverty-stricken parts of South Africa are as much as 59 percent female. Prolonged wars reduce the male population. And, in general, women tend to outlive men. The population pyramid is also useful in showing gender ratios. Note, for instance, the larger female populations in the upper bars for both the United States and Sun City, Arizona, in Figure 3.12 (page 90).

Beyond such general patterns, gender often influences demographic traits in specific ways. Often **gender roles**—culturally specific notions of what it means to be a man and what it means to be a woman—are closely tied to how many children are produced by couples. In many cultures, women are considered more womanly when they produce many offspring. By the same token, men are seen as more manly when they father many children. Because the raising of children often falls to women, the spaces that many cultures associate with women tend to be the private family spaces of the home. Public spaces such as streets, plazas, and the workplace, by contrast, are often associated with men. Some cultures go so far as to restrict where women and men may and may not go, resulting in a distinctive geography of gender, as **Figure 3.15**, page 94, illustrates (see also Seeing Geography, page 112). Falling fertility levels that coincide with higher levels of education for women, however, have resulted in numerous challenges to these cultural ideas of male and female spaces. As more

gender roles
What it means to be a man or a woman in different cultural and historical contexts.

Subject to Debate

Female: An Endangered Gender?

Does the simple fact of being female expose a person to demographic peril? In most societies, women are viewed as valuable, even powerful, particularly as mothers, nurturers, teachers, and spiritual leaders. Yet in other important ways, to be female is to be endangered. We will consider this controversial idea with an eye to how demographics and culture closely shape each other.

Many cultures demonstrate a marked preference for males. The academic term describing this is *androcentrism;* you may know it as *patriarchy, male bias,* or simply *sexism.* Whether a preference for males is a feature of all societies has been disputed. Some societies pass along forms of their wealth, property, and prestige from mother to daughter, rather than from father to son. This is rare, however, and it is clear that the roots of the cultural preference for males are historically far-reaching and widespread. In most societies, positions of economic, political, social, and cultural prestige and power are held largely by men. Men typically are considered to be the heads of households. Family names tend to pass from father to son, and with them, family honor and wealth. In traditional societies, when sons marry, they usually bring their wives to live in their parents' house and are expected to assume economic responsibility for aging parents.

For all of these reasons, in many places a cultural premium is placed on producing male children. Because couples often can choose to have more children, having a girl first is usually not a problem. However, in countries that have enacted strict population control programs, couples may not be given the chance to try again for a male child. This has resulted in severe pressures on couples to have boys in some countries, particularly in China and India. In both of these countries, female-specific infanticide or abortion has resulted in a growing gender imbalance. Ultrasound devices that allow gender identification of fetuses are now available even to rural peasants in China. About 100,000 such devices were in use as early as 1990 in China, and by the middle of that decade there were 121 males for every 100 females among children two years of age or younger. The sex ratio in China is radically changing, and a profound gender imbalance already exists there. In India, too, there were only 927 girls for each 1000 boys in 2001.

The cultural ramifications of such male-heavy populations are potentially profound. Men of marriageable age are increasingly unable to find female partners. Social analysts speculate that this will lead to human trafficking and violence against women. In China, the policy of one child per couple has resulted in the so-called four-two-one problem. This refers to the fact that the generational structure of many families now reflects four grandparents, two parents, and a single male child. This places enormous pressures on the shoulders of the male child to care for aging parents and grandparents. It also encourages parents and grandparents to lavish all of their attention, wealth, and hopes on the only child. For some families, this has led to the "little emperor syndrome," whereby the male heir becomes spoiled, unable to cope independently, and even obese.

Continuing the Debate

As noted, most societies value females. For a number of reasons, however, some societies show a clear preference for males. Keeping all of this in mind, consider these questions:

- Are Chinese families somewhat justified in emphasizing having a boy at all costs?

- According to a recent report, Americans using technology to select their baby's gender, unlike the Chinese, more often choose to have a girl. Why do you think there is a difference between male and female preference in these two societies?

This "little emperor" poses with his grandparents. *(Dennis Cox, LLC.)*

Youth and Old-Age Populations

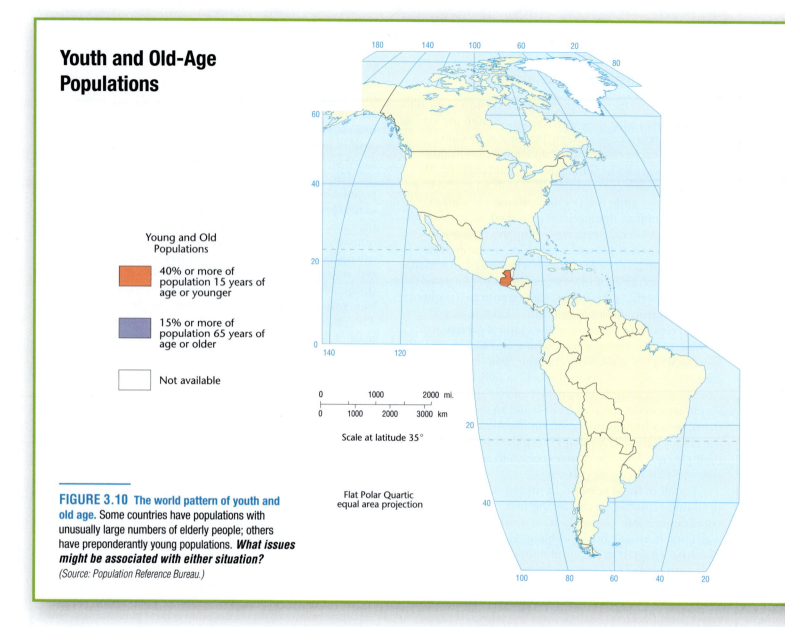

Young and Old Populations

- 40% or more of population 15 years of age or younger
- 15% or more of population 65 years of age or older
- Not available

0   1000   2000 mi.
0   1000   2000   3000 km
Scale at latitude 35°

Flat Polar Quartic equal area projection

FIGURE 3.10 The world pattern of youth and old age. Some countries have populations with unusually large numbers of elderly people; others have preponderantly young populations. *What issues might be associated with either situation?* (*Source: Population Reference Bureau.*)

and more women enter the workplace, for instance, ideas of where women should and should not go slowly become modified. Clearly, gender is an important factor to consider in our exploration of population geography (see also Subject to Debate, page 87, and Patricia's Notebook, page 91).

Standard of Living

Various demographic traits can be used to assess standard of living and analyze it geographically. **Figure 3.16**, pages 94–95, is a simple attempt to map living standards using the **infant mortality rate:** a measure of how many children per 1000 population die before reaching one year of age. Many experts believe that the infant mortality rate is the best single index of living standards because it is affected by many different factors: health, nutrition, sanitation, access to doctors, availability of clinics, education, ability to obtain medicines, and adequacy of housing. A vivid geographical pattern is revealed by the infant mortality rate.

Another good measure of quality of life is the United Nations Human Development Index (HDI), which combines measures of literacy, life expectancy, education, and wealth (see Figure 1.11, pages 14–15). The highest possible

infant mortality rate
The number of infants per 1000 live births who die before reaching one year of age.

(*continued on page 93*)

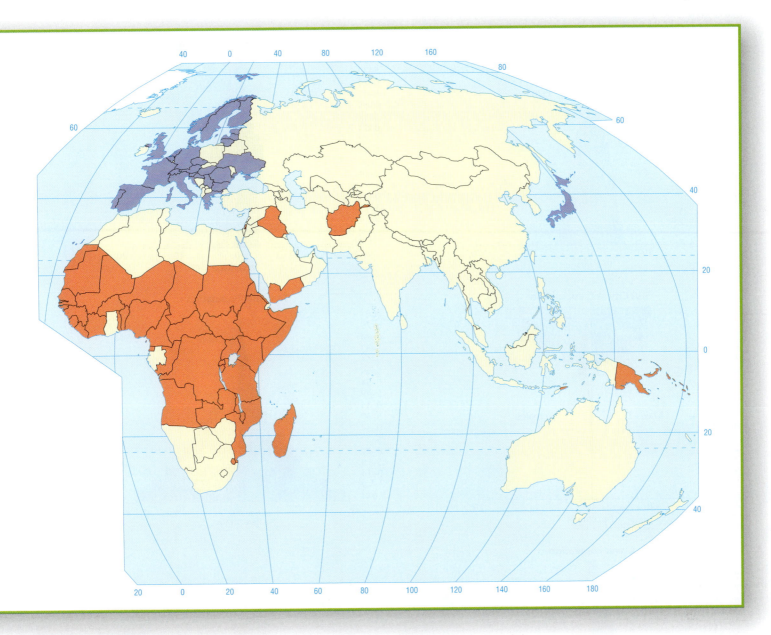

FIGURE 3.11 Residents of Sun City, Arizona, enjoy the many recreation opportunities provided in this planned retirement community, where the average age is 75. *(A. Ramey/PhotoEdit Inc.)*

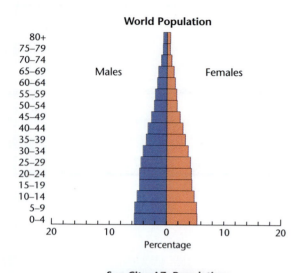

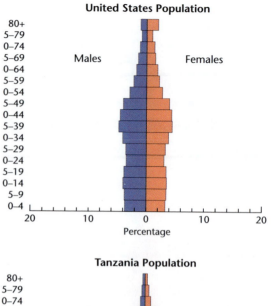

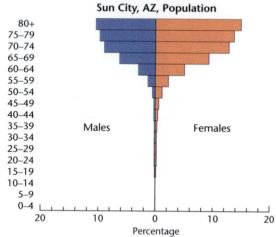

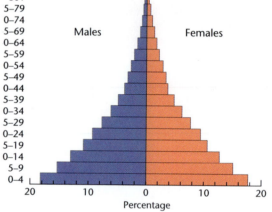

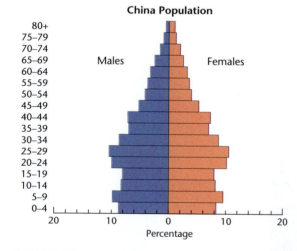

FIGURE 3.12 **Population pyramids for the world and selected countries and communities.**
Tanzania displays the classic stepped pyramid of a rapidly expanding population, whereas the U.S.
pyramid looks more like a precariously balanced pillar. China's population pyramid reflects the
lowered numbers of young people as a result of that country's one-child policy. *How do these*
pyramids help predict future population growth? (*Source: Population Reference Bureau.*)

Patricia's Notebook

Demystifying the Sunday Crowds in Hong Kong

Patricia Price

Filipina domestic servants in Hong Kong. On Sunday, their day off, these women congregate in covered walkways and other public spaces. They exchange gossip, play cards, cut each other's hair, and relax. *(Courtesy of Patricia L. Price.)*

During a recent trip to Hong Kong, I was struck by the dense crowds of women that can be seen on Sundays. Some of these gatherings are in public spaces, such as parks. Others are found in city spaces that are usually heavily trafficked during workdays, such as the steps outside large office buildings. The largest crowd I observed was under the covered walkway of a subway station. Initially, I thought that these women were gathered for a political demonstration of some kind. In Latin America—the world region I'm most familiar with—large gatherings of people almost always mean that a protest march, political rally, or labor strike is about to begin. But there were no men to be seen anywhere in these Hong Kong throngs! In addition, they didn't seem to be protesting anything. Rather, they happily chatted, ate, styled one another's hair, and played cards.

Later, I learned that these women were in fact maids enjoying their day off. For several decades now, women from the Philippines have migrated to Hong Kong to work as domestic servants, doing chores for and looking after the children of the families who hire them. Wages in Hong Kong are much higher than back home in the Philippines: a major pull factor. However, working conditions can be far from ideal, involving hard physical labor and long hours. Reports of abuse of Filipina servants by their employers are numerous. If a servant is fired, she will be deported if she cannot find another job quickly. The stress of long separations from husbands and children back in the Philippines has led to the breakup of families. In addition, some of these women find themselves coerced into Asia's booming sex industry. Knowing this makes me think twice about these happy-looking women. Though being a domestic servant isn't a pleasant or highly paid job anywhere in the world, Filipina maids in Hong Kong certainly seem to face a number of pressures, ranging from labor conditions, family circumstances, and the lack of good job alternatives. While the news coverage of domestic servants in the United States seems to be mostly about their status as documented or undocumented workers, there are obviously additional factors to keep in mind. These factors complicated my picture of these women, who, at first glance, seemed happy and carefree as they enjoyed their day off.

Posted by Patricia Price

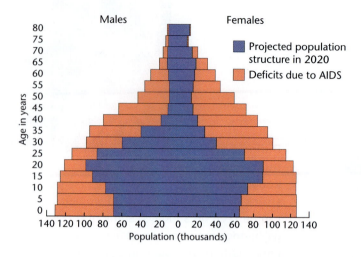

FIGURE 3.13 **Projected population structure with and without the AIDS epidemic in Botswana in 2020.** *(Sources: U.S. Bureau of Census; Food and Agriculture Organization of the United Nations.)*

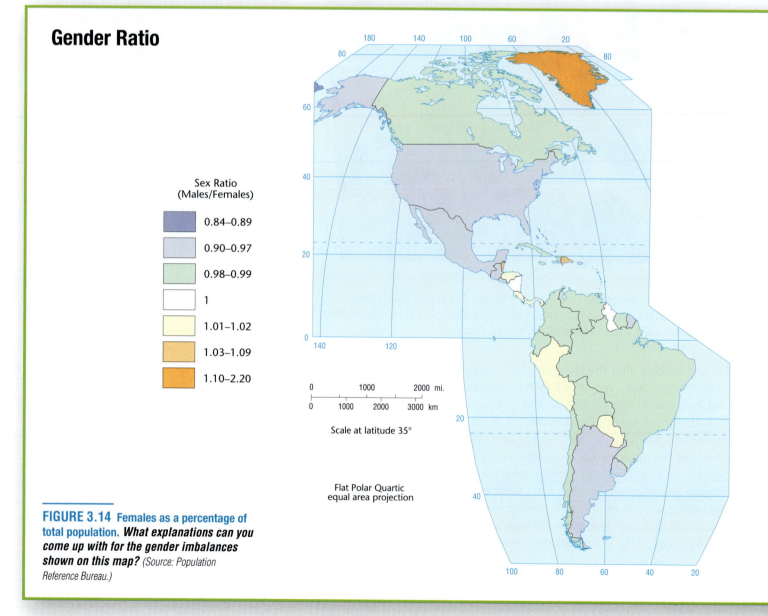

Gender Ratio

Sex Ratio
(Males/Females)

- 0.84–0.89
- 0.90–0.97
- 0.98–0.99
- 1
- 1.01–1.02
- 1.03–1.09
- 1.10–2.20

Scale at latitude 35°

Flat Polar Quartic
equal area projection

FIGURE 3.14 **Females as a percentage of total population.** *What explanations can you come up with for the gender imbalances shown on this map?* *(Source: Population Reference Bureau.)*

score is 1.000, and the three top-ranked countries in 2009 were Norway, Australia, and Iceland.

Examination of the HDI reveals some surprises. If all countries spent equally on those things that improve their HDI rankings, such as education and health care, then we would expect the wealthiest countries to place first on the list. According to the International Monetary Fund, the United States ranked 9th among 180 ranked nations in wealth as measured by the gross domestic product (GDP) per capita in 2009. Yet it ranked 13th in the world on the HDI. Compare this to Barbados, which ranked 45th by GDP per capita but came in much higher on the HDI at 37th. Why did Barbados rank higher in living standards than its monetary wealth would indicate, whereas the United States ranked lower? We would have to conclude that the government of Barbados places a relative priority on spending for education and health care, whereas the government of the United States does not.

Even more striking is the low standing of the United States when the Human Poverty Index (HPI) is used. The HPI measures social and economic deprivation by examining the prevalence of factors such as low life expectancy, illiteracy, poverty, and unemployment. Among the world's high-income countries, the United States ranks near the bottom, with only Italy and Ireland having larger numbers of deprived people.

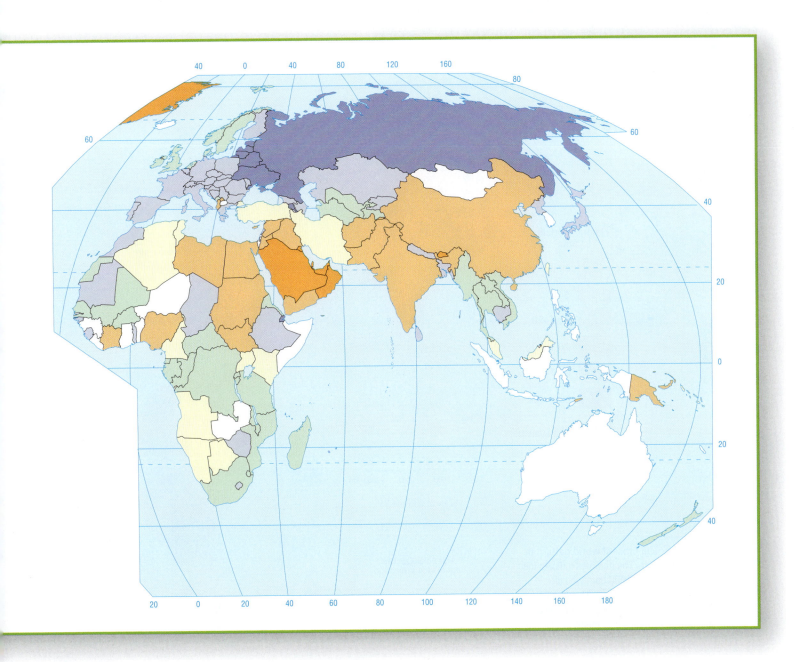

FIGURE 3.15 Segregated beach. This portion of the beachfront along the Israeli city of Tel Aviv allows women and men to swim on separate days. Young children of both genders accompany their mothers, whereas older boys visit with their fathers. Gender-based segregation is important for Israel's orthodox Jews. *(Courtesy of Patricia L. Price.)*

Infant Mortality Rate

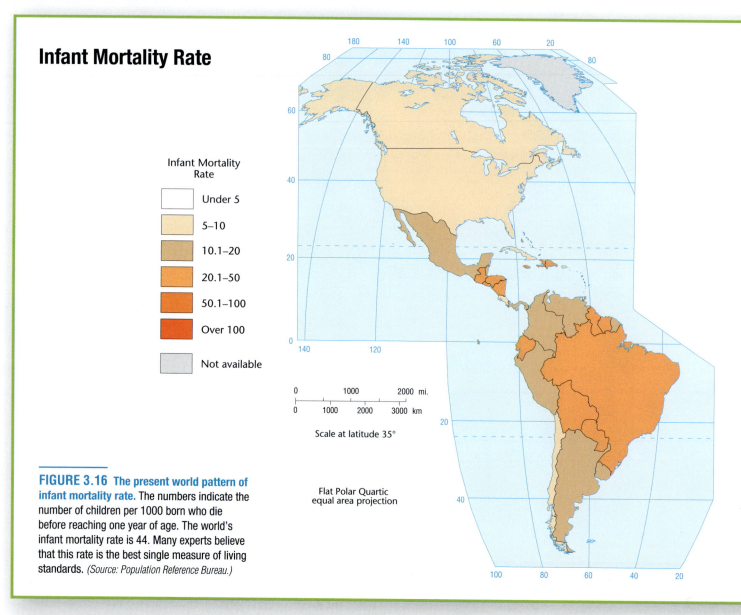

Infant Mortality Rate

- Under 5
- 5–10
- 10.1–20
- 20.1–50
- 50.1–100
- Over 100
- Not available

Scale at latitude 35°

0   1000   2000 mi.
0   1000   2000   3000 km

Flat Polar Quartic
equal area projection

FIGURE 3.16 The present world pattern of infant mortality rate. The numbers indicate the number of children per 1000 born who die before reaching one year of age. The world's infant mortality rate is 44. Many experts believe that this rate is the best single measure of living standards. *(Source: Population Reference Bureau.)*

Mobility

How does demography relate to the theme of mobility? After natural increases and decreases in populations, the second of the two basic ways in which population numbers are altered is the migration of people from one place to another, and it offers a straightforward illustration of relocation diffusion. When people migrate, they sometimes bring more than simply their culture; they can also bring disease. The introduction of diseases to new places can have dramatic and devastating consequences. Thus, the spread of diseases is also considered as part of the theme of mobility because diseases move with people.

Migration

Humankind is not tied to one locale. *Homo sapiens* most likely evolved in Africa, and ever since, we have proved remarkably able to adapt to new and different physical environments. We have made ourselves at home in all but the most inhospitable climates, shunning only such places as ice-sheathed Antarctica and the shifting sands of the Arabian Peninsula's "Empty Quarter." Our permanent habitat extends from the edge of the ice sheets to the seashores, from desert valleys below sea level to high mountain slopes. This far-flung distribution is the product of migration.

Early human groups moved in response to the migration of the animals they hunted for food and the ripening seasons of the plants they gathered. Indeed, the

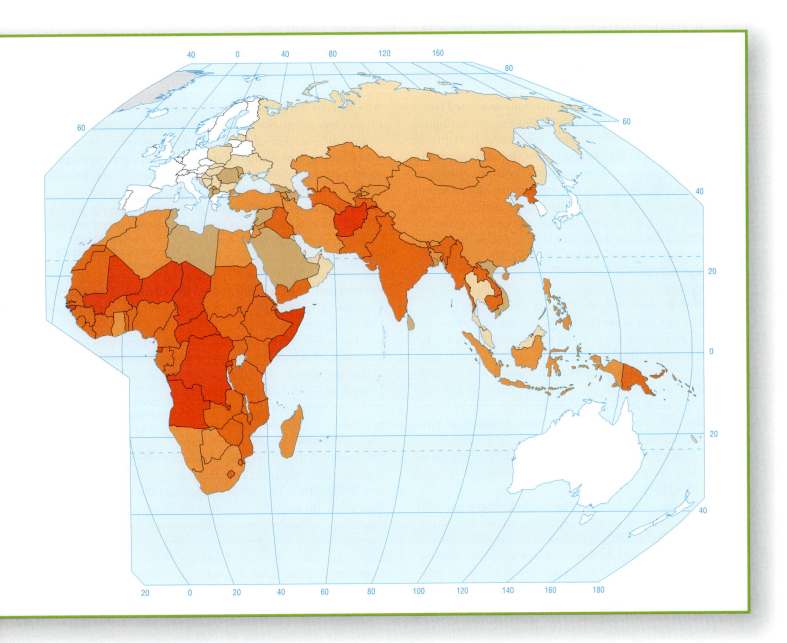

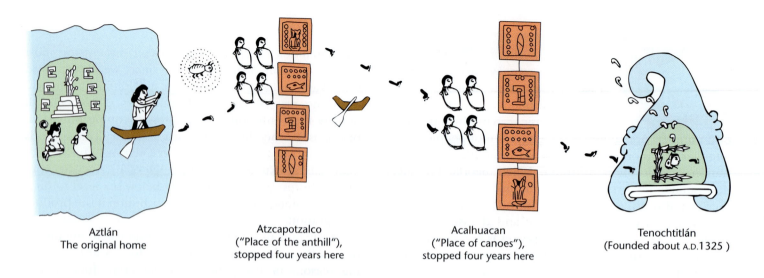

Aztlán
The original home

Atzcapotzalco
("Place of the anthill"),
stopped four years here

Acalhuacan
("Place of canoes"),
stopped four years here

Tenochtitlán
(Founded about A.D.1325)

FIGURE 3.17 Segments of an Aztec codex, depicting the prehistoric migration of the ancient Aztecs from an island, possibly in northwestern Mexico, to another island in a lake at the site of present-day Mexico City, where they founded their capital, Tenochtitlán. Clearly, the epic migration was a central event in their collective memory. *(After de Macgregor, 1984: 217.)*

agricultural revolution, whereby humans domesticated crops and animals and accumulated surplus food supplies, allowed human groups to stop their seasonal migrations. But why did certain groups opt for long-distance relocation? Some migrated in response to environmental collapse, others in response to religious or ethnic persecution. Still others—probably the majority, in fact—migrated in search of better opportunities. For those who migrate, the process generally ranks as one of the most significant events of their lives. Even ancient migrations often remain embedded in folklore for centuries or millennia (**Figure 3.17**).

The Decision to Move Migration takes place when people decide that moving is preferable to staying and when the difficulties of moving seem to be more than offset by the expected rewards. In the nineteenth century, more than 50 million European emigrants left their homelands in search of better lives. Today, migration patterns are very different (**Figure 3.18**). Europe, for example, now predominantly receives immigrants rather than sending out emigrants. International migration stands at an all-time high, much of it labor migration associated with the process of globalization. About 160 million people today live outside the country of their birth.

Historically, and to this day, forced migration also often occurs. The westward displacement of the Native American population of the United States; the dispersal of the Jews from Palestine in Roman times and from Europe in the mid-twentieth century; the export of Africans to the Americas as slaves; and the Clearances, or forced removal of farmers from Scotland's Highlands to make way for large-scale sheep raising—all provide depressing examples. Today, refugee movements are all too common, prompted mainly by despotism, war, ethnic persecution, and famine. Recent decades have witnessed a worldwide flood of **refugees:** people who leave their country because of persecution based on race, ethnicity, religion, nationality, or political opinion (note that economic persecution does not fall under the definition of *refugee*). Perhaps as many as 16 million people who live outside their native country are refugees.

> **refugees**
> Those fleeing from persecution in their country of nationality. The persecution can be religious, political, racial, or ethnic.

Every migration, from the ancient dispersal of human-kind out of Africa to the present-day movement toward urban areas, is governed by a host of **push-and-pull factors** that act to make the old home unattractive or unlivable and the new land attractive. Generally, push factors are the most central. After all, a basic dissatisfaction with the homeland is prerequisite to voluntary migration. The most important factor prompting migration throughout the thousands of years of human existence has been economic. More often than not, migrating people seek greater prosperity through better access to resources, especially land. Both forced migrations and refugee movements, however, challenge the basic assumption

> **push-and-pull factors**
> Unfavorable, repelling conditions (push factors) and favorable, attractive conditions (pull factors) that interact to affect migration and other elements of diffusion.

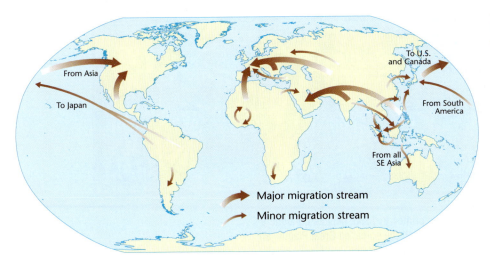

FIGURE 3.18 **Major and minor migration flows today.** Why have these flows changed so profoundly in the past hundred years? *(Source: Population Reference Bureau.)*

of the push-and-pull model, which posits that human movement is the result of choices.

Diseases on the Move

Throughout history, infectious disease has periodically decimated human populations. One has only to think of the vivid accounts of the Black Death episodes that together killed one-third of medieval Europe's population to get a sense of how devastating disease can be. Importantly for cultural geographers, diseases both move in spatially specific ways and occasion responses that are spatial in nature.

The spread of disease provides classic illustrations of both expansion and relocation diffusion. Some diseases are noted for their tendency to expand outward from their points of origin. Commonly borne by air or water, the viral pathogens for contagious diseases such as influenza and cholera spread from person to person throughout an affected area. Some diseases spread in a hierarchical diffusion fashion, whereby only certain social strata are exposed. The poor, for example, have always been much more exposed to the unsanitary conditions—rats, fleas, excrement, and crowding—that have occasioned disease epidemics. Other illnesses, particularly airborne diseases, affect all social and economic classes without regard for human hierarchies: their diffusion pattern is literally contagious. Indeed, disease and migration have long enjoyed a close relationship. Diseases have spread and relocated thanks to human movement. Likewise, widespread human migrations have occurred as the result of disease outbreaks.

When humans engage in long-distance mobility, their diseases move along with them. Thus, cholera—a waterborne disease that had for centuries been endemic to the Ganges River region of India—broke out in Calcutta (now called Kolkata) in the early 1800s. Because Calcutta was an important node in Britain's colonial empire, cholera quickly began to relocate far beyond India's borders. Soldiers, pilgrims, traders, and travelers spread cholera throughout Europe, then from port to port across the world through relocation diffusion, making cholera the first disease of global proportions.

Early responses to contagious diseases were often spatial in nature. Isolation of infected people from the healthy population—known as quarantine—was practiced. Sometimes only the sick individuals were quarantined, whereas at other times households or entire villages were shut off from outside contact. Whole shiploads of eastern European immigrants to New York City were routinely quarantined in the late nineteenth century. Another early spatial response was to flee the area where infection had occurred. During the time of the plague in fourteenth-century Europe, some healthy individuals abandoned their ill neighbors, spouses, and even children. Some of them formed altogether separate communities, whereas others sought merely to escape the city walls for the countryside (**Figure 3.19**, page 98).

Targeted spatial strategies could be implemented once it became known that some diseases spread through specific means. The best-known example is John Snow's 1854 mapping of cholera outbreaks in London, illustrated in **Figure 3.20** (page 98), which allowed him to trace the source of the infection to one water pump. Thanks to his detective work, and the development of a broader medical understanding of the role of germs in the spread of disease, cholera was discovered to be a waterborne disease that could be controlled by increasing the sanitary conditions of water delivery.

The threat of deadly disease is hardly a thing of the past. Today, geographers play vital roles in understanding the diffusion of HIV/AIDS, SARS (severe acute respiratory

FIGURE 3.19 **Fleeing from disease.** This woodcut, which was printed in 1630, depicts Londoners leaving the pestilent city in a cart. In 1665–1666, London experienced the so-called Great Plague, an outbreak of bubonic plague that killed one-fifth of the city's population. *(Bettmann/Corbis.)*

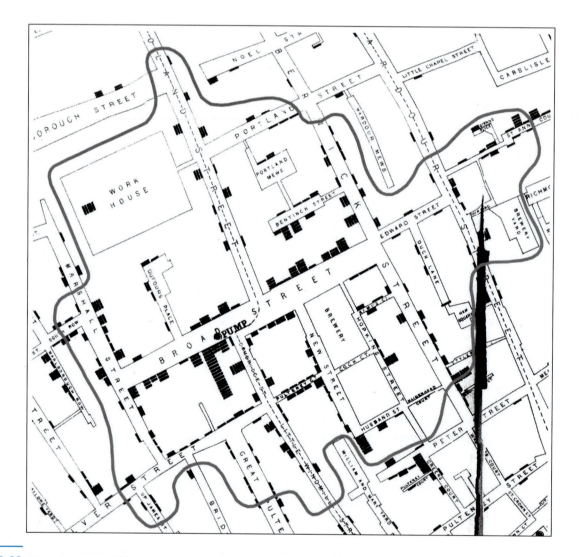

FIGURE 3.20 **Mapping disease.** This map was constructed by London physician John Snow. Snow was skeptical of the notion that "bad air" somehow carried disease. He interviewed residents of the Soho neighborhood stricken by cholera to construct this map of cholera cases and used it to trace the outbreak to the contaminated Broad Street pump. Snow is considered to be a founder of modern epidemiology. *(Courtesy of The John Snow Archive.)*

syndrome), and the so-called swine flu in order to better address outbreaks and halt the spread of these global scourges. By mapping the spread of these contemporary diseases, understanding the pathways traveled by their carriers, and developing appropriate spatial responses, it is hoped that mass epidemics and disease-related panics can be avoided.

Globalization

How much is enough? How much is too much? One of the great debates today involves the population of the Earth. How many people can the Earth support? Should humans limit their reproduction, and who should decide? There are no precise numbers involved in this debate, but these questions are so hotly contested because addressing them in a conscientious way may well determine the long-term fate of the human race.

Population Explosion?

population explosion
The rapid, accelerating increase in world population since about 1650 and especially since about 1900.

One of the fundamental issues of the modern age is the **population explosion**: a dramatic increase in world population since 1900 (**Figure 3.21**).

The crucial element triggering this explosion has been a steep decline in the death rate, particularly for infants and children, in most of the world, without an accompanying universal decline in fertility. At one time in traditional cultures, only two or three offspring in a family of six to eight children might live to adulthood, but when improved health conditions allowed more children to survive, the cultural norm encouraging large families persisted.

Until very recently, the number of people in the world has been increasing geometrically, doubling in shorter and shorter periods of time. It took from the beginning of human history until A.D. 1800 for the Earth's population to grow to 1 billion people. But from 1800 to 1930, it grew to 2 billion, and in only 45 more years it doubled again (**Figure 3.22**, page 100). The overall effect of even a few population doublings is astonishing. As an illustration of a simple geometric progression, consider the following legend. A king was willing to grant any wish to the person who could supply a grain of wheat for the first square of his chessboard, two grains for the second square, four for the third, eight for the fourth, and so on. To cover all 64 squares and win, the candidate would have had to present a cache of wheat larger than today's worldwide wheat crop. Looking at the population explosion in another way, it is estimated that 61 billion humans have lived in the entire 200,000-year period since *Homo sapiens* originated. Of these, 6.8 billion (roughly 10 percent) are alive today. One

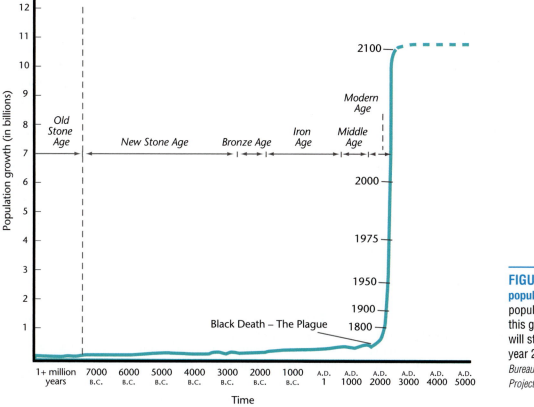

FIGURE 3.21 S-shaped world population curve. Is the global population explosion nearing its end? If this graph is right, the world's population will stabilize at nearly 11 billion by the year 2100. *(Adapted from Population Reference Bureau and United Nations, World Population Projections 2100, 1998.)*

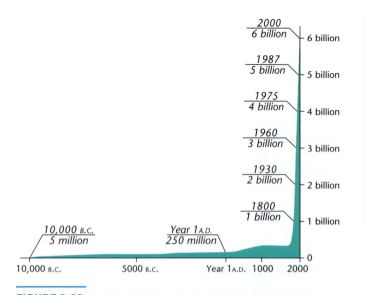

FIGURE 3.22 **World population doubling timeline.** This graph illustrates the ever-faster doubling times of the world's population. Whereas accumulating the first billion people took all of human history until about A.D. 1800, the next billion took slightly more than a century to add, the third billion took only 30 years, and the fourth took only 15 years. *(Adapted from Sustainablescale.org.)*

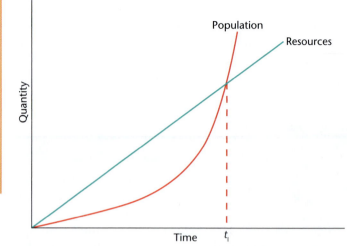

FIGURE 3.23 **Malthus's dismal equation.** Thomas Malthus based his theory of population growth on the simple notion that resources (food) grow in a linear fashion, whereas population grows geometrically. Beyond time t_1, population outstrips resources, resulting in famine, conflict, and disease.

of every 10 humans who ever lived on Earth is alive today. If we were to consider only those humans who survived into adulthood, the proportion alive today would come closer to one in five!

Some scholars foresaw long ago that an ever-increasing global population would eventually present difficulties. The most famous pioneer observer of population growth, the English economist and cleric Thomas Malthus, published *An Essay on the Principle of Population*—known as the "dismal essay"—in 1798. He believed that the human ability to multiply far exceeded our ability to increase food production. Consequently, Malthus maintained that "a strong and constantly operating check on population" would necessarily act as a natural control on numbers. Malthus regarded famine, disease, and war as the inevitable outcome of the human population's outstripping the food supply (**Figure 3.23**). He wrote, "Population, when unchecked, increases in a geometrical ratio. Subsistence only increases in an arithmetical ratio. A slight acquaintance with numbers will show the immensity of the first power in comparison of the second."

Malthusian
Those who hold the views of Thomas Malthus, who believed that overpopulation is the root cause of poverty, illness, and warfare.

The adjective **Malthusian** entered the English language to describe the dismal future Malthus foresaw. Being a cleric as well as an economist, however, Malthus believed that if humans could voluntarily restrain the "passion between the sexes," they might avoid their otherwise miserable fate.

Or Creativity in the Face of Scarcity?

But was Malthus right? From the very first, his ideas were controversial. The founders of communism, Karl Marx and Friedrich Engels, blamed poverty and starvation on the evils of capitalist society. Taking this latter view might lead one to believe that the miseries of starvation, warfare, and disease are more the result of maldistribution of the world's wealth than of overpopulation. Indeed, severe food shortages in the late 2000s in several regions point to precisely this issue of inequitable food distribution and its catastrophic consequences.

Malthus did not consider that when faced with conundrums such as scarce food supplies, human beings are highly creative. This has led critics of Malthus and his modern-day followers to point out that although the global population has doubled three times since Malthus wrote his essay, food supplies have doubled five times. Scientific innovations such as the green revolution have led to food increases that have far outpaced population growth (see Chapter 8). Other measures of well-being, including life expectancy, air quality, and average education levels, have all improved, too. Some of Malthus's critics, known as **cornucopians,** argue that human beings are, in fact, our greatest resource and that attempts to curb our numbers misguidedly cheat us out of geniuses who could devise creative solutions to our resource

cornucopians
Those who believe that science and technology can solve resource shortages. In this view, human beings are our greatest resource rather than a burden to be limited.

shortages. Modern-day followers of Malthus, known as **neo-Malthusians,** counter that the Earth's support systems are being strained beyond their capacity by the widespread adoption of wasteful Western lifestyles.

The fact is that the world's population is growing more slowly than before. The world's TFR has fallen to 2.6; one demographer has declared that "the population explosion is over"; and Figure 3.21 (page 99) depicts a leveling-off of global population at about 11 billion by the year 2100, suggesting that the world's current "population explosion" is merely a stage in a global demographic transition. Relatively stable population totals worldwide, however, mask the population declines and aging of the population structure already under way in some regions of the world.

It is difficult to speculate about the state of the world's population beyond the year 2050. What is clear, however, is that the lifestyles we adopt will affect how many people the Earth can ultimately support. In particular, whether wealthy countries continue to use the amount of resources they currently do, and whether developing countries decide to follow a Western-style route toward more and more resource consumption, will significantly determine whether we have already overpopulated the planet.

The Rule of 72

A handy tool for calculating the doubling time of a population is called the Rule of 72: take a country's rate of annual increase, expressed as a percent, and divide it into the number 72. The result is the number of years a population, growing at a given rate, will take to double.

For example, the natural annual growth of the United States in 2007 was 0.6 percent. Dividing 72 by 0.6 yields 120, which means that the population of the United States will double every 120 years. This does not, however, factor in the relatively high levels of immigration experienced by the United States, which will cause its population to double faster than every 120 years.

What about countries with faster rates of growth? Consider Guatemala, which is growing at 2.8 percent per year. Doing the math, we find that Guatemala's population is doubling every 25.7 years!

Percentages such as 0.6 or 2.8 don't sound like such high rates. If we were discussing your bank account rather than the populations of countries, you would hope that your money would double more quickly than every 25 years! (Incidentally, you can apply the Rule of 72 to your bank account or to any other figure that grows at a steady annual rate.) You may be tempted to say, "Look, there are only 12 million people in Guatemala, so it doesn't really matter if its population is doubling quickly. What really matters is that at an annual increase of 1.7 percent, India's 1 billion people

will double to 2 billion in 42 years, and China's 1.3 billion will double as fast as the U.S. population—every 120 years—but that will add another 1.3 billion to the world's population, more than four times what the United States will add by doubling!"

This assessment is partly right and partly wrong, depending on your vantage point. Viewed at the global scale, it does indeed make a significant difference when China's or India's population doubles. But if you are a resident of Guatemala or an official of the Guatemalan government, a doubling of your country's population every 25 years means that health care, education, jobs, fresh water, and housing must be supplied to twice as many people every 25 years. As mentioned in the introduction to this chapter, the scale at which population questions are asked is vital for the answers that are given.

Population Control Programs

Though the debates may occur at a global scale, most population control programs are devised and implemented at the national level. When faced with perceived national security threats, some governments respond by supporting pronatalist programs that are designed to increase the population. Labor shortages in Nordic countries have led to incentives for couples to have larger families. Most population programs, however, are antinatalist: they seek to reduce fertility. Needless to say, this is an easier task for a nonelected government to carry out because limiting fertility challenges traditional gender roles and norms about family size.

Although China certainly is not the only country that has sought to limit its population growth, its so-called one-child policy provides the best-known modern example. Mao Zedong, the longtime leader of the People's Republic of China (1949–1976), did not initially discourage large families in the belief that "every mouth comes with two hands." In other words, he believed that a large population would strengthen the country in the face of external political pressures. He reversed his position in the 1970s when it became clear that China faced resource shortages as a result of its burgeoning population. In 1980, the one-child-per-couple policy was adopted. With it, Chinese authorities sought not merely to halt population growth but, ultimately, to decrease the national population. All over China today one sees billboards and posters admonishing the citizens that "one couple, one child" is the ideal family (**Figure 3.24**, page 102). Violators face huge monetary fines, cannot request new housing, lose the rather generous benefits provided to the elderly by the government, forfeit their children's access to higher education, and may even lose their jobs. Late marriages are encouraged. In response, between 1970 and 1980, the TFR in China plummeted from 5.9 births per woman to only 2.7, then to 2.2 by 1990,

FIGURE 3.24 **Population control in the People's Republic of China.** China has aggressively promoted a policy of "one couple, one child" in an attempt to relieve the pressures of overpopulation. These billboards convey the government's message. Violators—those with more than one child—are subject to fines, loss of job and old-age benefits, loss of access to better housing, and other penalties. *How effective would such billboards be in influencing people's decisions? Why is one of the signs in English?* (Courtesy of Terry G. Jordan-Bychkov.)

2.0 by 1994, 1.7 by 2007, and 1.5 by 2010 (see Figure 3.3, pages 76–77).

China achieved one of the greatest short-term reductions of birthrates ever recorded, thus proving that cultural changes can be imposed from above, rather than simply waiting for them to diffuse organically. In recent years, the Chinese population control program has been less rigidly enforced as economic growth has eroded the government's control over the people. This relaxation has allowed more couples to have two children instead of one; however, the increase in economic opportunity and migration to cities have led some couples to have smaller families voluntarily. Because of China's population size, its actions will greatly affect the global context as well as its own national one (see Subject to Debate on page 87 for a discussion of the gender implications of strict population control policies).

Reflecting on Geography

Would a global population control policy be desirable or even feasible?

Nature-Culture

Why do human populations form such a diverse mosaic? The reasons are often cultural. Though the differences may have started out as adaptations to given physical conditions, when repeated from generation to generation these patterns become woven tightly into the cultural fabric of places. Thus, demographic practices such as living in crowded settlements or having large families may well have deep roots in both nature and culture.

Environmental Influence

Local population characteristics are often influenced in a possibilistic manner by the availability of resources. In the middle latitudes, population densities tend to be greatest where the terrain is level, the climate is mild and humid, the soil is fertile, mineral resources are abundant, and the sea is accessible. Conversely, population tends to thin out with excessive elevation, aridity, coldness, ruggedness of terrain, and distance from the coast.

Climatic factors influence where people settle. Most of the sparsely populated zones in the world have, in some respect, "defective" climates from the human viewpoint (see Figure 3.1, pages 72–73). The thinly populated northern edges of Eurasia and North America are excessively cold, and the belt from North Africa into the heart of Eurasia matches the major desert zones of the Eastern Hemisphere. Humans remain creatures of the humid and subhumid tropics, subtropics, or midlatitudes and have not fared well in excessively cold or dry areas. Small populations of Inuit (Eskimo), Sami (Lapps), and other peoples live in some of the less hospitable areas of the Earth, but these regions do not support large populations. Humans have proven remarkably adaptable, and our cultures contain strategies that allow us to live in many different physical environments; as a species, however, perhaps we have not entirely moved beyond the adaptive strategies that suited us so well to the climatic features of sub-Saharan Africa, where we began.

Humankind's preference for lower elevations is especially true for the middle and higher latitudes. Most mountain ranges in those latitudes stand out as sparsely populated regions. By contrast, inhabitants of the tropics often prefer to live at higher elevations, concentrating in dense clusters in mountain valleys and basins (see Figure 3.1). For example, in tropical portions of South America, more people live in the Andes Mountains than in the nearby Amazon lowlands. The capital cities of many tropical and subtropical nations lie in mountain areas above 3000 feet (900 meters) in elevation. Living at higher elevations allows residents to escape the hot, humid climate and diseases of the tropic lowlands. In addition, these areas were settled because the fertile volcanic soils of these mountain valleys and basins were able to support larger populations in agrarian societies.

Humans often live near the sea. The continents of Eurasia, Australia, and South America resemble hollow shells, with the majority of the population clustered around the rim of each continent (see Figure 3.1). In Australia, half the total population lives in only five port cities, and most of the remainder is spread out over nearby coastal areas. This preference for living by the sea stems partly from the trade and fishing opportunities the sea offers. At the same time, continental interiors tend to be regions of climatic extremes. For example, Australians speak of the "dead heart" of their continent, an interior land of excessive dryness and heat.

Disease also affects population distribution. Some diseases attack valuable domestic animals, depriving people of food and clothing resources. Such diseases have an indirect effect on population density. For example, in parts of East Africa, a form of sleeping sickness attacks livestock. This particular disease is almost invariably fatal to cattle but not to humans. The people in this part of East Africa depend heavily on cattle, which provide food, represent wealth, and serve a religious function in some tribes. The spread of a disease fatal to cattle has caused entire tribes to migrate away from infested areas, leaving those areas unpopulated.

Environmental Perception and Population Distribution

Perception of the physical environment plays a major role in a group's decision about where to settle and live. Different cultural groups often "see" the same physical environment in different ways. These varied responses to a single environment influence the distribution of people. A good example appears in a part of the European Alps shared by German- and Italian-speaking peoples. The mountain ridges in that area—near the point where Switzerland, Italy, and Austria border one another—run in an east-west direction, so that each ridge has a sunny, south-facing slope

and a shady, north-facing one. German-speaking people, who rely on dairy farming, long ago established permanent settlements some 650 feet (200 meters) higher on the shady slopes than the settlements of Italians, who are culturally tied to warmth-loving crops, on the sunny slopes. This example demonstrates how contrasting cultural attitudes toward the physical environment and land use affect settlement patterns.

Sometimes, the same cultural group changes its perception of an environment over time, with a resulting redistribution of its population. The coalfields of western Europe provide a good case in point. Before the industrial age, many coal-rich areas—such as the Midlands of England, southern Wales, and the lands between the headwaters of the Oder (or Odra) and Vistula rivers in Poland—were only sparsely or moderately settled. The development of steam-powered engines and the increased use of coal in the iron-smelting process, however, created a tremendous demand. Industries grew up near the European coalfields, and people flocked to these areas to take advantage of the new jobs. In other words, once a technological development gave a new cultural value to coal, many sparsely populated areas containing that resource acquired large concentrations of people.

Recent studies indicate that much of the interregional migration in the United States today is prompted by a desire for a pleasant climate and other desirable physical environmental traits, such as beautiful scenery. Surveys of immigrants to Arizona revealed that its sunny, warm climate is a major reason for migration. An attractive environment provided the dominant factor in the growth of the population and economy of Florida. The most desirable environmental traits that serve as stimulants for American migration include (1) mild winter climate and mountainous terrain, (2) a diverse natural vegetation that includes forests and a mild summer climate with low humidity, (3) the presence of lakes and rivers, and (4) nearness to the seacoast. Different age and cultural groups often express different preferences, but all are influenced by their perceptions of the physical environment in making decisions about migration.

Reflecting on Geography

What is your ideal climate? Do you now live in a place with such a climate? If not, do you eventually intend to migrate for this reason?

Population Density and Environmental Alteration

People modify their habitats through their adaptive strategies. Particularly in areas where population density is high, radical alterations often occur. This can happen in fragile environments even at relatively low population densities because, as discussed earlier in this chapter, the Earth's carrying capacity varies greatly from one place to another and from one culture to another.

Many of our adaptive strategies are not sustainable. Population pressures and local ecological crises are closely related. For example, in Haiti, where rural population pressures have become particularly severe, the trees in previously forested areas have been stripped for fuel, leaving the surrounding fields and pastures increasingly denuded and vulnerable to erosion (**Figure 3.25**). In short, overpopulation relative to resource availability can precipitate environmental destruction—which, in turn, results in a downward cycle of worsening poverty, with an eventual catastrophe that is both ecological and demographic. Thus, many cultural ecologists believe that attempts to restore the balance of nature will not succeed until we halt or even reverse population growth, although they recognize that other causes are also at work in ecological crises.

The worldwide ecological crisis is not solely a function of overpopulation. A relatively small percentage of the Earth's population controls much of the industrial technology and consumes a disproportionate percentage of the world's resources each year. Americans, who make up less than 5 percent of the global population, account for about 25 percent of the natural resources consumed globally each year. New houses built in the United States in 2002 were, on average, 38 percent bigger than those built in 1975, despite a shrinking average household size. If everyone in the world had an average American standard of living, the Earth could support only about 500 million people—only 8 percent of the present population. As the economies of large countries such as India and China continue to surge, the resource

FIGURE 3.25 Overpopulation and deforestation. This aerial photograph depicts the border between Haiti and the Dominican Republic. Both nations share the Caribbean island of Hispaniola. Population pressures in Haiti, on the left side of the photograph, have led to deforestation; the Dominican Republic is the greener area to the right. The political border is also an environmental border. *(NASA.)*

consumption of their populations is likely to rise as well because the human desire to consume appears to be limited only by the ability to pay for it. Indeed, in mid-2010, China surpassed the United States to become the world's largest consumer of energy resources.

CULTURAL LANDSCAPE

What are the many ways in which demographic factors are expressed in the cultural landscape? Population geographies are visible in the landscapes around us. The varied densities of human settlements and the shapes these take in different places provide clues to the intertwined cultural and demographic strategies implemented in response to diverse local factors. These clues are evident wherever people have settled, be they urban, suburban, or rural locations. Less visible, but no less important, factors are also at work on and throughout the cultural landscape. In this final part of the chapter we examine the diverse, and often creative, ways in which different places are organized spatially in order to accommodate the populations that live there.

Diverse Settlement Types

Human settlements range in density from the isolated farmsteads found in some rural areas to the teeming streets of megacities such as Kolkata, India (see Seeing Geography, page 112). The size and pattern of human settlements depend in part on the size and needs of the population to be accommodated. Small rural populations and those engaged in subsistence farming don't inhabit large, dense cities. Nomadic peoples require dwellings that can be easily packed up and moved, and they tend to have few household possessions (**Figure 3.26**). On the other end of the spectrum, dense cities make more sense for populations employed in the service and information sectors. These people benefit from being as close as possible to their jobs, and higher densities—at least in theory—reduce commuting distances. People in postindustrial societies also tend to have smaller families, which can be more easily accommodated in apartments (**Figure 3.27**).

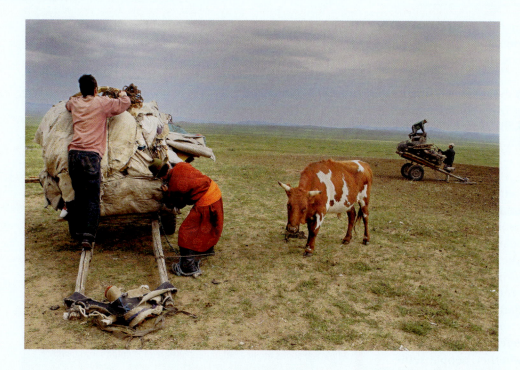

FIGURE 3.26 Nomadic family moving on. These Mongol herders are nomadic; they move often in search of the best pasture for their livestock. Here, a family packs their yurt, a traditional Mongolian tent structure, and their few household possessions. *(Bruno Morandi/The Image Bank/ Getty Images.)*

But there is also a significant cultural aspect at work, such that similar populations may live in settlements that look and feel vastly different from each other. Take rural settlements, for instance.

In many parts of the world, farming people group themselves together in clustered settlements called **farm villages.** These tightly bunched settlements vary in size from a few dozen inhabitants to several thousand. Contained in the village **farmstead** are the house, barn, sheds, pens, and garden. The fields, pastures, and meadows lie out in the country beyond the limits of the village, and farmers must journey out from the village each day to work the land.

Farm villages are the most common form of agricultural settlement in much

farm villages
Clustered rural settlements of moderate size, inhabited by people who are engaged in farming.

farmstead
The center of farm operations, containing the house, barn, sheds, and livestock pens.

FIGURE 3.27 High-density dwelling in Amsterdam. The Borneo-Sporenburg development provides high-density, low-rise housing to accommodate Holland's small families in an urban setting. Instead of yards, these dwellings feature rooftop terraces. The waterfront also provides a kind of "green space" for residents, who tie their boats up in front of their houses. *(© Iain Masterton/Alamy.)*

of Europe, in many parts of Latin America, in the densely settled farming regions of Asia (including much of India, China, and Japan), and among the sedentary farming peoples of Africa and the Middle East.

In many other parts of the world, the rural population lives on dispersed, isolated farmsteads, often some distance from the nearest neighbors (**Figure 3.28**). These dispersed rural settlements grew up mainly in Anglo America, Australia, New Zealand, and South Africa—that is, in the lands colonized by emigrating Europeans.

Why do so many farm people settle together in villages? Historically, the countryside was unsafe, threatened by roving bands of outlaws and raiders. Farmers could better defend themselves against such dangers by grouping together in villages. In many parts of the world, the populations of villages have grown larger during periods of insecurity and shrunk again when peace returned. Many farm villages occupy the most easily defended sites in their vicinity, what geographers call *strong-point settlements*.

In addition to concerns about defense, the quality of the environment helps determine whether people settle in villages. In deserts and in limestone areas, where the ground absorbs moisture quickly, farmsteads are built around the few sources of water. Such *wet-point villages* cluster around oases or deep wells. Conversely, a superabundance of water—in marshes, swamps, and areas subject to floods—often prompts people to settle in villages on available dry points at higher elevations.

Various communal ties strongly bind villagers together. Farmers linked to one another by blood relationships, religious customs, communal landownership, or other similar bonds usually form clustered villages. Mormon farm villages in the United States provide an excellent example of the clustering force of religion. Communal or state ownership of the land—as in China and parts of Israel—encourages the formation of farm villages.

The conditions encouraging dispersed settlement are precisely the opposite of those favoring village development. These include peace and security in the countryside, eliminating the need for defense; colonization by individual pioneer families rather than by socially cohesive groups; private agricultural enterprise, as opposed to some form of communalism; and well-drained land where water is readily available. Most dispersed farmsteads originated rather recently, dating primarily from the colonization of new farmland in the past two or three centuries.

Reflecting on Geography

> What challenges might a settlement pattern of isolated farmsteads present for the people who live there?

Landscapes and Demographic Change

Population change in a place can occur rapidly or more slowly over time. Rapid **depopulation** can come about as the result of sudden catastrophic events, such as natural disasters, disease epidemics, and warfare (**Figure 3.29**). Or depopulation can take place at a slower pace. For example, places may lose population over time because of the gradual out-migration of people in search of opportunities elsewhere, as a cumulative result of declining fertility rates, or as a response to climate change.

depopulation
A decrease in population that sometimes occurs as the result of sudden catastrophic events, such as natural disasters, disease epidemics, and warfare.

Populations can also grow in more or less rapid fashions. A place might experience an abrupt influx of people who have been displaced from other areas. Population increases commonly occur more gradually, as the result of demographic improvements such as longer life spans or lower infant mortality rates, or the accumulation of steady streams of immigrants to attractive areas over time.

Depopulation in one area and population increase in another are often linked processes. For instance, you could easily envision a scenario in which the same natural disaster victims exiting one place become

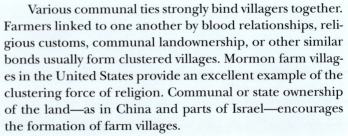

FIGURE 3.28 A truly isolated farmstead, in the Vesturland region of western Iceland. This type of rural settlement dominates almost all lands colonized by Europeans who migrated overseas. Iceland was settled by Norse Vikings a thousand years ago. *(© Jay Dickman/Corbis.)*

FIGURE 3.29 Refugees flee fighting in eastern Congo. This area of the Democratic Republic of Congo, near the Rwandan border, has experienced the repeated movement of people displaced by civil strife. *(Stephen Morrison/epa/Corbis.)*

the refugees suddenly flooding into a neighboring area. As a longer-term example, you could equally well imagine how broad economic changes or technological innovations could lead to regional population shifts whereby some areas become depopulated and others gain population.

Indeed, shifting economic and technological conditions have influenced the demographic landscape in profound ways. The process of industrialization during the past 200 years has caused the greatest voluntary relocation of people in world history. Within industrial nations, people moved from rural areas to cluster in manufacturing regions (see Chapter 10). Agricultural changes have also influenced population density. For example, the complete mechanization of cotton and wheat cultivation in mid-twentieth-century America allowed those crops to be raised by a much smaller labor force (see Chapter 8). As a result, profound depopulation occurred, to the extent that many small towns serving these rural inhabitants ceased to exist.

Regardless of the reasons, population changes must be accommodated, and these adaptations are invariably reflected in the cultural landscape.

Depopulation in History: Ancient Rome There are as many theories accounting for the decline of the Roman Empire as there are historians theorizing about it. Lead poisoning, overextension of the empire's reach, conquest by Germanic tribes, and at least 200 other theories have been proposed.

Rome, the empire's capital city, suffered this process acutely. For nearly a millennium, Rome had been the world's wealthiest, most powerful, and most populous city (**Figure 3.30**, page 108). At the close of the first century A.D., Rome's population surpassed 1 million. As the Roman Empire's economic, political, and military might waned across the third, fourth, and fifth centuries, the demographic balance shifted as well. Under the emperor Constantine in A.D. 330, the empire was split into eastern and western halves, and its capital was relocated to Constantinople (now Istanbul, in modern-day Turkey).

As barbarian invasions further weakened the divided empire, Rome's physical infrastructure—its famed roads, bridges, aqueducts, and monumental buildings—crumbled, as did its impressive administrative infrastructure. With no access to former amenities, from running water to educational opportunities, Rome's population had shrunk to around 20,000 by A.D. 550. The Roman landscape had become a ghost of its former glory: dilapidated, uninhabited, and in ruins (see Figure 3.30).

Bogotá Rising Dilapidated-looking landscapes can result from population decline, as the previous example of ancient Rome illustrates, but they can also arise from rapid population influx. Many of the world's shantytowns exemplify the sort of chaotic landscape that can result from rapid population increases. Los Altos de Cazucá, a neighborhood in the Colombian capital of Bogotá, is but one example. In this case, the population influx comes mostly from people arriving in the capital after being displaced by armed conflict in the countryside.

FIGURE 3.30 Rome at its height and Rome in ruins. The image on the left, from a nineteenth-century engraving, shows Rome at its height, as seen from Mt. Palatine, one of Rome's seven hills. The image on the right, from a nineteenth-century painting, shows the ruins of the Roman Forum from the Capitoline Hill, another of Rome's seven hills. (*Both photos: The Granger Collection, New York/The Granger Collection.*)

The United Nations High Commissioner for Refugees (UNHCR) estimates that, by 2009, there were 26 million internally displaced persons (IDPs) due to armed conflict worldwide. Colombia is one of the globe's hotspots for IDPs, alongside Sudan. Colombia's estimated 5 million IDPs have been uprooted largely as the result of ongoing civil warfare between the paramilitary group known as the FARC (Revolutionary Armed Forces of Colombia) and government forces bent on eradicating them. The UNHCR has identified this as the "worst humanitarian crisis in the western hemisphere."

Colombia's armed conflict has depopulated the rural areas where it occurs, forcing its victims into the cities. Terrorized, landless, and impoverished, the mainly indigenous and Afro-Colombian IDPs settle in the outskirts of cities like Colombia's capital, Bogotá.

Los Altos de Cazucá is one of the settlements inhabited by Colombia's internally displaced persons (**Figure 3.31**). Known as shantytowns, areas like Los Altos de Cazucá arise for different reasons, but they exist in all large cities throughout the developing world. Housing is constructed by the residents themselves, using found materials like cardboard, tin panels, and old tires. Some shantytowns are located far away from downtown areas where wealthy people reside and where jobs are, while others are

FIGURE 3.31 Los Altos de Cazucá. This is a shantytown on the outskirts of Colombia's capital city, Bogotá. Many of its approximately 50,000 residents are displaced people from other parts of the country. (*imagebroker.net/SuperStock.*)

FIGURE 3.32 Landscapes of poverty and wealth.
This scene from Bogotá is a common Latin American urban landscape, with wealthy high-rises located next to impoverished shantytowns. *(Victor Englebert.)*

literally pressed up against wealthier neighborhoods (**Figure 3.32**).

Most shantytowns arise spontaneously, in order to address population influx to cities that lack the resources to plan systematically for rapid growth. But are shantytowns temporary, disappearing as their residents become incorporated into city life? Hardly. Instead, shantytowns gradually become part of the urban fabric. Indeed, most of the spatial expansion of Latin American cities like Bogotá occurs precisely thanks to growth of the shantytowns surrounding them. Over time, the dwellings are constructed of more-permanent materials such as concrete blocks. Roads are paved and running water installed. Power and phone lines are extended. The once-temporary areas slowly become visually, economically, and culturally integrated into the permanent fabric of the city. New shantytowns then arise beyond their borders, to accommodate recent arrivals.

Reflecting on Geography

Many cities in the United States have seen shantytowns arise in the wake of the home foreclosure crisis of the late 2000s. Are these comparable to Los Altos de Cazucá? In what significant ways are they similar, and in what ways are they different? (Figure 3.33)

FIGURE 3.33 Mortgage foreclosure tent city in Reno, Nevada. Nevada was one of the hardest-hit states during the economic crisis in which many people lost their jobs and their homes; some even became homeless. Similar tent cities could be found in many U.S. cities in the late 2000s. *(Max Whittaker/ Getty Images.)*

CONCLUSION

In our study of population geography, we have seen that humankind is unevenly distributed over the Earth. Spatial variations in fertility, death rates, rates of population change, age groups, gender ratios, and standards of living also exist: these patterns can be depicted as demographic culture regions. The principles of mobility prove useful in analyzing human migration and also help explain the spread of factors influencing demographic characteristics such as disease.

Population geography proves particularly intriguing when scale is taken into account. Although most geodemographic issues are experienced locally, and policies shaping them are usually set at the national level, the important debates about the world's population are truly global in nature.

The theme of nature-culture shows how the natural environment and people's perception of and engagement with it influence demographic factors. Nature-culture interactions shape the spatial distribution of people and sometimes help guide migrations. In addition, population density is linked to the level of environmental alteration, and overpopulation can have a destructive impact on the environment.

The cultural landscape visually expresses the varied ways in which societies accommodate their populations. How people distribute themselves over the Earth's surface finds a vivid expression in the cultural landscape. The look and feel of places are constantly adapting to demographic change.

DOING GEOGRAPHY

Public Space, Personal Space: Too Close for Comfort?

Culture can condition people to accept or reject crowding. Personal space—the amount of space that individuals feel "belongs" to them as they move about their everyday business—varies from one cultural group to another. As the Seeing Geography section for this chapter notes, different people seem to require different amounts of personal space. One's comfort zone varies with social class, gender, ethnicity, the situation at hand, and what one has grown accustomed to over one's life. Some Arabs, for example, consider it appropriate and even polite to be close enough for another to smell his or her breath during conversation. Those from cultures that have not developed a high tolerance for personal contact might experience such closeness as intrusive. Americans conducting business in Japan are often surprised at the level of physical

closeness expected in their dealings. Such closeness might well be interpreted as overstepping one's bounds, literally, in the United States!

In this exercise, you will gather some data on the amount of personal space needed by those around you. Observe and record your findings, and discuss them in class as a group.

Steps to Understanding Personal Space

Step 1: Observe your professors as they lecture in class. Is your class a large one that meets in a lecture hall? If so, where does your professor sit or stand in relation to the students? Does the professor have his or her own designated space in the classroom? Where is it located, and how big is it? Does the professor ever step outside of it? How does this professor's use of space compare to that of other professors you have, and why do you think this is so? If you have a smaller class,

Population densities in parks. These two images depict similarly designed public playgrounds featuring a large sandbox. The park on the top is located in Minnesota in the United States; the park on the bottom is located in Taipei, Taiwan. **Where would you rather play? Why?** (Top: James Shaffer/PhotoEdit; Bottom: Christian Klein/Alamy.)

compare the use of space by that professor. Is it different from the behavior of the professors in large lecture halls? Under what circumstances, if any, do your professors get close to students, and how close do they get?

Step 2: How close can you get to friends? Strike up a conversation with a same-gender friend standing next to you. Discreetly move closer and closer to your friend until he or she moves away or says something about your proximity. How much space separated you when this happened? What do you think would happen if you tried this with a stranger? With a friend or a stranger of a different gender? With a friend or a stranger from a different culture?

Step 3: Discuss your findings. Did all your classmates have similar experiences, or were your findings notably different? Are all students in your class from similar economic or ethnic backgrounds? If not, that may explain some of the differences that emerge.

There has been some talk lately about the future of the ever-huger "McMansions" and sprawling suburban developments that for several decades have characterized middle-class life in the United States. Clearly, both waste resources. McMansions are expensive to heat and cool, and their landscaping is often out-of-step with the local environment. Living in a suburb often means commuting long distances to jobs or school, in cars occupied by just one or two people. But think, too, about the proxemics involved and consider these questions:

- Is there such a thing as too much space?
- Do you know or have you heard of people who have opted out of living large in the 'burbs? Why do you think they made this decision?
- In what ways do planners and architects take culturally diverse preferences about personal space into account when they design cities, streets, buildings, homes, and classrooms?

Key Terms

Population Geography on the Internet

You can learn more about population geography on the Internet at the following web sites:

Population Reference Bureau, Inc., Washington, D.C.
http://www.prb.org
This organization is concerned principally with overpopulation and standard of living. The "World Population Data Sheet" provides up-to-date basic demographic information at a glance, and the "Datafinder" section has a wealth of images on all aspects of global population for use in presentations and reports. Some of the maps in this chapter were adapted from PRB maps.

United Nations High Commissioner for Refugees
http://www.unhcr.org
This United Nations web site provides basic information about refugee situations worldwide. The site includes regularly updated maps showing refugee locations and populations as well as photos of refugee life.

U.S. Census Bureau Population Clocks
http://www.census.gov/main/www/popclock.html
Check real-time figures here for the population of the United States and the population of the world. The main web site, http://www.census.gov, hosts the most important and comprehensive data sets available on the U.S. population.

World Health Organization, Geneva, Switzerland
http://www.who.int/en
Learn about the group that distributes information on health, mortality, and epidemics as it seeks to improve health conditions around the globe. The "Global Health Atlas" allows you to create detailed maps from WHO data.

Worldwatch Institute, Washington, D.C.
http://www.worldwatch.org
This organization is concerned with the ecological consequences of overpopulation and the wasteful use of resources. It seeks sustainable ways to support the world's population and brings attention to ecological crises.

SEEING GEOGRAPHY Street in Kolkata, India

Would you feel comfortable walking here? If not, why not?

A street scene in the large city of Kolkata, India.

Do you need your "personal space"? Most Americans and Canadians do. If so, Kolkata (formerly Calcutta), India, is a place you might want to avoid. West Bengal state, where Kolkata is located, has the highest population density in the country.

Why do people form such dense clusters? The theme of cultural interaction would tell us of push factors that encourage people to leave their farms and move to the city. Some can no longer make a living or feed their families with the food provided by the tiny plots of land they work. Others are forced off the land by landlords who want to convert their farms to use mechanized Western methods of agriculture that use far less labor (see Chapter 8). But cultural interaction also tells us of pull factors exerted by cities such as Kolkata—the hope or promise of better-paying jobs, the encouragement of friends and relatives who came to the city earlier, or the greater availability of government services. And so, pushed and pulled, they come to the teeming, overcrowded city, to jostle and elbow their way through the streets.

But it is not only cities in the developing world that become so dense. If you have ever visited, or lived in, Manhattan, New York, you are all too familiar with dense crowds of people. In fact, there are some who grow up in these environments and find that the relative solitude of rural areas verges on terrifying. They prefer the bustle of activity and the sounds of the city, and they feel at home in a crowd.

Being a woman is another reason that you might feel uncomfortable in this Kolkata street environment. Notice the nearly complete absence of women in this crowd. Many societies have strict norms that dictate where women, and men, may and may not go. Harassment or even violence may be the result of violating these norms.

The study of the size and shape of people's envelopes of personal space is called **proxemics.** Anthropologist Edward T. Hall, whose book *The Hidden Dimension* is listed in Ten Recommended Books on Population Geography at the end of the chapter, is the founder of this science. Urban planners, architects, psychologists, and sociologists, as well as geographers, use proxemics to explain why some people need more space than others and how this varies culturally.

> **proxemics**
> The study of the size and shape of people's envelopes of personal space.

Sources

Coclanis, Peter. 2010. "Russia's Demographic Crisis and Gloomy Future," *The Chronicle of Higher Education,* 19 February, pp. B9–B11.

de Macgregor, María T. de Gutiérrez. 1984. "Population Geography in Mexico," in John J. Clarke (ed.), *Geography and Population.* Oxford: Pergamon.

D'Emilio, Frances. 2004. "Italy's Seniors Finding Comfort with Strangers," *Miami Herald,* 30 October, pp. 1A–2A.

Gould, Peter. 1993. *The Slow Plague: A Geography of the AIDS Pandemic.* Oxford: Blackwell.

Hooper, Edward. 1999. *The River: A Journey to the Source of HIV and AIDS.* Boston: Little, Brown.

Malthus, Thomas R. 1989 [1798]. *An Essay on the Principle of Population.* Patricia James (ed.). Cambridge: Cambridge University Press.

Paul, Bimal K. 1994. "AIDS in Asia." *Geographical Review* 84: 367–379.

Shannon, Gary W., Gerald F. Pyle, and Rashid L. Bashshur. 1991. *The Geography of AIDS: Origins and Course of an Epidemic.* New York: Guilford.

Rosin, Hanna. 2010. "The End of Men." *Atlantic Monthly,* July/August. Available online at http://www.theatlantic.com/magazine/archive/2010/07/the-end-of-men/8135/.

United Nations High Commissioner for Refugees. 2009. *Global Report 2008.* New York: United Nations.

World Bank, "Statistics in Africa." Available online at http://www.worldbank.org/afr/stats.

World Population Data Sheet. 2007. Washington, D.C.: Population Reference Bureau.

Ten Recommended Books on Population Geography

(For additional suggested readings, see *The Human Mosaic* web site: www.whfreeman.com/domosh12e)

Castles, Stephen, and Mark J. Miller. 1998. *The Age of Migration: International Population Movements in the Modern World,* 2nd ed. New York: Guilford. A global perspective on migrations, why they occur, and the effects they have on different countries, in an age of an unprecedented volume of migration. Explores how migration has led to the formation of ethnic minorities in numerous countries as well as its impact on domestic politics and economics.

Diamond, Jared. 2005. *Collapse: How Societies Choose to Fail or Succeed.* New York: Penguin. Why do thriving civilizations die out? Ecology, biology, and geography all play a role in Diamond's analysis.

Hall, Edward T. 1966. *The Hidden Dimension.* Garden City, N.Y.: Doubleday. This is the classic study of proxemics conducted by an anthropologist. Hall argues that culture, above all else, shapes our criteria for defining, organizing, and using space.

Johnson, Steven. 2006. *The Ghost Map: The Story of London's Most Terrifying Epidemic—And How It Changed Science, Cities, and the Modern World.* New York: Riverhead Books. In this gripping historical narrative, physician John Snow's tracing of the Soho cholera outbreak of 1854 is placed in the larger context of the history of science and urbanization.

Mann, Charles C. 2006. *1491.* New York: Vintage. This highly readable account of the American landscape on the eve of European conquest challenges long-held notions. Contending that the indigenous population of the Americas was in fact much larger and well-off than assumed, Mann paints a picture of a preconquest landscape that was culturally rich, technologically sophisticated, and environmentally compromised.

Meade, Melinda S., and Robert J. Earickson. 2005. *Medical Geography,* 2nd ed. New York: Guilford. Surveys the perspectives, theories, and methodologies that geographers use in studying human health; a primary text that undergraduates can readily understand.

Newbold, K. Bruce. 2007. *Six Billion Plus: World Population in the Twenty-First Century,* 2nd ed. Lanham, Md.: Rowman & Littlefield. The impact of increased global population levels across the next century is assessed with respect to interaction with environmental, epidemiological, mobility, and security issues.

Roberts, Brian K. 1996. *Landscapes of Settlement.* London: Routledge. Discusses the role and significance of rural settlements, drawing from global case studies. Outlines the formation of different spatial arrangements at the farmstead, hamlet, and village scales.

Seager, Joni, and Mona Domosh. 2001. *Putting Women in Place: Feminist Geographers Make Sense of the World.* New York: Guilford. A highly readable account of why paying attention to gender is crucial to understanding the spaces in which we live and work.

Tone, Andrea. 2002. *Devices and Desires: A History of Contraceptives in America.* New York: Hill & Wang. This social history of birth control in the United States details the fascinating relationship between the state and the long-standing attempts of men and women to limit their fertility.

Journals in Population Geography

Gender, Place and Culture: A Journal of Feminist Geography. Published by the Carfax Publishing Co., P.O. Box 2025, Dunnellon, Fla. 34430. Volume 1 appeared in 1994.

Population and Environment: A Journal of Interdisciplinary Studies. Volume 1 was published in 1996.

U.S. Army veterans at a recruiting table in Miami, Florida.
(Jeff Greenberg/AgeFotostock.)

Aquí se habla Spanglish. What does this sign tell you about who lives in this city?

Go to "Seeing Geography" on page 144 to learn more about this image.

4 THE GEOGRAPHY OF LANGUAGE

Building the Spoken Word

Language is one of the primary features that distinguishes humans from other animals. Many animals, including dolphins, whales, and birds, do indeed communicate with one another through patterned systems of sounds, movements, or scents and other chemicals. Some nonhuman primates have been taught to use sign language to communicate with humans. In turn, we have attempted to translate animal noises into words in human languages, with results that vary across cultures (**Figure 4.1**). However, the complexity of human language, its ability to convey nuanced emotions and ideas, and its importance for our existence as social beings set it apart from the communication systems used by other animals. In many ways, language is the essence of culture. It provides the single most common variable by which different cultural groups are identified and by which groups assert their unique identity. Language not only facilitates the cultural diffusion of innovations but it also helps to shape the way we think about, perceive, and name our environment. **Language,** a mutually agreed-on system of symbolic communication, offers the main means by which learned belief systems, customs, and skills pass from one generation to the next.

One of our first and most long-lasting ties to place is forged through language. The ability to name, or rename, a place is a key step in claiming a place as one's own, as shown, for example, by **Figure 4.2**, page 116, a political map of Antarctica. Notice how many places are named after Antarctic explorers and monarchs of countries that claim portions of the continent: the Ross Sea and Ross Ice Shelf are named after James Clark Ross, the English explorer who charted much

language
A mutually agreed-on system of symbolic communication that has a spoken and usually a written expression.

	English	Indonesian	Japanese	Greek
Dog	bow-wow	gonggong	wanwan	gav
Cat	meow	ngeong	nyaa	niaou
Bird	tweet-tweet	kicau	chunchun	tsiou tsiou
Rooster	cock-a-doodle-doo	kikeriku	kokekokkoo	ki-kiriki

FIGURE 4.1 Translating animal sounds. Although a cat obviously sounds the same in any part of the world, the speakers of different human languages render that sound in diverse ways. This chart shows how the sounds of various animals "translate" into different human languages. *(Compiled using information from a project by Dr. Catherine N. Ball in Georgetown University's Department of Linguistics, titled "Sounds of the World's Animals.")*

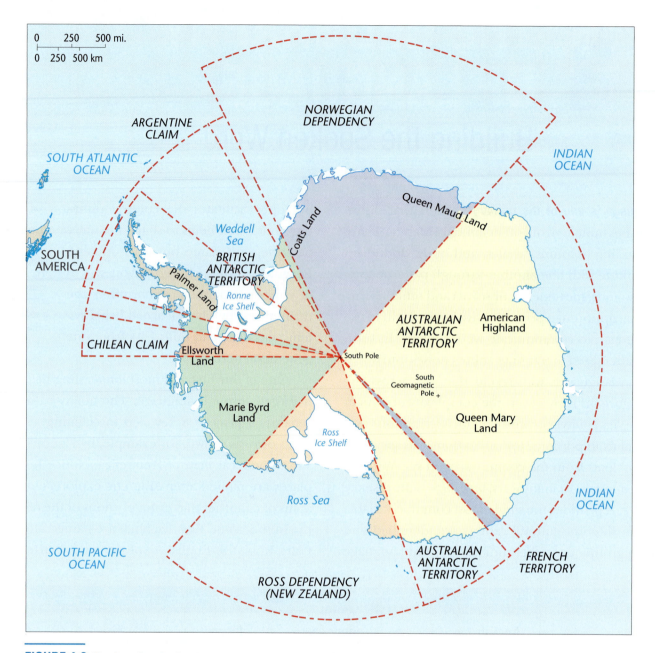

FIGURE 4.2 Naming place is closely related to claiming place. This map shows the pie-shaped land claims of various nations. Notice how place-names reflect the names of Antarctic explorers or political rulers on the political map of this continent. *(Adapted from Latrimer Clarke Corporation Pty Ltd.)*

of Antarctica's coastline, and Queen Maud Land is named after Norway's Queen Maud. Interestingly, Roald Amundsen, the Norwegian explorer who gave Queen Maud Land its name in 1939, took part in an earlier trip that highlights the importance of language. In 1898, Amundsen was part of an ill-fated expedition to locate the magnetic south pole when his ship be-

came stuck in the ice for more than one year. Several of Amundsen's shipmates, including men from Poland, Romania, Norway, the United States, and Belgium, went insane. This was not only because of the long winter nights and the cold weather; in great part, language differences among the men became their biggest problem. They simply could not understand one another!

Region

What is the geographical patterning of languages? Do various languages provide the basis for formal and functional culture regions? The spatial variation of speech is remarkably complicated, adding intricate patterns to the human mosaic (**Figure 4.3**, pages 118–119). Because language is such a central component of culture, understanding the spatiality of language, and how and why its patterns change over time, provides a particularly valuable window into cultural geography more generally. The logical place to begin our geographical study of language is with the regional theme.

Separate languages are those that cannot be mutually understood. In other words, a monolingual speaker of one language cannot comprehend the speaker of another. **Dialects,** by contrast, are variant forms of a language where mutual comprehension is possible. A speaker of English, for example, can generally understand that language's various dialects, regardless of whether the speaker comes from Australia, Scotland, or Mississippi. Nevertheless, a dialect is distinctive enough in vocabulary and pronunciation to label its speaker as hailing from one place or another, or even from a particular city. About 6000 languages and many more dialects are spoken in the world today.

dialect
A distinctive local or regional variant of a language that remains mutually intelligible to speakers of other dialects of that language; a subtype of a language.

When different linguistic groups come into contact, a **pidgin** language, characterized by a very small vocabulary derived from the languages of the groups in contact, often results. Pidgins primarily serve the purposes of trade and commerce: they facilitate exchange at a basic level but do not have complex vocabularies or grammatical structures. An example is Tok Pisin, meaning "talk business." Tok Pisin is a largely English-derived pidgin spoken in Papua New Guinea, where it has become the official national language in a country where many native Papuan tongues are spoken. Although New Guinea pidgin is not readily intelligible to a speaker of Standard English, certain common words such as *gut bai* ("good-bye"), *tenkyu* ("thank you"), and *haumas* ("how much") reflect the influence of English. When pidgin languages acquire fuller vocabularies and become native languages of their speakers, they are called creole languages. Obviously, deciding precisely *when* a pid-

pidgin
A composite language consisting of a small vocabulary borrowed from the linguistic groups involved in commerce.

creole
A language derived from a pidgin language that has acquired a fuller vocabulary and become the native language of its speakers.

gin becomes a creole language has at least as much to do with a group's political and social recognition as it does with what are, in practice, fuzzy boundaries between language forms.

Another response to the need for speakers of different languages to communicate with one another is the elevation of one existing language to the status of a **lingua franca.** A lingua franca is a language of communication and commerce spoken across a wide area where it is not a mother tongue. The Swahili language enjoys lingua franca status in much of East Africa, where inhabitants speak a number of other regional languages and dialects. English is fast becoming a global lingua franca. Finally, regions that have linguistically mixed populations may be characterized by **bilingualism,** which is the ability to speak two languages with fluency. For example, along the U.S.-Mexico border, so many residents speak both English and Spanish (with varying degrees of fluidity) that bilingualism in practice—even if not in policy—means there is no need for a lingua franca.

lingua franca
An existing, well-established language of communication and commerce used widely where it is not a mother tongue.

bilingualism
The ability to speak two languages fluently.

Language Families

One way in which geolinguists often simplify the mapping of languages is by grouping them into **language families:** tongues that are related and share a common ancestor. Words are simply arbitrary sounds associated with certain meanings. Thus, when words in different languages are alike in both sound and meaning, they may well be related. Over time, languages interact with one another, borrowing words, imposing themselves through conquest, or organically diverging from a common ground. Languages and their interrelations can thus be graphically depicted as a tree with various branches (**Figure 4.4**, page 120). This classification makes the complicated linguistic mosaic a bit easier to comprehend.

language family
A group of related languages derived from a common ancestor.

Indo-European Language Family The largest and most widespread language family is the Indo-European, which is spoken on all the continents and is dominant in Europe, Russia, North and South America, Australia, and parts of southwestern Asia and India (see Figure 4.3). Romance, Slavic, Germanic, Indic, Celtic, and Iranic are all Indo-European subfamilies. These subfamilies are in turn divided into individual languages. For example, English is a Germanic Indo-European language. Seven Indo-European

Languages

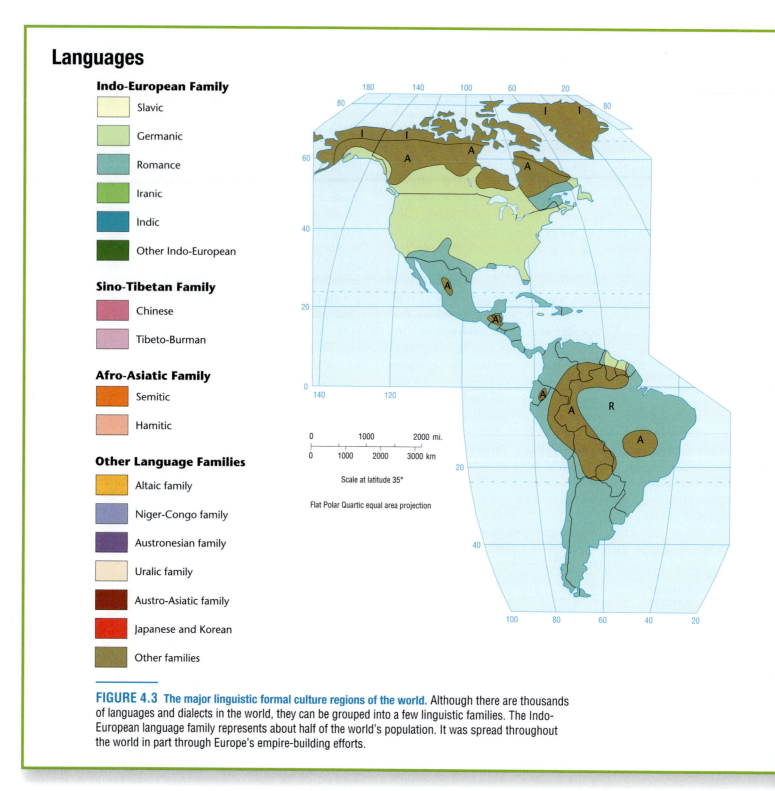

Indo-European Family

- Slavic
- Germanic
- Romance
- Iranic
- Indic
- Other Indo-European

Sino-Tibetan Family

- Chinese
- Tibeto-Burman

Afro-Asiatic Family

- Semitic
- Hamitic

Other Language Families

- Altaic family
- Niger-Congo family
- Austronesian family
- Uralic family
- Austro-Asiatic family
- Japanese and Korean
- Other families

0   1000   2000 mi.
0   1000   2000   3000 km

Scale at latitude 35°

Flat Polar Quartic equal area projection

FIGURE 4.3 **The major linguistic formal culture regions of the world.** Although there are thousands of languages and dialects in the world, they can be grouped into a few linguistic families. The Indo-European language family represents about half of the world's population. It was spread throughout the world in part through Europe's empire-building efforts.

tongues, including English, are among the 10 most spoken languages in the world as classified by the number of native speakers (**Table 4.1**, page 121).

Comparing the vocabularies of various Indo-European tongues reveals their kinship. For example, the English word *mother* is similar to the Polish *matka*, the Greek *meter*, the Spanish *madre*, the Farsi *madar* in Iran, and the Sinhalese *mava* in Sri Lanka. Such similarities demonstrate that these languages have a common ancestral tongue.

Sino-Tibetan Family Sino-Tibetan is another of the major language families of the world and is second only to Indo-

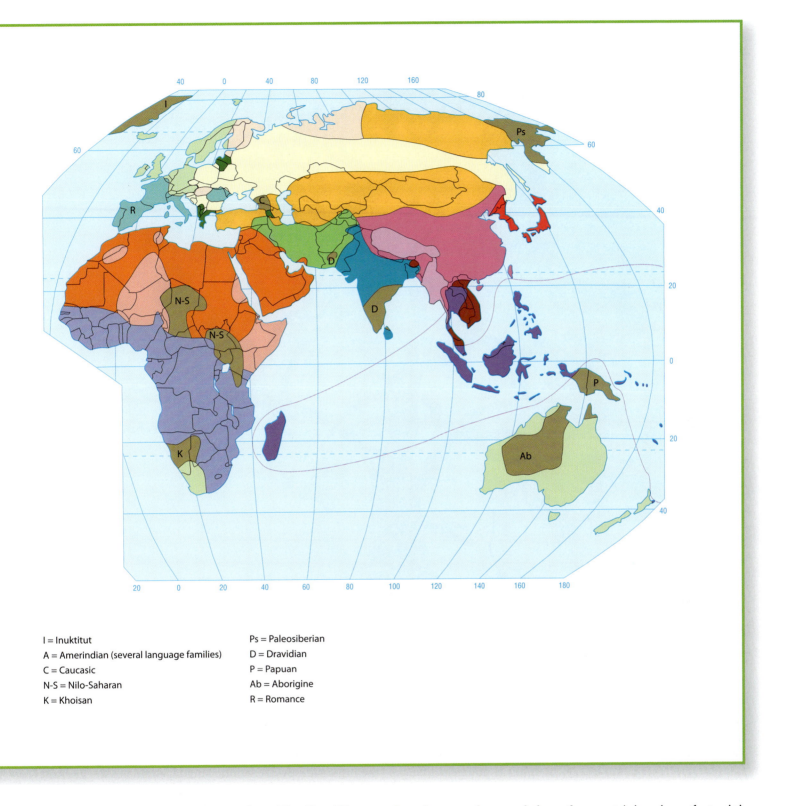

I = Inuktitut

A = Amerindian (several language families)

C = Caucasic

N-S = Nilo-Saharan

K = Khoisan

Ps = Paleosiberian

D = Dravidian

P = Papuan

Ab = Aborigine

R = Romance

European in numbers of native speakers. The Sino-Tibetan region extends throughout most of China and Southeast Asia (see Figure 4.3). The two language branches that make up this group, Sino- and Tibeto-Burman, are believed to have had a common origin some 6000 years ago in the Himalayan Plateau; speakers of the two language groups subsequently moved along the great Asian rivers that originate in this area. "Sino" refers to China and in this context indicates the various languages spoken by more than 1.3 billion people in China. Han Chinese (Mandarin) is spoken in a variety of dialects and serves as the official language of China. The nearly 400 languages and dialects

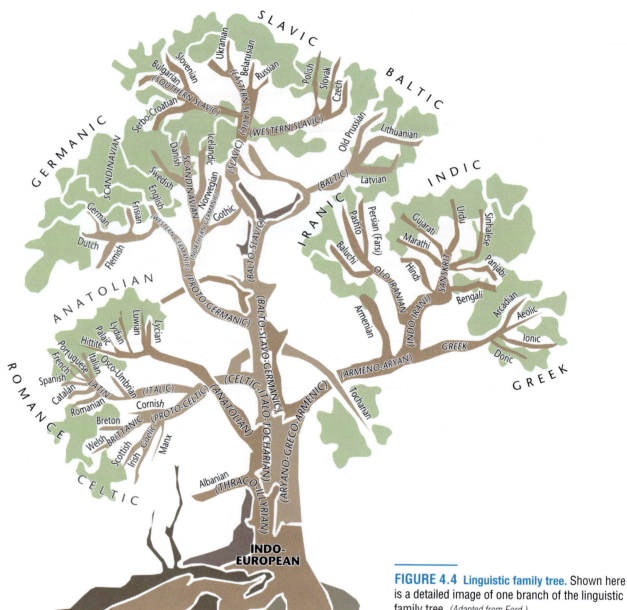

FIGURE 4.4 Linguistic family tree. Shown here is a detailed image of one branch of the linguistic family tree. *(Adapted from Ford.)*

that make up the Burmese and Tibetan branch of this language family border the Chinese language region on the south and west. Other East Asian languages, such as Vietnamese, have been heavily influenced by contact with the Chinese and their languages, although it is not clear that they are linguistically related to Chinese at all.

Afro-Asiatic Family The third major language family is the Afro-Asiatic. It consists of two major divisions: Semitic and Hamitic. The Semitic languages cover the area from the Arabian Peninsula and the Tigris-Euphrates river valley of Iraq westward through Syria and North Africa to

the Atlantic Ocean. Despite the considerable size of this region, there are fewer speakers of the Semitic languages than you might expect because most of the areas that Semites inhabit are sparsely populated deserts. Arabic is by far the most widespread Semitic language and has the greatest number of native speakers, about 235 million. Although many different dialects of Arabic are spoken, there is only one written form.

Hebrew, which is closely related to Arabic, is another Semitic tongue. For many centuries, Hebrew was a "dead" language, used only in religious ceremonies by millions of Jews throughout the world. With the creation of the state of

TABLE 4.1 The 10 Leading Languages in Numbers of Native Speakers*

Language	Family	Speakers (in millions)	Main Areas Where Spoken
Han Chinese (Mandarin)	Sino-Tibetan	885	China, Taiwan, Singapore
Hindi/Urdu	Indo-European	426	Northern India, Pakistan
Spanish	Indo-European	358	Spain, Latin America, southwestern United States
English	Indo-European	343	British Isles, United States, Caribbean, Australia, New Zealand, South Africa, Philippines, former British colonies in tropical Asia and Africa
Arabic	Afro-Asiatic	235	Middle East, North Africa
Bengali	Indo-European	207	Bangladesh, eastern India
Portuguese	Indo-European	176	Portugal, Brazil, southern Africa
Russian	Indo-European	167	Russia, Kazakhstan, parts of Ukraine and other former Soviet republics
Japanese	Japanese and Korean	125	Japan
German	Indo-European	100	Germany, Austria, Switzerland, Luxembourg, eastern France, northern Italy

*"Native speakers" means mother tongue.
(*Sources:* Encyclopaedia Britannica, *2000;* World Almanac Books, *2001.*)

Israel in 1948, a common language was needed to unite the immigrant Jews, who spoke the languages of their many different countries of origin. Hebrew was revived as the official national language of what otherwise would have been a **polyglot,** or multi-language, state (**Figure 4.5**). Amharic, a third major Semitic tongue, today claims 18 million speakers in the mountains of East Africa.

polyglot
A mixture of different languages.

FIGURE 4.5 Sign in Arab Quarter of Nazareth. Many of Israel's cities are home to diverse populations. This sign at a child-care center reflects Israel's polyglot population, with its English, Arabic, and Hebrew wording. (*Courtesy of Patricia L. Price.*)

Smaller numbers of people who speak Hamitic languages share North and East Africa with the speakers of Semitic languages. Like the Semitic languages, these tongues originated in Asia but today are spoken almost exclusively in Africa by the Berbers of Morocco and Algeria, the Tuaregs of the Sahara, and the Cushites of East Africa.

Other Major Language Families Most of the rest of the world's population speak languages belonging to one of six remaining major families. The Niger-Congo language family, also called Niger-Kordofanian, which is spoken by about 325 million people, dominates Africa south of the Sahara Desert. The greater part of the Niger-Congo culture region belongs to the Bantu subgroup. Both Niger-Congo and its Bantu constituent are fragmented into a great many different languages and dialects, including Swahili. The Bantu and their many related languages spread from what is now southeastern Nigeria about 4000 years ago, first west and then south in response to climate change and new agricultural techniques (**Figure 4.6**, page 122).

Flanking the Slavic Indo-Europeans on the north and south in Asia are the speakers of the Altaic language family, including Turkic, Mongolic, and several other subgroups. The Altaic homeland lies largely in the inhospitable deserts, tundra, and coniferous forests of northern and central Asia. Also occupying tundra and grassland areas adjacent to the Slavs is the Uralic family. Finnish and

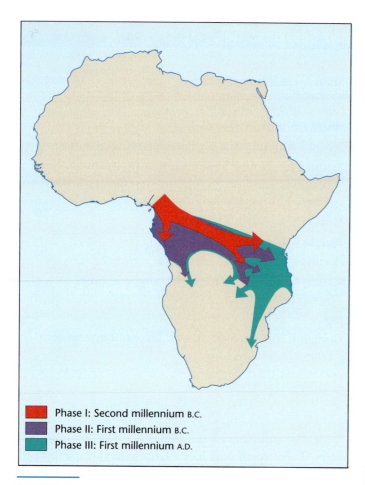

FIGURE 4.6 Bantu expansionism. The Bantu and their many languages spread from what is today southeastern Nigeria to the west and south. *(Adapted from Dingemanse/Wikipedia.)*

Legend:
- Phase I: Second millennium B.C.
- Phase II: First millennium B.C.
- Phase III: First millennium A.D.

Hungarian are the two more widely spoken Uralic tongues, and both enjoy the status of official languages in their respective countries.

One of the most remarkable language families in terms of distribution is the Austronesian. Representatives of this group live mainly on tropical islands stretching from Madagascar, off the east coast of Africa, through Indonesia and the Pacific islands, to Hawaii and Easter Island. This longitudinal span is more than half the distance around the world. The north-south, or latitudinal, range of this language area is bounded by Hawaii and Taiwan in the north and New Zealand in the south. The largest single language in this family is Malay-Indonesian, with 58 million native speakers, but the most widespread is Polynesian.

Japanese and Korean, with about 200 million speakers combined, probably form another Asian language family. The two perhaps have some link to the Altaic family, but even their kinship to each other remains controversial and unproven.

In Southeast Asia, the Vietnamese, Cambodians, Thais, and some tribal peoples of Malaya and parts of India speak languages that constitute the Austro-Asiatic family. They occupy an area into which Sino-Tibetan, Indo-European, and Austronesian languages have all encroached.

Mobility

How did the mosaic of languages and dialects come to exist? How is the spatial patterning of language changing today? Different types of cultural diffusion have helped shape the linguistic map. Relocation diffusion has been extremely important because languages spread when groups, in whole or in part, migrate from one area to another. Some individual tongues or entire language families are no longer spoken in the regions where they originated, and in certain other cases the linguistic hearth is peripheral to the present distribution (compare Figures 4.3, pages 118–119, and 4.7). Today, languages continue to evolve and change based on the shifting locations of peoples and on their needs as well as on outside forces.

Indo-European Diffusion

According to a widely accepted new theory, the earliest speakers of the Indo-European languages lived in southern and southeastern Turkey, a region known as Anatolia, about 9000 years ago (**Figure 4.7**). According to the so-called **Anatolian hypothesis,** the initial diffusion of these Indo-European speakers was facilitated by the innovation of plant domestication. As sedentary farming became adopted throughout Europe, a gradual and peaceful expansion diffusion of Indo-European languages occurred. As these people dispersed and lost contact with one another, different Indo-European groups gradually developed variant forms of the language, causing fragmentation of the language family. The Anatolian hypothesis has been criticized by scholars who note that specific terms related to animals (particularly horses), as opposed to agriculture, link Indo-European languages to a common origin. The so-called **Kurgan hypothesis** places the rise of Indo-European languages in the central Asian steppes only 6000 years ago, positing that the spread of Indo-European languages was both swifter and less peaceful than is maintained by those who subscribe to the Anatolian hypothesis. No one theory has been definitively proven to be correct.

Anatolian hypothesis
A theory of language diffusion holding that the movement of Indo-European languages from the area in contemporary Turkey known as Anatolia followed the spread of plant domestication technologies.

Kurgan hypothesis
A theory of language diffusion holding that the spread of Indo-European languages originated with animal domestication in the central Asian steppes and grew more aggressively and swiftly than proponents of the Anatolian hypothesis maintain.

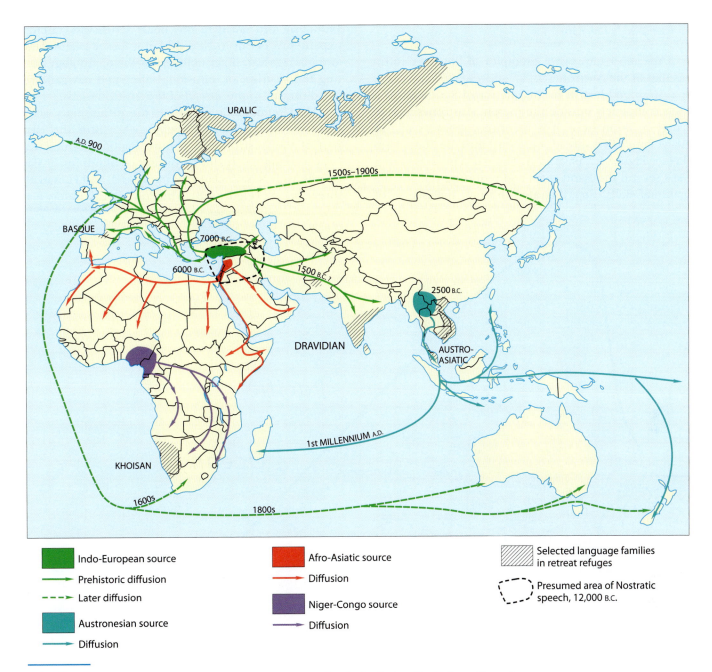

FIGURE 4.7 Origin and diffusion of four major language families in the Eastern Hemisphere. The early diffusion of Indo-European speech to the west and north probably occurred in conjunction with the diffusion of agriculture from the Middle Eastern center, as did the early spread of the Afro-Asiatic family. All such origins, lost in time, are speculative. As these and other groups advanced, certain linguistic families retreated to refuges in remote places, where they hold out to the present day. Sources, dates, and routes are mostly speculative. *(Sources: Krantz, 1988; Renfrew, 1989.)*

What is more certain is that in later millennia, the diffusion of certain Indo-European languages—in particular Latin, English, and Russian—occurred in conjunction with the territorial spread of great political empires. In such cases of imperial conquest, relocation and expansion diffusion were not mutually exclusive. Relocation diffusion occurred as a small number of conquering elites came to rule an area. The language of the conqueror, implanted by relocation diffusion, often gained wider acceptance through expansion diffusion. Typically, the conqueror's language spread hierarchically—adopted first by the more important and influential persons and by city dwellers. The diffusion of Latin with Roman conquests, and Spanish with the conquest of Latin America, occurred in this manner.

Austronesian Diffusion

One of the most impressive examples of linguistic diffusion is that of the Austronesian languages, 5000 years ago, from a presumed hearth in the interior of Southeast Asia that was completely outside the present Austronesian culture region. From here, it is theorized, speakers of this language family first spread southward into the Malay Peninsula (see Figure 4.7, page 123). Then, in a process lasting perhaps several thousand years and requiring remarkable navigational skills, they migrated through the islands of Indonesia and sailed in tiny boats across vast, uncharted expanses of ocean to New Zealand, Easter Island, Hawaii, and Madagascar. If agriculture was the technology permitting Indo-European diffusion, sailing and navigation provided the key to the spread of the Austronesians.

Most remarkable of all was the diffusionary achievement of the Polynesian people, who form the eastern part of the Austronesian culture region. Polynesians occupy a triangular realm consisting of hundreds of Pacific islands, with New Zealand, Easter Island, and Hawaii at the three corners (**Figure 4.8**). The Polynesians' watery leap of

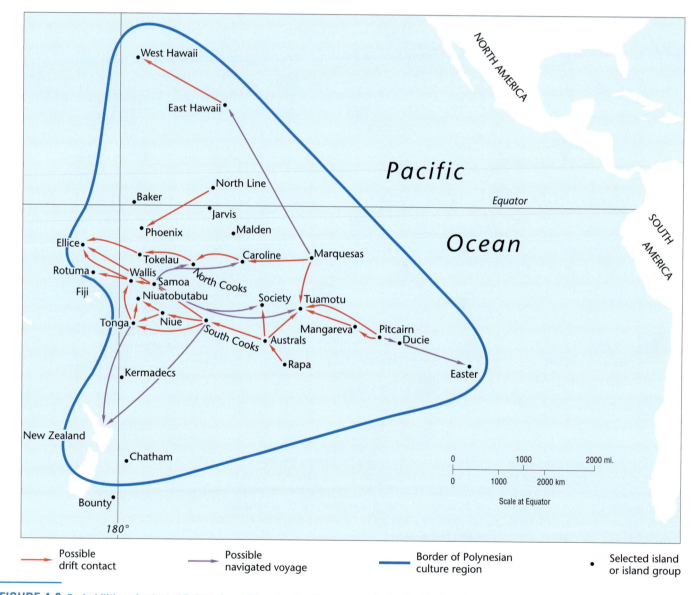

FIGURE 4.8 Probabilities of selected Polynesian drift and navigation voyages in the Pacific Ocean.
According to a computer model, the outer arc of Polynesia, represented by Hawaii, Easter Island, and New Zealand, could have been reached only by navigated voyages. The earliest known Polynesian pottery shards were found in Tonga by archaeologists David Burley and William Dickinson, suggesting that the first distinctively Polynesian culture originated there. *(Adapted from Levison, Ward, and Webb, 1973: 5, 33, 35, 43, 61.)*

2500 miles (4000 kilometers) from the South Pacific to Hawaii, a migration in outrigger canoes against prevailing winds into a new hemisphere with different navigational stars, must rank as one of the greatest achievements in seafaring. No humans had previously found the isolated Hawaiian Islands, and the Polynesian sailors had no way of knowing ahead of time that land existed in that quarter of the Pacific.

The relocation diffusion that produced the remarkable present distribution of the Polynesian languages has long been the subject of controversy. How, and by what means, could a traditional people have achieved the diffusion? Geolinguists Michael Levison, Gerard Ward, and John Webb answered these questions by developing a computer model incorporating data on winds, ocean currents, vessel traits and capabilities, island visibility, and duration of voyage. Both *drift* voyages, in which the boat simply floats with the winds and currents, and *navigated* voyages were considered.

After performing more than 100,000 voyage simulations, they concluded that the Polynesian triangle had probably been entered from the west, from the direction of the ancient Austronesian hearth area, in a process of "island hopping"—that is, migrating from one island to another one visible in the distance. The core of eastern Polynesia was probably reached in navigated voyages, but once this was attained, drift voyages easily explain much of the internal diffusion. According to Levison and his colleagues, a peripheral region, the "outer arc from Hawaii through Easter Island to New Zealand," could be reached only by means of intentionally navigated voyages: truly astonishing and daring feats (see Figure 4.8).

In carrying out their investigation, Levison and his colleagues employed the themes of region (present distribution of Polynesians) and nature-culture (currents, winds, visibility of islands) to help describe and explain the workings of a third theme, mobility.

Religion and Linguistic Mobility

Cultural interaction creates situations in which language is linked to a particular religious faith or denomination, a linkage that greatly heightens cultural identity. Perhaps Arabic provides the best example of this cultural link. It spread from a core area on the Arabian Peninsula with the expansion of Islam. Had it not been for the evangelical success of the Muslims, Arabic would not have diffused so widely. The other Semitic languages also correspond to particular religious groups. Hebrew-speaking people are of the Jewish faith, and the Amharic speakers in Ethiopia tend to be Coptic, or Eastern, Christians. Indeed, we can attribute the preservation and recent revival of Hebrew to

its active promotion by Jewish nationalists who believe that teaching and promoting Hebrew to diasporic Jews facilitates unity.

Certain languages have even acquired a religious status. Latin survived mainly as the ceremonial language of the Roman Catholic Church and Vatican City. In non-Arabic Muslim lands, such as Iran, where people consider themselves Persians and speak Farsi, Arabic is still used in religious ceremonies. Great religious books can also shape languages by providing them with a standard form. Martin Luther's translation of the Bible led to the standardization of the German language, and the Qur'an is the model for written Arabic. Because they act as common points of frequent cultural reference and interaction, great religious books can also aid in the survival of languages that would otherwise become extinct. The early appearance of a hymnal and the Bible in the Welsh language aided the survival of that Celtic tongue, and Christian missionaries in diverse countries have translated the Bible into local languages, helping to preserve them. In Fiji, the appearance of the Bible in one of the 15 local dialects elevated it to the dominant native language of the islands.

Language's Shifting Boundaries

Dialects, as well as the language families discussed previously, reveal a vivid geography. Their boundaries—what separates them from other dialects and languages—shift over time, both spatially and in terms of what elements they contain or discard.

isogloss
The border of usage of an individual word or pronunciation.

Geolinguists map dialects by using **isoglosses,** which indicate the spatial borders of individual words or pronunciations. The dialect boundaries between Latin American Spanish speakers using *tú* and those using *vos* are clearly defined in some areas, as shown on the map in **Figure 4.9**, page 126. The choice of *tú* or *vos* for the second-person singular carries with it a cultural indication. *Vos* represents a usage closer to the original Spanish but is considered by many in the Spanish-speaking world to be rather archaic and, in fact, has died out in Spain itself. In other regions, particularly throughout Argentina, *vos* has long been used by the media; often, it is considered to reflect the "standard" dialect and usage for the area in which it is used. Because certain words or dialects can fall out of fashion or simply become overwhelmed by an influx of new speakers, isogloss boundaries are rarely clear or stable over time. Indeed, in Central America, the media are increasingly using *vos*—long used in conversation in the region—thus elevating *vos* to a more official status covering a larger territory. Because of this, geolinguists often disagree about how many dialects are present in an area or exactly where isogloss borders should

ATLANTIC OCEAN

PACIFIC OCEAN

tú
vos
tú/vos
Not applicable

FIGURE 4.9 **Dialect boundaries in Latin America.** Spanish speakers in the Americas use either *vos* or *tú* as the second-person singular verb form. They represent dialects of Spanish: both are correct, linguistically speaking, but the *vos* form is older. Some regions use *vos* and *tú* interchangeably. *(Adapted from Pountain, 2005.)*

be drawn. The language map of any place is a constantly shifting kaleidoscope.

The dialects of American English provide another good example. At least three major dialects, corresponding to major culture regions, had developed in the eastern United States by the time of the American Revolution: the Northern, Midland, and Southern dialects (**Figure 4.10**). As the three subcultures expanded westward, their dialects spread and fragmented. Nevertheless, they retained much of their basic character, even beyond the Mississippi River. These culture regions have unusually stable boundaries. Even today, the "r-less" pronunciation of words such as *car* ("cah") and *storm* ("stohm"), characteristic of the East Coast Midland regions, is readily discernible in the speech of its inhabitants.

Although we are sometimes led to believe that Americans are becoming more alike, as a national culture overwhelms regional ones, the current status of American English dialects suggests otherwise. Linguistic divergence is still under way, and dialects continue to mutate on a regional level, just as they always have. Local variations in grammar and pronunciation proliferate, confounding the

proponents of standardized speech and defying the homogenizing influence of the Internet, television, and other mass media.

Shifting language boundaries involve content as well as spatial reach, and this, too, changes over time. Today, for example, some of the unique vocabulary of American English dialects is becoming old-fashioned. For instance, the term *icebox,* which was literally a wooden box with a compartment for ice that was used to cool food, was used widely throughout the United States to refer to the precursor of the refrigerator. Although the modern electric refrigerator is ubiquitous in the United States today, some people, particularly those of older generations and in the South, still use the term *icebox.* Many young people, by contrast, no longer pause to say the entire word *refrigerator,* shortening it instead to *fridge.*

As illustrated by the birth of the new word *fridge,* slang terms are quite common in most languages, and American English is no exception. **Slang** refers to words and phrases that are not a part of a standard, recog-

slang
Words and phrases that are not part of a standard, recognized vocabulary for a given language but that are nonetheless used and understood by some of its speakers.

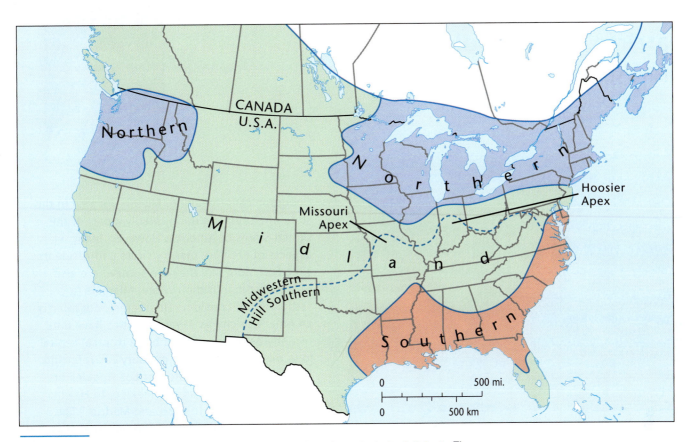

FIGURE 4.10 **Major dialects of North American English, with a few selected subdialects.** These dialects had developed by the time of the American Revolution and have remained remarkably stable over time. *(After Carver, 1986; Kurath, 1949.)*

nized vocabulary for a given language but are nonetheless used and understood by some or most of its speakers. Often, subcultures—for example, youth, drug dealers, and nightclub-goers—have their own slang that is used within that community but is not readily understood by nonmembers. Slang words tend to be used for a period of time and then discarded as newer terms replace them. For example, *cool* and *groovy* were used to refer to desirable, attractive, or fashionable things in the 1970s. Although many people today still recognize these words, a new generation of young people is much more likely to use a new set of words. Their children, in turn, will more than likely use yet another set of words. Slang illustrates another way in which American English changes over time.

Some African-Americans speak a distinctive form of English. African-American Vernacular English (AAVE) shares characteristics with the Southern dialect and also displays considerable African influence in pitch, rhythm, and tone. Some linguists understand AAVE as a creole language that grew out of a pidgin that developed on the early slave plantations and is today spoken by some African-Americans. Indeed, AAVE shares many characteristics with other English

creole languages worldwide. Today, it is considered a dialect, or variation, of Standard American English. It is also considered an **ethnolect:** a dialect spoken by an ethnic group, in this case, African-Americans. The popularity of AAVE's distinctive vocabulary and syntax among some white Americans, however, calls into question its ethnic exclusivity and serves to point out the instability of any language boundaries, including ethnic ones.

Some distinguishing characteristics of AAVE's speech patterns include the use of double negatives ("She don't like nothing"), omission of forms of the verb *to be* ("He my friend"), and nonconjugation of verbs ("She give him her paper yesterday"). There is some controversy over the place of Ebonics in the U.S. educational system. Does Ebonics constitute a distinctive language, rather than simply a dialect of English? Should it be taught—with its attendant grammar, structure, and literature—to American schoolchildren? Are those who speak Ebonics and Standard English technically bilingual?

In the United States today, many descendents of Spanish speakers have adapted their speech to include words

ethnolect
A dialect spoken by a particular ethnic group.

Patricia's Notebook
Miami State of Mind

Patricia Price

Little Havana commercial landscape. The lingua franca of Miami's Little Havana neighborhood is Spanish. This is reflected in the names and descriptions of local businesses. English is seldom seen (or heard). *(Courtesy of Patricia L. Price.)*

Geographically speaking, Miami—located in the state of Florida—is undoubtedly a part of the United States of America. However, many aspects of life here make visitors and residents alike feel more like they are somewhere else. As novelist Joan Didion wrote, Miami is "a settlement of considerable interest, not exactly an American city as American cities have until recently been understood but a tropical capital: long on rumor, short on memory . . . referring not to New York or Boston or Los Angeles or Atlanta but to Caracas and Mexico, to Havana and to Bogotá and to Paris and Madrid." The climate is subtropical, with the only real seasonal variation being between the rainy (summer) and dry (winter) months. Even after 15 years, I still miss autumn! Demographically, about 65 percent of residents are Latino, while the rest are roughly evenly split between African- and Anglo-Americans. And the United Nations has ranked Miami number one in the world in terms of the proportion of foreign-born residents (just over 50 percent).

Perhaps the biggest transition for a new arrival to Miami from another U.S. city, however, involves language. Spanish is the lingua franca of Miami. You can get by here just fine without speaking a word of English, though the reverse is not true: monolingual English speakers will struggle to understand and be understood in many areas of their daily lives. In other words, not speaking Spanish is an impediment here, whether in one's personal transactions or even in the ability to get a job, since many require applicants to understand and speak some Spanish.

I wonder whether Miami will remain this way or whether—as with other so-called immigrant gateways such as New York, San Francisco, and Chicago—immigrants will eventually acculturate, and the lingua franca of this town will transition yet again to English. I can already see this happening with my students, most of whose parents or grandparents are immigrants. Already their Spanish (or Portuguese, or Haitian Kreyol) is limited to what I call "kitchen" status. In other words, their vocabularies are limited, and they speak Spanish or other languages mostly with non-English-speaking relatives. Their children might not learn any language beyond English.

Posted by Patricia Price

and variants of words in both Spanish and English, in a dialect known as "Spanglish" (see Seeing Geography, page 144). Although acceptance of AAVE and Spanglish as legitimate language forms is hardly without conflict (see Patricia's Notebook and Subject to Debate), they illustrate the fluidity of languages and how they are constantly evolving and changing as the needs and experiences of their users change.

Subject to Debate
Imposing English

English-only laws are nothing new in the United States. Its history as a nation of immigrants has led to a population that, at any one point in time, has spoken a variety of languages besides English. In its early days as a colony, one could hear German, Dutch, French, and a multitude of Native American languages spoken alongside English. This prompted both Benjamin Franklin and John Adams to propose enforcing English as the sole acceptable language, and Theodore Roosevelt once said, "The one absolutely certain way of bringing this nation to ruin or preventing all possibility of its continuing as a nation at all would be to permit it to become a tangle of squabbling nationalities. We have but one flag. We must also learn one language, and that language is English."

Citing concerns that providing official documents and services in multiple languages would simply be too expensive, contemporary advocates of English-only legislation claim that mandating one language is one way to reduce the cost of government. Some proponents also believe that English-only laws encourage immigrants to assimilate by learning the official language of the United States. Opponents accuse the laws of being racist and suggest that supporters of English-only legislation are threatened by cultural diversity. Linguistic unity, they say, does not lead to political or cultural unity. Furthermore, providing official documents and services only in English in effect denies these services and information to those who do not understand English. Debates such as these bring up questions of the legal, social, and political status of minority groups and their languages, debates that exist in many countries besides the United States.

Today, most of those who wish to legislate English as the official language of the country target Spanish-speaking immigrants as the object of their concern. Language is often at the heart of the immigration debates that appear so often on the nightly news. Anxieties about being culturally "overwhelmed" by Spanish speakers who refuse to learn English culminate in claims that Latino immigrants are dividing the nation in two: one English speaking and culturally "American," and the other Spanish speaking and unable to assimilate into the mainstream (see Patricia's Notebook).

Historically, most immigrants to the United States sooner or later abandon their native tongues. As late as 1910, one out of every four Americans could speak some language other than English with the skill of a native (as compared to 14 percent in 1990). This was a result of the mass immigrations from Germany, Poland, Italy, Russia, China, and many other foreign lands. Much of this linguistic diversity has given way to English, partly because these other languages lacked legal status, partly because of the monolingual educational system in the United States, and partly because of social pressures.

Continuing the Debate

As this discussion illustrates, many cultures do not want to assimilate into English-speaking society, choosing instead to actively assert pride in their language. Keeping this in mind, consider these questions:

- Do you think the wave of immigrants today, with their pride in language, is different from earlier waves? How so?

- Will monoglot English speakers become a dwindling minority as more and more people become bilingual through either choice or necessity?

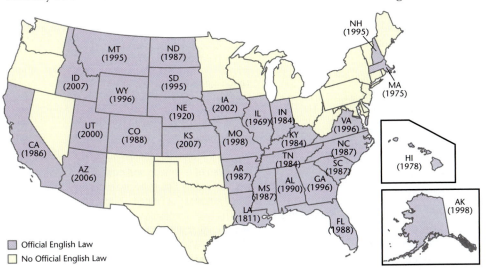

States that have some form of official English-only laws. Dates show the years English-only laws were enacted. In the 1980s, the influx of immigrants to the United States from Asia and Latin America prompted many of these laws. Typically, English-only laws require that state documents be published in English only. Some states' laws, however, also prohibit the state from doing business in a language other than English or providing services such as multilingual emergency medical hotlines. (*Source: Adapted from U.S. English, Inc.*)

Globalization

What is the relationship between technology and language? How has the global reach of empires past and present affected the spatial distribution of language? Why is the number of spoken languages on the decline worldwide?

More often than not, the diffusion of some languages has come at the expense of many others. Ten thousand years ago, the human race consisted of only 1 million people, speaking an estimated 15,000 languages. Today, a population 6000 times larger speaks only 40 percent as many tongues. Only 1 percent of all languages have as many as 500,000 speakers. Some experts believe that all but 300 languages will be extinct or dying by the year 2100. Clearly, cultural diffusion has worked to favor some languages and eliminate others. Each passing decade witnesses the extinction of more minor languages. Languages die out when their speakers do; often the entire cultural world associated with a language vanishes as well. Thus, globalization both presents the opportunity for more people to communicate directly with one another and, at the same time, threatens to extinguish the cultural diversity that goes hand-in-hand with linguistic diversity.

Technology, Language, and Empire

Technological innovations affecting language range from the basic practice of writing down spoken languages to the sophisticated information superhighway provided by the Internet. Technological innovations have in the past facilitated the spread and proliferation of multiple languages, but more recently they have encouraged the tendency of only a few languages—especially English—to dominate all others. Particular language groups achieve cultural dominance over neighboring groups in a variety of ways, often with profound results for the linguistic map of the world. Technological superiority is usually involved. Earlier, we saw how plant and animal domestication—the technology of the "agricultural revolution"—aided the early diffusion of the Indo-European language family.

An even more basic technology was the invention of writing, which appears to have developed as early as 5300 years ago in several hearth areas, including in Egypt, among the Sumerians in what is today Iraq, and in China. Writing helped civilizations develop and spread, giving written languages a major advantage over those that remained spoken only. Written languages can be published and distributed widely, and they carry with them the status of standard, official, and legal communication.

Written language facilitates record keeping, allowing governments and bureaucracies to develop. Thus, the languages of conquerors tend to spread with imperial expan-

sion. The imperial expansion of Britain, France, the Netherlands, Belgium, Portugal, Spain, and the United States across the globe altered the linguistic practices of millions of people (**Figure 4.11**). This empire building superimposed Indo-European tongues on the map of the tropics and subtropics. The areas most affected were Asia, Africa, and the Austronesian island world. A parallel case from the ancient world is China, also a formidable imperial power that spread its language to those it conquered. During the Tang dynasty (A.D. 618–907), Chinese control extended to Tibet, Mongolia, Manchuria (in contemporary northeastern China), and Korea. The 4000-year-old written Chinese language proved essential for the cohesion and maintenance of its far-flung empire. Although people throughout the empire spoke different dialects or even different languages, a common writing system lent a measure of mutual intelligibility at the level of the written word.

Even though imperial nations have, for the most part, given up their colonial empires, the languages they trans-

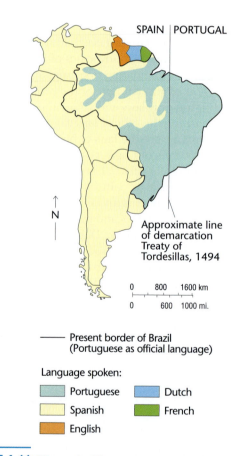

FIGURE 4.11 The mesh of language and empire in South America. Latin America was colonized by Spanish speakers and Portuguese speakers as well as by speakers of French, Dutch, and English in northern areas. The Treaty of Tordesillas, signed by Spain and Portugal in 1494, established the political basis for the present linguistic pattern in South America. Portugal was awarded the eastern part of the continent and Spain, the rest. *(Courtesy of Terry G. Jordan-Bychkov.)*

planted overseas survive. As a result, English still has a foothold in much of Africa, South Asia, the Philippines, and the Pacific islands. French persists in former French and Belgian colonies, especially in northern, western, and central Africa; Madagascar; and Polynesia (**Figure 4.12**). In most of these areas, English and French function as the languages of the educated elite, often holding official legal status. They are also used as a lingua franca of government, commerce, and higher education, helping hold together states with multiple native languages.

Transportation technology also profoundly affects the geography of languages. Ships, railroads, and highways all serve to spread the languages of the culture groups that build them, sometimes spelling doom for the speech of less technologically advanced peoples whose lands are suddenly opened to outside contacts. The Trans-Siberian Railroad, built about a century ago, spread the Russian language eastward to the Pacific Ocean. The Alaska Highway, which runs through Canada, carried English into Native American refuges. The construction of highways in Brazil's remote Amazonian interior threatens the native languages of that region.

Another example is the predominance of English on the Internet, which can be understood as a contemporary information highway (**Figure 4.13**). What will happen when other languages begin to challenge the dominance of English on the Internet? This will inevitably happen sooner or later, although exactly when English will be surpassed by another language is anyone's guess. For example, from 2000 to 2010 there was a truly impressive 1277 percent growth in the number of Chinese speakers on the Internet. If this trend continues, and when—not if—the 77 percent of

FIGURE 4.12 French, the colonial language of the empire, shares this sign on the isle of Bora Bora in French Polynesia with the native variant of the Polynesian tongue. Until recently, French rulers allowed no public display of the Polynesian language and tried to make the natives adopt French. *(Courtesy of Terry G. Jordan-Bychkov.)*

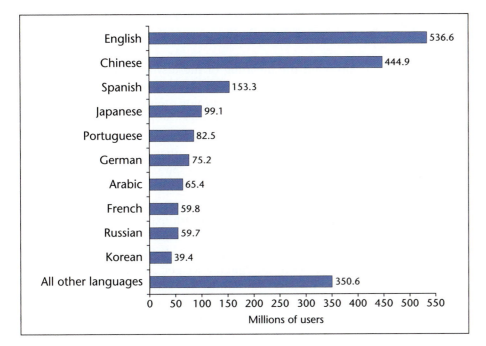

FIGURE 4.13 The 10 most prevalent languages on the Internet as of 2010, measured as a percentage of total users. English is the second most widely spoken language on Earth, after Mandarin Chinese. And English is the most widely spoken second language in the world. It dominates the Internet as well. *(Adapted from http://www.internetworldstats.com/stats7.htm.)*

Chinese speakers who do not now use the Internet begin to log on, we can expect Chinese to surpass English as the most popular language on the Internet. Speakers of other languages grew even more impressively over this time period. For instance, Arabic-speaking Internet users grew 2501 percent from 2000 to 2010, while speakers of Russian grew 1826 percent over the same period.

Reflecting on Geography

Can we view the Internet as a principal transportation route responsible for spreading English throughout the world today? If so, what does the spread of Internet access to areas formerly isolated by their physical landscape mean for the survival of linguistic minorities?

Texting and Language Modification

Though English may dominate the Internet (see Figure 4.13, page 131), much of what comes across our computer and cell phone screens isn't a readily recognizable form of English. As with the diffusion of spoken English to far-flung regions of the British Empire, the English language that is spread via electronic correspondence is subject to significant modification. E-mailing, instant messaging, and text messaging Standard English on cell phones requires a lot of typing, and text is notoriously deficient in conveying emotions when compared to the spoken word. For these reasons, users often use abbreviations and symbols to shorten the number of keystrokes used, to add emotional punctuation to their correspondence, and to make electronic communication hard to monitor by those who don't understand the language—particularly parents and teachers! (See the cartoon below.)

English is an alphabetic writing system, where letters represent discrete sounds that must be strung together to form a word's complete sound. A second major writing system is syllabic, where characters represent blocks of word

"FIRST PAPER I'VE HAD TO GRADE WRITTEN ENTIRELY IN EMOTICONS."

(Sidney Harris/ScienceCartoonsPlus.com.)

sounds. This type of writing, prevalent throughout the Middle East and Southeast Asia, includes Arabic, Hebrew, Hindi, and Thai. The third form of writing is logographic, where characters represent entire words. Chinese is the only major language in this category.

English-speaking texters quickly learn to take shortcuts around the lengthiness inherent to alphabetic writing systems. The simplest and most commonly used shortcut involves using acronyms instead of whole words to convey common phrases. *POS* (parent over shoulder), *AFK* (away from keyboard), *VBG* (very big grin), *LOL* (laughing out loud), and *GMTA* (great minds think alike) are some examples of this technique. Another shortcut involves using characters that sound like words. For example, the number "8" can substitute for the word or sound "ate," so "h8" is "hate," and "i 8" is "I ate." In the first example, a syllabic approach to writing is used because "8" condenses the syllable "ate" into one character. Another technique involves using symbols to substitute for entire words. Using "8" for the word "ate" is an example of this technique, called rebus writing, where a symbol is used for what it sounds like as opposed to what it stands for. As your instructor for this class will likely confirm, such abbreviations and symbols have even worked their way into term papers at the college level, much to the consternation or delight of language scholars.

The use of symbols such as ☺ and ♥, which are called pictograms, or word pictures, is similar to Chinese writing. Their popularity has resulted in a vast lexicon of pictograms as well as simpler symbol groupings, called emoticons, used to convey entire words, ideas, and emotions in a compact and often humorous form. Here are some examples:

:-)   (user is smiling or joking)
:-@   (user is screaming or cursing)
:-#   (well, shut my mouth!)

Although the meaning of these symbols is understood by speakers of many languages because of their ubiquity, non-English languages also employ their own symbol combinations. In Korean, for instance, ^^ is used instead of :) to convey a smiling face and -_- is used instead of :(to depict a sad face. In Chinese, the number "5" is pronounced in a way that resembles crying, so "555" is the Chinese texter's way of conveying sadness.

Spoken and written English are quickly picking up these cyberspeech patterns with expressions such as *PITA*, which is used to indicate that someone is a "pain in the ass" without actually saying so, and *G2G*, which is uttered instead of the marginally longer sounds "got to go" for which this acronym stands. As speakers of non-English languages become more heavily involved in texting-based activities, their spoken and written languages also doubtlessly will become modified.

Language Proliferation: One or Many?

Could all the world's languages have derived from one single mother tongue? It may seem a large leap from the primordial tongue to a consideration of globalization and languages, but, in fact, the two are related. If we humans began with one language, why shouldn't we return to that condition? If one language became 15,000 and the 6000 or so that remain will dwindle to 300 within a century, then why not end up with one again?

Are the forces of modernization working to produce, through cultural diffusion, a single world language? And if so, what will that language be? English? Worldwide, about 343 million people speak English as their mother tongue and perhaps another 350 million speak it well as a second, learned language. Adding other reasonably competent speakers who can "get by" in English, the world total reaches about 1.5 billion, more than for any other language. What's more, the Internet is one of the most potent agents of diffusion, and its language, overwhelmingly, is English.

English earlier diffused widely with the British Empire and U.S. imperialism, and today it has become the de facto language of globalization. Consider the case of India, where the English language imposed by British rulers was retained (after independence) as the country's language of business, government, and education. It provided some linguistic unity for India, which had 800 indigenous languages and dialects. This is why today many of India's nearly 1 billion people speak English well enough to provide customer support services over the telephone for clients in the United States. Even so, many people resent its use and wish India to be rid of this hated linguistic colonial legacy once and for all (see Figure 4.25, page 141). Although English is not likely to be driven out of India any time soon, it is true that the spoken English of India has drifted away from Standard British English. The same holds true for the English of Singapore, which is now a separate language called Singlish. Many other regional, English-based languages have developed, languages that could not be readily understood in London or Chicago.

But is the diffusion of English to the entire world population likely? Will globalization and cultural diffusion produce one world language? Probably not. More likely, the world will be divided largely among 5 to 10 major languages.

Language and Cultural Survival

Because language is the primary way of expressing culture, if a language dies out, there is a good chance that the culture of its speakers will, too. Languages, like animal species, are classified as endangered or extinct. Endangered languages are those that are not being taught to children by their parents and are not being used actively in everyday matters.

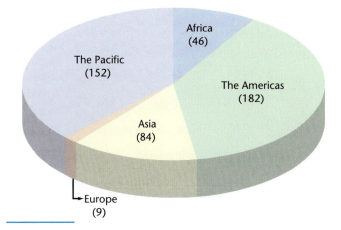

FIGURE 4.14 **Distribution of the world's nearly extinct languages.** This chart shows the regional distribution of the world's nearly extinct languages, or those languages that have only a few elderly speakers still living. Nearly extinct languages account for 6 percent of the world's total existing languages. (*Source: Adapted from Ethnologue, "Nearly Extinct Languages."*)

Some linguists believe that more than half of the world's roughly 6000 languages are endangered. Ethnologue, an online language resource (see Linguistic Geography on the Internet, page 145) for this web site address, considers 473 world languages to be nearly extinct. Languages that have only a few elderly speakers still living fall into this category. The Americas and the Pacific regions together account for more than three-quarters of the world's current nearly extinct languages, thanks to their many and varied indigenous language groups (**Figure 4.14**). In Argentina, for example, only five families speak Vilela, and only seven or eight speakers of Tuscarora remain in Canada. When these speakers die, it is likely that their language will die out with them.

As Figure 4.14 shows, almost 40 percent of the world's nearly extinct languages are found in the Americas. They represent a wealth of Native American languages that are slowly becoming suffocated by English, Spanish, and Portuguese. Other **language hotspots**—places with the most unique, misunderstood, or endangered tongues—are located around the globe (**Figure 4.15**, page 134). Two of the world's five most vulnerable regions—northern Australia, central South America, North America's upper Pacific coast, eastern Siberia, and Oklahoma—are located in the United States.

Languages can also be used to keep cultural traditions alive by speaking them repeatedly. Keith Basso, an anthropologist who has written an intriguing book titled *Wisdom Sits in Places,* discusses the landscape of distinctive place-names used by the Western Apache of New Mexico. The people Basso studied use place-names to invoke stories that help the Western Apache remember their collective history. According to Nick Thompson, one of Basso's interviewees, "White

language hotspots
Those places on Earth that are home to the most unique, misunderstood, or endangered languages.

FIGURE 4.15 Global "language hotspots."
The Enduring Voices Project and the National Geographic Society have teamed up to document endangered languages and thereby attempt to prevent language extinction. *(Source: Anderson and Harrison, 2007.)*

men need paper maps. . . . We have maps in our minds." Thompson goes on to assert that calling up the names of places can guard against forgetting the correct way of living, or adopting the bad habits of white men, once Western Apaches move to other areas. "The names of all these places are good. They make you remember how to live right, so you want to replace yourself again." One of the places Basso heard about is called Shades of Shit. Here is what he was told:

> *It happened here at Shades of Shit.*
>
> *They had much corn, those people who lived here, and their relatives had only a little. They refused to share it. Their relatives begged them but still they refused to share it.*
>
> *Then their relatives got angry and forced them to stay at home. They wouldn't let them go anywhere, not even to defecate. So they had to do it at home. Their shades [shelters] filled up with it. There was more and more of it! It was very bad! Those people got sick and nearly died.*
>
> *Then their relatives said, "You have brought this on yourselves. Now you live in shades of shit!" Finally, they agreed to share their corn.*
>
> *It happened at Shades of Shit. (p. 24)*

Today, merely standing at this place or speaking its graphic name reminds the Western Apache that stinginess is a vice that can threaten the survival of the entire community.

Related to linguistic extinction is the existence of so-called remnant languages that survive, typically in small linguistic islands that are surrounded by the dominant language. One example is Khoisan, found in the Kalahari Desert of southwestern Africa and characterized by distinc-

tive clicking sounds. The pockets of Khoisan seen in Figure 4.3 (pages 118–119) survived after the expansion of the Bantu, discussed earlier (see Figure 4.6, page 122). Other remnant languages include Dravidian, spoken by hundreds of millions of people in southern India, adjacent northern Sri Lanka, and a part of Pakistan, as well as Australian Aborigine, Papuan, Caucasic, Nilo-Saharan, Paleosiberian, Inuktitut, and a variety of Native American language families. In a few cases, individual minor languages represent the sole survivors of former families. Basque, spoken in the borderland between Spain and France, is such a survivor, unrelated to any other language in the world.

Nature-Culture

What relationships exist between language and the physical environment? Language interacts with the environment in two basic ways. First, the specific physical habitats in which languages evolve help shape their vocabularies. Second, the environment can guide the migrations of linguistic groups or provide refuges for languages in retreat. The following section, from the viewpoint of possibilism—the notion that the physical environment shapes, but does not fully determine, cultural phenomena—illustrates how the physical environment influences vocabulary and the distribution of language.

Habitat and Vocabulary

Humankind's relationship to the land played a strong role in the emergence of linguistic differences, even at the level

TABLE 4.2 Some Spanish Words Describing Mountains and Hills

Spanish Word	English Meaning
Candelas	Literally "candles"; a collection of *peñas*
Ceja	Steep-sided breaks or escarpments separating two plains of different elevations
Cejita	A low escarpment
Cerrillo or *cerrito*	A small *cerro;* a hill
Cerro	A single eminence, intermediate in size between English *hill* and *mountain*
Chiquito	Literally "small," describing minor secondary fringing elevations at the base of and parallel to a *sierra* or *cordillera*
Cordillera	A mass of mountains, as distinguished from a single mountain summit
Cuchilla	Literally "knife"; the comblike secondary crests that project at right angles from the sides of a *sierra*
Cumber	The highest elevation or peak within a *sierra* or *cordillera;* a summit
Eminencia	A mountainous or hilly protuberance
Loma	A hill in the midst of a plain
Lomita	A small hill in the midst of a plain
Mesa	Literally "table"; a flat-topped eminence
Montaña	Equivalent to English *mountain*
Pelado	A barren, treeless mountain
Pelon	A bare conical eminence
Peloncilla	A small *pelon*
Peña	A needlelike eminence
Picacho	A peaked or pointed eminence
Pico	A summit point; English *peak*
Sandia	Literally "watermelon"; an oblong, rounded eminence
Sierra	An elongated mountain mass with a serrated crest
Tinaja	A solitary, hemispheric mountain shaped like an inverted bowl

(Source: Hill, 1986.)

of vocabulary. For example, the Spanish language—which originated in Castile, Spain, a dry and relatively barren land rimmed by hills and high mountains—is especially rich in words describing rough terrain, allowing speakers of this tongue to distinguish even subtle differences in the shape and configuration of mountains, as **Table 4.2** reveals. Similarly, Scottish Gaelic possesses a rich vocabulary to describe types of topography; this terrain-focused vocabulary is a common attribute of all the Celtic languages spoken by hill peoples. In the Romanian tongue, also born of a rugged landscape, words relating to mountainous features emphasize use of that terrain for livestock herding. English, by contrast, which developed in the temperate wet coastal plains of northern Europe, is relatively deficient in words describing mountainous terrain (**Figure 4.16**). However, English

FIGURE 4.16 A scene in the desert of the western United States. "Mountains," yes, but what kind of mountains? (See Table 4.2.) The English language cannot describe such a place adequately because it is the product of a very different, humid and cool, physical landscape. As a result, the ability of English speakers to name dryland environmental features will be less precise in such places. *(David Muench/Corbis.)*

abounds with words describing flowing streams and wetlands: typical physical features found in northern Europe. This vocabulary transferred well to the temperate East Coast of the United States. In the rural American South alone, one finds *river, creek, branch, fork, prong, run, bayou,* and *slough.* This vocabulary indicates that the area is a well-watered land with a dense network of streams.

Clearly, then, language serves an adaptive strategy. Vocabularies are highly developed for those features of the environment that involve livelihood. Without such detailed vocabularies, it would be difficult to communicate sophisticated information relevant to the community's livelihood, which in most places is closely bound to the physical landscape.

The Habitat Helps Shape Language Areas

Environmental barriers and natural routes have often guided linguistic groups onto certain paths. The wide distribution of the Austronesian language group, as we have seen, was profoundly affected by prevailing winds and water currents in the Pacific and Indian oceans. The Himalayas and the barren Deccan Plateau deflected migrating Indo-Europeans entering the Indian subcontinent into the rich Ganges-Indus river plain. Even today in parts of India, according to Charles Bennett, the Indo-European/Dravidian "language boundary seems to approximate an ecological boundary" between the black soils of the plains and the thinner, reddish Deccan soils.

Because such physical barriers as mountain ridges can discourage groups from migrating from one area to another, they often serve as linguistic borders as well. In parts of the Alps, speakers of German and Italian live on opposite sides of a major mountain ridge. Portions of the mountain rim along the northern edge of the Fertile Crescent in the Middle East form the border between Semitic and Indo-European tongues. Linguistic borders that follow such physical features generally tend to be stable, and they often endure for thousands of years. By contrast, language borders that cross plains and major routes of communication are often unstable.

The Habitat Provides Refuge

The environment also influences language insofar as inhospitable areas provide protection and isolation. Such

linguistic refuge area
An area protected by isolation or inhospitable environmental conditions in which a language or dialect has survived.

areas often provide minority linguistic groups refuge from aggressive neighbors and are, accordingly, referred to as **linguistic refuge areas.** Rugged hilly and mountainous areas, excessively cold or dry climates, dense

FIGURE 4.17 **The environment provides a linguistic refuge in the Caucasus Mountains.** The rugged mountainous region between the Black and Caspian seas—including parts of Armenia, Russia, Georgia, and Azerbaijan—is peopled by a great variety of linguistic groups, representing three major language families. Mountain areas are often linguistic mosaics because the rough terrain provides refuge and isolation. For more information about this fascinating and diverse region, see Wixman, 1980.

forests, remote islands, and extensive marshes and swamps can all offer protection to minority language groups. For one thing, unpleasant environments rarely attract conquerors. Also, mountains tend to isolate the inhabitants of one valley from those in adjacent ones, discouraging contact that might lead to linguistic diffusion.

Examples of these linguistic refuge areas are numerous. The rugged Caucasus Mountains and nearby ranges in central Eurasia are populated by a large variety of peoples and languages (Figure 4.17). In the Rocky Mountains of northern New Mexico, an archaic form of Spanish survives, largely as a result of isolation that ended only in the early 1900s. Similarly, the Alps, the Himalayas, and the highlands of Mexico form fine-grained linguistic mosaics, thanks to the mountains that provide both isolation and protection for multitudinous languages. The Dhofar, a mountain tribe in Oman, preserves Hamitic speech, a language family otherwise vanished from all of Asia. Bitterly cold tundra climates of the far north have sheltered Uralic and Inuktitut speakers, and a desert has shielded Khoisan speakers from Bantu invaders. In short, rugged, hostile, or isolated environments protect linguistic groups that might otherwise be eclipsed by more dominant languages.

Still, environmental isolation is no longer the vital linguistic force it once was. Fewer and fewer places are so isolated that they remain little touched by outside influences. Today, inhospitable lands may offer linguistic refuge, but it is no longer certain that they will in the future. Even an island situated in the middle of the vast Pacific Ocean does not offer reliable refuge in an age of airplanes, satellite-transmitted communications, and global tourism. Similarly, marshes and forests provide refuge only if they are not drained and cleared by those who wish to use the land more intensively. The nearly 10,000 Gullah-speaking descendants of African slaves have long nurtured their distinctive African-influenced culture and language, in part because they reside on the Sea Islands of South Carolina, Georgia, and North Florida. Today, the development of these islands for tourism and housing for wealthy nonlocals, as well as the out-migration of Gullah youth in search of better economic opportunities, threatens the survival of the Gullah culture and language. The reality of the world is no longer isolation, but contact.

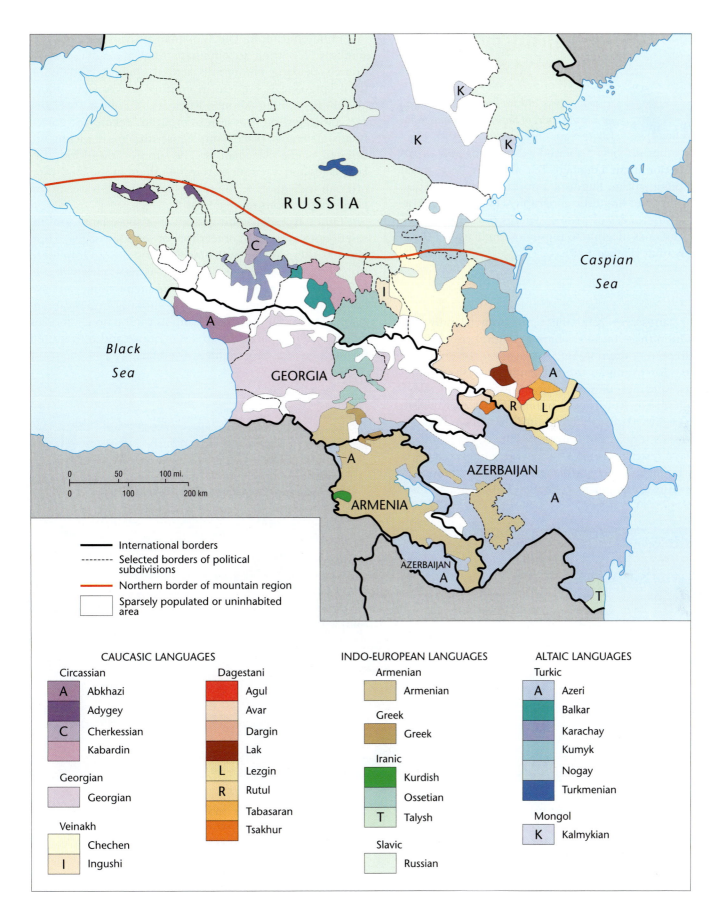

International borders

Selected borders of political subdivisions

Northern border of mountain region

Sparsely populated or uninhabited area

CAUCASIC LANGUAGES

Circassian
A Abkhazi
Adygey
C Cherkessian
Kabardin

Georgian
Georgian

Veinakh
Chechen
I Ingushi

Dagestani
Agul
Avar
Dargin
Lak
L Lezgin
R Rutul
Tabasaran
Tsakhur

INDO-EUROPEAN LANGUAGES

Armenian
Armenian

Greek
Greek

Iranic
Kurdish
Ossetian
T Talysh

Slavic
Russian

ALTAIC LANGUAGES

Turkic
A Azeri
Balkar
Karachay
Kumyk
Nogay
Turkmenian

Mongol
K Kalmykian

CULTURAL LANDSCAPE

In what ways are languages visible and, as a result, part of the cultural landscape? Road signs, billboards, graffiti, placards, and other publicly displayed writings not only reveal the locally dominant language but also can be a visual index to bilingualism, linguistic oppression of minorities, and other facets of linguistic geography (**Figure 4.18**). Furthermore, differences in writing systems render some linguistic landscapes illegible to those not familiar with these forms of writing (**Figure 4.19**).

Messages

Linguistic landscapes send messages, both friendly and hostile. Often these messages have a political content and deal with power, domination, subjugation, or freedom. In Turkey, for example, until recently Kurdish-speaking minorities were not allowed to broadcast music or television programs in Kurdish, to publish books in Kurdish, or even to give their children Kurdish names. Because Turkey wishes to join the European Union, these minority language restrictions have come under intense outside scrutiny. In 2002, Turkey reformed its legal restrictions to allow the Kurdish language to be used in daily life but not

in public education. The Canadian province of Québec, similarly, has tried to eliminate English-language signs. French-speaking immigrants settled Québec, and its official language is French, in contrast to Canada's policy elsewhere of bilingualism in English and French. As **Figure 4.20** indicates, there is a movement in Ireland to replace English-language place-name signs with signs depicting the original Gaelic place-names. The suppression of minority languages and attempts to reinstate them in the landscape offer an indication of the social and political status of minority populations more generally.

Other types of writing, such as gang-related graffiti, can denote ownership of territory or send messages to others that they are not welcome (**Figure 4.21**). Only those who understand the specific gang symbols used will be able to decipher the message. Misreading such writing can have dangerous consequences for those who stray into unfriendly territory. In this way, gang symbols can be understood as a dialect that is particular to a subculture and transmitted through symbols or a highly stylized script.

FIGURE 4.18 Samoan, a Polynesian language belonging to the Austronesian family, becomes part of the linguistic landscape of Apia, the capital of independent Western Samoa in the Pacific Ocean. When this area was still a British colony, such a visual display of the native language would not have been permitted. *(Courtesy of Terry G. Jordan-Bychkov.)*

FIGURE 4.19 Linguistic landscapes can be hard to read for those who are not familiar with the script used for writing. For many English-speaking monoglots, who are visually accustomed to the Latin alphabet, the linguistic landscape of countries such as Korea appears illegible. *(Courtesy of Terry G. Jordan-Bychkov.)*

Toponyms

Language and culture also intersect in the names that people place on the land, whether they are given to settlements, terrain features, streams, or various other aspects of their surroundings. These place-names, or **toponyms,** often directly reflect the spatial patterns of language, dialect, and ethnicity. Toponyms become part of the cultural landscape when they appear on signs and placards. Toponyms can be very revealing because, as geographer Stephen Jett said, they often provide insights into "linguistic origins, diffusion, habitat, and environmental perception." Many place-names consist of two parts— the generic and the specific. For example, in the American place-names Huntsville, Harrisburg, Ohio River, Newfound Gap, and Cape Hatteras, the specific segments are *Hunts-, Harris-, Ohio, Newfound,* and *Hatteras.* The generic parts, which tell what *kind* of place is being described, are *-ville, -burg, River, Gap,* and *Cape.*

Generic toponyms are of greater potential value to the cultural geographer than specific names because they appear again and again throughout a culture region. There are literally thousands of generic place-names, and every culture or subculture has its own distinctive set of them. They are particularly valuable both in tracing the spread of a culture and in reconstructing culture regions of the past. Sometimes generic toponyms provide information about changes people wrought long ago in their physical surroundings.

> **toponym**
> A place-name, usually consisting of two parts, the generic and the specific.

> **generic toponym**
> The descriptive part of many place-names, often repeated throughout a culture area.

Generic Toponyms of the United States

The three dialects of the eastern United States (see Figure 4.10)—Northern, Midland, and Southern—illustrate the value of generic toponyms in cultural geographical detective work. For example, New Englanders, speakers of the Northern dialect, often used the terms *Center* and *Corners* in the names of the towns or hamlets. Outlying settlements frequently bear the prefix *East, West, North,* or *South,* with the specific name of the township as the suffix. Thus, in Randolph Township, Orange County, Vermont, we find settlements named Randolph Center, South Randolph, East Randolph, and North Randolph (**Figure 4.22**). A few miles away lies Hewitts Corners.

These generic usages and duplications are peculiar to New England, and we can locate areas settled by New Englanders as they migrated westward by looking for such place-names in other parts of the country. A trail of

FIGURE 4.20 Road sign in Dublin, Ireland, shows directions in Irish Gaelic on top and English underneath. *(JoeFoxDublin/Alamy.)*

FIGURE 4.21 Graffiti is used to mark gang territory. This wall in the Polanco neighborhood of Guadalajara, Mexico, is covered with graffiti. Gangs use stylized scripts that are often unintelligible to nonmembers to mark their territory. *(Courtesy of Patricia L. Price.)*

FIGURE 4.22 This map depicts a portion of the state of Vermont, displaying the generic toponyms that abound in New England. *(Courtesy of Bing.)*

"Centers" and name duplications extending westward from New England through upstate New York and Ontario and into the upper Midwest clearly indicates their path of migration and settlement (**Figure 4.23**). Toponymic evidence of New England exists in areas as far afield as Walworth County, Wisconsin, where Troy, Troy Center, East Troy, and Abels Corners are clustered; in Dufferin County, Ontario, where one finds places such as Mono Centre; and even in distant Alberta, near Edmonton, where the toponym Michigan Centre doubly suggests a particular cultural diffusion. Similarly, we can identify Midland American areas by such terms as *Gap, Cove, Hollow, Knob* (a low, rounded hill), and *-burg,* as in Stone Gap, Cades Cove, Stillhouse Hollow, Bald Knob, and Fredericksburg. We can recognize southern speech by such names as *Bayou, Gully,* and *Store* (for rural hamlets), as in Cypress Bayou, Gum Gully, and Halls Store.

Toponyms and Cultures of the Past

Place-names often survive long after the culture that produced them vanishes from an area, thereby preserving traces of the past. Australia abounds in Aborigine toponyms, even in areas from which the native peoples disappeared long ago (**Figure 4.24**). No toponyms are more permanently established than those identifying physical geographical features, such as rivers and mountains. Even the most absolute conquest, exterminating an aboriginal people, usually does not entirely destroy such names. Quite the contrary, in fact. Geographer R. D. K. Herman speaks of anticonquest, in which the defeated people finds its toponyms venerated and perpetuated by the conqueror, who at the same time denies the people any real power or cultural influence. The abundance of Native American toponyms in the United States provides an example (see Doing Geography

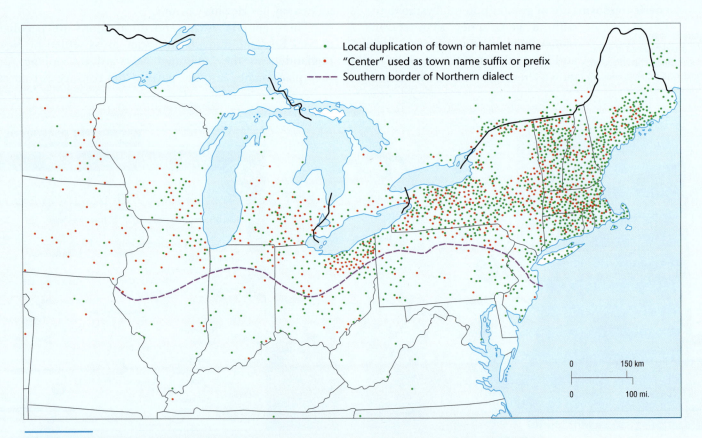

FIGURE 4.23 Generic place-names reveal the migration of Yankee New Englanders and the spread of the Northern dialect. Two of the most typical place-name characteristics in New England are the use of *Center* in the names of the principal settlements in a political subdivision and the tendency to duplicate the names of local towns and villages by adding the prefixes *East, West, North,* and *South* to the subdivision's name. As the concentration of such place-names suggests, these two Yankee traits originated in Massachusetts, the first New England colony. Note how these toponyms moved westward with New England settlers but thinned out rapidly to the south, in areas not colonized by New Englanders.

FIGURE 4.24 An Australian Aborigine specific toponym joined to an English generic name, near Omeo in Victoria state, Australia. Such signs give a special, distinctive look to the linguistic landscape and speak of a now-vanished culture region. *(Courtesy of Terry G. Jordan-Bychkov.)*

on page 143). India, however, has recently decided to revert to traditional toponyms, after many Indian place-names had been Anglicized under British colonial rule (**Figure 4.25**).

In Spain and Portugal, seven centuries of Moorish rule left behind a great many Arabic place-names (**Figure 4.26**, page 142). An example is the prefix *guada-* on river names (as in Guadalquivir and Guadalupejo). The prefix is a corruption of the Arabic *wadi,* meaning "river" or "stream." Thus, Guadalquivir, corrupted from Wadi-al-Kabir, means

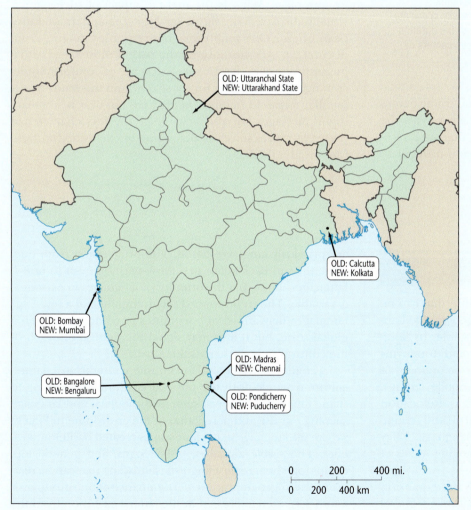

OLD: Uttaranchal State
NEW: Uttarakhand State

OLD: Calcutta
NEW: Kolkata

OLD: Bombay
NEW: Mumbai

OLD: Madras
NEW: Chennai

OLD: Bangalore
NEW: Bengaluru

OLD: Pondicherry
NEW: Puducherry

| 0 | 200 | 400 mi. |
| 0 | 200 | 400 km |

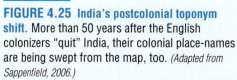

FIGURE 4.25 India's postcolonial toponym shift. More than 50 years after the English colonizers "quit" India, their colonial place-names are being swept from the map, too. *(Adapted from Sappenfield, 2006.)*

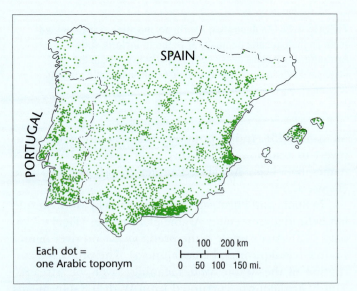

FIGURE 4.26 Arabic toponyms in Iberia. Arabic, a Semitic language, spread into Spain and Portugal with the Moors more than a thousand years ago. Romance Indo-European speakers subsequently reconquered the Arabic-speaking North Africans in Iberia. A reminder of the Semitic language survives in Iberian toponyms, or place-names. Using this map, you can easily speculate about the direction of the Moorish invasion and retreat, the duration of Moorish rule in different parts of Iberia, and the main centers of former Moorish power. *(Source: Houston, 1967.)*

"the great river." The frequent occurrence of Arabic names in any particular region or province of Spain reveals the remnants of Moorish cultural influence in that area, rather than anticonquest. Many such names were brought to the Americas through Iberian conquest, so that Guadalajara, for example, appears on the map as an important Mexican city.

New Zealand, too, offers some intriguing examples of the subtle messages that can be conveyed by archaic toponyms. The native Polynesian people of New Zealand are the Maori. As cultural geographer Hong-key Yoon has observed, the survival rate of Maori names for towns varies according to the size of their populations. The smaller the town, the more likely it is to bear a Maori name. The four largest New Zealand cities all have European names, but of the 20 regional centers, with populations of 10,000 to 100,000, 40 percent have Maori names. Almost 60 percent of the small towns, with fewer than 10,000 inhabitants, bear Maori toponyms. Similarly, whereas only 20 percent of New Zealand's provinces have Maori names, 56 percent of the counties do. Nearly all streams, hills, and mountains retain Maori names. The implication is that the British settlement of New Zealand was largely an urban phenomenon.

These Maori toponyms, which are heard and seen as one drives across New Zealand, help make the country a unique place. What is the mental impact of such names, visually displayed on signs, on New Zealanders of European origin? One might imagine responses ranging from discomfort, even hostility, on the part of those faced with a linguistically alien landscape, to a sense of comfort, homecoming, and belonging for those to whom this landscape is familiar terrain. Linguistic landscapes not only bear meaningful messages but also help shape the very character of places, as well as senses of belonging or exclusion for those who inhabit them.

CONCLUSION

Language, then, is an essential part of culture that can be studied using the five themes of cultural geography. Language is firmly enmeshed in the cultural whole. Its families, dialects, vocabulary, pronunciation, and toponyms display distinct spatial variations that are shown on maps of linguistic culture regions. Languages are mobile entities, ebbing and flowing across geographical areas. Relocation and expansion diffusion, both hierarchical and contagious, are apparent in the movement of language. They are also shaped and reshaped with the changing needs of their users.

The globalization of language, through the expansion of ancient empires as well as today's interlinked global exchanges, underscores the fact that human interactions are primarily language based. The progression from one, to many, and back again to a few—perhaps even to one—languages shows that the number of languages in existence is variable. The trend toward the dominance of a few "big" languages may afford opportunities to communicate on a global scale, but it also may signal the demise of much of the cultural richness across the Earth.

Language and physical environment interact in a nature-culture dynamic, with the physical environment helping to shape linguistic elements, such as vocabulary, and language shaping our use and perception of the environment. Finally, we can see language in the landscapes created by literate societies. The visible alphabet, public signs, and generic toponyms together create a linguistic landscape that can be read using one's "geographic eyes." Dominance of one group over another is often expressed in the latter's exclusion from the linguistic cultural landscape.

DOING GEOGRAPHY

Toponyms and Roots of Place

As you recall from this chapter, toponyms can give us important clues about the historical, social, political, and physical geography of a place. One example of this is the prevalence of indigenous place-names throughout the Americas, from Canada to Chile. You may say a place-name on a daily basis without being aware of its roots in an indigenous language. According to Charles Cutler, European settlers simply appropriated many of the Native American words for plants, animals, foods, and places with little or no modification in their pronunciation. These words are known as *loanwords*. For example, *Milwaukee* comes from an Algonquin word meaning "good spot or place," and *Chicago,* also Algonquin in origin, probably means "garlic field." The commonly used derogatory place-name *Podunk* is also indigenous in origin, from the Natick word for "swampy place." In fact, the names of more than half of the states in the United States are of Native American origin.

For this exercise, you will explore the theme of toponyms in more detail.

Steps to Exploring Toponyms

Step 1: Choose (or your instructor will assign) a state in the United States or a province in Canada on which to focus.

Step 2: Find a map of your chosen or assigned state or province. It should be a map that is detailed enough to show the names of political and physical features, such as cities, towns, counties or parishes, rivers, mountains, lakes, and so on. You can find such maps in the reference section of your library; in printed or CD-ROM atlases; and online at sites such as the University of Texas's Perry-Castañeda map collection, located at http://www.lib.utexas.edu/maps/. Many maps in atlases also have an index of place-names that can be useful to you.

Step 3: Examine your map with an eye to the different categories of toponyms. Make a list of at least five place-names for each category:

- Historical people or events
- Non-English place-names (excluding Native American names)
- Native American place-names
- Place-names transplanted from elsewhere (e.g., "New" York)
- Descriptions of physical features (landforms, elevation, and so on)
- Descriptions of natural resources

Based on what you learned, answer the following questions:

- Did you find at least five examples for each category? If not, why do you think you didn't?
- For which category of toponyms did you find the most examples? Why?

- Were toponyms of one or more categories clustered spatially on the map? If so, where and why?
- What do the names say about the history, culture, and physical geography of the state or province you examined?

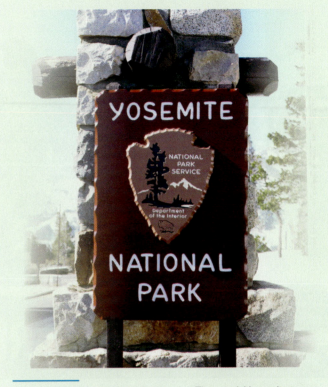

Yosemite National Park in California. The word *Yosemite* comes from the name of the indigenous group that inhabited this area. Notice the arrowhead symbol, which underscores the park's Native American history. *(Adamsmith/Taxi/Getty Images.)*

Key Terms

U.S. Army veterans at a recruiting table in Miami, Florida.

some U.S. cities, Hispanics constitute more than half of the population, a fact that brings into question the designation "minority." For example, the population of Miami, Florida, is two-thirds Hispanic, and more than three-fourths of the residents of El Paso and San Antonio, Texas, are Hispanic. In the United States today, Spanish-speaking peoples from Latin America provide the largest flow of immigrants into the United States. Even midsize and smaller towns in the midwestern and southern United States are becoming destinations for Spanish-speaking immigrants, sharply changing the ethnic composition of cities such as Shelbyville, Tennessee; Dubuque, Iowa; and Siler City, North Carolina.

Languages are fluid, always being altered and reinvented as the needs and experiences of their users change. Thanks to relocation diffusion resulting from the conquest of the East Coast of North America by the British in the seventeenth century, the primary language of the United States is English. The language then expanded westward as English-speaking peoples conquered more and more of the continent's territory. But the English spoken in Britain's overseas colonies has never been "the Queen's English." Rather, British colonies in North America, Australia, the Caribbean, Africa, and eastern and southern Asia all developed their own distinctive dialects. Indigenous words have been incorporated into English vocabularies, as the sections on toponyms in this chapter show. Pidgins, creoles, and distinctive dialects have resulted in places of high multilingual exposure, such as Singapore and the Anglophone Caribbean islands. Waves of immigrants have added further to the linguistic richness of English-speaking areas.

In 2002, Hispanics surpassed African-Americans as the nation's numerically most significant minority group. In

It is logical to assume that the Spanish language spoken by these new immigrants, and by the families of Hispanic-Americans, will have a growing impact on American English. As the opening photograph for this chapter illustrates, Spanish words have become common sights in U.S. cities, appearing frequently on street signs. But Spanish and English have combined in a rich, complex fashion as well, to produce a hybrid language called Spanglish. The phrase "Vámonos al downtown a tomar una bironga after work hoy" is an excellent example. It translates into Standard English as "Let's go downtown and have a beer after work today." *Vámonos* ("let's go"), *tomar* ("to drink"), and *hoy* ("today") are Spanish words that are combined in the same sentence with the English words *downtown* and *after work*. Linguists refer to this as code-switching. But the noun *bironga*, which means "beer" in English, is a Spanglish invention: it exists in neither English nor Spanish. This is quite common in Spanglish, and neologisms such as *hanguear* ("to hang out"), *deioff* ("day off"), and *parquear* ("to park" a vehicle) abound.

In the image that opens this chapter, the phrase "Yo soy el Army" employs both Spanish and English to recruit. The tendency to shift between languages in the same sentence is known as code-switching. It is a very common practice among bilingual speakers of Spanish and English. This image also underscores that people of Hispanic descent can be of any race.

Spanglish reflects the growing Spanish-English bilingualism of many U.S. residents, the flexibility of language, and the enduring creativity of human beings as we attempt to communicate with one another. Ilan Stavans, author of *Spanglish: The Making of a New American Language,* likens Spanglish to jazz. "Yes," Stavans writes, "it is the tongue of the uneducated. Yes, it's a hodgepodge. . . . But its creativity astonished me. In many ways, I see in it the beauties and achievements of jazz, a musical style that sprung up [*sic*] among African-Americans as a result of improvisation and lack of education. Eventually, though, it became a major force in America, a state of mind breaching out of the ghetto into the middle class and beyond. Will Spanglish follow a similar route?"

Linguistic Geography on the Internet

You can learn more about linguistic geography on the Internet at the following web sites:

Dictionary of American Regional English

http://dare.wisc.edu/

Discover a reference web site that describes regional vocabulary contrasts of the English language in the United States and includes numerous maps.

Enduring Voices

http://www.nationalgeographic.com/mission/enduringvoices/

This flash map allows you to explore the world's "language hotspots," those regions that are home to the most linguistic diversity, the highest levels of linguistic endangerment, and the least studied tongues.

Ethnologue

http://www.ethnologue.com

This site provides information on how languages change over time as well as on endangered and nearly extinct languages.

Language Log

http://itre.cis.upenn.edu/~myl/languagelog/

Search the fascinating posts on this language-themed blog, run by University of Pennsylvania phonetician Mark Liberman and featuring guest linguists. Themes range from aversion to the word *moist,* the peculiar naming of drink sizes at Starbucks, and a broader range of insulting words than you can imagine.

Sources

Anderson, Greg, and David Harrison. 2007, "Language Hotspots." Model developed at the Living Tongues Institute for Endangered Languages, available online at http://www.nationalgeographic.com/mission/enduringvoices/.

Basso, Keith H. 1996. *Wisdom Sits in Places: Landscape and Language Among the Western Apache.* Albuquerque: University of New Mexico Press.

Bennett, Charles J. 1980. "The Morphology of Language Boundaries: Indo-Aryan and Dravidian in Peninsular India," in David E. Sopher (ed.), *An Exploration of India: Geographical Perspectives on Society and Culture.* Ithaca, N.Y.: Cornell University Press, 234–251.

Carver, Craig M. 1986. *American Regional Dialects: Word Geography.* Ann Arbor: University of Michigan Press.

Cutler, Charles. 1994. *O Brave New Words! Native American Loanwords in Current English.* Norman: University of Oklahoma Press.

Didion, Joan. 1987. *Miami.* New York: Simon & Schuster.

Dingemanse, Mark. Wikipedia. Available online at http://en.wikipedia.org/wiki/Image:Bantu_expansion.png

Encyclopaedia Britannica. 2000. *2000 Britannica Book of the Year.* Chicago: Encyclopaedia Britannica.

Ethnologue, "Nearly Extinct Languages." Available online at http://www.ethnologue.com/nearly_extinct.asp.

Ford, Clark. "Early World History: Indo-Europeans to the Middle Ages." Available online at http://www.public.iastate.edu/~cfford/342worldhistoryearly.html.

Herman, R. D. K. 1999. "The Aloha State: Place Names and the Anti-Conquest of Hawaii." *Annals of the Association of American Geographers* 89: 76–102.

Hill, Robert T. 1986. "Descriptive Topographic Terms of Spanish America." *National Geographic* 7: 292–297.

Houston, James M. 1967. *The Western Mediterranean World.* New York: Praeger.

Jett, Stephen C. 1997. "Place-Naming, Environment, and Perception Among the Canyon de Chelly Navajo of Arizona." *Professional Geographer* 49: 481–493.

Krantz, Grover S. 1988. *Geographical Development of European Languages.* New York: Peter Lang.

Kurath, Hans. 1949. *Word Geography of the Eastern United States.* Ann Arbor: University of Michigan Press.

Latrimer Clarke Corporation Pty Ltd. Available online at http://www.altapedia.com.

Levison, Michael R., Gerard Ward, and John W. Webb. 1973. *The Settlement of Polynesia: A Computer Simulation.* Minneapolis: University of Minnesota Press.

Pountain, Chris. 2005. "Varieties of Spanish." Available online at http://www.qmul.ac.uk/~mlw058/varspan/varspanla.pdf (Queen Mary School of Modern Languages).

Renfrew, Colin. 1989. "The Origins of Indo-European Languages." *Scientific American* 261(4): 106–114.

Sappenfield, Mark. 2006. "Tear Up the Maps: India's Cities Shed Colonial Names." *Christian Science Monitor,* September 7, online edition.

Stavans, Ilan. 2003. *Spanglish: The Making of a New American Language.* New York: Rayo.

U.S. English, Inc. Available online at http://www.us-english.org/inc/official/states.asp.

Wixman, Ronald. 1980. *Language Aspects of Ethnic Patterns and Processes in the North Caucasus.* Research Paper No. 191. University of Chicago, Department of Geography.

World Almanac Books. 2001. *World Almanac and Book of Facts 2001.* New York: World Almanac Books.

Yoon, Hong-key. 1986. "Maori and Pakeha Place Names for Cultural Features in New Zealand," in *Maori Mind, Maori Land: Essays on the Cultural Geography of the Maori People from an Outsider's Perspective.* Bern, Switzerland: Peter Lang, 98–122.

Ten Recommended Books and Special Issues on the Geography of Language

(For additional suggested readings, see *The Human Mosaic* web site: www.whfreeman.com/domosh12e)

Abley, Mark. 2005. *Spoken Here: Travels Among Threatened Languages.* New York: Mariner Books. A fascinating journey across the global map of dwindling languages, exploring their diversity, their speakers, and their vulnerability to eclipse by the ever-more-dominant major tongues, especially English.

Cassidy, Frederic C. (ed.). 1985–2002. *Dictionary of American Regional English.* 4 vols. Cambridge, Mass.: Harvard University Press. A massive compilation of words used only regionally within the United States, with maps showing distributions.

Kenneally, Christine. 2007. *The First Word: The Search for the Origins of Language.* New York: Penguin. Do languages really "evolve," increasing in a linear fashion from the first grunts of cavemen millennia ago to today's cacophony of 6000 different tongues? Kenneally traces both the science of language evolution and the history of its study.

Laponce, J. A. 1987. *Languages and Their Territories.* Anthony Martin-Sperry (trans.). Toronto: University of Toronto Press. Treats the themes that (1) languages protect themselves by territoriality and (2) the modern political state typically acts overtly to destroy minority languages.

MacNeil, Robert, and William Cran. 2005. *Do You Speak American?* New York: Mariner Books. Spoken English in the United States displays a wealth of regional dialects, urban and rural differences, and historical as well as contemporary ties to languages spoken in other countries. The authors traveled the country documenting this fascinating language mosaic. A three-volume DVD of the same title is also available.

Moseley, Christopher, and R. E. Asher (eds.). 1994. *Atlas of the World's Languages.* London: Routledge. A wonderfully detailed color map portrait of the world's complex linguistic mosaic. Thumb through it at your library and you will come to appreciate how complicated the patterns and spatial distributions of languages remain, even in the age of globalization.

Ostler, Nicholas. 2005. *Empires of the Word: A Language History of the World.* New York: HarperCollins. This fascinating book explores the spread and evolution of languages through conquest, with many maps accompanying the text.

Pinker, Steven. 2007. *The Language Instinct: How the Mind Creates Language.* New York: Harper Perennial. Ever notice how linguistically inventive the average toddler is? This best seller looks at many angles of language development in humans, from its origins, to its social functions, to the instinct humans share to learn, understand, and speak language.

Rose-Redwood, Reuben, and Alderman, Derek, guest editors. 2011. New Directions in Political Toponymy. Special Thematic Interventions Section of *ACME: An International E-Journal for Critical Geographies* 10(1): 1–41. Available online at http://www.acme-journal.org/Volume10-1.htm. A collection of the latest geographic scholarship on political place names.

Withers, Charles W. J. 1988. *Gaelic Scotland: The Transformation of a Culture Region.* London: Routledge. A geographer analyzes one of the dying Celtic languages within the framework of the models presented in this chapter

A Journal in the Geography of Language

World Englishes. Published by the International Association for World Englishes, the journal documents the fragmentation of English into separate languages around the world. Edited by Margie Berns and Daniel R. Davis.

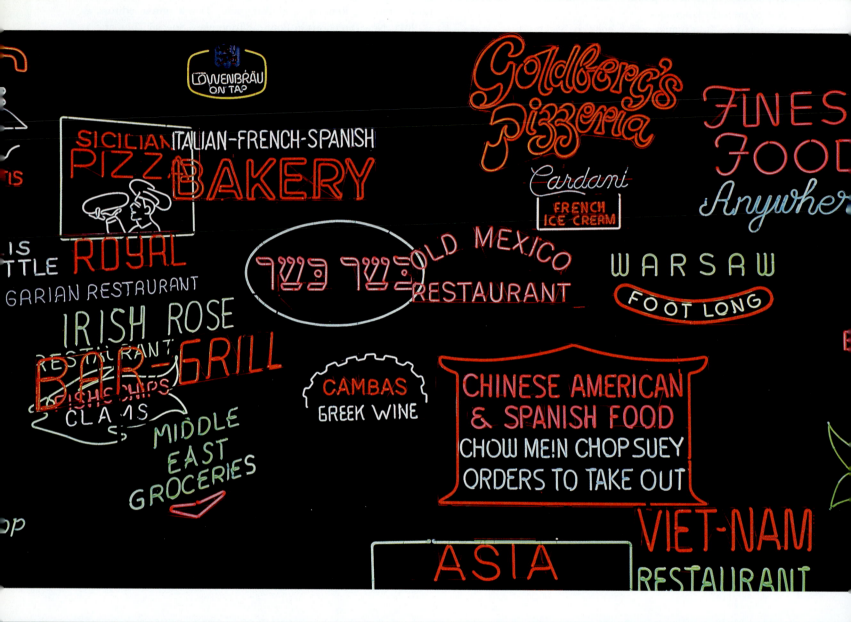

When does cuisine cease to be ethnic and become simply "American"? What role does cultural diffusion play in the process?

Go to "Seeing Geography" on page 184 to learn more about this image.

5 GEOGRAPHIES OF RACE AND ETHNICITY

Mosaic or Melting Pot?

One of the enduring stories that the people of the United States proudly tell themselves is that "ours is an immigrant nation." This story is on display during annual festivals celebrating the mosaic of ethnic traditions in countless cities, towns, and villages across the nation. For example, the midwestern town of Wilber, settled by Bohemian immigrants beginning about 1865, bills itself as "The Czech Capital of Nebraska" and annually invites visitors to attend a "National Czech Festival." Celebrants are promised Czech foods, such as *koláce, jaternice,* poppy seed cake, and *jelita;* Czech folk dancing; "colored Czech

postcards and souvenirs" imported from Europe; and handicraft items made by Nebraska Czechs (bearing an official seal and trademark to prove authenticity). Thousands of visitors attend the festival each year. Without leaving Nebraska, these tourists can move on to "Norwegian Days" at Newman Grove, the "Greek Festival" at Bridgeport, the Danish "Grundlovs Fest" in Dannebrog, "German Heritage Days" at McCook, the "Swedish Festival" at Stromsburg, the "St. Patrick's Day Celebration" at O'Neill, several Native American powwows, and assorted other ethnic celebrations (**Figure 5.1**).

FIGURE 5.1 The town of Stromsburg, Nebraska (*left*). Proud of its Swedish heritage, Stromsburg holds a Swedish Festival each year in June. **Hispanic Heritage Festival in Nebraska (*right*).** Hispanic immigrants have expanded the range of ethnic pride festivals throughout the Midwest. *(Right: Steve Skjold/Alamy.)*

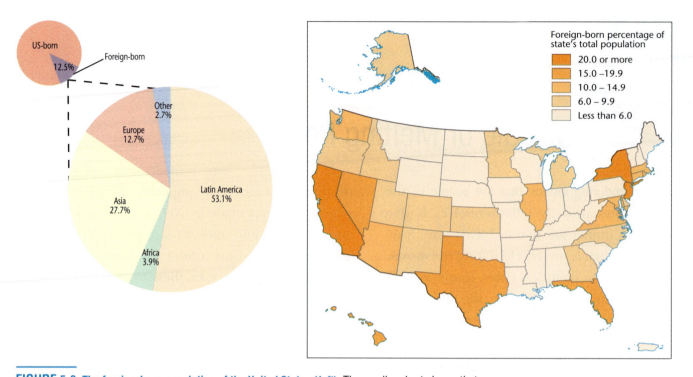

FIGURE 5.2 **The foreign-born population of the United States (*left*).** The smaller chart shows that 12.5 percent, or 38.5 million, of the total U.S. population in 2009 was born abroad. The larger chart shows that the majority of these people came from Latin America and Asia. **The foreign-born population by state and Puerto Rico (*right*).** This map shows where foreign-born people reside throughout the United States and Puerto Rico. *(Sources: U.S. Census Bureau, American Community Survey, 2009; Puerto Rico Community Survey, 2009.)*

Today, Nebraska is still a magnet for immigrants, but since the 1990s, the state's new arrivals have been overwhelmingly non-European. In particular, Mexican immigrants employed in Nebraska's meat-processing industry find destinations such as Nebraska and other upper midwestern states attractive. In general, immigrants to the United States today are more likely to come from Asia or Latin America than from Europe, and they are changing the face of ethnicity in the United States (**Figure 5.2**). Indeed, ethnicity is a central aspect of the cultural geography of most places, forming one of the brightest motifs in the human mosaic.

Race is often used interchangeably with *ethnicity*, but the two terms have very different meanings, and one must be careful in choosing between them. **Race** can be understood as a genetically significant difference among human populations. A few biologists today support the view that human populations do form racially distinct groupings, arguing that race explains phenomena such as the susceptibility to certain diseases. In contrast, many social scientists (and many biologists) have noted the fluidity of definitions of *race* across time and space, suggesting

race
A classification system that is sometimes understood as arising from genetically significant differences among human populations, or from visible differences in human physiognomy, or as a social construction that varies across time and space.

that race is a social construct rather than a biological fact (**Figures 5.3** and **5.4**).

Because race is a social construction, it takes different forms in different places and times. In the United States, for example, the so-called one-drop rule meant that anyone with any African-American ancestry at all was considered black. This law was intended to prevent interracial marriage. It also meant that moving out of the category "black" was, and still is today, extremely difficult because one's racial status is determined by ancestry. Yet the notion that "black blood" somehow makes a person completely black is challenged today by the growing numbers of young people with diverse racial backgrounds. The rise in interracial marriages in the United States, along with the fact that—for the first time in 2000—one could declare multiple races on the census form, means that more and more people identify themselves as "racially mixed" or "biracial" instead of feeling they must choose only one facet of their ancestry as their sole identity. Thus, golfer Tiger Woods calls himself a "Cablinasian" to describe his mixed Caucasian, black, American Indian, and Asian background. Race and ethnicity arise from multiple sources: your own definition of yourself; the way others see you; and the way society treats you, particularly legally. All three of these are subject to change over time. Still, President Barack Obama, whose individuals formerly classified as black. However, one must keep in mind that Brazil was also the last country in the Americas to abolish slavery, and discrimination against darker-skinned Brazilians is common today.

Studies of genetic variation have demonstrated that there is far more variability within so-called racial groups than between them, which has led most scholars to believe that all human beings are, genetically speaking, members of only one race: *Homo sapiens sapiens.* In fact, some social scientists have dropped the term *race* altogether in favor of *ethnicity.* This is not to say, however, that racism, the belief that human capabilities are determined by racial classification and that some races are superior to others, does not exist. In this chapter, we use the verb *racialize* to refer to the processes whereby these socially constructed differences are understood—usually, but not always, by the powerful majority—to be impervious to assimilation. Across the world, hatreds based in racism are at the root of the most incendiary conflicts imaginable (see Subject to Debate, page 152).

What exactly is an ethnic group? The word *ethnic* is derived from the Greek word *ethnos,* meaning a "people" or "nation," but that definition is too broad. For our purposes, an **ethnic group** consists of people of common ancestry and cultural tradition. A strong feeling of group identity characterizes ethnicity. Membership in an ethnic group is largely involuntary, in the sense that a person cannot simply decide to join; instead, he or she must be born

> **ethnic group**
> A group of people who share a common ancestry and cultural tradition, often living as a minority group in a larger society.

FIGURE 5.3 Castas painting. These paintings were common in colonial Mexico. They depict the myriad racial and ethnic combinations perceived to arise from intermixing among Europeans, indigenous peoples, and African slaves in the New World. They provided a way for those in power to keep track of the confusing racial hierarchy they had created. *(Schalkwijk/Art Resource, NY.)*

mixed-race ancestry has led him to identify with both his black and his white relatives, struggled during the 2008 presidential campaign with the claims that he was at once "too black" and "not black enough" to be a viable presidential candidate.

In Brazil, by contrast, a range of physiognomic features, such as skin pigmentation, eye color, and hair texture, are used to identify a person racially, with the result that many racial categories exist in Brazil. Siblings are frequently classified in significantly different racial terms depending on their appearance. The same person can be put into multiple racial categories by different people, and individuals' own racial self-designations can depend on such variables as their mood at the time they are asked. Increased economic or educational status can "whiten"

White   Black   Hispanic   Asian

FIGURE 5.4 Mug shots of people from different races. Phenotype variations—visible bodily differences such as facial features, skin color, and hair texture—are considered indicative of "race" by the U.S. Federal Bureau of Investigation (FBI). "Race" is one of several visible characteristics—which also include tattoos, scars, height, and weight—and is often used by law enforcement to identify suspects. These are photos from the FBI's 10 Most Wanted list. *How is this idea of race as defined by visible differences different from race as it is commonly understood in the United States? Do you see any similarities between this image and Figure 5.3?* *(Courtesy of the Federal Bureau of Investigation.)*

Subject to Debate

Racism: An Embarrassment of the Past, or Here to Stay?

Most modern societies are quick to claim that they have moved beyond the scourge of racism. Brazil, for example, bills itself as a "racial democracy." Swedes see theirs as a nation that is particularly tolerant in matters of race. And the "land of opportunity" privileges of the United States are supposedly extended to all, regardless of race, creed, color, and so forth.

Yet we know through our day-to-day experiences—studying, working, and socializing in specific places—that although "postracism" may be the ideal of modern societies, it is not always the fact in real life. In France in 2005, for instance, working-class Paris suburbs such as Clichy-sous-Bois—home to mostly Muslim immigrants from North Africa—became the site of rioting by youths who felt persecuted and excluded from French society because of their religion, culture, skin color, and immigrant status. A group of teenage boys who had been playing soccer were returning to the housing projects where they lived when they spotted the police. They scattered in order to avoid the detainment, questioning, and requests to show identity papers that such youths often face. Thinking that they were being chased in relation to a reported burglary at a construction site, three of the teenagers hid in a power substation; two of them were accidentally electrocuted. This incident tapped into a well of unrest over what many see as the systematic exclusion in France of those with Arabic- or African-sounding names and darker skins, exclusion that has led to high levels of unemployment among *banlieue* (the French equivalent of a ghetto) residents. Months of rioting ensued throughout France. Nearly 10,000 cars were set on fire, property damage totaled in the hundreds of millions of euros, and 3000 arrests were made.

The following year in the United States, widespread immigration protests filled city streets across the nation. In the spring of 2006, an estimated half a million people hit the streets of Los Angeles, Chicago, and Washington, D.C., and groups numbering in the tens of thousands marched in Denver, Milwaukee, Charlotte, Atlanta, Phoenix, and many other cities across the nation. Hispanics and their supporters took to the streets to protest legislation that threatened to deport undocumented workers and to classify anyone who helped an undocumented individual as a felon. As in the French incidents, Hispanic immigrants to the United States said they felt discriminated against and excluded from society. Even those who merely "look" or "sound" Hispanic are often the targets of racism. Rather than being violent terrorists or drains on society, protesters argued that Hispanics are, in fact, the backbone of low-wage labor in the United States. "When did you ever see a Mexican blow up the World Trade Center? Who do you think built the World Trade Center? I'm in my homeland!" exclaimed 22-year-old David González, whose mother immigrated to the United States from Mexico (quoted in the Associated Press).

These incidents highlight the fact that race-related tensions are not only about ancestry or skin pigmentation but also about religious differences, immigration, and social opportunities. As anxieties arise over ever-higher levels of contact in a globalizing world, racialized intolerance is all too often the response. Indeed, A. Sivanandan has argued that "poverty is the new black" in a globalizing world that has become increasingly hostile toward involuntary migrants regardless of their skin color.

Continuing the Debate

Possible responses to racism lie along a continuum, from violent protest to waiting for time to erase racism. Consider these questions:

- Is racism really a serious problem, or are minority groups simply agitating as they always have?

- Will racism finally go away by itself as societies modernize and move beyond intolerance?

- Can legislation adequately and appropriately address the fundamental causes of racism, or are more radical measures—protests or work stoppages, for example—necessary given the deeply rooted nature of race-based discrimination?

Protesters numbering around half a million rally for immigrant rights in Los Angeles, part of a wave of protests occurring throughout U.S. cities in the spring of 2006. *(J. Emilio Flores/Getty Images.)*

into the group. In some cases, outsiders can join an ethnic group by marriage or adoption.

Reflecting on Geography

Are ethnic groups always minorities, or do majority groups also have an ethnicity? What different sorts of traits might a majority group use to define its identity, as compared to a minority group? Is it possible for a racialized minority group to racialize other groups, or even itself?

Different ethnic groups may base their identities on different traits. For some, such as Jews, ethnicity primarily means religion; for the Amish, it is both folk culture and religion; for Swiss-Americans, it is national origin; for German-Americans, it is ancestral language; for African-Americans, it is a shared history stemming from slavery. Religion, language, folk culture, history, and place of origin can all help provide the basis of the sense of "we-ness" that underlies ethnicity.

As with race, ethnicity is a notion that is at once vexingly vague yet hugely powerful (see Patricia's Notebook, page 154). The boundaries of ethnicity are often fuzzy and shift over time. Moreover, ethnicity often serves to mark minority groups as different, yet majority groups also have an ethnicity, which may be based on a common heritage, language, religion, or culture. Indeed, an important aspect

of being a member of the majority is the ability to decide how, when, and even if one's ethnicity forms an overt aspect of one's identity. Finally, some scholars question whether ethnicity even exists as intrinsic qualities of a group, suggesting instead that the perception of ethnic difference arises only through contact and interaction.

Apropos of this last point is the distinction between *immigrant* and *indigenous* (sometimes called *aboriginal*) groups. Many, if not most, ethnic groups around the world originated when they migrated from their native lands and settled in a new country. In their old home, they often belonged to the host culture and were not ethnic, but when they were transplanted by relocation diffusion to a foreign land, they simultaneously became a minority and ethnic. Han Chinese are not ethnic in China (**Figure 5.5**), but if they come to North America they are. Indigenous ethnic groups that continue to live in their ancient homes become ethnic when they are absorbed into larger political states. The Navajo, for example, reside on their traditional and ancient lands and became ethnic only when the United States annexed their territory. The same is true of Mexicans, who, long resident in what is today the southwestern United States, found themselves labeled ethnic minorities when the border was moved after the 1848 U.S.–Mexican War. "We did not jump the border, the border jumped us!" is a common local comeback to this sudden shift in ethnic status.

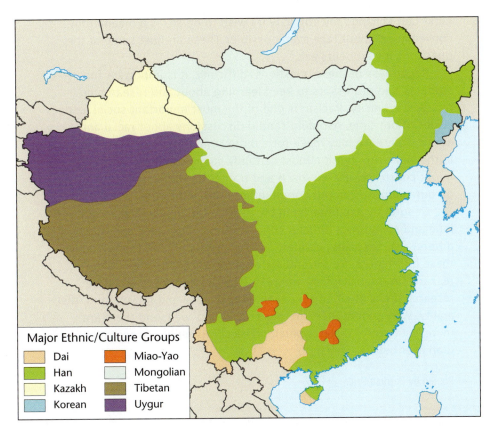

Major Ethnic/Culture Groups
- Dai
- Han
- Kazakh
- Korean
- Miao-Yao
- Mongolian
- Tibetan
- Uygur

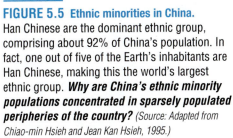

FIGURE 5.5 Ethnic minorities in China. Han Chinese are the dominant ethnic group, comprising about 92% of China's population. In fact, one out of five of the Earth's inhabitants are Han Chinese, making this the world's largest ethnic group. ***Why are China's ethnic minority populations concentrated in sparsely populated peripheries of the country?*** (Source: Adapted from Chiao-min Hsieh and Jean Kan Hsieh, 1995.)

Patricia's Notebook

Teaching Race and Ethnicity

Patricia Price

Anti-immigrant rally in Washington, D.C. This protest was organized by the Minutemen Project. Named after the American Revolutionary militiamen, this activist group attempts to police the border between the United States and Mexico. They and their supporters view this border as a source of drugs, crime, and undocumented immigration into the United States. *Critics claim that Mexicans and Mexican-Americans are stigmatized, discriminated against, and persecuted by groups like this. What do you think?* (Paul J. Richards/AFP/Getty Images.)

In my own research, I explore the changing geographies of race in the United States. In particular, I examine new immigration flows and cultural debates about immigration, race, and Latinos that have arisen in response. There is no question that Latino immigrants, particularly those who are poorest and brownest, are treated as a racial group in this country. True, the U.S. Census recognizes that "Latino" or "Hispanic" is an ethnicity and that those who identify as Latino or Hispanic can be of any (or mixed) race. But in practice most Americans, in their day-to-day engagement with race, view Latinos as one "point" in what David Hollinger has termed the ethno-racial pentagon: black, white, red, yellow, and brown.

We academics tend to teach what we research, and so I teach a great deal on race and ethnicity. Despite my affection for the topic, however, I'm the first to admit that it's hard to think of a subject that is more controversial or more difficult to teach. Students are often angry or confused about race, based on their own, often bitter, experiences. Race and ethnicity are hard to define, and their borders seem to be perennially under siege. Are Hispanics a separate race? Can I change my ethnicity if I wish to do so? Is a meaningful conversation about race even possible? These are just some of the questions my students ask.

Regardless of these questions' difficulty, it is important for students and professors to keep learning about and researching race and ethnicity. Racism is one of the most important social justice issues of our time. Avoiding candid discussions about racial and ethnic difference may make for a more congenial classroom climate in the short run, but it almost certainly stymies engagement that might constructively approach and transform the injustices of racism on our campuses, in our communities, and across the country.

Geography itself has been transformed thanks in part to our long and sometimes difficult conversations about race and other differences: sexuality, gender, and (dis)ability among them. Today these differences are the subject of some of the most exciting research in the discipline, whereas 20 years ago there was little research if any at all on sexuality, gender, race and racism, and (dis)ability. Who geographers are has changed as well, and there are more "minority"—racial, sexual, physical (dis)ability—geographers than ever before.

Posted by Patricia Price

This is not to say that ethnic minorities remain unchanged by their host culture. **Acculturation** often occurs, meaning that the ethnic group adopts enough of the ways of the host society to be able to function economically and socially. Stronger still is **assimilation,** which implies a complete blending with the host culture and may involve the loss of many distinctive ethnic traits. Intermarriage is perhaps the most effective way of encouraging assimilation. Many students of American culture long assumed that all ethnic groups would eventually be assimilated into the American melting pot, but relatively few have been; instead they use acculturation as their way of survival. The past three decades, in fact, have witnessed a resurgence of ethnic identity across the globe (see Subject to Debate, page 152).

Ethnic geography is the study of the spatial aspects of ethnicity. Ethnic groups are the keepers of distinctive cultural traditions and the focal points of various kinds of social interaction. They are the basis not only of group identity but also of friendships, marriage partners, recreational outlets, business success, and political power bases. These interactions can offer cultural security and reinforcement of tradi-

acculturation
The adoption by an ethnic group of enough of the ways of the host society to be able to function economically and socially.

assimilation
The complete blending of an ethnic group into the host society, resulting in the loss of all distinctive ethnic traits.

tion. Ethnic groups often practice unique adaptive strategies and usually occupy clearly defined areas, whether rural or urban. In other words, the study of ethnicity has built-in geographical dimensions. The geography of race is a related field of study that focuses on the spatial aspects of how race is socially constructed and negotiated. Cultural geographers who study race and ethnicity tend to always have an eye on the larger economic, political, environmental, and social power relations at work when race is involved. This chapter draws on insights and examples from both ethnic geography and the geography of race.

Region

How are ethnic groups distributed geographically? Do ethnic culture regions have a special spatial character? Formal ethnic culture regions exist in most countries (see Figure 5.5, page 153). To map these regions, geographers rely on data as diverse as probable origin of surnames in telephone directories and census totals for answers to questions on ancestry, primary racial identification, or language spoken at home. Given the cultural complexity of the real world, each method produces a slightly different map (**Figure 5.6**). Regardless of the mapping method,

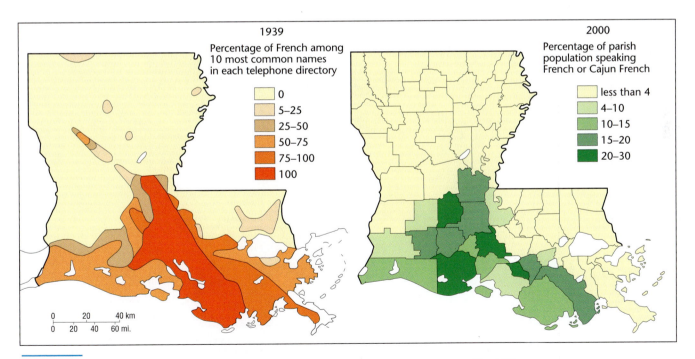

FIGURE 5.6 Acadiana, the Louisiana French homeland, as mapped by two different methods. The 1939 map was compiled by sampling the surnames in telephone directories. The 10 most common names in each directory were determined, and the percentage of these 10 that were of French origin was recorded. When no telephone directories were available, surnames on mailboxes were used. The 2000 map is based on census data for the percentage of those residing in the parish who speak French or Cajun French. *(After Meigs, 1941: 245; U.S. Bureau of the Census, 2000.)*

ethnic culture regions reveal a vivid mosaic of minorities in most countries of the world.

Ethnic Homelands and Islands

There are four types of ethnic culture regions: the rural ethnic homelands and islands, and urban ethnic neighbor-hoods and ghettos. The difference between ethnic homelands and islands is their size, in terms of both area and population. Rural **ethnic homelands** cover large areas, often have overlapping municipal borders, and

ethnic homelands
Sizable areas inhabited by an ethnic minority that exhibits a strong sense of attachment to the region and often exercises some measure of political and social control over it.

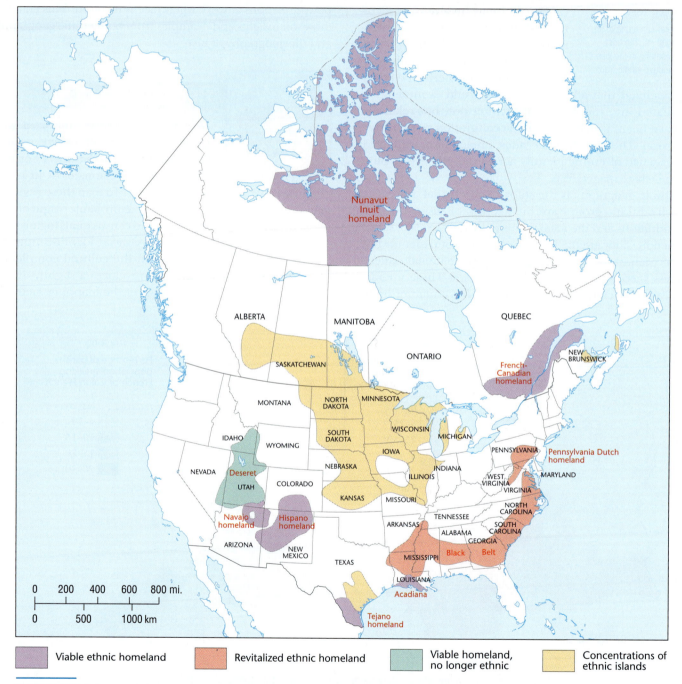

FIGURE 5.7 Selected ethnic homelands in North America, past and present, and concentrations of rural ethnic islands. The Hispano homeland is also referred to as the Spanish-American homeland. With the return migration of African-Americans from northern industrial cities such as Chicago to rural southern areas, the long-moribund Black Belt homeland appears to be undergoing a revitalization.

(Sources: Arreola, 2002; Carlson, 1990; Meinig, 1965; Nostrand and Estaville, 2001; U.S. Bureau of the Census, 2000.)

have sizable populations. Because of their size, the age of their inhabitants, and their geographical segregation, they tend to reinforce ethnicity. The residents of homelands typically seek or enjoy some measure of political autonomy or self-rule. Homeland populations usually exhibit a strong sense of attachment to the region. Most homelands belong to indigenous ethnic groups and include special, venerated places that serve to symbolize and celebrate the region—shrines to the special identity of the ethnic group. In its fully developed form, the homeland represents that most powerful of geographical entities, one combining the attributes of both formal and functional culture regions. By contrast, **ethnic islands** are small dots in the countryside, usually occupying an area smaller than a county and serving as home to several hundred to several thousand people (at most). Because of their small size and isolation, they do not exert as powerful an influence as homelands do.

> **ethnic islands**
> Small ethnic areas in the rural countryside; sometimes called folk islands.

North America includes a number of viable ethnic homelands (**Figure 5.7**), including Acadiana, the Louisiana French homeland now increasingly identified with the Cajun people and also recognized as a vernacular region; the Hispano or Spanish-American homeland of highland New Mexico and Colorado; the Tejano homeland of south Texas; the Navajo reservation homeland in Arizona, Utah, and New Mexico; and the French-Canadian homeland centered on the valley of the lower St. Lawrence River in Québec. Some geographers would also include Deseret, a Mormon homeland in the Great Basin of the intermontane West. Some ethnic homelands have experienced decline and decay. These include the Pennsylvania Dutch homeland, weakened to the point of extinction by assimilation, and the southern Black Belt, diminished by the collapse of the plantation-sharecrop system and the resulting African-American relocation to northern urban areas. Mormon absorption into the American cultural mainstream has eroded Mormon ethnic status, whereas nonethnic immigration has diluted the Hispano homeland. At present, the most vigorous ethnic homelands are those of the French-Canadians and south Texas Mexican-Americans.

If residents of ethnic homelands assimilate and are absorbed into the host culture, then, at the very least, a geographical residue, or **ethnic substrate,** remains. The resulting culture region, though no longer ethnic, nevertheless retains some distinctiveness, whether in local cuisine, dialect, or traditions. Thus, it differs from surrounding regions in a variety of ways. In seeking to explain its distinctiveness, geographers often discover an ancient, vanished ethnicity. For example, the Italian province of Tuscany

> **ethnic substrate**
> Regional cultural distinctiveness that remains following the assimilation of an ethnic homeland.

owes both its name and some of its uniqueness to the Etruscan people, who ceased to be an ethnic group 2000 years ago, when they were absorbed into the Latin-speaking Roman Empire. More recently, the massive German presence in the American heartland (**Figure 5.8**, page 158), now largely nonethnic, helped shape the cultural character of the Midwest, which can be said to have a German ethnic substrate.

Ethnic islands are much more numerous than homelands or substrates, peppering large areas of rural North America, as Figures 5.7 and 5.8 suggest. Ethnic islands develop because, in the words of geographer Alice Rechlin, "a minority group will tend to utilize space in such a way as to minimize the interaction distance between group members," facilitating contacts within the ethnic community and minimizing exposure to the outside world. People are drawn to rural places where others of the same ethnic background are found. Ethnic islands survive from one generation to the next because most land is inherited. Moreover, land is typically sold within the ethnic group, which helps to preserve the identity of the island. Social stigma is often attached to those who sell land to outsiders. Even so, the smaller size of ethnic islands makes their populations more susceptible to acculturation and assimilation.

Ethnic Neighborhoods and Racialized Ghettos

Formal ethnic culture regions also occur in cities throughout the world, as minority populations initially create, or are consigned to, separate ethnic residential quarters. Two types of urban ethnic culture regions exist. An **ethnic neighborhood** is a voluntary community where people of common ethnicity reside by choice. Such neighborhoods are, in the words of Peter Matwijiw, an Australian geographer, "the results of preferences shown by different ethnic groups . . . toward maintaining group cohesiveness." An ethnic neighborhood has many benefits: common use of a language other than that of the majority culture, nearby kin, stores and services specially tailored to a certain group's tastes, the presence of employment that relies on an ethnically based division of labor, and institutions important to the group—such as churches and lodges—that remain viable only when a number of people live close enough to participate frequently in their activities. Miami's orthodox Jewish population clusters in Miami Beach–area neighborhoods in part because the proximity of synagogues and kosher food establishments makes religious observance far easier than it would be in a neighborhood that did not have a sizable orthodox Jewish population.

> **ethnic neighborhood**
> A voluntary community where people of like origin reside by choice.

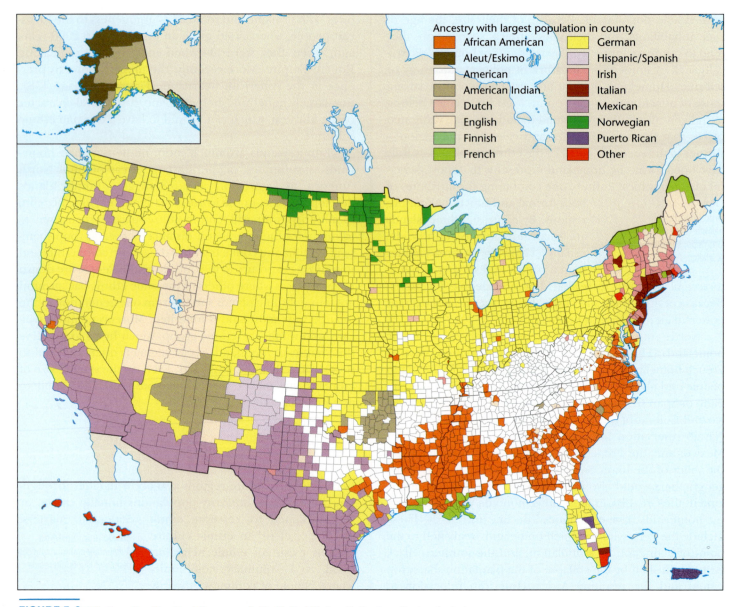

FIGURE 5.8 Ethnic and national-origin groups in the United States. Notice how the border between Canada and the United States generally also forms a cultural boundary. Several ethnic homelands appear on this map, as do many ethnic islands. *(Sources: Wikipedia/U.S. Census Bureau.)*

ghetto
Traditionally, an area within a city where an ethnic group lives, either by choice or by force. Today in the United States, the term typically indicates an impoverished African-American urban neighborhood.

The second type of urban ethnic region is a **ghetto.** Historically, the term dates from thirteenth-century medieval Europe, when Jews lived in segregated, walled communities called ghettos (**Figure 5.9**). Ethnic residential quarters have, in fact, long been a part of urban cultural geography. In ancient times, conquerors often forced the vanquished native population to live in ghettos. Religious minorities usually received similar treatment. Islamic cities, for example, had Christian dis-tricts. If one abides by the origin of the term, ghettos need to be neither composed of racialized minorities nor impoverished. Typically, however, the term *ghetto* is commonly used in the United States today to signal an impoverished, urban, African-American neighborhood. A related term, *barrio,* refers to an impoverished, urban, Hispanic neighborhood. Ghettos and barrios are as much functional culture regions as formal ones.

Coinciding with the urbanization and industrialization of North America, ethnic neighborhoods became typical in the northern United States and in Canada about 1840.

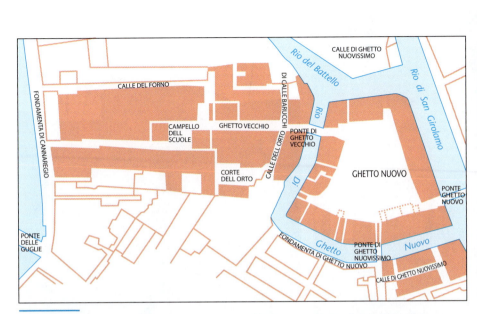

FIGURE 5.9 **Venetian ghetto.** In the sixteenth century, Venice's Jewish population lived in a segregated, walled neighborhood called a ghetto. On the left is a map of this early ghetto. Though most of Venice's Jews do not reside in the ghetto today (*right*), many attend religious services there, and the ghetto continues to be at the heart of Venetian Jewish life. *(Left: Adapted from the Jewish Museum of Venice; Right: LusoItaly/Alamy.)*

Instead of dispersing throughout the residential areas of the city, immigrant groups clustered together. To some degree, ethnic groups that migrated to cities came from different parts of Europe from those who settled in rural areas. Whereas Germany and Scandinavia supplied most of the rural settlers, the cities attracted those from Ireland and eastern and southern Europe. Catholic Irish, Italians, and Poles, along with Jews from eastern Europe, became the main urban ethnic groups, although lesser numbers of virtually every nationality in Europe also came to the cities of North America. These groups were later joined by French-Canadians, southern blacks, Puerto Ricans, Appalachian whites, Native Americans, Asians, and other non-European groups.

Regardless of their particular history, the neighborhoods created by ethnic migrants tend to be transitory. As a rule, urban ethnic groups remain in neighborhoods while undergoing acculturation. As a result, their central-city ethnic neighborhoods experience a life cycle in which one group is replaced by another, later-arriving one. We can see this process in action in the succession of groups that resided in certain neighborhoods and then moved on to more desirable areas. The list of groups that passed through one Chicago neighborhood from the nineteenth century to the present provides an almost complete history of American migra-

tory patterns. First came the Germans and Irish, who were succeeded by the Greeks, Poles, French-Canadians, Czechs, and Russian Jews, who were soon replaced by the Italians. The Italians, in turn, were replaced by Chicanos and a small group of Puerto Ricans. As this succession occurred, the established groups had often attained enough economic and cultural capital to move to new areas of the city. In many cities, established ethnic groups moved to the suburbs. Even when ethnic groups relocated from inner-city neighborhoods to the suburbs, residential clustering survived (**Figure 5.10**, page 160). The San Gabriel Valley, about 20 miles (32 kilometers) from downtown Los Angeles, has developed as a major Chinese suburb. These suburban ethnic neighborhoods, called **ethnoburbs,** can house relatively affluent immigrant populations. Ethnic neighborhoods will receive additional attention in Chapter 11.

ethnoburbs
A suburban ethnic neighborhood, sometimes home to relatively affluent immigration populations.

Recent Shifts in Ethnic Mosaics

In the United States, immigration laws have changed during the past 40 years, shifting in 1965 from the quota system based on national origins to one that allowed a certain number of immigrants from the Eastern and Western hemispheres, as well as giving preference to certain

FIGURE 5.10 Two Chinese ethnic neighborhoods. The image on the left depicts Los Angeles's traditional urban Chinatown, which is a popular tourist destination as well. The image on the right illustrates the suburban location and flavor of new ethnoburbs. This image is from the San Gabriel Valley, outside Los Angeles. *(Left: Susan Seubert/drr.net; Right: Ron Lim.)*

categories of migrants, such as family members of those already residing in the United States. These changes, and the rising levels of undocumented immigration, have led to a growing ethnic variety in North American cities.

Asia and Latin America, rather than Europe, is now the principal source of immigrants to North America. In 2010, Mexico alone supplied 23.7 percent of immigrants to the United States; China, the Philippines, India, and Vietnam were next, with a combined total of 15.9 percent. Asian and Pacific Islanders are now the fastest-growing group in the United States, projected to grow from only 4 percent of the U.S. population today to 10 percent by 2050. People of Japanese ancestry form the largest national-origin group in Hawaii, and Washington State elected the first Chinese-American governor in the country's history in 1996. Many West Coast cities, from Vancouver to San Diego, have acquired very sizable Asian populations. Vancouver, already 11 percent Asian in 1981, has since absorbed many more Asian immigrants, particularly from Hong Kong, which again became part of China in 1997. The 2006 Canadian census shows Chinese are by

far the largest non-European ethnic group in Vancouver, accounting for 29.4 percent of that city's visible minority population. In the United States, the West Coast is home to about 40 percent of the Asian population, mostly in California, whereas the urban corridor that stretches from New York City to Boston houses another concentration (**Figure 5.11**). Spatially speaking, Asians are less segregated than African-American or Hispanic populations.

Latin America, including the Caribbean countries, has also surpassed Europe as a source of immigrants to North America (**Figure 5.12**, page 162). The two largest national-origin groups of Hispanic immigrants—Mexicans and Cubans—are still the most prevalent, but they have been supplemented since the 1980s by waves of Central American and, increasingly, South American arrivals. East Coast cities have absorbed large numbers of immigrants from the West Indies. The two largest national-origin groups coming to New York City as early as the 1970s were from the Dominican Republic and Jamaica, displacing Italy as the leading source of immigrants. For the United States as a whole, in 2002, Latinos narrowly surpassed African-Americans as the largest

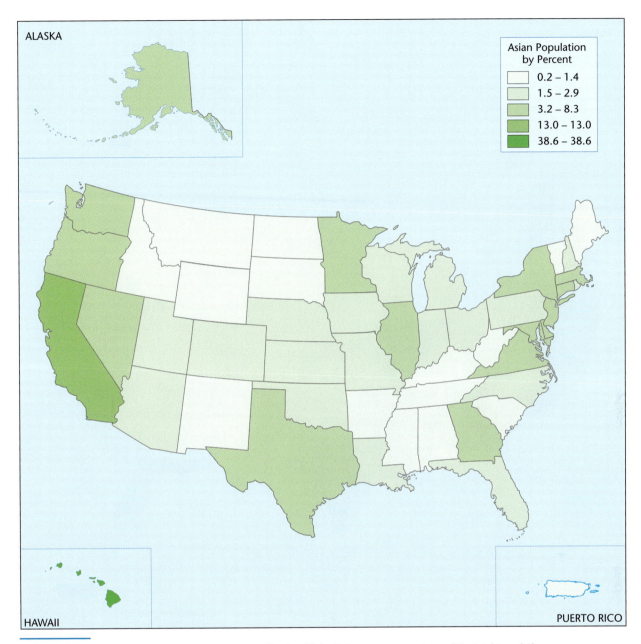

FIGURE 5.11 **Asian population by state.** People indicating "Asian" alone as a percentage of the total population by state. *(Source: U.S. Bureau of the Census, 2010.)*

ethnic group, after non-Hispanic whites. In some cities that are popular destinations for Hispanic immigrants, Hispanics constitute majority populations (**Table 5.1**, page 162). The Latino influx into the United States has rewoven significant portions of the cultural mosaic, influencing food preferences, music, fashion, and language.

As demographer William Frey has noted, ethnic populations continue to concentrate in established ports of entry, most of them in coastal locations. There is a large swath of the United States that has remained largely non-Hispanic white (**Figure 5.13**, page 163). **Figure 5.14**

(page 164) shows that black Americans, too, remain spatially concentrated. The 2010 census reported that 79% of whites live in white-majority neighborhoods, while 44% of African-Americans live in black-majority neighborhoods and 45% of Hispanics reside in Hispanic-majority neighborhoods. As a nation, we may be becoming more diverse—a more colorful mosaic—but we are nowhere near the melting pot we sometimes portray ourselves to be.

Although the United States may boast of its immigrant background, only one U.S. city—Miami—ranked in the top 10 list of the world's cities that have significant proportions

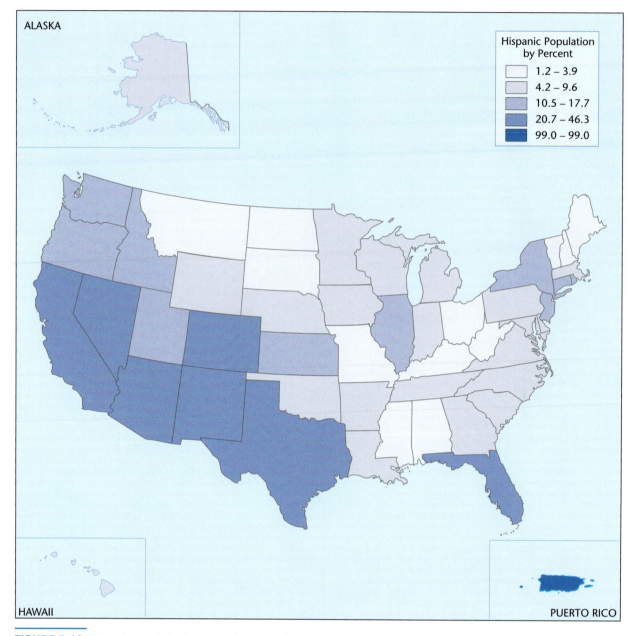

FIGURE 5.12 Hispanic population by state. Hispanic or Latino population as a percentage of the total population by state. "Hispanic" is an ethnic category and can encompass several racial designations. *(Source: U.S. Bureau of the Census, 2010.)*

TABLE 5.1 U.S. Cities of 100,000 or More with the Highest Ethnic Concentrations

	Largest Concentration	Second-Largest Concentration	Third-Largest Concentration
Black	Gary, Indiana (84%)	Detroit, Michigan (81.6%)	Birmingham, Alabama (73.5%)
White	Livonia, Michigan (95.5%)	Cape Coral, Florida (93%)	Boise, Idaho (92.2%)
Hispanic	East Los Angeles, California (96.8%)	Laredo, Texas (94.1%)	Brownsville, Texas (91.3%)
Asian	Honolulu, Hawaii (55.9%)	Daly City, California (50.7%)	Fremont, California (37%)

The figures for black, white, and Asian are for that category alone, not in combination with other races. The figures for Hispanic can include any racial designation.
(Source: U.S. Bureau of the Census, 2000.)

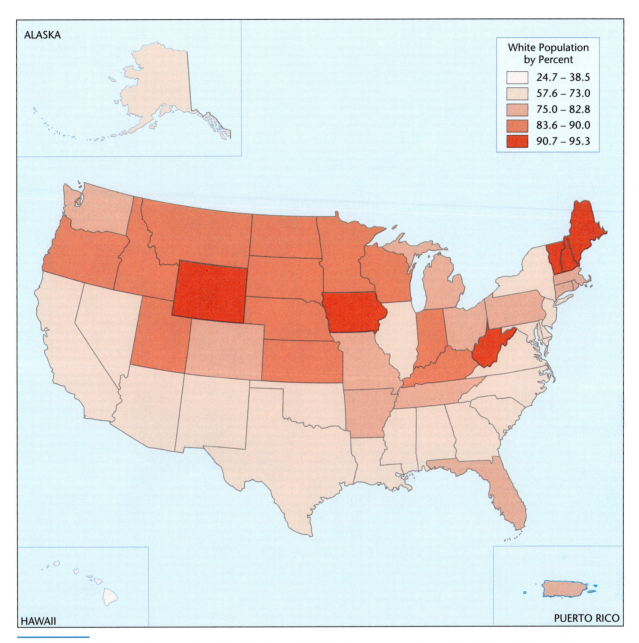

ALASKA

White Population
by Percent

	24.7 – 38.5
	57.6 – 73.0
	75.0 – 82.8
	83.6 – 90.0
	90.7 – 95.3

HAWAII

PUERTO RICO

FIGURE 5.13 **White population by state.** People indicating "white" alone as a percentage of the total population by state. *(Source: U.S. Bureau of the Census, 2010.)*

of foreign-born residents (**Table 5.2**, page 164). Three of the top 10 cities are located in the Middle East, three in Europe, two in Canada, and one in the Pacific. In addition, national-origin populations prevalent in the United States are even more prevalent in other countries. Twenty-eight million ethnic Chinese reside outside China and Taiwan. Most of these overseas Chinese live not in North America but in Southeast Asian countries and even Polynesia (**Figure 5.15**, page 165). Indonesia has more than 7 million, Thailand nearly 6 million, and Malaysia more than 5 million. Pacific Islanders exhibit a similar pattern. Auckland,

New Zealand, has the largest Polynesian population of any city in the world and ranks number seven in the world by percentage of foreign-born (see Table 5.2). Australia, Argentina, and Brazil also have large ethnic populations. Parts of East Africa have long been home to relatively affluent South Asian Indian populations, whereas Lebanese populations in West Africa enjoy a similarly privileged position. Even European countries such as Germany, the United Kingdom, Italy, and Spain—long known as sources rather than destinations of migrants—today are home to millions of Africans, Turks, and Asians. Immigration-based

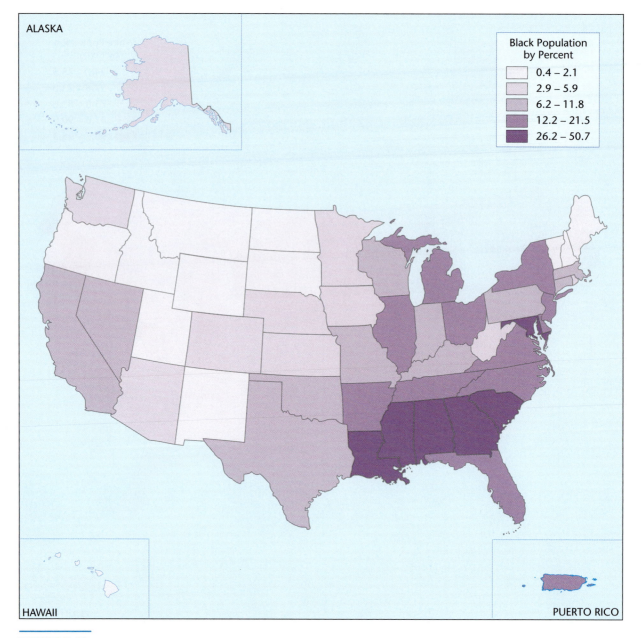

FIGURE 5.14 Black population by state. People indicating "black or African-American" alone as a percentage of the total population by state. *(Source: U.S. Bureau of the Census, 2010.)*

TABLE 5.2 Top 10 Cities by Percentage of Foreign-Born

	Percentage		Percentage
Dubai, United Arab Emirates	82	Vancouver, Canada	39
Miami, Florida	51	Auckland, New Zealand	39
Amsterdam, the Netherlands	47	Geneva, Switzerland	38
Toronto, Canada	45	Mecca, Saudi Arabia	38
Muscat, Oman	45	The Hague, the Netherlands	37

(Source: Adapted from Benton-Short, Price, and Friedman, 2005, p. 953.)

FIGURE 5.15 Store owned by a prosperous ethnic Chinese retailer on the island of Bora Bora in French Polynesia. The population of the island is overwhelmingly Polynesian. Only 0.8 percent of the people are Europeans, 6.6 percent are "Demis" (a mixture of white and Polynesian), and less than 0.5 percent are Asian. Yet this store and many others in the archipelagoes of the Pacific are owned by persons of Chinese heritage. Why don't Polynesians own such stores? Because (1) they have no tradition of retailing and (2) theirs is a communal society that shares wealth. If a Polynesian were to open a store, all of his or her relatives would have the right to come and take merchandise for free, sharing the wealth. The store would fail within a month. And so the way was left open for the Chinese, who have a very different culture. *(Courtesy of Terry G. Jordan-Bychkov.)*

ethnicity is far from being a phenomenon limited to North America (see Subject to Debate, page 152).

Mobility

How do the various reasons behind mobility help to shape the complicated geographical patterns of ethnicity and race? When groups move, how do cultural patterns reassemble in new places? As noted in the introduction to this chapter, it is often through migration that a group formerly in the majority becomes, in a new land, different from the mainstream and thus labeled as ethnic or racialized. The many motives for mobility can have different results in terms of where a group chooses to go, which parts of its original culture relocate and which do not, and who may become its neighbors in its new home.

Migration and Ethnicity

Much of the ethnic pattern in many parts of the world is the result of relocation diffusion. The migration process itself often creates ethnicity, as people leave countries where they belonged to a nonethnic majority and become a minority in a new home. *National Geographic's* Genographic Project uses DNA samples from volunteers worldwide to substantiate the claim that all humans descended from a group of Africans who began to migrate out of Africa about 60,000 years ago. The contemporary global mosaic of ethnic populations can be traced to their specific journeys. Today's voluntary migrations have also produced much of the ethnic diversity in the United States and Canada, and the involuntary migration of political and economic refugees has always been an important factor in ethnicity worldwide and is becoming ever more so in North America.

Chain migration may be involved in relocation diffusion. In chain migration, an individual or small group decides to migrate to a foreign country. This decision typically arises from negative conditions in the home area, such as political persecution or lack of employment, and the perception of better conditions in the receiving country. Often ties between the sending and receiving areas are preexisting, such as those formed when military bases of

chain migration
The tendency of people to migrate along channels, over a period of time, from specific source areas to specific destinations.

receiving countries are established in sending countries. The first emigrants, or "innovators," may be natural leaders who influence others, particularly friends and relatives, to accompany them in the migration. The word spreads to nearby communities, and soon a sizable migration is under way from a fairly small district in the source country to a comparably small area or neighborhood in the destination country (**Figure 5.16**). In village after village, the first emigrants often rank high in the local social order, so that hierarchical diffusion also occurs. That is, the *decision* to migrate spreads by a mixture of hierarchical and contagious diffusion, whereas the actual migration itself represents relocation diffusion.

involuntary migration
Also called forced migration, refers to the forced displacement of a population, whether by government policy (such as a resettlement program), warfare or other violence, ethnic cleansing, disease, natural disaster, or enslavement.

Involuntary migration also contributes to ethnic diffusion and the formation of ethnic culture regions. African slavery constituted the most demographically significant involuntary migration in human history and has strongly shaped the ethnic mosaic of the Americas. Refugees from Cambodia and Vietnam created ethnic groups in North America, as did Guatemalans and Salvadorans fleeing political repression in Central America. Often, such forced migrations may result from policies of ethnic cleansing, whereby countries expel or massacre minorities outright to produce cultural homogeneity in their populations. In Europe, the newly independent country of Croatia has systematically expelled its Serb minority in a campaign of **ethnic cleansing.** Following forced migration, the relocated group often engages in voluntary migration to concentrate in some new locality. Cuban political refugees, scattered widely throughout the United States in the 1960s, reconvened in south Florida, and Vietnamese refugees continue to gather in Southern California, Louisiana, and Texas.

ethnic cleansing
The removal of unwanted ethnic minority populations from a nation-state through mass killing, deportation, or imprisonment.

Outright warfare, too, displaces populations. During World War II, Polish populations were deported from the lands annexed into the German Reich, in an attempt to "Germanize" this region. In all, some 2 million Poles were expelled from their homes. In Iraq, more than 3 million people have been displaced by war-related violence there. **Figure 5.17** shows that about half of them remain in the country, whereas the rest have migrated across Iraq's borders, with most fleeing to neighboring nations.

Return migration represents another type of ethnic diffusion and involves the voluntary movement of a group back to its ancestral homeland or native country. The large-scale return since 1975 of African-Americans from the cities of the northern and

return migration
A type of ethnic diffusion that involves the voluntary movement of a group of migrants back to their ancestral or native country or homeland.

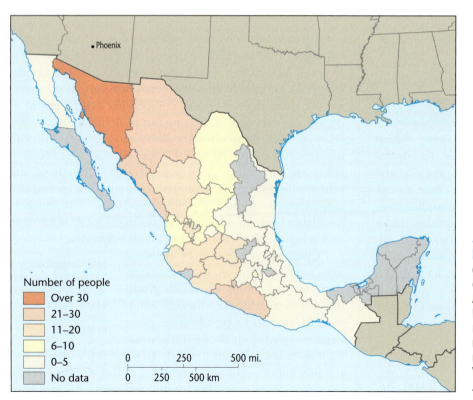

Number of people

- Over 30
- 21–30
- 11–20
- 6–10
- 0–5
- No data

0   250   500 mi.

0   250   500 km

FIGURE 5.16 Origins of migrants. This map depicts the Mexican state of origin for the residents of Garfield, an ethnic neighborhood in Phoenix, Arizona, that is composed mostly of Mexican and Mexican-American residents. Note that most of the residents trace their homeland to only a few northern Mexican states; immigrants from Mexico's more southern states, such as Oaxaca and Veracruz, are relatively rare. *(Adapted from Oberle and Arreola, 2008.)*

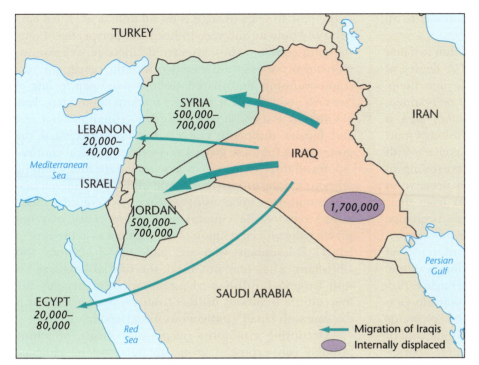

FIGURE 5.17 **Iraqis displaced by conflict.** This map shows the war-related displacement of more than 3 million Iraqis, who constitute involuntary migrants. Roughly half have left Iraq for neighboring countries. There, they may face discrimination and poverty as unwanted minority populations. In Syria, nearly one-third of Iraqi refugee children do not attend school because the infrastructure is not able to accommodate them. The UN refugee agency claims that this is the largest movement of displaced people in the Middle East since the Palestinian exodus after the establishment of Israel in 1948. *(Adapted from BBC News, 2007.)*

western United States to the Black Belt ethnic homeland in the South is one of the most notable such movements now under way. This type of ethnic migration is also channelized. Channelization is a process whereby a specific source region becomes linked to a particular destination, so that neighbors in the old place became neighbors in the new place as well. The 2010 census found that, thanks to migration from cities like Los Angeles and Philadelphia, the South is home to 57% of the black population. This proportion is higher than that at any time in the preceding 50 years. Black Americans are returning to the South, particularly to cities like Atlanta, because of greater economic opportunities there than elsewhere in the country. Cultural factors play a role as well, particularly the historic attachment to place and the perception that racial prejudice has lessened in the region. Return migration has recently revitalized the once-moribund Black homeland (see Figure 5.7, page 156).

Similarly, many of the 200,000 or so expatriate Estonians, Latvians, and Lithuanians left Russia and other former Soviet republics to return to their newly independent Baltic home countries in the 1990s, losing their ethnic status in the process. Clearly, migration of all kinds turns the ethnic mosaic into an ever-changing kaleidoscope.

Reflecting on Geography

Why might African-Americans have begun return migration to the South after 1975, and why did this movement accelerate in the 1990s?

Simplification and Isolation

When groups migrate and become ethnic in a new land, they have, in theory at least, the potential to introduce the totality of their culture by relocation diffusion. Conceivably, they could reestablish every facet of their traditional way of life in the area where they settle. However, ethnic immigrants never successfully introduce the totality of their culture. Rather, profound **cultural simplification** occurs. As geographer Cole Harris noted, "Europeans established overseas drastically simplified versions of European society." This happens, in part, because of chain migration: only fragments of a culture diffuse overseas, borne by groups from particular places migrating in particular eras. In other words, some simplification occurs at the point of departure. Moreover, far more cultural traits are implanted in the new home than actually survive. Only selected traits are successfully introduced, and others undergo considerable modification before becoming established in the new homeland. In other words, absorbing barriers prevent the diffusion of many traits, and permeable barriers cause changes in many other traits, greatly simplifying the migrant cultures. In addition, choices that did not exist in the old home become available to immigrant ethnic groups. They can borrow novel ways from those they encounter in the new land, invent new techniques better suited to the adopted place, or modify existing approaches

> **cultural simplification**
> The process by which immigrant ethnic groups lose certain aspects of their traditional culture in the process of settling overseas, creating a new culture that is less complex than the old.

as they see fit. Most immigrant ethnic groups resort to all these devices in varying degrees.

The displacement of a group and its relocation to a new homeland can have widely differing results. The degree of isolation an ethnic group experiences in the new home helps determine whether traditional traits will be retained, modified, or abandoned. If the new settlement area is remote and contacts with outsiders are few, diffusion of traits from the sending area is more likely. Because contacts with groups in the receiving area are rare, little borrowing of traits can occur. Isolated ethnic groups often preserve in archaic form cultural elements that disappear from their ancestral country; that is, they may, in some respects, change less than their kinfolk back in the mother country.

Language and dialects offer some good examples of this preservation of the archaic. Germans living in ethnic islands in the Balkan region of southeastern Europe preserve archaic South German dialects better than do Germans living in Germany itself, and some medieval elements survive in the Spanish spoken in the Hispano homeland of New Mexico. The highland location of Taiwanese aboriginal peoples helped maintain their archaic Formosan language, belonging to the Austronesian family, despite attempts by Chinese conquerors to acculturate them to the Han Chinese culture and language.

Globalization

What is the future of ethnicity and race? Will the potent forces of globalization erase ethnic differences and wipe out racism? Or do ethnicity and race provide enduring constructs that predate—and will outlast—the global era? The perception of differences among human groups, whether based in visible physical traits, cultural practices, language differences, or religion, is probably as old as human civilization itself. For at least a century, however, the demise of ethnic groups has been predicted. The idea of capitalist America as a melting pot has been used to describe the process wherein the mixing of diverse peoples would eventually absorb everyone into American mainstream culture. In communist lands, Marxist doctrine preached that racism would vanish in the golden age of socialist egalitarianism. The fact that ethnic differences and racial strife still persist—and in some instances have worsened—points to their deeply rooted nature.

A Long View of Race and Ethnicity

In keeping with the notion that globalization is a process that has unfolded over several centuries, as opposed to only in the past 40 years or so, it is useful to look at global concepts of ethnicity and race in historical perspective. Long-standing cohesion through shared language, religion, or ethnicity provides group members with the perception of a common history and shared destiny—that deep feeling of "we-ness"—that is the basis of the modern nation-state. Israel's identity as a Jewish nation, Turkey's common language, and Japan's distinctive cultural traditions facilitated the groundwork for these groups to cohere as political entities.

Yet all three of these nation-states harbor long-standing tensions among groups that reside within the national borders but who are not considered (or do not consider themselves) part of the cultural mainstream. They may not practice the official religion of the country, as in the case of Israel's Palestinians, who are made up of Muslims and Christians. They may not speak the official language, as with Turkey's Kurdish-speaking minority. Or they may follow cultural traditions distinct from the mainstream, as is the case with Japan's indigenous Ainu population.

Occupying a minority religious, linguistic, or ethnic position can expose a group to persecution. Violence against racial or ethnic minorities dates back to ancient times. The persecution of Jews, for example, began with pogroms, or widespread anti-Semitic rioting, during the Roman Empire some two millennia ago. Across medieval Europe, violence against Jews was episodic, occurring in Spain in the eleventh century, England in the twelfth century, Germany in the fourteenth century, and Russia in the nineteenth century. Anti-Semitic hate crimes still occur today throughout Europe and the United States. Yet when Israel was founded in 1948 as the only nation-state to have Judaism as its official religion, non-Jewish minorities were, in turn, marginalized. Even minority Jewish populations, such as the Mizrahim (Arab) Jews residing in Israel, "are excluded and marginalized not only from social and political centers, but . . . from the very definition of what it is to be 'Israeli'," claims Israeli anthropologist Pnina Motzafi-Haller. Thus, it appears that no place or time is free of tensions based in ethnic or racialized difference.

Reflecting on Geography

Why do you think racialized and ethnic differences are so deeply rooted in human cultures? What—if anything—might cause them to disappear?

Race and European Colonization

Europe's colonization of vast territories in Africa, Asia, and Latin America was predicated on the drawing of sharp distinctions between the colonizers and the colonized. These distinctions were often depicted in racial terms. Indeed, conquered peoples were viewed as so different

FIGURE 5.18 La Leyenda Negra. Illustrations such as this gruesome depiction of Spanish conquistadores hanging and roasting indigenous people, while dashing their babies against a wall, were used to construct La Leyenda Negra, or "The Black Legend." La Leyenda Negra was circulated in the sixteenth century by the Dutch and the British, Spain's religious and colonial rivals. They wished to depict the Spaniards as cruel barbarians who had no moral business building empires in the New World. *(akg-images.)*

from Europeans that their very status as human beings was debated. The famous exchanges in the mid-sixteenth century between Caribbean-based Spanish priests and intellectuals in Spain is a case in point. Observing the abuses of indigenous populations firsthand led to an outcry by these clerics, who argued that the subjugated peoples were not animals and should not be subject to such treatment. "Are these not men? Have they not rational souls? Must you not love them as you love yourselves?" asked Antonio de Montesinos in 1511, in a sermon delivered on the island of Hispaniola (**Figure 5.18**). These same clerics, however, saw no problem with owning black slaves, simply because they did not view them as human beings.

European colonialism frequently drew on existing ethnic and racial cleavages in colonized societies. Rwanda provides a tragic example of this practice. Rwanda is composed of two major ethnic groups: the Hutu, who, at 85 percent of the population, constitute the majority, and the minority Tutsi. Although differences in status have long existed between the two groups, it was not until the Belgian takeover of the territory in 1918 that the distinction between Hutu and Tutsi became racialized. Tutsis were given positions of power in the colonial government, educational system, and economic structure of colonial Rwanda under Belgian rule (**Figure 5.19**). Hutus' physical differences from Tutsis were examined under a European racial lens. The shorter and darker Hutus were considered ugly, intellectually inferior, and natural slaves, whereas the taller, lighter

FIGURE 5.19 Identification card of a Rwandan Tutsi. In Rwanda, cards like this one were used to distinguish racialized Tutsis from Hutus. To show a card with "Tutsi" written on it was tantamount to a death sentence during the Rwandan genocide. Whether to instigate a mandatory national identification card is a hot topic in many nations faced with undocumented immigration today. Because these cards can serve to mark some groups as undesirable in some potentially dangerous way, you can understand why there is usually vigorous opposition to a national identification card. *(Antony Njuguna/Reuters/Landov.)*

FIGURE 5.20 Rwandan genocide. These skulls are the remains of some 500,000 to 1,000,000 Tutsis, and their Hutu sympathizers, who were massacred over a period of four months in 1994 by Hutu militias. (Peter Van Agtmael/Polaris.)

Tutsis were extolled for their physical beauty and virtues of cultural refinement and leadership.

In 1959, the oppressed Hutu majority rebelled, leading to the deaths of some 20,000 Tutsis and the displacement of many more. Rwanda became independent from Belgium in 1961. Displaced Tutsi refugees regrouped in neighboring Uganda and in 1990 launched an invasion of Rwanda, demanding an end to racial discrimination and a reinstatement of their citizenship. Rapid deterioration of Hutu-Tutsi relations ensued, and plans to exterminate Tutsis and their Hutu sympathizers were promoted as the only

genocide
The systematic killing of a racial, ethnic, religious, or linguistic group.

way to rid Rwanda of its problems. In 1994, this **genocide** was conducted, resulting in the massacre of as many as 1 million people in the span of a few months (**Figure 5.20**).

As these examples illustrate, ethnic and racial distinctions provided central features of European colonization and the rise of nation-states. Yet, as we have seen, tensions based on race and ethnicity predate the modern era. Racial and ethnic conflict is far from erased from the face of today's global world and may in fact become aggravated as the bonds of the nation-state are loosened (see Subject to Debate, page 152).

Indigenous Identities in the Face of Globalization

Differences do not always cause conflict, tension, and oppression. Indeed, pride in one's ethnic or racial distinctiveness is common. Yet as revolutions in communications

and transportation bring diverse peoples into heightened contact with one another the world over, distinctive markers of ethnicity come under pressure. Exposure to seductive U.S.-based cultural practices—in music, cuisine, fashion, lifestyles, and values—is thought to encourage people to drop their distinctive ways in favor of adopting a homogeneous modern, Westernized culture.

Late in the twentieth century, however, an ethnic resurgence, especially among indigenous groups, became evident in many countries around the world. In a very real way, many ethnic groups and their geographical territories have become bulwarks of resistance to globalization. Yet it is too simplistic to romanticize indigenous groups as simply the keepers of ethnic distinctiveness in a globalizing world (**Figure 5.21**).

FIGURE 5.21 Native Huichol girl. Most contemporary indigenous peoples do not wear head-to-toe ethnic garb on a daily basis. Those who live in cities or regularly interact with the majority society may adopt Western-style dress full time. Others, perhaps those who live in smaller villages or those who do not interact often with the majority society, may blend indigenous and Western styles. Some indigenous peoples view indigenous clothing as a mark of ethnic pride and wear it daily regardless of their place of residence or level of interaction with the majority society. This Huichol girl blends a baseball cap with traditional dress, itself a blend of European and indigenous materials and styles. The Huicholes are residents of western Mexico. (Susana Valadez/Huichol Center.)

Members of indigenous groups must interpret the challenges as well as the opportunities offered by globalization and can be expected to modify their identities in complex ways.

In the Andean region of South America, for example, indigenous populations have both interpreted and resisted globalization. The Andes are a long, high mountain range that runs parallel to South America's Pacific coast. Their highland valleys, called *altiplano*, have been home to dense indigenous populations for millennia. Lynn Meisch describes the preservation of traditional forms of dress among the highland peoples of central Ecuador's Otavalo Valley. On conquest, the indigenous men were quickly put to work by the Spaniards in weaving workshops. Because they no longer had time to weave their family's clothing on traditional backstrap looms, and because they were legally prohibited from dressing like whites, they developed a distinctive style of dress in the sixteenth century. Although these garments are produced on European-style treadle looms rather than by traditional methods, their form and color mark those who wear them as *indígenas*, or natives, of the Otavalo Valley region.

Dress styles, like all other things cultural, change and evolve over time through conquest, diffusion of new techniques, and the whims of fashion. For Ecuador's indigenous highlanders, conquest by the Incas in 1495 meant a shift from the *manta*, or wrap, for men to the *camiseta*, or shirt. For women, typical dress has evolved from the *anaco*, a square cotton wrap held closed at the shoulders by copper or silver pins and bound at the waist by a belt called a *faja*, described by the Spaniard Sancho Paz Ponce just after Spanish conquest of the region in 1582. Today, women dress quite conservatively and retain many pre-Hispanic elements, including the red backstrap-woven *mama chumbi*, or mother belt, which is held shut by the *wawa chumbi*, or baby belt.

Meisch notes that community pressure discourages drastic change in styles of dress. "In the summer of 1989 a young woman wearing a green *anaku* [wrap skirt] was met by stares and murmurs of disapproval from other *indígenas* as she walked in Otavalo." Young men who appear in public without poncho and hat are referred to by older men as *lluchu*, meaning "naked." Meisch quotes 19-year-old Breenan Conteron: "When I leave my village to visit other cities in my country of Ecuador I always wear this costume because in this way I value and respect my ancestors, who fought to maintain their culture, traditions and customs. And I am proud that through my inheritance and in my blood I am a bearer of this culture." So much for globalization eradicating tradition! In this case, wearing traditional clothing has become a visible marker of ethnic pride.

Nature-Culture

How do ethnic groups interact with their habitats? Is there a special bond between ethnic groups and the land they inhabit that helps to form their self-identity? Do ethnic groups find shelter in certain habitats? Are the areas inhabited by racialized groups targets of environmental racism? Ethnicity is very closely linked to the nature-culture theme. The possibilistic interplay between people and physical environment is often evident in the pattern of ethnic culture regions, in ethnic migration, and in ethnic persistence or survival. Ethnicity and race are sometimes also linked to the environment in harmful ways, particularly when the places they inhabit become the targets of pollution.

Cultural Preadaptation

For those ethnic groups created by migration or relocation diffusion, the concept of cultural preadaptation provides an interesting approach. **Cultural preadaptation** involves a complex of adaptive traits possessed by a group in advance of migration that gives them the ability to survive and a competitive advantage in colonizing the new environment. Most often, preadaptation occurs in groups migrating to a place environmentally similar to the one they left behind. The adaptive strategy they had pursued before migration works reasonably well in the new home.

> **cultural preadaptation**
> A complex of adaptive traits and skills possessed in advance of migration by a group, giving it survival ability and competitive advantage in occupying the new environment.

The preadaptation may be accidental, but in some cases the immigrant ethnic group deliberately chooses a destination area that physically resembles their former home. In Africa, the Bantu expansion mentioned in Chapter 4 was probably initially driven by climate change and expansion of the Sahara (see Figure 4.6, page 122). The Bantu spread south and west in search of forested lands similar to those they had previously inhabited. But their progress southward was finally inhibited because their agricultural techniques and cattle were not adapted to the drier Mediterranean climate of southern Africa.

The state of Wisconsin, dotted with scores of ethnic islands, provides some fine examples of preadapted immigrant groups that sought environments resembling their homelands. Particularly revealing are the choices of settlement sites made by the Finns, Icelanders, English, and Cornish who came to Wisconsin (**Figure 5.22**, page 172). The Finns—coming from a cold, thin-soiled, glaciated, lake-studded, coniferous forest zone in Europe—settled the North Woods of Wisconsin, a land similar in almost every

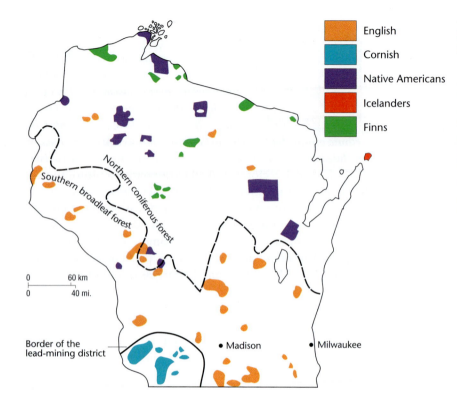

English
Cornish
Native Americans
Icelanders
Finns

FIGURE 5.22 **The ecology of selected ethnic islands in Wisconsin.** Notice that Finnish settlements are concentrated in the infertile North Woods section, as are the Native American reservations. The Finns went there by choice, and the Native Americans survived there because few white people were interested in such land. The English, by contrast, are found more often in the better farmland south of the border of the North Woods. Some of the English were miners from Cornwall, and they were drawn to the lead-mining country of southwestern Wisconsin, where they could practice the profession already known to them. Icelanders, an island people, chose an island as their settlement site in Wisconsin. *(After Hill, 1942.)*

respect to the one from which they had migrated. Icelanders, from a bleak, remote island in the North Atlantic, located their only Wisconsin colony on Washington Island, an isolated outpost surrounded by the waters of Lake Michigan. The English, accustomed to good farmland, generally founded ethnic islands in the better agricultural districts of southern and southwestern Wisconsin. Cornish miners from the Celtic highlands of Cornwall in southwestern England sought out the lead-mining communities of southwestern Wisconsin, where they continued their traditional occupation.

Such ethnic niche-filling has continued to the present day. Cubans have clustered in southernmost Florida, the only part of the U.S. mainland to have a tropical savanna climate identical to that in Cuba, and many Vietnamese have settled as fishers on the Gulf of Mexico, especially in Texas and Louisiana, where they could continue their traditional livelihood. Yet historical and political patterns, as well as the factors driving chain migration discussed earlier, are also at work in these contemporary patterns of ethnic clustering. They act to temper the influence of the physical environment on ethnic residential selection, making it only one of many considerations.

This deliberate site selection by ethnic immigrants represents a rather accurate environmental perception of the new land. As a rule, however, immigrants tend to perceive the ecosystem of their new home as more like that of their abandoned native land than is actually the case. Their per-

ceptions of the new country emphasize the similarities and minimize the differences. Perhaps the search for similarity results from homesickness or an unwillingness to admit that migration has brought them to a largely alien land. Perhaps growing to adulthood in a particular kind of physical environment inhibits one's ability to perceive a different ecosystem accurately. Whatever the reason, the distorted perception occasionally caused problems for ethnic farming groups (for examples, see Chapter 8). A period of trial and error was often necessary to come to terms with the New World environment. Sometimes crops that thrived in the old homeland proved poorly suited to the new setting. In such cases, **cultural maladaptation** is said to occur.

cultural maladaptation
Poor or inadequate adaptation that occurs when a group pursues an adaptive strategy that, in the short run, fails to provide the necessities of life or, in the long run, destroys the environment that nourishes it.

Reflecting on Geography

Can you think of examples of cultures that were particularly maladapted to their new surroundings and experienced spectacular ecological failures as a result?

Habitat and the Preservation of Difference

Certain habitats may act to shelter and protect ethnically or racially distinct populations. The high altitudes and rugged terrain of many mountainous regions make these areas relatively inaccessible and thus can provide refuge for minority populations while providing a barrier to outside influences. As we saw in the previous chapter on language, mountain dwellers sometimes speak archaic dialects or preserve their unique tongues thanks to the refuge provided by their habitat. In more general terms, the ways of life—including language—that are associated with ethnic distinctiveness are often preserved by a mountainous location.

The ethnic patchwork in the Caucasus region, for example, persists thanks in no small measure to the mountainous terrain found there. As **Figure 5.23** shows, distinct groups occupy the rugged landscape of valleys, plateaus, peaks, foothills, steppes, and plains in this region. Religions and languages overlap in complex ways with ethnic identities and are made even more confounding by the geopolitical boundaries in this region, which do not necessarily follow the contours of the Caucasian ethnic mosaic. Thus, although this is one of the most ethnically diverse places on Earth, the Caucasus region has seen more than its fair share of conflict as well, much of it ethnically motivated.

Islands, too, can provide a measure of isolation and protection for ethnically or racially distinct groups. The Gullah, or Geechee, people are descendants of African slaves brought to the United States in the eighteenth and nineteenth centuries to work on plantations. Many of their nearly 10,000 descendants today inhabit coastal islands of South Carolina, Georgia, and north Florida. Their island location has allowed much of the original African cultural roots to be preserved, so much so that the Gullah are sometimes called the most African-American community in the

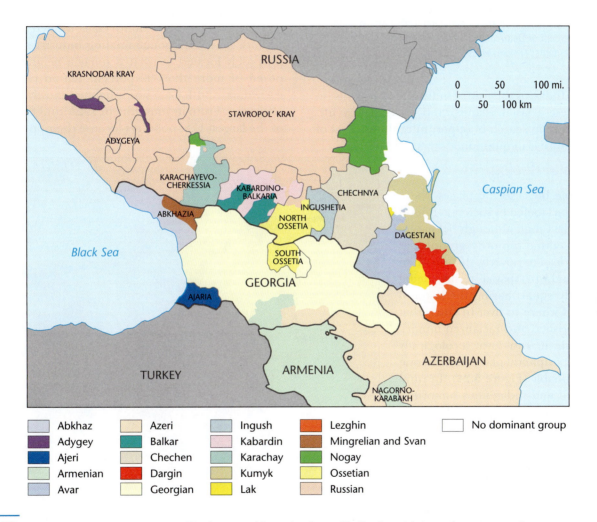

FIGURE 5.23 Ethnic pluralities in the Caucasus. The Caucasus Mountains, located in Southwest Asia, are home to one of the world's most ethnically diverse populations. They are also the site of much ethnically based conflict. Because ethnic territories often overlap, this map depicts pluralities. The populations shown comprise at least 40 percent of that place's ethnic population and in most cases also constitute a majority population, although some of the more heterogeneous areas have no dominant ethnic population. The racialized term *Caucasian* is derived from this area's name, although, in fact, it has little to do with the people actually living in this region. (Adapted from O'Loughlin et al., 2007.)

FIGURE 5.24 Arab slum in Delhi, India. This ethnic neighborhood is home to low-income Arabs, who are an ethnic minority population in India. Nearly 50 percent of Delhi's residents live in slums or unauthorized settlements. *(Viviane Dalles/REA/Redux.)*

United States. Today, as the tourist industry sets it sights on developing these coastal islands, and as Gullah youth migrate elsewhere in search of opportunity, the survival of this unique culture is in question.

Environmental Racism

Not only people but also the places where they dwell can be labeled by society as minority. And it is a fact that, although belonging to a racial or ethnic minority group does not necessarily lead to poverty, the two often go hand-in-hand. In spatial terms, being poor frequently equals having the last and worst choice of where to live. In cities, that means impoverished and racialized minorities often reside in run-down, ecologically precarious, or peripheral places that no one else wants to inhabit (**Figure 5.24**). In rural areas, racialized and indigenous minorities work the smallest and least fertile lands, or—more commonly—they work the plots of others, having become dispossessed of their own lands.

environmental racism
The targeting of areas where ethnic or racial minorities live with respect to environmental contamination or failure to enforce environmental regulations.

Environmental racism refers to the likelihood that a racialized minority population inhabits a polluted area. As Robert Bullard asserts, "Whether by conscious design or institutional neglect, communities of color in urban ghettos, in rural 'poverty pockets,' or on economically impoverished Native American reservations face some of the worst environmental devastation in the nation." In the United States, Hispanics, Native Americans, and blacks are more likely than whites to reside in places where toxic wastes are dumped, polluting industries are located, or environmental legislation is not enforced. Inner-city populations, which in many U.S. cities are disproportionately made up of minorities, occupy older buildings contaminated with asbestos and toxic lead-based paints. Farmworkers, who in the United States are disproportionately Hispanic, are exposed to high levels of pesticides, fertilizers, and other agricultural chemicals. Native American reservation lands are targeted as possible sites for disposing of nuclear waste.

As **Figure 5.25** depicts, there is an overlap of nonwhite populations and toxic release facilities that is difficult to explain away as simply a random coincidence. Rather, areas such as the city of South Gate, located in Los Angeles County, have seen contaminating industries and activities purposely located in what is today a neighborhood composed of more than 90 percent Latino residents. In its early days, South Gate was inhabited mostly by white non-Hispanics. From its founding in 1923 until the 1960s, some of the largest U.S. polluters located their operations there:

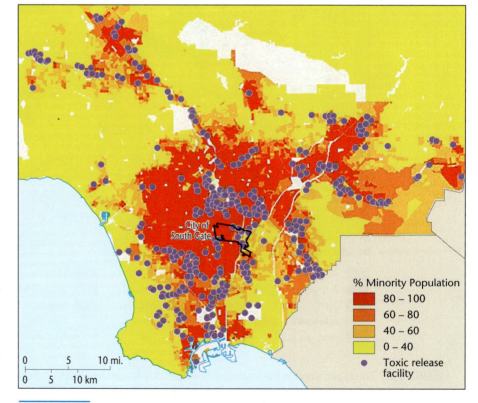

% Minority Population
- 80 – 100
- 60 – 80
- 40 – 60
- 0 – 40
- Toxic release facility

FIGURE 5.25 Environmental racism in Los Angeles. The map of Los Angeles County's nonwhite population concentrations overlaps closely with the map of toxic release facility locations in Los Angeles County. It is hard to believe that this is simply a coincidence. *(Sources: Adapted from Toxic Release Inventory [U.S. EPA], 1996; U.S. Bureau of the Census, 1990.)*

Firestone Tire and Rubber Company, A. R. Maas Chemical Company, Bethlehem Steel, Alcoa Aluminum, and hundreds of smaller operations were allowed to set up shop in a period when few were aware of the health consequences of their activities. These industries attracted labor arriving from Mexico. During this period, the neighborhood was encircled by four major freeways and a flight path was routed overhead. Pollution resulting from the transportation-related fossil fuels added to the industrial waste in this community. As the deleterious health impacts of these environmental factors became more evident, white residents began to move away, being replaced by Hispanics, some 30 percent of whom are believed to be recent undocumented immigrants. Although many of the industries are now closed, their contaminants remain in the soil and water.

Racialized minority populations are not powerless in the face of environmental racism. Organizations such as Communities for a Better Environment work with local residents, such as those living in South Gate, to promote research, legal assistance, and activism. *Environmental justice* is the general term for the movement to redress environmental racism.

Environmental racism extends to the global scale. Industrial countries often dispose of their garbage, industrial waste, and other contaminants in poor countries. Industries headquartered in wealthy nations often outsource the dangerous and polluting phases of production to countries that do not have—or do not enforce—strict environmental legislation. Whether areas inhabited by poor, ethnic, or racialized minorities are deliberately targeted for unhealthy conditions, or whether they are simply the unfortunate recipients of the effects of larger detrimental processes, the connections among race, ethnicity, and the environment pose central questions for cultural geographers.

CULTURAL LANDSCAPE

What traces do ethnicity and race leave on the landscape? Ethnic and racialized landscapes often differ from mainstream landscapes in the styles of traditional architecture, in the patterns of surveying the land, in the distribution of houses and other buildings, and in the degree to which they "humanize" the land. Often the imprint is subtle, discernible only to those who pause and look closely. Sometimes it is quite striking, flaunted as an **ethnic flag:** a readily visible marker of ethnicity on the landscape that strikes even the untrained eye (**Figure 5.26**). Sometimes the distinctive markers of race and ethnicity are not visible at all but rather are audible, tactile, olfactory, or—as we explore shortly with culinary landscapes—tasty.

ethnic flag
A readily visible marker of ethnicity on the landscape.

Urban Ethnic Landscapes

Ethnic cultural landscapes often appear in urban settings, in both neighborhoods and ghettos. A fine example is the

FIGURE 5.26 An "ethnic flag" in the cultural landscape. This maize granary, called a cuezcomatl, is unique to the indigenous population of Tlaxcala state, Mexico. The structure holds shelled maize. In Mexico, even the cultivation of maize long remained an indigenous trait because the Spaniards preferred wheat. (*www.photographersdirect.com.*)

FIGURE 5.27 Mexican-American exterior mural, Barrio Logan, San Diego, California. This mural bears an obvious ideological/political message and helps create a sense of this place as a Mexican-American neighborhood. See also Arreola, 1984. *(Courtesy of Terry G. Jordan-Bychkov.)*

brightly colored exterior mural typically found in Mexican-American ethnic neighborhoods in the southwestern United States (**Figure 5.27**). These began to appear in the 1960s in Southern California, and they exhibit influences rooted in both the Spanish and the indigenous cultures of Mexico according to geographer Daniel Arreola (see Practicing Geography). A wide variety of wall surfaces, from apartment house and store exteriors to bridge abutments, provide the space for this ethnic expression. The subjects portrayed are also wide-ranging, from religious motifs to political ideology, from statements about historical wrongs to ones about urban zoning disputes. Often they are specific to the site,

incorporating well-known elements of the local landscape and thus heightening the sense of place and ethnic "turf." Inscriptions can be in either Spanish or English, but many Mexican murals do not contain a written message, relying instead on the sharpness of image and vividness of color to make an impression.

Usually, the visual ethnic expression is more subtle. Color alone can connote and reveal ethnicity to the trained eye. Red, for example, is a venerated and auspicious color to the Chinese, and when they established Chinatowns in Canadian and American cities, red surfaces proliferated (**Figure 5.28**). Light blue is a Greek ethnic color, derived from

FIGURE 5.28 Two urban ethnic landscapes. (*Left*) Houses painted red reveal the addition of a Toronto residential block to the local Chinatown. (*Right*) The use of light blue trim, the Greek color, coupled with the planting of a grape arbor at the front door of a dwelling in the Astoria district of Queens, New York City, marks the neighborhood as Greek. *(Courtesy of Terry G. Jordan-Bychkov.)*

PRACTICING GEOGRAPHY

For Daniel Arreola, practicing geography is a family affair. He traces his interest in cultural geography to his grandfathers. His paternal grandfather owned a ranch outside of Escondido, California, where Professor Arreola spent several summers during his youth. "The trip to this part of Southern California from my hometown in Santa Monica was always an adventure and surely imprinted my later desire to travel. The experience of the out-of-doors in the San Diego backcountry was a wonderland of nature during the 1950s and helped cultivate my attraction to ranch living Mexican style." His grandfather on his mother's side took him on frequent walks through the neighborhood and to the beach. "As part of those wanderings, he would tell me stories about the people and places we passed through, and this experience no doubt branded me as an observer and cemented my passion for walking as exploration."

(Courtesy of Daniel Arreola.)

Daniel Arreola

Professor Arreola has written a number of books, including the award-winning *Tejano South Texas: A Mexican American Cultural Province.* In this book, he examines the physical, historical, and social aspects of south Texas that make it an ethnic homeland distinct from other areas of Mexican-American population in the United States. Professor Arreola's concern with illuminating the diverse cultural geographic patterns within Latino communities in the United States is taken up again in his newest edited collection, titled *Hispanic Places, Latino Places* (2004).

For Professor Arreola, research is actually a two-step process involving re-search *and* search. "To do any type of investigation, one must first re-search—in other words, examine and come to some understanding of what others have written about a subject." Typically, this involves reading materials in the library and in archives where written or visual records are stored. The most interesting part of the project—the search—comes next. "Search as opposed to re-search means discovery of new information, and the best way to discover is to go into the field. Geographers, since ancient times, have been entrusted with the responsibility to describe the surface of the Earth." Typically, he uses the case study method, which he describes as "an empirical inquiry that enables me to examine a contemporary place in its real-life context, and where the boundaries between the place I am studying and its geographical context are not clearly evident. As part of my case study method, I depend on diachronic inquiry, or understanding of past situations relevant to the place I am studying, because to understand a contemporary place almost always necessitates an appreciation of a past for that place." Professor Arreola does interviews, draws maps, and takes photographs of the places where he works.

His interest in the way places look and how places change over time has shaped Professor Arreola's current project. He began to collect historical postcards of the borderlands region between Mexico and the United States about 15 years ago, and now he has thousands of them. He has visited the places where these pictures were originally taken and rephotographed about 100 of them from the same perspectives as the originals in order to assess how these border places have changed during three separate time periods: the 1920s, the 1930s/1940s, and the 1950s/1960s. As part of the fieldwork for this project, he interviewed local residents of different generations to understand how residents remember place. He says, "I hope to combine my understanding of landscape change through my visual documentation with the oral histories of place residents to tell the story of changing border communities."

---

the flag of their ancestral country. Not only that, but Greeks also avoid red, which is perceived as the color of their ancient enemy, the Turks. Green, an Irish Catholic color, also finds favor in Muslim ethnic neighborhoods (**Figure 5.29**, page 178) in countries as far-flung as France and China because it is the sacred color of Islam (see Chapter 7).

Indeed, urban ethnic and racial landscapes are visible in cities across the world. This is true even when urban planners try their best to prevent the emergence of such landscapes. When Brazil's capital was moved from Rio de Janeiro to Brasília in 1960, for example, very few pedestrian-friendly public spaces were incorporated into the urban plan of architects Lúcio Costa and Oscar Niemeyer. Rather, residences were concentrated in high-rise apartments called *superquadra;* streets were designed for high-speed motor traffic only; and no smaller plazas, cafés, or

FIGURE 5.29 Ethnic storefront in Coney Island, New York. The liberal use of the color green in this storefront design might indicate an Islamic presence. Indeed, this is a Pakistani ethnic grocery. *(© David Grossman/Alamy.)*

sidewalks were included—all of which discouraged informal socialization (**Figure 5.30**). As historian James C. Scott notes in his discussion of Brasília, the lack of human-scale public gathering places was intentional. "Brasília was to be an exemplary city, a center that would transform the lives of the Brazilians who lived there—from their personal habits and household organization to their social lives, leisure, and work. The goal of making over Brazil and Brazilians necessarily implied a disdain for what Brazil had been." But because the housing needs of those building the city and serving the government workers had not been planned, Brasília soon developed surrounding slums that

did not follow an orderly layout. And because the desires of wealthier residents were not met by the uniform *superquadra* apartments, unplanned but luxurious residences and private clubs were also built. By 1980, 75 percent of Brasília's population lived in unanticipated settlements. And because, as previously mentioned, Brazil's social structure is in part based on skin pigmentation, it is the poorer and darker-skinned workers who live in the peripheral slums, whereas the wealthier and lighter-skinned residents tend to live in isolated enclaves for the rich. Thus, a racialized landscape of wealth and poverty emerged in Brasília despite all efforts to the contrary.

FIGURE 5.30 The planned urban landscape of Brasília. As you can see from this aerial view, Brasília is a city of straight roadways, high-rise residential superquadra, and green spaces in between. It is a city that was literally built from scratch in an area of cleared jungle in Brazil's Amazonian interior. This allowed the planners to follow a modern urban design blueprint down to the last detail. Residents of Brasília complain that its larger-than-life public spaces, high-speed vehicle traffic, and impersonal housing blocks inhibit socialization and make Brasília a difficult place to live. *(Cassio Vasconcellos/SambaPhoto/Getty Images.)*

The Re-Creation of Ethnic Cultural Landscapes

As mentioned previously in this chapter, migration is the principal way in which groups not ethnic in their homelands become ethnic or racialized, through relocation to a new place. As we have seen, ethnic groups may choose to relocate to places that remind them of their old homelands. Once there, they set about re-creating some—although, because of cultural simplification, not all—of their particular landscapes in their new homes.

Cuban-American immigrants, for example, have reconstructed aspects of their prerevolutionary Cuban homeland in Miami. In the early 1960s, the first wave of Cuban refugees came to the United States on the heels of Castro's communist revolution on the island. Some went north, to places such as Union City, New Jersey, where today there is a sizable Cuban-American population. Others, however, came (or eventually relocated) to Miami. Once a predominantly Jewish neighborhood called Riverside, what is today called the Little Havana neighborhood became the heart of early Cuban immigration. Shops along the main thoroughfare, Calle Ocho (or Southwest Eighth Street), reflect the Cuban origins of the residents: grocers such as Sedanos and La Roca cater to the Spanish-inspired culinary traditions of Cuba; cafés selling small cups of strong, sweet *café Cubano* exist on every block; and famed restaurants such as Versailles are social gathering points for Cuban-Americans.

Because ethnicity and its expression on the landscape are fluid and ever-changing, the landscape of Calle Ocho reflects the current demographic changes under way in Little Havana. Although the neighborhood is still a Latino enclave, with a Hispanic population of over 95 percent, only half identify as "Cuban identity." More than one-quarter of Little Havana's residents are recent arrivals from Central American countries, particularly Nicaragua and Honduras, whereas more and more Argentineans and Colombians are arriving—like the Cubans before them—on the heels of political and economic chaos in their home countries. Today, you are just as likely to see a Nicaraguan *fritanga* restaurant as a Cuban coffee shop along Calle Ocho. This has prompted some to suggest that the neighborhood's name be changed from Little Havana to The Latin Quarter (**Figure 5.31**).

Geographer Christopher Airriess has studied Vietnamese refugees in the United States. These refugees were initially brought to one of four reception centers, and then further relocated to rural and urban locations throughout the country, on the theory that spatial dispersal would hasten their assimilation into mainstream U.S. society. Airriess's work reveals that a process of secondary, chain migration later led many of them to cluster in selected urban areas that offer warm weather and proximity to Vietnamese friends and relatives. Focusing on the Versailles neighborhood of New Orleans, Airriess found that the fact that the neighborhood is surrounded by swamps, canals, or bayous

FIGURE 5.31 Restaurant in Little Havana, Miami. *How many different national groups are identified in this restaurant façade?*
(Courtesy of Patricia L. Price.)

on three sides affords a degree of cultural isolation from other ethnic groups in the city. This has led to Versailles becoming an *ethnic neighborhood* where Vietnamese refugees re-create elements of their home landscape in the United States. The most prominent ethnic flag of the Vietnamese in Versailles is the vegetable gardens found on the perimeter of the neighborhood. Cultural preadaptation has led to the reproduction of Vietnamese rural landscapes here, sometimes with plants sent directly from Vietnam. It is the older Vietnamese who plant and tend the gardens. Such gardening provides a therapeutic activity, allows them to retain traditional dietary habits, enables them to produce folk medicines from plants, and reduces household food expenditures. Airriess notes that "the image of conical-hat-wearing gardeners leaning over plants within a multi-textured and moisture-soaked, vibrant green agricultural scene, coupled with frequent flyovers of military helicopters, is an eerie scene of wartime Vietnam reproduced."

After Hurricane Katrina devastated sections of New Orleans—including Versailles (today more commonly called Village de L'est)—in 2005, the ethnic cultural landscape underwent a transformation. Many Vietnamese immigrants moved away from Versailles after the storm, whereas many Hispanics moved into the area, attracted by the many post-Katrina construction jobs. A 2006 survey found nearly 7000 Asians in New Orleans after Katrina, compared to nearly 12,000 before the storm. By contrast, Latinos have grown from about 14,000 before the hurricane to 16,000 post-Katrina: the only ethnic group, in fact, to have increased after 2005. Today, the commercial landscape reflects this transition. The first arrivals on the scene were the *loncheras,* or mobile lunch trucks (**Figure 5.32**). Then came the Hispanic-oriented restaurants, supermarkets, and services. Today, the remaining Vietnamese-American residents are learning to adapt to the newcomers. Sara Catania recounts how Mai Thi Nguyen, a business development director in Village de L'est, reports that her aunt, a market owner, "is learning Spanish. She's learning to say hello, how to tell customers how much something costs. It's wild. I love it. It's exciting."

Reflecting on Geography

What common ground might Vietnamese-American and Hispanic immigrants have that could help them to bridge their differences?

Ethnic Culinary Landscapes

"Tell me what you eat, and I'll tell you who you are." This oft-quoted phrase arises from the connections between identity and **foodways,** or the customary behaviors associated with food preparation and consumption that vary from place to place and from ethnic group to ethnic group (**Figure 5.33**). As geographers Barbara and James Shortridge point out, "Food is a sensitive indicator of identity and change." Immigration, intermarriage, technological innovation, and the availability of certain ingredients mean that modifications and simplifications of traditional foods are inevitable over time. An examination of culinary cultural landscapes reveals that, although landscapes are typically understood to be visual entities, they can be constructed from the other four senses as well: touch, smell, sound, and—in this case—taste.

foodways
Customary behaviors associated with food preparation and consumption.

Although Singapore occupies the southern tip of the Malaysian Peninsula in extreme Southeast Asia, its cuisine draws heavily from southern Chinese cooking. This is because three-quarters of Singapore's multiethnic population is of Chinese ancestry. Intermarriage between Chinese men, who came to Singapore as traders and settled there, and local Malay women resulted in a distinctive spicy cuisine called *nonya.* As we have already seen, there are far more ethnic Chinese in Asia than in other parts of the world, principally because of China's proximity to other Asian countries.

FIGURE 5.32 Lunch truck in New Orleans. Loncheras, or mobile lunch trucks, are found in areas of New Orleans that suffered devastation from Hurricane Katrina in 2005. Mexican workers were attracted to New Orleans after 2005 because of opportunities to work in the construction sector as neighborhoods rebuilt. Lunch trucks cater to Mexican immigrant tastes and budgets.
(Russell McCulley/AFP/Getty Images.)

FIGURE 5.33 Market in Chinatown, New York City. Grocery stores selling distinctive produce items are one of the most visible landscape markers of an ethnic presence. Here, the spiny fruits hanging from the bags are called durian, also known as "stinky fruit." Durian is sometimes banned from public places in its native Asia because of its pungent smell. *(AP Photo/Kathy Willens.)*

The corn tortilla remains a staple of central Mexican foodways and is consumed at every meal by some families. But many middle-class urban women in Mexico no longer have the time to soak, hull, and grind corn by hand to make corn tortillas from scratch. Rather, children are often sent every afternoon to the corner *tortillería* to purchase *masa*, or corn dough, or even hot stacks of the finished product. Convenience notwithstanding, old-timers complain that the uniform machine-made tortillas will never come close to the flavor and texture of a handmade tortilla. In some regions, notably in the northern part of Mexico and southern Texas, the labor-intensive corn tortilla was replaced altogether by wheat tortillas, which are much easier to make.

Geographer Daniel Arreola has plotted the Taco-Burrito and the Taco-Barbecue isoglosses in southwestern Texas shown in **Figure 5.34**. These lines show the transition between different Mexican culinary influences (tacos versus burritos) and between Mexican and European foodways (tacos versus barbecue), with the barbecue sandwich preferred by the German, Czech, Scandinavian, Anglo, and black Texan populations to the north and east of the line.

In the rural, mountainous Appalachian region of the eastern United States, distinct ethnic foodways persist and literally flavor this region. Geographer John Rehder observed that, during and after World War II, Appalachians maintained their cultural distinctiveness in the face of migration to northern cities in part through "care packages" of lard, dried beans, grits, cornmeal, and other foods unavailable in the North. Grocery stores in the Little Appalachia ethnic neighborhoods that formed in cities such as Detroit,

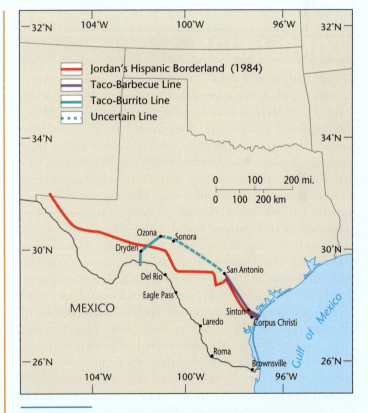

FIGURE 5.34 The Taco-Burrito and Taco-Barbecue lines. The transition between different styles of Mexican, and Mexican and European, foodways is depicted on this map of Texas. Note that the lines coincide with the borders of the Hispanic homeland as defined by Terry Jordan. In this map, Daniel Arreola builds on that work. *(Sources: After Arreola, 2002: 175; Jordan, 1984.)*

Cleveland, and Chicago soon began to cater to the distinct food preferences of the population. Thus, Appalachian foodways were maintained as Appalachians moved north.

Rehder tells of sending students on a scavenger hunt during a class field trip to Pikeville, Tennessee. To one young man, he handed a card that read, "What are cat head biscuits and sawmill gravy?"

About a half an hour later, my "biscuit man" returned with a scared look on his face and tears welling up in his eyes. I asked, "Where did you go and what happened?" He replied, "Well, . . . I . . . went to the feed store and asked an old man there, 'What are cat head biscuits and sawmill gravy?' just like you told me to do. Only the old man just growled at me and said, 'Son, if you don't know, HELL, I ain't goin' to tell you!' And with that our lad left the feed store dejected and thoroughly upset. To calm him down, I said, "Why don't you just go on down to the café.

Get a little something to drink and maybe calmly run your question by them down there." After a sufficient amount of time, I decided to go check on him. He was sitting on a red upholstered stool at the counter. As I drew closer, he looked up and, with fresh and quite different tears in his eyes, pointed to his plate: "This," he said proudly, "is cat head biscuits and sawmill gravy!" (p. 208)

Rehder explains that cat head biscuits are flour biscuits made in the exact size and shape of a cat's head and baked golden in a wood stove. Sawmill gravy is a southern-style white gravy made in a cast-iron skillet. Legend has it that the cook at the Little River Lumber Company logging camp ran out of flour for the gravy and substituted cornmeal. The loggers complained about the texture, referring to it as "sawmill gravy." For further discussion of ethnicity and food, see Doing Geography and Seeing Geography (page 184).

CONCLUSION

The inclusion of a critical focus on race into the long-standing field of ethnic geography provides contemporary, fruitful inroads into understanding human diversity. Here, the five themes of cultural geography have provided new perspectives on ethnicity and race. Ethnicity and race help shape the pattern of rural homelands and islands as well as urban ethnic neighborhoods and ghettos. Immigration is actively reshaping the ethnic mosaic of many places, including the United States. Human mobility routes—whether voluntary or involuntary, undertaken in response to persecution in the home region or opportunities beckoning beyond one's borders—help to shape the patterns that ethnicity and race assume in the new land. Sometimes similarities between home and host habitats direct migrants to certain places rather than others. Globalization appears to intensify age-old tendencies to distinguish, and discriminate, among groups. Racism seems particularly resistant to change. Certain natural habitats can protect ethnically distinct groups. Yet environmental racism illustrates how minority peoples and the places they inhabit can become the targets of undesirable practices, such as pollution. On the flip side, ethnic pride and the active construction of ethnically and racially distinct landscapes are evident the world over.

By now, we hope you are beginning to think and see as geographers do. The world is patterned in terms of cul-

ture. What these patterns are, why they change, how they change, and how these changes affect people living in places are the basic focus of cultural geography. Ethnicity, along with language and religion, is at the heart of many of today's most pressing political geography questions.

DOING GEOGRAPHY

Tracing Ethnic Foodways Through Recipes

At some point in our family histories, we all trace our roots back to migrants. Perhaps your ancestors walked here some 30,000 years ago. Perhaps they arrived on slave ships in the seventeenth century. Perhaps they were traders who settled and married locals. Were they part of the waves of Europeans in the eighteenth, nineteenth, and early twentieth centuries, or have you yourself only recently immigrated? More than likely, your ethnic inheritance results from a combination of different immigrant groups. As the geographer Doreen Massey wrote, "In one sense or another most places have been 'meeting places'; even their 'original inhabitants' usually came from somewhere else." In other words, if you dig into the history of any place, you'll find layers on layers of people coming in from other places and bringing their cultural baggage—recipes and all—with them.

For this exercise, you will analyze one of the most commonplace, yet revealing, items of ethnic geography: a recipe. Certainly, you are what you eat, but you also eat where you are, and the foodways in which you take an active part are very revealing of both who you are and where you are. Even though you may not be conscious of it, the simple act of cooking a meal sets into motion all sorts of cultural geography elements: regional identity, ethnic heritage, place-specific agricultural traditions, and so on. Together these form important components of identity and place.

Steps to Analyzing a Recipe

Step 1: Identify a recipe to analyze. Most of you grew up or are now living in households where meals are cooked on site at least some of the time. Choose a recipe that is used often and has been around for a while. The best candidate is a favorite family recipe that has been passed down through the generations (see note, below). If you don't have a copy of the recipe, you will need to interview a person who does—if necessary, by phone or e-mail.

Step 2: With a written recipe now in front of you, answer the following questions. You can interview someone in your family about this recipe, too.

- From where does this recipe come? With what country, or region, is it identified?
- Do any of the recipe's ingredients give clues about the origins of the recipe? Do the ingredients draw on particular animal or plant ingredients that are, or were, produced where the recipe originated?
- Has the recipe been modified to substitute ingredients that are no longer available, either because the person who used the recipe migrated or because the ingredients went out of production?
- Are there ingredients used in the recipe that are identified with particular ethnic groups and perhaps aren't consumed by others living in the same place?
- Do elements of the recipe's preparation give additional clues about the foodways of the people who developed the recipe?
- Are there special occasions, such as holidays, when this recipe is always used?
- Do any of the ingredients or methods of preparation have symbolic meanings or stories associated with them?

In sharing the results of the recipe analysis, the class can list the various places and ethnicities that together comprise the foodways of the students. The class might want to create a cookbook of their recipes and map the places from which they come.

Note: If you are attending school in a foreign country, use a family recipe from your native homeland. E-mail or call a family member to discuss this recipe using the guidelines for this exercise. Also, some of you may have grown up in an institutional setting, where you consumed food prepared in a cafeteria. Institutional foods can be very revealing of local ethnic influences, so for this exercise choose a dish that is commonly prepared for the evening meal.

Recipe box. Back when most Americans cooked their meals at home, a personalized recipe collection like this one—written on 3 × 5 cards and safeguarded in a recipe box—held treasured family recipes, some passed down for generations. *(Edward Pond/Masterfile.)*

Key Terms

acculturation	p. 155
assimiliation	p. 155
chain migration	p. 165
cultural maladaptation	p. 172
cultural preadaptation	p. 171
cultural simplification	p. 167
environmental racism	p. 174
ethnic cleansing	p. 166
ethnic flag	p. 175
ethnic group	p. 151
ethnic homelands	p. 156
ethnic islands	p. 157
ethnic neighborhood	p. 157
ethnic substrate	p. 157
ethnoburbs	p. 159
foodways	p. 180
genocide	p. 170
ghetto	p. 158
involuntary migration	p. 166
race	p. 150
return migration	p. 166

Ethnic Geography on the Internet

You can learn more about ethnic geography and the geography of race on the Internet at the following web sites:

2010 Census of the United States
http://www.census.gov

Go to the American FactFinder section of the web page. Here you will find a wonderful selection of maps that

SEEING GEOGRAPHY American Restaurant Neon Signs

When does cuisine cease to be ethnic and become simply "American"? What role does cultural diffusion play in the process?

Neon signs collected from ethnic restaurants in the United States by the Smithsonian Institution in Washington, D.C.

This remarkable image comes not from a cultural landscape but from a montage of neon signs collected for an exhibit some years ago at the Smithsonian Institution in Washington, D.C. The exhibit represented a small sampling of the kinds of ethnic foods available commercially in the United States. It showed how ethnically diverse America had become.

What, more precisely, can these diverse, artificially assembled fragments from many cultural landscapes tell us? That we are a multiethnic society? Of course. As cultural geographer Richard Pillsbury recently said, America has "no foreign food" because we have accepted every possible foreign cuisine and made it our own.

Cultural interaction is also revealed here. Massive changes in U.S. immigration laws in the 1960s had the effect of greatly diversifying the immigrant stream, allowing such a food diversity to become established.

Implicit in the photo, too, are culture regions—in the form of ethnic neighborhoods. Each of these signs comes from an ethnic neighborhood, and the further regional implication is that such neighborhoods are proliferating.

Cultural diffusion is also obvious. How could these different cuisines have reached our shores other than by relocation diffusion?

So, in this manner, landscape images demand cultural interactive explanations, imply cultural regions, and require cultural diffusion. The various themes of cultural geography work together, are inseparable, and constitute a functioning whole. Are there any ethnic cuisines you think should be added to this montage, thanks to new immigration flows? Any that should be deleted, either because there are no sizable immigrant populations from these places any longer or because their ethnicity has since melted away and become mainstream?

show themes (thematic maps) generated from census data. Some of the maps used in this chapter were found here. Data from the 2010 Census will be released through September 2013.

Ethnic Geography Specialty Group, Washington, D.C.
http://www.unl.edu/geography/ethnic
This web site provides information about a specialty group within the Association of American Geographers,

whose membership includes nearly all U.S. specialists in the study of ethnic geography.

Food: Past and Present
http://www.teacheroz.com/food.htm
Here you will find a list of links to hundreds of web pages that deal with all aspects of food: diets of historical and contemporary cultures, recipes, and foodways of regions and ethnic groups in the United States.

New American Media

http://newamericamedia.org/

This rich web site is dedicated to "expanding the news lens through ethnic media." You can select news stories by ethnicity, visit blogs, or view polls designed to capture the opinions of those typically excluded from mainstream surveys.

Racialicious

http://www.racialicious.com

Feel the need to know the top 10 trends in race and pop culture? Want to follow celebrity gaffes, politicians' missteps, and questionable media representations? Log on to Racialicious, a no-holds-barred blog about the intersection of race and pop culture.

Sources

Airriess, Christopher. 2002. "Creating Vietnamese Landscapes and Place in New Orleans," in Kate A. Berry and Martha L. Henderson (eds.), *Geographical Identities of Ethnic America: Race, Space, and Place.* Reno: University of Nevada Press, 228–254.

Allen, James, and Eugene Turner. 1988. *We the People: An Atlas of America's Ethnic Diversity.* New York: Macmillan.

Arreola, Daniel D. 1984. "Mexican-American Exterior Murals." *Geographical Review* 74: 409–424.

Arreola, Daniel D. 2002. *Tejano South Texas: A Mexican American Cultural Province.* Austin: University of Texas Press.

Associated Press, 2006 (March 26). "500,000 Rally Immigration Rights in Los Angeles." Available online at http://www.msnbc.com.

BBC News. 2007 (February 7). "UN Warns of Iraq Refugee Disaster." Available online at http://newsvote.bbc.co.uk.

Benton-Short, Lisa, Marie D. Price, and Samantha Friedman. 2005. "Globalization from Below: The Ranking of Immigrant Cities." *International Journal of Urban and Rural Research* 29(4): 945–959.

Bullard, Robert D. (ed.). 1993. *Confronting Environmental Racism: Voices from the Grassroots.* Boston: South End Press.

Carlson, Alvar W. 1990. *The Spanish American Homeland: Four Centuries in New Mexico's Río Arriba.* Baltimore: Johns Hopkins University Press.

Catania, Sara. 2006 (October 16). "From Fish Sauce to Salsa— New Orleans Vietnamese Adapt to Influx of Latinos." Available online at *New American Media,* http://news.newamericamedia.org/news/.

Dawson, C. A. 1936. *Group Settlement: Ethnic Communities in Western Canada.* Toronto: Macmillan.

Frey, William H. 2001. "Micro Melting Pots." *American Demographics* (June): 20–23.

Grieco, Elizabeth M., and Edward N. Trevelyan. 2010. "Place of Birth of the Foreign-Born Population: 2009." American Community Survey Briefs. Washington, D.C.: U.S. Census Bureau.

Available online at http://www.census.gov/prod/2010pubs/acsbr09-15.pdf.

Harris, R. Colebrook. 1977. "The Simplification of Europe Overseas." *Annals of the Association of American Geographers* 67: 469–483.

Hill, G. W. 1942. "The People of Wisconsin According to Ethnic Stocks, 1940." *Wisconsin's Changing Population.* Bulletin, Serial No. 2642. Madison: University of Wisconsin.

Hollinger, David. 1995. *Postethnic America: Beyond Multiculturalism.* New York: Basic Books.

Hsieh, Chiao-min and Jean Kan Hsieh. 1995. *China: A Provincial Atlas.* New York: Macmillan

Johnson, James H., Jr., and Curtis C. Roseman. 1990. "Recent Black Outmigration from Los Angeles: The Role of Household Dynamics and Kinship Systems." *Annals of the Association of American Geographers* 80: 205–222.

Jordan, Terry G., with John L. Bean, Jr. and William M. Holmes. 1984. *Texas: A Geography.* Boulder, Colo.: Westview Press.

Massey, Doreen. 1994. *Space, Place, and Gender.* Minneapolis: University of Minnesota Press.

Matwijiw, Peter. 1979. "Ethnicity and Urban Residence: Winnipeg, 1941–71." *Canadian Geographer* 23: 45–61.

Meigs, Peveril, III. 1941. "An Ethno-Telephonic Survey of French Louisiana." *Annals of the Association of American Geographers* 31: 243–250.

Meinig, Donald W. 1965. "The Mormon Culture Region." *Annals of the Association of American Geographers* 55: 191–220.

Meisch, Lynn A. 1991. "We Are Sons of Atahualpa and We Will Win: Traditional Dress in Otavalo and Saraguro, Ecuador," in Margot Blum Schevill, Janet Catherine Berlo, and Edward B. Dwyer (eds.), *Textile Traditions of Mesoamerica and the Andes: An Anthology.* New York: Garland Publishing, 145–177.

Motzafi-Haller, Pnina. 1997. "Writing Birthright: On Native Anthropologists and the Politics of Representation," in Deborah E. Reed-Danahay (ed.), *Auto/Ethnography: Rewriting the Self and the Social.* Oxford: Berg, 195–222.

National Geographic, "The Genographic Project." Available online at http://www3.nationalgeographic.com/genographic/.

Nostrand, Richard L., and Lawrence E. Estaville, Jr. (eds.). 2001. *Homelands: A Geography of Culture and Place Across America.* Baltimore: Johns Hopkins University Press.

Oberle, Alex P., and Daniel D. Arreola. 2008. "Resurgent Mexican Phoenix." *Geographical Review* 98(2): 171–196.

O'Loughlin, John, Frank Witmer, Thomas Dickinson, Nancy Thorwardson, and Edward Holland. 2007. "Preface to Special Issue and Caucasus Map Supplement." *Eurasian Geography and Economics,* 48(1): 127–134.

Pillsbury, Richard. 1998. *No Foreign Food: American Diet in Time and Place.* Boulder, Colo.: Westview Press.

Rechlin, Alice. 1976. *Spatial Behavior of the Old Order Amish of Nappanee, Indiana.* Geographical Publication No. 18. Ann Arbor: University of Michigan.

Rehder, John B. 2004. *Appalachian Folkways*. Baltimore: Johns Hopkins University Press.

Scott, James C. 1998. *Seeing Like a State: How Certain Schemes to Improve the Human Condition Have Failed*. New Haven, Conn.: Yale University Press.

Shortridge, Barbara G., and James R. Shortridge (eds.). 1998. *The Taste of American Place: A Reader on Regional and Ethnic Foods*. Lanham, Md.: Rowman & Littlefield.

Sivanandan, A. 2001 (August 17). "Poverty Is the New Black." Available online at http://www.guardian.co.uk.

Toxic Release Inventory (U.S. EPA). 1996. Available online at http://www.inmotionmagazine.com/auto/mapofla.html.

U.S. Bureau of the Census. 2000. MLA Language Map Data Center. Available online at http://www.mla.org/map_main.

Ten Recommended Books on Ethnic Geography

(For additional suggested readings, see *The Human Mosaic* web site: www.whfreeman.com/domosh12e)

Berry, Kate A., and Martha L. Henderson (eds.). 2002. *Geographical Identities of Ethnic America: Race, Space, and Place*. Reno: University of Nevada Press. Eighteen different experts give their views on American ethnic geography, explaining how place shapes ethnic/racial identities and, in turn, how these groups create distinctive spatial patterns and ethnic landscapes.

Jordan, Terry G., and Matti E. Kaups. 1989. *The American Backwoods Frontier: An Ethnic and Ecological Interpretation*. Baltimore: Johns Hopkins University Press. The authors use the concepts of cultural preadaptation and ethnic substrate to reveal how the forest colonization culture of the American eastern woodlands developed and helped shape half a continent.

Lee, Jennifer 8. 2008. *The Fortune Cookie Chronicles: Adventures in the World of Chinese Food*. New York: Twelve. Many "ethnic" cuisines common in the United States are in fact largely invented and popularized here, rather than in the home country of the immigrants in question. Such is the case with American "Chinese food" favorites, such as fortune cookies, General Tsao chicken, chop suey, and other dishes. Lee traces the fascinating cultural, nutritional, migratory, interethnic, and economic histories at work behind this American tradition.

McKee, Jesse O. (ed.). 2000. *Ethnicity in Contemporary America: A Geographical Appraisal,* 2nd ed. Lanham, Md.: Rowman & Littlefield. This clear and thoughtful text offers a geographical analysis of U.S. immigration patterns and the development of selected ethnic minority groups, focusing especially on their origin, diffusion, socioeconomic characteristics, and settlement patterns within the United States; many well-known geographers contributed chapters.

Miyares, Ines M., and Christopher A. Airriess. 2007. *Contemporary Ethnic Geographies in America.* Lanham, Md.: Rowman & Littlefield. An edited collection featuring chapters authored by renowned contemporary ethnic geographers on a variety of topics ranging from Central American soccer leagues to Muslim immigrants from Lebanon and Iran; also includes a geographer's view of the ethnic festivals with which we began this chapter.

Noble, Allen G. (ed.). 1992. *To Build in a New Land: Ethnic Landscapes in North America.* Baltimore: Johns Hopkins University Press. A superb study of widely differing traditional ethnic cultural landscapes in the United States and Canada, revealing what diverse countries we inhabit.

Nostrand, Richard L., and Lawrence E. Estaville, Jr. (eds.). 2001. *Homelands: A Geography of Culture and Place Across America.* Baltimore: Johns Hopkins University Press. A collection of essays on an array of North American ethnic homelands, together with in-depth treatment of the geographical concept of homeland.

Rehder, John B. 2004. *Appalachian Folkways.* Baltimore: Johns Hopkins University Press. An exploration of the folk culture of the Appalachian region of the United States, including its distinctive settlement history, folk architecture, cuisine, speech, and belief systems.

Schein, Richard (ed.). 2006. *Landscape and Race in the United States.* London and New York: Routledge. Contributions by leading geographers studying the cultural geographies of race provide an updated and critical insight into this growing subfield.

Yoon, Hong-key. 1986. *Maori Mind, Maori Land: Essays on the Cultural Geography of the Maori People from an Outsider's Perspective.* Bern, Switzerland: Peter Lang. A highly readable study of the indigenous Polynesian ethnic group in New Zealand; also useful as a comparison to the situation in North America.

Are these border fences or walls?

Go to "Seeing Geography" on page 226 to learn more about these images.

6 POLITICAL GEOGRAPHY
A Divided World

From the breakup of empires to regional differences in voting patterns, from the drawing of international boundaries to congressional redistricting in the U.S. electoral system, from the resurgence of nationalism to separatist violence, human political behavior is inherently geographical. As geographer Gearóid

political geography
The geographic study of politics and political matters.

Ó Tuathail has said, **political geography** "is about power, an ever-changing map revealing the struggle over borders, space, and authority."

Region

How is political geography revealed in regions? The theme of region is essential to the study of political geography because an array of both formal and functional political regions exist. Among these, the most important and influential is the **state.**

state
A centralized authority that enforces a single political, economic, and legal system within its territorial boundaries. Often used synonymously with "country."

A World of States

The fundamental political-geographical fact is that the Earth is divided into nearly 200 independent countries or states, creating a diverse mosaic of functional regions (**Figure 6.1**, pages 190–191). The state is a political institution that has taken a variety of forms over the centuries, ranging from Greek city-states, to Chinese dynastic states,

to European feudal states. When we talk about states today, however, we mean something very specific and historically recent. States are independent political units with a centralized authority that makes claims to sole jurisdiction over a bounded territory. Within that territory, the central authority controls and enforces a single system of political and legal institutions. Importantly, in the modern international system, states recognize each other's **sovereignty.** That is, virtually every state recognizes every other state's right to exist and control its own affairs within its territorial boundaries.

sovereignty
The right of individual states to control political and economic affairs within their territorial boundaries without external interference.

Closer inspection of Figure 6.1 reveals that some parts of the world are fragmented into many different states, whereas others exhibit much greater unity. The United States occupies about the same amount of territory as Europe, but the latter is divided into 45 independent countries. The continent of Australia is politically united, whereas South America has 12 independent entities and Africa has 54.

The modern state is a tangible geographical expression of one of the most common human tendencies: the need to belong to a larger group that controls its own piece of the Earth, its own territory. So universal is this trait that scholars coined the term **territoriality** to describe it. Most geographers view territoriality as a learned cultural response. Robert Sack, for example, regards territoriality as a cultural strategy that uses power to control an area and communicate that control, thereby subjugating the inhabitants and acquiring resources. He argues, for example, that the precise marking of borders is a practice originally unique to modern Western culture. The modern territorial state, he

territoriality
A learned cultural response, rooted in European history, that produced the external bounding and internal territorial organization characteristic of modern states.

Independent Countries

Abbreviations

A	AUSTRIA
AL	ALBANIA
B	BELGIUM
BA	BOSNIA-HERZEGOVINA
BF	BURKINA FASO
BG	BULGARIA
BOTS	BOTSWANA
BY	BELARUS
CH	SWITZERLAND
CZ	THE CZECH REPUBLIC
D	GERMANY
EG	EQUATORIAL GUINEA
EST	ESTONIA
GBI	GUINEA BISSAU
H	HUNGARY
HR	CROATIA
IC	IVORY COAST
K	KOSOVO
L	LUXEMBOURG
LT	LITHUANIA
LV	LATVIA
MK	MACEDONIA
NL	NETHERLANDS
RCA	CENTRAL AFRICAN REPUBLIC
RL	LEBANON
RO	ROMANIA
RU	RUSSIA
SK	SLOVAKIA
SLO	SLOVENIA
S	SERBIA
M	MONTENEGRO
TC	TURKISH CYPRUS
TL	TIMOR-LESTE
TM	TURKMENISTAN
UAE	UNITED ARAB EMIRATES
WAG	THE GAMBIA
WAL	SIERRA LEONE
ZW	ZIMBABWE

Scale at latitude 35°

Flat Polar Quartic
equal area projection

Most of the countries on this map have homepages on the Worldwide Web. Visit The Human Mosaic *online to learn more about them.*

colonialism
The building and maintaining of colonies in one territory by people based elsewhere.

claims, emerged rather recently in sixteenth-century Europe and diffused around the globe through European **colonialism.**

Political territoriality, then, is a thoroughly cultural-geographical phenomenon. The sense of collective identity that we call nationalism (see page 195) springs from a learned or acquired attachment to region and place. Geography and national identity cannot be separated.

Distribution of National Territory An important geographical aspect of the modern state is the shape and configuration of the national territory. Theoretically, the more compact the territory, the easier is national governance. Circular

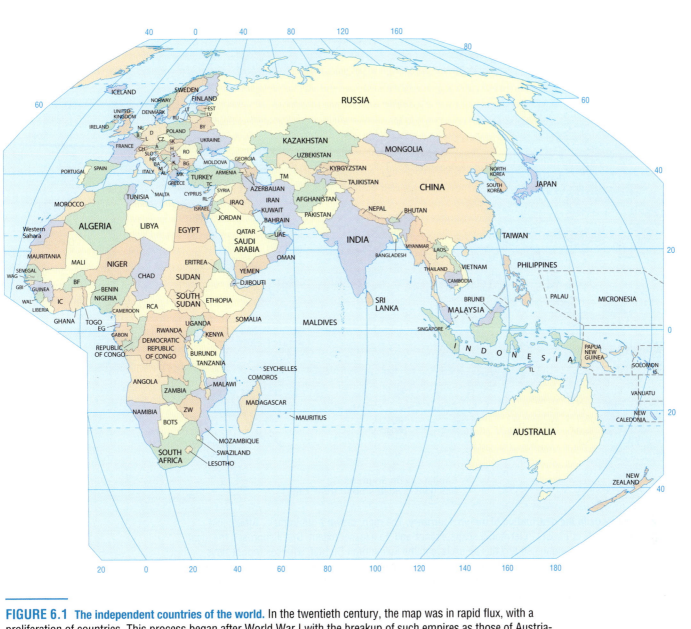

FIGURE 6.1 The independent countries of the world. In the twentieth century, the map was in rapid flux, with a proliferation of countries. This process began after World War I with the breakup of such empires as those of Austria-Hungary and Turkey, then intensified after World War II when the overseas empires of the British, French, Italians, Dutch, Americans, and Belgians collapsed. More recently, the Russian-Soviet Empire disintegrated.

or hexagonal forms maximize compactness, allow short communication lines, and minimize the amount of border to be defended. Of course, no country actually enjoys this ideal degree of compactness, although some—such as France and Brazil—come close (**Figure 6.2**, page 192).

Any one of several unfavorable territorial distributions can inhibit national cohesiveness. Potentially most damaging to a country's stability are enclaves and exclaves. An **enclave** is a district surrounded by a country but not ruled by it. Enclaves can be either self-governing (e.g., Lesotho in Figure 6.2) or an exclave of another country. In either case, its presence can pose problems for the surrounding country. Potentially just as disruptive is the

enclave
A piece of territory surrounded by, but not part of, a country.

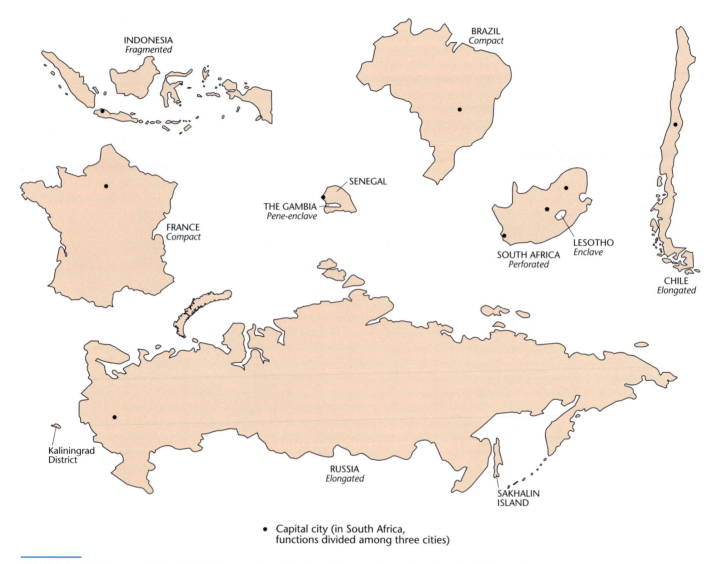

• Capital city (in South Africa, functions divided among three cities)

FIGURE 6.2 Differences in the distribution of national territory. The map, drawn from Eurasia, Africa, and South America, shows wide contrasts in territorial shape. France and, to a lesser extent, Brazil approach the ideal hexagonal shape, but Russia is elongated and has an exclave in the Kaliningrad District, whereas Indonesia is fragmented into a myriad of islands. The Gambia intrudes as a pene-enclave into the heart of Senegal, and South Africa has a foreign enclave, Lesotho. Chile must overcome extreme elongation. *What problems can arise from elongation, enclaves, fragmentation, and exclaves?*

pene-enclave, an intrusive piece of territory with only the smallest of outlets (e.g., The Gambia in Figure 6.2).

exclave
A piece of national territory separated from the main body of a country by the territory of another country.

Exclaves are parts of a national territory separated from the main body of the country to which they belong by the territory of another (e.g., the Kaliningrad District in Figure 6.2; **Figure 6.3**). Exclaves are particularly undesirable if a hostile power holds the intervening territory because defense of such an isolated area is difficult and makes substantial demands on national resources. Moreover, an exclave's inhabitants, isolated from their compatriots, may develop separatist feelings, thereby causing additional problems. Pakistan provides a good example of the national instability created by exclaves. Pakistan was created in 1947 as two main bodies of territory separated from each other by almost 1000 miles (1600 kilometers) of territory in northern India. West Pakistan had the capital and most of the territory, but East Pakistan was home to most of the people. West Pakistan hoarded the country's wealth, exploiting East Pakistan's resources but giving little in return. Ethnic differences between the peoples of the two sectors further complicated matters. In 1971, a quarter of a century after its founding, Pakistan broke apart. The distant exclave seceded and became the independent country of Bangladesh (see Figure 6.1). Even when a national territory is geographically united,

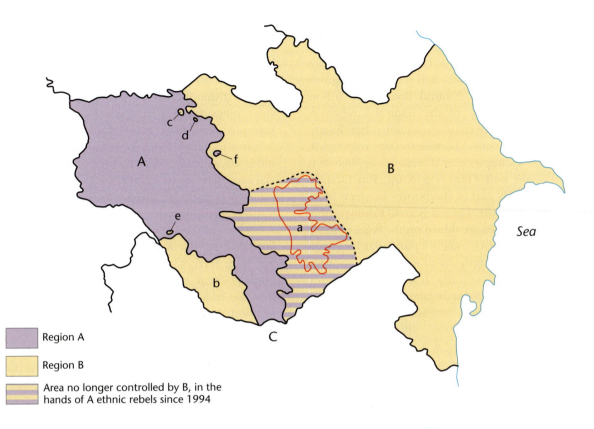

Region A

Region B

Area no longer controlled by B, in the hands of A ethnic rebels since 1994

FIGURE 6.3 Two independent countries, A and B. A seeks to liberate region a, where a population speaking the same language and adhering to the same religion as the people of A live. In a war lasting from 1990 to 1994, the people of region a seceded from B, and a tenuous ceasefire was arranged. B, meanwhile, has an exclave, b, on the opposite, western side of A, and the people of region b form another ethnic minority, unrelated to B. Country B also possesses several much smaller enclaves—c, d, and e, the last of which is regarded as part of b (an exclave of an exclave!). A also has a tiny exclave, f. In other words, the distribution of national territories is troublesome to both A and B, particularly given the hostile relations between them. These are real countries. Using an atlas, try to identify them. If you fail, you can find the answer at the end of this chapter on page 229. **Have any recent events occurred here?** (Sources: Office of the Geographer, U.S. Department of State, personal communication, 1997; Smith et al., 1997: 37.)

instability can develop if the shape of the state is awkward. Narrow, elongated countries, such as Chile, The Gambia, and Norway, can be difficult to administer, as can nations consisting of separate islands (see Figures 6.1 and 6.2). In these situations, transportation and communications are difficult, causing administrative problems. Several major secession movements have challenged the multi-island country of Indonesia; one of these—in East Timor—succeeded in 2002 in creating the first new sovereign state of the twenty-first century (see Figure 6.1). Similarly, the three-island country of Comoros, in the Indian Ocean, is troubled today by a separatist movement on Anjouan Island (see Figure 6.1).

The shape and configuration of a country can make it harder or easier to administer, but they do not determine its stability. We can think of exceptions for all of the territorial forms in Figure 6.2. For example, Alaska is an exclave of the United States, but it is unlikely to produce a secessionist movement similar to the one in Bangladesh. Con-

versely, the Democratic Republic of the Congo and Nigeria both have compact territories and both have suffered through secessionist wars. Geography is an important factor in national governance, but it is not the only factor.

Boundaries   Political territories have different types of boundaries. Until fairly recent times, many boundaries were not sharp, clearly defined lines but instead were referred to as zones called **marchlands.** Today, the nearest equivalent to the marchland is the **buffer state,** an independent but small and weak country lying between two powerful, potentially belligerent countries. Mongolia, for example, is a buffer state between Russia and China; Nepal occupies a similar position between India and China (see Figure 6.1). If one of the neighboring countries assumes control of the buffer

marchland
A strip of territory, traditionally one day's march for infantry, that served as a boundary zone for independent countries in premodern times.

buffer state
An independent but small and weak country lying between two powerful countries.

satellite state
A small, weak country dominated by one powerful neighbor to the extent that some or much of its independence is lost.

natural boundary
A political border that follows some feature of the natural environment, such as a river or mountain ridge.

ethnographic boundary
A political boundary that follows some cultural border, such as a linguistic or religious border.

geometric boundary
A political border drawn in a regular, geometric manner, often a straight line, without regard for environmental or cultural patterns.

relic boundary
A former political border that no longer functions as a boundary.

state, it loses much of its independence and becomes a **satellite state.**

Most modern boundaries are lines rather than zones, and we can distinguish several types. **Natural boundaries** follow some feature of the natural landscape, such as a river or mountain ridge. Examples of natural boundaries are numerous, for example, the Pyrenees lie between Spain and France, and the Rio Grande serves as part of the border between Mexico and the United States. **Ethnographic boundaries** are drawn on the basis of some cultural trait, usually a particular language spoken or religion practiced. The border that divides India from predominantly Islamic Pakistan is one case. **Geometric boundaries** are regular, often perfectly straight lines drawn without regard for physical or cultural features of the area. The U.S.–Canada border west of the Lake of the Woods (about 93° west longitude) is a geometric boundary, as are most county, state, and province borders in the central and western United States and Canada. Not all boundaries are easily categorized; some boundaries are of mixed type, composites of two or more of the types listed.

Finally, **relic boundaries** are those that no longer exist as international borders. Nevertheless, they often leave behind a trace in the local cultures. With the reunification of Germany in the autumn of 1990, the old Iron Curtain border between the former German Democratic Republic in the east and the Federal Republic of Germany in the west was quickly dismantled (**Figure 6.4**). In a remarkably short time span, measured in weeks, the Germans reopened severed transport lines and created new ones, knitting the enlarged country together. Even so, remnants and reminders of the old border remained, as it continued to function as provincial boundaries within Germany. Furthermore, it still separated two parts of the country with strikingly different levels of prosperity.

Spatial Organization of Territory States differ greatly in the way their territory is organized for purposes of administration. Political geographers recognize two basic types of spatial organization: **unitary** and **federal.** In unitary countries, power is concentrated centrally, with little or no provincial authority. All major decisions come from the central government, and policies are applied uniformly throughout the national territory. France and China are unitary in structure, even though one is democratic and the other totalitarian. A federal government, by contrast, is a more geographically expressive political system. That is, it acknowledges the existence of regional cultural differences and

unitary state
An independent state that concentrates power in the central government and grants little authority to the provinces.

federal state
An independent country that gives considerable powers and even autonomy to its constituent parts.

FIGURE 6.4   A boundary disappears. In Berlin, the view toward the Brandenburg Gate changed radically between 1989 and 1991, when the Berlin Wall was destroyed and Germany reunited. *(Courtesy of Terry G. Jordan-Bychkov.)*

provides the mechanism by which the various regions can perpetuate their individual characters. Power is diffused, and the central government surrenders much authority to the individual provinces. The United States, Canada, Germany, Australia, and Switzerland, though exhibiting varying degrees of federalism, provide examples. The trend in the United States has been toward a more unitary, less federal government, with fewer states' rights. By contrast, federalism remains vital in Canada, representing an effort to counteract French-Canadian demands for Québec's independence. That is, by emphasizing federalism, the central government allows more latitude for provincial self-rule, thus reducing public support for the more radical option of secession.

Whether federal or unitary, a country functions through some system of political subdivisions. In federal systems, these subdivisions sometimes overlap in authority, with confusing results. For example, the Native American reservations in the United States occupy a unique and ambiguous place in the federal system of political subdivisions. These semiautonomous enclaves are legally sanctioned political territories that, theoretically, only indigenous Americans can possess. In reality, ownership is often fragmented by non–Native American land holdings within the reservations. Although not completely sovereign, Native Americans do have certain rights to self-government that differ from those of surrounding local authorities. For example, many reservations can (and do) build casinos on their land even though gambling may be illegal in surrounding political jurisdictions. Reservations do not fit neatly into the American political system of states, counties, townships, precincts, and incorporated municipalities.

Centrifugal and Centripetal Forces Although physical factors such as the spatial organization of territory, degree of compactness, and type of boundaries can influence an independent country's stability, other forces are also at work. In particular, cultural factors often make or break a country. The most viable independent countries, those least troubled by internal discord, have a strong feeling of group solidarity among their population. Group identity is the key.

nationalism
The sense of belonging to and self-identifying with a national culture.

In the case of the modern state, the primary source of group identity is **nationalism.** Nationalism is the idea that the individual derives a significant part of his or her social identity from a sense of belonging to a nation. We can trace the origins of modern nationalism to the late eighteenth century and the emergence of the modern nation-state. Political geographer John Agnew cautions us, however, that the meaning and form of nationalism are unstable, making it difficult to generalize. One version is state nationalism, wherein the nation-state is exalted and individuals are called to sacrifice

for the good of the greater whole. The twentieth-century history of state nationalism in Europe is marked by two horrific world wars in which millions of lives were sacrificed. Referring to World War I, Agnew points out that the "war itself was also the outcome of a mentality in which the individual person had to sacrifice for the good of the greater whole: the nation-state."

When a specific ethnic group or ethnically based political party controls the state's military, we sometimes see forced deportations and even attempted genocides in efforts to create an ethnically homogeneous national citizenry (see Chapter 5). This happened most recently in Rwanda in 1994 and in the Balkans in the 1990s. The most notorious modern example is Nazi Germany's attempt in the 1930s and 1940s to clear German territory of non-Germans, especially Jews and Roma (Gypsies) (see Rod's Notebook, page 196). Another form is sub-state nationalism, in which ethnic or linguistic minority populations seek to secede from the state or to alter state territorial boundaries to promote cultural homogeneity and political autonomy.

Geographers refer to factors that promote national unity and solidarity as **centripetal forces.** By contrast, anything that disrupts internal order and furthers the break-up of the territory is called a **centrifugal force.** Many states encourage centripetal forces that help fuel nationalistic sentiment. Such things as an official national language, national history museums, national parks and monuments, and sometimes even a national religion are actively promoted and supported by the state. We consider an array of centripetal and centrifugal forces later, in the sections on globalization and nature-culture.

centripetal force
Any factor that supports the internal unity of a country.

centrifugal force
Any factor that disrupts the internal unity of a country.

Political Boundaries in Cyberspace

What happens to political boundaries in cyberspace? E-mail, the Internet, and the World Wide Web can cross borders in ways previously not possible, although radio, telephone, and fax machines possess some of the same border-defying qualities. In Germany, for example, Nazi and neo-Nazi propaganda is prohibited, yet the First Amendment right of freedom of speech protects the dissemination of this material in the United States. What happens when this propaganda originates in the United States and is posted electronically to online bulletin boards worldwide? Can the German government prosecute the originator of the message—an American—for violating a German law?

How can countries impose their laws and boundaries in the computer age? One way is to restrict citizens' access to technological hardware such as satellite dishes, computers, phone lines, and cell phones. This is becoming

Rod's Notebook

Places of a Genocidal State

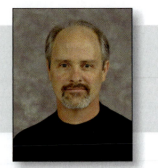

Rod Neumann

For years I have known of the Holocaust in broad historical outline. I knew that 6 million Jews were systematically murdered by the Nazi-controlled German state. The ideology of the Nazi Party was a rabid form of German nationalism founded on racial hatred, especially anti-Semitism. For Nazi leader Adolf Hitler, the "final solution" to the "problem" of European Jews was their extermination. By the 1940s, Nazi Germany had become the world's first "genocidal state."

Although I had long been aware of these historical events, I had not fully grasped the nature of a genocidal state and how it functioned until I traveled to Poland and visited the Auschwitz-Birkenau State Museum. Auschwitz-Birkenau is a Nazi death camp complex where 1.1 million people were murdered during World War II. The abstraction of 6 million murders was suddenly made concrete as I saw firsthand how the German state organized the collection of Jews from as far away as the Netherlands and Greece and brought them to this concentration camp in southern Poland. I saw how the rail lines were built to efficiently unload thousands of prisoners and quickly send them to slave labor or quick death. I saw how the Birkenau camp had been built close to manufacturing facilities of German companies in order to more efficiently exploit prisoners' labor. Walking through the gas chambers where hundreds of thousands had died, I perceived the unspeakable horror of state-run mass murder.

Birkenau camp in the Auschwitz complex. These chimneys are all that remain of hundreds of prisoner barracks, which once stretched for many kilometers and held a total of approximately 90,000 people. *(Courtesy of Roderick Neumann.)*

I left Auschwitz-Birkenau thinking how important the experience of place is to knowledge. I could not have grasped fully the scale and magnitude of industrial mass genocide had I not looked out across the miles of rail lines and hundreds of hectares of barracks where people were stored until they could be murdered. I could never have understood how critical spatial relations were to the conduct of mass murder without physically moving through the spaces of the death camps.

I also came away realizing again how memory and place are so tightly bound. We memorialize places so that we do not forget. The United Nations declared Auschwitz-Birkenau a World Heritage Site in 1979 in hopes that the world will remember and prevent future atrocities. I know I will not forget my encounter with the places of a genocidal state.

Posted by Rod Neumann

---

increasingly difficult for even the most totalitarian regimes. Another tactic deployed by states is to allow citizens' access to the Web but to control the flow of information they receive. The most well-known example is the deal cut between the search engine company Google and the People's Republic of China in 2006. In order to gain access to

China's rapidly expanding market for Internet services, estimated at 162 million customers in 2007, Google agreed to create a self-censoring search engine. Its Chinese-language version was specially designed to match the Chinese government's censorship laws. Web sites promoting ideas that the state views as threatening to its ideological dominance, such as free speech and the practice of unauthorized religions, are filtered out by Google's search engine.

Google, based in Mountain View, California, has been widely criticized in the United States and has even been called to testify before Congress. In their own defense, Google officials argue that they can do more good for China's citizens by participating in their Internet business than by withdrawing completely. This case illustrates that political boundaries are far from irrelevant in the age of the Internet. The Internet is, to a degree, eroding political boundaries by allowing information and ideas to diffuse more rapidly and completely. As a result, political barriers to cultural diffusion have become more fragile. However, countries still have some power to control what ideas are allowed inside their territorial boundaries.

Supranational Political Bodies

In addition to independent countries and their governmental subdivisions, the third major type of political functional region is the **supranational organization** (**Figure 6.5**). **Supranationalism** exists when countries voluntarily give up some portion of their sovereignty to gain the advantages of a closer political, economic, and cultural association with their neighbors. Sometimes supranational organizations take the form of **regional trading blocs,**

supranational organization
A group of independent countries joined together for purposes of mutual interest.

supranationalism
A situation that occurs when states willingly relinquish some degree of sovereignty in order to gain the benefits of belonging to a larger political-economic entity.

regional trading blocs
Agreements made among geographically proximate countries that reduce trade barriers in order to better compete with other regional markets.

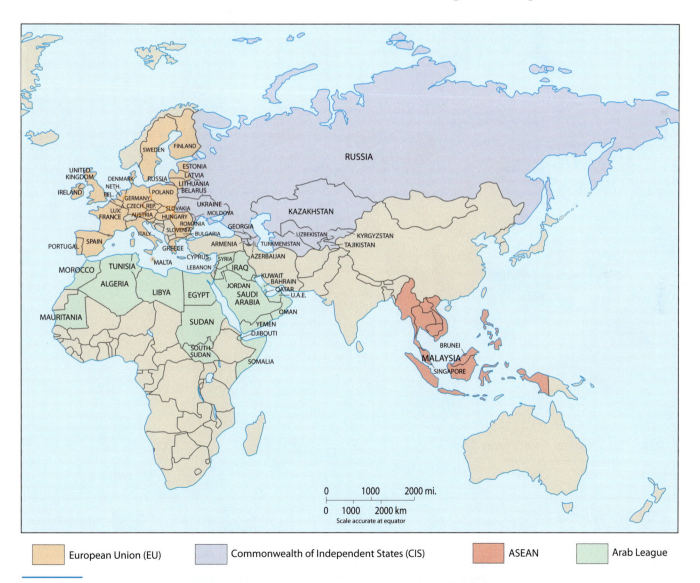

European Union (EU)   Commonwealth of Independent States (CIS)   ASEAN   Arab League

FIGURE 6.5 Some supranational political organizations in the Eastern Hemisphere. These organizations vary greatly in purpose and cohesion. ASEAN stands for the Association of Southeast Asian Nations, and its purposes are both economic and political. *What might this map indicate about globalization?*

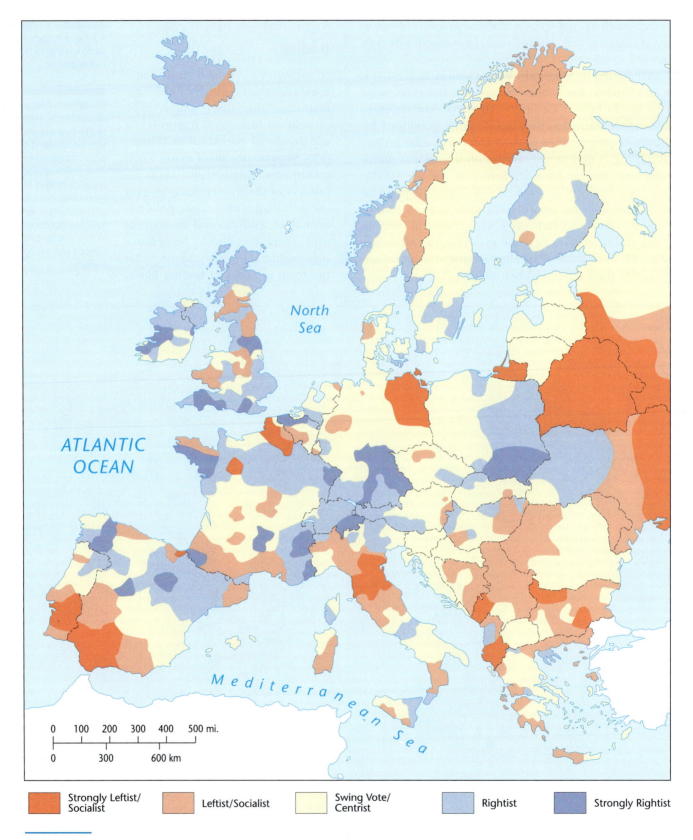

| Strongly Leftist/ Socialist | Leftist/Socialist | Swing Vote/ Centrist | Rightist | Strongly Rightist |

FIGURE 6.6 **The electoral geography of Europe.** A conservative-rightist core contrasts with a socialist-leftist periphery. Data are based on elections held in the period 1950 to 1995. In the formerly communist countries, the record of free elections began only in 1990. (*Source: Jordan-Bychkov and Jordan, 2002: 226.*)

such as the North American Free Trade Agreement (NAFTA, which comprises Canada, the United States, and Mexico), that promote the freer flow of goods and services across international borders.

In the twentieth century, supranational organizations grew in numbers and importance, coincident with and counterbalancing the proliferation of independent countries. Some represent the vestiges of collapsed empires, such as the British Commonwealth, French Community, and Commonwealth of Independent States (CIS)—the latter a shadow of the former Soviet Union. Most supranationals, such as the Arab League or the Association of Southeast Asian Nations (ASEAN), possess little cohesion.

The European Union, or EU, is by far the most powerful, ambitious, and successful supranational organization in the world (see Figure 6.5, page 197). It grew from a central core area of six countries in the 1950s to its present membership of 27. At first, the EU was merely a customs union whose purpose was to lower or remove tariffs that hindered trade, but it gradually took on more and more of a political and cultural role. An underlying motivation was to weaken the power of its member countries to the point that they could never again wage war against one another—a response to the devastation of Europe in two world wars.

The member countries have all sacrificed some of their sovereign powers to the EU administration. A single monetary currency, the euro, has been adopted by most EU members. Most international borders within the EU are now completely open, requiring no passport checks. More importantly for citizenship and national identity, the EU is standardizing a range of social norms related to the tolerance of religious and ethnic difference, human rights, and gender relations. In effect, the EU is pushing supranationalism to its logical conclusion. That is, at some future date nationalism as a focus of identity will be obsolete and people will think of themselves as European rather than as, say, Italian, or Latvian, or Polish. We will have to wait to learn whether centripetal or centrifugal forces will win out in the EU's grand experiment.

Reflecting on Geography

What will motivate the existing and prospective member countries of the European Union to continue to sacrifice aspects of their independence to create a stronger union?

Electoral Geographical Regions

Voting in elections creates another set of political regions. A free vote of the people on some controversial issue provides one of the purest expressions of culture, revealing attitudes on religion, ethnicity, and ideology. Geographers can devise formal culture regions based on voting patterns, giving rise to the subspecialty known as **electoral geography.**

Mapping voting tendencies over many decades shows deep-rooted, formal electoral behavior regions. In Europe, for example, some districts and provinces have a long record of rightist sentiment, and many of these lie toward the center of Europe. Peripheral areas, especially in the east, are often leftist strongholds (**Figure 6.6**). Every country where free elections are permitted has a similarly varied electoral geography. In other words, cumulative voting patterns typically reveal sharp and pronounced regional contrasts. Electoral geographers refer to these borders as cleavages.

Electoral geographers also concern themselves with functional regions, in this case the voting district or precinct. Their interests are both scholarly and practical. For example, following each census, political redistricting takes place in the United States. In redistricting, new boundaries are drawn for congressional districts to reflect the population changes since the previous census. The goal is to establish voting areas of more or less equal population and to increase or reduce the number of districts depending on the amount and direction of change in total population. These districts form the electoral basis for the U.S. House of Representatives. State legislatures are based on similar districts. Geographers often assist in the redistricting process.

The pattern of voting precincts or districts can influence election results. If redistricting remains in the hands of legislators, then the majority political group or party will often try to arrange the voting districts geographically in such a way as to maximize and perpetuate its power. Cleavage lines will be crossed to create districts that have a majority of voters favoring the party in power or some politically important ethnic group. This practice is called **gerrymandering** (**Figure 6.7**, page 200), and the resultant voting districts often have awkward, elongated shapes. Gerrymandering can be accomplished by one of two methods. One is to draw district boundaries so as to concentrate all of the opposition party into one district, thereby creating an unnecessarily large majority while also ensuring that it cannot win elsewhere. A second is to draw the district boundaries so as to dilute the opposition's vote so that it does not form a simple majority in any district.

electoral geography
The study of the interactions among space, place, and region and the conduct and results of elections.

gerrymandering
The drawing of electoral district boundaries in an awkward pattern to enhance the voting impact of one constituency at the expense of another.

Red States, Blue States

In recent years, various commentators in academia and the popular media have used maps of national election results to suggest that the United States is divided into

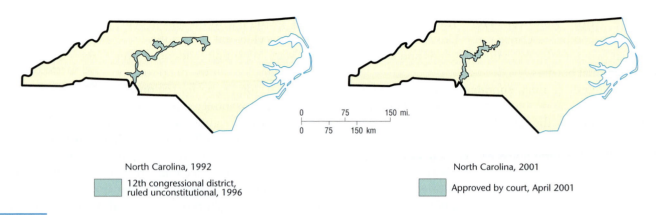

North Carolina, 1992

12th congressional district,
ruled unconstitutional, 1996

North Carolina, 2001

Approved by court, April 2001

FIGURE 6.7 Gerrymandering of a congressional district in North Carolina. The North Carolina Twelfth District was created in 1992 to ensure an African-American majority so that an additional minority candidate could be elected to Congress. Note the awkward shape of these districts, often a sign of gerrymandering. In 1996, the North Carolina district and others gerrymandered for racial purposes were declared unconstitutional. The lines were redrawn, and although the district still looks very much gerrymandered, the courts approved the revised version. *(Source: New York Times.)*

distinct culture regions. We can trace this argument to the postelection analysis of the controversial 2000 presidential election, when, for the first time, news media adopted a universal color scheme for mapping voter preferences. States where the majority of voters favored the Democratic Party's presidential candidate were assigned the color blue, whereas those favoring the Republican candidate were assigned red. When cartographers mapped the state-by-state results of the 2000 and subsequently the 2004 and 2008 presidential elections, the solid blocks of starkly contrasting colors gave the appearance of a deeply divided country (**Figure 6.8**).

The terms *red state* and *blue state* entered the popular lexicon in reference to the division of the United States into regions of "conservative" and "liberal" cultures that

these maps implied. Figure 6.8 suggests, for example, that the U.S. South is politically and culturally conservative, whereas the West Coast is liberal. These maps, so common in the popular media, are overly simplistic and, therefore, misleading. For example, the solid red coloring of a large portion of the country in Figure 6.8 implies that most of the United States is culturally and politically conservative. However, this map gives inordinate visual importance to states of larger areal extent, no matter what their population. A cartogram provides a better representation of both the true significance of states in the presidential election and the relative proportion of voters favoring Democrats or Republicans (**Figure 6.9**). When states' populations are

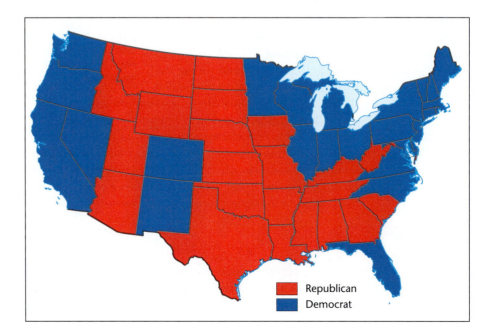

Republican
Democrat

FIGURE 6.8 2008 presidential election results. News media have adopted a standard color coding for political parties—red for Republican, blue for Democrat—in election result maps. It is now common to hear of the United States being politically and culturally divided into "red state" and "blue state" regions. *(Source: Courtesy of Mark Newman, http://www-personal.umich.edu/~mejn/election/2008/.)*

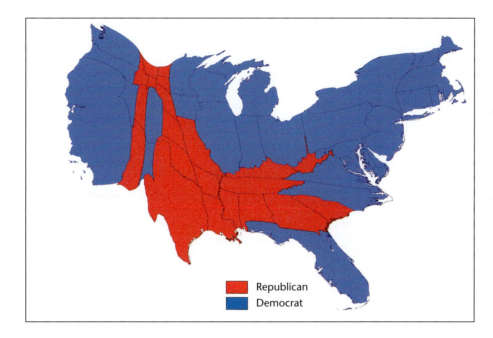

FIGURE 6.9 Cartogram of the 2008 presidential election results. This map sizes the states in proportion to their respective populations. *How does this cartogram compare to the map in Figure 6.8 in its representation of the relative popularity of the Republican and Democratic candidates?* (Source: Courtesy of Mark Newman, http://www-personal.umich.edu/~mejn/election/2008/.)

taken into account, U.S. voters look much more favorable toward Democratic rather than Republican candidates.

The red state/blue state designation also exaggerates the appearance of sharp geographic divisions within the country. Whether a state is designated "red" or "blue" is based on the winner-take-all system that the United States uses for presidential elections. That is, the candidate that receives a simple majority in the popular vote takes all the Electoral College votes for that state. George W. Bush, for example, defeated Al Gore in Florida by only 537 votes in 2000, but he received all 25 of the state's electoral votes. This was an unusually close result, but it is common in presidential elections for only a small percentage of a state's popular vote to separate the winner and the loser. Designating a state such as Florida as "red" thus masks the narrow margin of the Republican victory and exaggerates the existence of regional polarization.

The limitations of the red state/blue state designation have led to the introduction of the term *purple state* to signify closely divided states. If we apply the purple state idea (i.e., shadings of color rather than stark contrasts) to county-level election results, we find that the simplistic division of the United States into liberal and conservative regions begins to break down (**Figure 6.10**). There is a

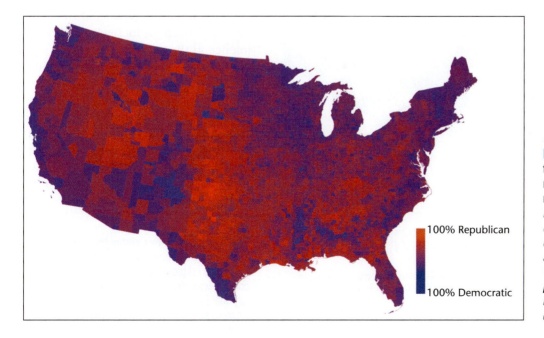

100% Republican

100% Democratic

FIGURE 6.10 Purple America. Rather than using stark color contrasts to represent the 2008 presidential election results, this map uses color shading. *How does the use of shading affect our understanding of a sharply divided American electorate that appears so prominent in Figure 6.8? What other kinds of geographic voting patterns emerge?* (Source: Courtesy of Mark Newman, http://www-personal.umich.edu/~mejn/election/2008/.)

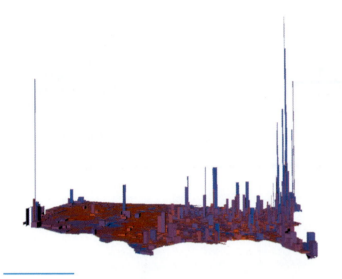

FIGURE 6.11 Presidential election results in 3-D. In this map height represents voters per square mile (for the 2004 presidential election), so urban areas stand out in contrast to suburban and rural areas. *What political-geographic patterns are noticeable here?* (Robert J. Vanderbei, Princeton University.)

great deal of county-by-county variation within states across the United States. New geographic patterns emerge in this map that belie the idea of large regional divisions in the country.

Using shades of color and county-level results—rather than using contrasting colors and state-level results—provides a very different picture of U.S. regional political and cultural differences. Even this nuanced and finely grained cartographic presentation does not reveal all of the geo-

graphic complexities of U.S. presidential elections. With three-dimensional cartographic techniques, another pattern emerges. In the map in **Figure 6.11**, height represents voters per square mile, so that volume represents total number of voters. Here the core geographic difference appears to be between cities and rural areas, not between regions. The general pattern we see is that rural areas are predominantly Republican and cities are predominantly Democratic.

A closer analysis of the red state/blue state phenomenon reveals some of the complexities involved in electoral geography. Choices about the scale at which data are aggregated, the cartographic techniques employed, and even the map colors influence the interpretation of election results. This is something to keep in mind when you evaluate claims about a divided America.

Islamic Law in Nigerian Politics

As we will see in Chapter 7, politics is often intertwined with religion. Some countries function as theocracies. In many others, internal religious divisions are expressed politically. A good example is Nigeria, which is divided between a Muslim north and a Christian/animist south (**Figure 6.12**). In recent years, all of the states of northern Nigeria added criminal law to the jurisdiction of sharia courts. Sharia is Islamic law, which is detailed in the Qur'an and has long been applied in personal and civil law in northern Nigeria, where the majority of the population is Muslim. Although Muslims welcomed the change, it has caused tensions between the Muslim majority and local non-Muslim minorities. Non-Muslims are not tried in sharia courts, but they have complained that their political and economic marginalization has increased since the change. There has been an increase in religious violence in the northern states and even reports that some Christian populations have chosen to flee the region. It is unlikely that the country will fragment along this politico-religious divide in the near future. At the very least, this movement strengthens the centrifugal forces at work in this troubled country, which has already experienced one bloody, but unsuccessful, secessionist war.

Nigerian states that have adopted Islamic law

0   100   200 mi.
0 100 200 km

FIGURE 6.12 States within Nigeria that have adopted Islamic law as of January 1, 2002. The religious tension in this African country was already intense before the regional imposition of a religion-based legal system. *Can Nigeria hold together?* (Source: Embassy of Nigeria, Washington, D.C.)

Mobility

Is mobility an important factor in understanding political geography? Absolutely! Political events and developments can trigger massive human migrations, sometimes spanning continents. Moreover, the forces of globalization have accentuated the importance of mobility across political boundaries. However, political boundaries also act as barriers to the movement of people, ideas, knowledge, and resources.

Movement Between Core and Periphery

core area
The territorial nucleus from which a country grows in area and over time, often containing the national capital and the main center of commerce, culture, and industry.

Many independent states sprang fully formed into the world, but some expanded outward when powerful political entities emerged from a small nucleus called a **core area.** These core political powers enlarged their territories by annexing adjacent lands, often over many centuries. Generally, core areas possess a particularly attractive set of resources for human life and culture. Larger numbers of people cluster there than in surrounding districts, especially if the area has some measure of natural defense against aggressive neighboring political entities. This denser population, in turn, may produce enough wealth to support a large army, which then provides the base for further expansion outward from the core area. Resources, people, and capital investment begin to flow between the core and peripheral territories, usually resulting in a further consolidation of the core's wealth and power.

In states formed in this way, the core area typically remains the country's single most important district, housing the capital city and the cultural and economic heart of the nation. The core area is the node of a functional region. France expanded to its present size from a small core area around the capital city of Paris. China grew from a nucleus in the northeast, and Russia originated in the small principality of Moscow (**Figure 6.13**). The United States grew westward from a core between Massachusetts and Virginia on the Atlantic coastal plain, an area that still has the national capital and the densest population in the country.

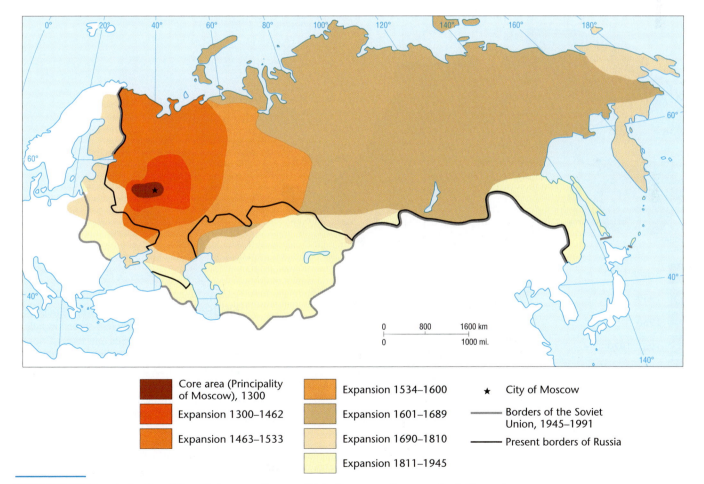

FIGURE 6.13 Russia developed from a core area. *Can you think of reasons why expansion to the east was greater than expansion to the west? What environmental goals might have motivated Russian expansion?*

The evolution of independent countries in this manner produces the core-periphery configuration, described in Chapter 1 as typical of both functional and formal regions. Although the core dominates the periphery, a certain amount of friction exists between the two. Peripheral areas generally display pronounced, self-conscious regionality and occasionally provide the settings for secession movements. Even so, countries that diffused from core areas are, as a rule, more stable than those created all at once to fill a political void. The absence of a core area can blur or weaken citizens' national identity and make it easier for various provinces to develop strong local or even foreign allegiances. Belgium and the Democratic Republic of the Congo offer examples of countries without political core areas. In the case of the Congo, this situation partly accounts for the history of secessionist conflicts and internecine wars since the country gained independence from colonial rule in 1960.

Countries with multiple, competing core areas are potentially the least stable of all. This situation often develops when two or more independent countries are united. The main threat is that one of the competing cores will form the center of a separatist movement and break apart the country. In Spain, Castile and Aragon united in 1479, but tensions in the union remain more than five centuries later—in part because the old core areas of the two former countries, represented by the cities of Madrid and Barcelona, continue to compete for political control and cultural influence. That these two cities symbolize two language-based cultures, Castilian and Catalan, compounds the division.

Mobility, Diffusion, and Political Innovation

One of the consequences of colonialism was the creation of new patterns of mobility among the conquered populations. One pattern in particular contained the seeds of colonialism's demise. Colonial rule in Africa depended on the establishment of a relatively small cadre of educated African civil servants. The colonial rulers selected those whom they considered the best and the brightest (and most politically cooperative) among their African subjects and sent them to the best universities in Europe. In addition to learning the skills required of colonial functionaries,

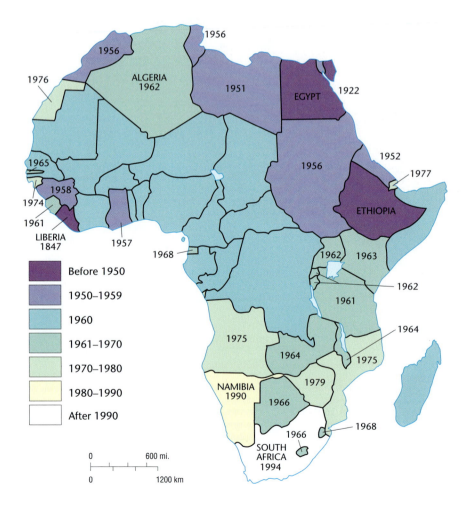

FIGURE 6.14 Independence from European colonial or white minority rule spread rapidly through Africa in the 1950s and 1960s. Prior to the 1950s, there were only three independent countries in Africa. Between 1951 and 1968, most attained self-rule. The remaining few gained freedom over the next three decades. By 1994, independence had spread from the Mediterranean to the Cape of Good Hope. Many of the ideas about national liberation were spread by African leaders educated in Europe. *Why do you think it took a few countries so much longer than most to gain independence?*

they absorbed the ideals of Western political philosophy, such as the right of self-determination, independent self-rule, and individual rights. Educated Africans returned to their countries armed with these ideas and organized new nationalist movements to oppose colonial rule.

The consequences can be seen in the spread of political independence in Africa. In 1914, only two African countries—Liberia and Ethiopia—were fully independent of European colonial or white minority rule, and even Ethiopia later fell temporarily under Italian control. Influenced by developments in India and Pakistan, the Arabs of North Africa began a movement for independence. Their movement gained momentum in the 1950s and swept southward across most of the African continent between 1960 and 1965. Many of Africa's great nationalist leaders of this period had obtained university degrees in England, Scotland, France, and other European countries. By 1994, African nationalists had helped spread the idea of independence into all remaining parts of the continent, eventually reaching the Republic of South Africa, formerly under white minority rule (**Figure 6.14**).

Despite its rapid spread, diffusion of African self-rule occasionally encountered barriers. Portugal, for example, clung tenaciously to its African colonies until 1975, when a change in government in Lisbon reversed a 500-year-old policy, allowing the colonies to become independent. In colonies containing large populations of European settlers, independence came slowly and usually with blood-shed. France, for example, sought to hold on to Algeria because many European colonists had settled there. The country nonetheless achieved independence in 1962, but only after years of violence. In Zimbabwe, a large population of European settlers refused London's orders to move toward majority rule, resulting in a bloody civil war and delaying the country's independence until 1979.

On a quite different scale, political innovations also spread within independent countries. American politics abounds with examples of cultural diffusion. A classic case is the spread of suffrage for women, a movement that culminated in 1920 with the ratification of a constitutional amendment (**Figure 6.15**). Opposition to women's suffrage was strongest in the U.S. South, a region that later exhibited the greatest resistance to ratification of the Equal Rights Amendment and displayed the most reluctance to elect women to public office.

Federal statutes permit, to some degree, laws to be adopted in the individual functional subdivisions. In the United States and Canada, for example, each state and province enjoys broad lawmaking powers, vested in the legislative bodies of these subdivisions. The result is often a patchwork legal pattern that reveals the processes of cultural diffusion at work. A good example is the clean air movement, which began in California with the initiation of state legislation regulating automotive and industrial emissions. It later spread to other states and provided the model for clean air legislation at the federal level.

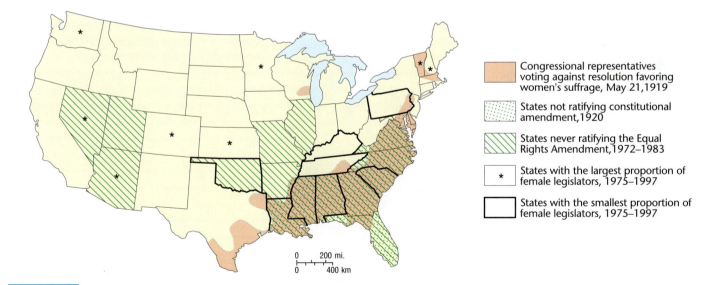

FIGURE 6.15 The diffusion of suffrage for women in the United States and of the Equal Rights Amendment. The suffrage movement achieved victory through a constitutional amendment in 1920. Both the suffrage movement and the campaign for an Equal Rights Amendment (ERA) for women failed to gain approval in the Deep South, an area that also lags behind most of the remainder of the country in the election of women to public office. *What might be the barriers to diffusion in the Deep South?* The states failing to ratify the ERA lay mostly in the same area. The ERA movement did not succeed, in contrast to the earlier suffrage movement. *(Adapted in part from Paulin and Wright, 1932.)*

Politics and Migration

Very often political events provide the motivation for migration, both voluntary and forced. An excellent example is provided by the breakup of the Soviet Union into 15 independent countries in 1991. During the long period of Soviet unity, the country's various national groups migrated into one another's territories in large numbers. Ethnic Russians, for instance, settled in all parts of the Soviet Union, even though only one of the 15 constituent republics was Russian by identity. Likewise, many members of the other 14 major ethnic groups had relocated outside their republics.

When the Soviet Union dissolved in 1991, a profound reverse migration set in almost immediately. Ethnic Russians returned to Russia, Estonians to Estonia, Azeris to Azerbaijan, and so on. By 1995, about 2 million ethnic Russians had migrated to Russia. Fully half a million of these came from recently independent Kazakhstan in central Asia. In that same period, 150,000 ethnic Kazakhs returned to Kazakhstan from Russia and other former Soviet republics. This process continues to the present day. (For other results of migration, see Practicing Geography.)

Globalization

How does globalization affect the world of states? The effects of globalization are complex. On the one hand, the idea of national self-determination continues to spread with the increase in global communications. Thus, new states appear on the world map almost annually. On the other hand, the forces of globalization may be rendering some functions of the territorial state obsolete. As we will learn in this section, a great deal of speculation exists regarding the impact of globalization on political geography (see Subject to Debate).

The Nation-State

The link between political and cultural patterns is epitomized by the **nation-state,** created when a nation—a people of common heritage, memories, myths, homeland, and culture; speaking the same language;

> **nation-state**
> An independent country dominated by a relatively homogeneous culture group.

PRACTICING GEOGRAPHY

Professor Katharyne Mitchell spent her undergraduate years studying literature, music, and art, all but oblivious to the discipline of geography. It was not until graduate school, working toward a degree in architecture, that she discovered the discipline. "I think I hardly even knew what geography was until Manuel Castells told me that all of the questions I had been pestering him with concerning space and power were questions then being pursued by geographers." She switched majors, earned her doctorate in geography, and never looked back. "It was like coming home." Professor Mitchell is now in the Department of Geography at the University of Washington, specializing in the study of cultural identity, citizenship, and transnationalism related to immigration.

Research for her sometimes "feels like detective work," where one painstakingly gathers evidence and "suddenly all of the different kinds of data click together." Professor Mitchell is convinced that "the more types of methods one can bring to bear on answering a research question, the better." If, on the one hand, she is addressing macroeconomic and political questions, she concentrates on examining statistics and quantitative data of various kinds. If, on the other hand, the questions are cultural and social, her methods are more qualitative. "I examine newspapers, photographs, flyers, advertisements, journals, postcards, and other documents from the time period I'm analyzing," she explains, "and I always interview people and read what they have to say in letters to the editor and other media." Part of her methodology is spending time walking through the relevant spaces and taking photographs.

She is currently studying how immigrants and second-generation children are educated to become "cultural" citizens of a particular nation. "By this I mean how do kids come to form a particular kind of allegiance to a particular country—especially in the contemporary time period when so much migration is transnational and there are so many global pressures on immigrant individuals and families." She was led to this research through her interest in differing explanations of immigrant integration, from multiculturalism to assimilation. One of the ways through which she approaches the question is to focus on national education systems. According to Professor Mitchell, national systems of education are important sites "in which to study the ways in which individuals are constituted or 'made' into certain kinds of national subjects or 'citizens.'"

(Courtesy of Katharyne Mitchell.)

Katharyne Mitchell

Subject to Debate
Whither the Nation-State?

Globalization presents powerful new challenges to the nation-state's dominant position in political geography. Undeniably, globalization has redefined the nation-state's role in regulating cultural, economic, and political life within its territorial boundaries. For instance, some scholars suggest that the increasing mobility of people on a global scale means that the ideal of a culturally homogeneous national identity is less and less attainable. In the economic sphere, production, trade, and finance are organized through transnational networks that operate as if national borders did not exist. Politically, the power of the nation-state is weakened by the rise of global institutions, such as the World Trade Organization, that have the power to penalize countries through trade sanctions.

There is, however, counterevidence suggesting that the nation-state is stronger than ever under globalization. In response to the greater transnational mobility of people, nationalist opposition has grown through so-called nativist movements that seek to staunch the flow of immigrants at their countries' borders. Although finance and trade operate through global networks, transnational flows remain constrained and directed by the laws and policies of nation-states. Capital investment remains solidly grounded in the legal and institutional structures of the nation-state. The international pecking order, in place for more than a century, remains intact because globalization has done little to shift the structural inequalities among nation-states. Clearly globalization's implications for the continued supremacy of the nation-state are complex and unresolved.

Continuing the Debate

Based on the discussion above, consider these questions:

- Is globalization a new phenomenon operating independently of the nation-state, or is it a set of processes structured by an underlying foundation of sovereign countries?

- Are the territorial barriers erected by nation-states being dissolved or reinforced by globalization?

- Under what circumstances and because of what cultural, economic, or political phenomenon might globalization strengthen or weaken nation-states?

- Is a major reorganization of the world's political geography still unfolding, with the consequences of globalization yet to be fully evident?

- What new forms of political-geographic organization might emerge to supplant the territorial nation-state?

The Hungarian parliament building, built in 1904, is the seat of the Hungarian national government. The monumental size and historic architectural styles of many national capitols are meant to associate the nation-state with the qualities of stability, permanence, and antiquity *(MIVA Stock/eStock Photo.)*

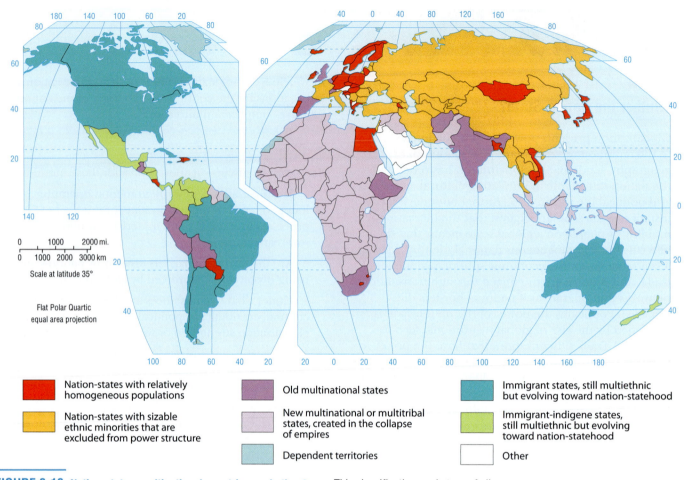

FIGURE 6.16 Nation-states, multinational countries, and other types. This classification, as is true of all classifications, is arbitrary and debatable. *How would you change it, and why?*

and/or sharing a particular religious faith—achieves independence as a separate country. Nationality is culturally based in the nation-state, and the country's raison d'être lies in that cultural identity. The more the people have in common culturally, the more stable and potent is the resultant nationalism. Examples of modern nation-states include Germany, Sweden, Japan, Greece, Armenia, and Finland (**Figure 6.16**).

Scholars generally trace the nation-state model to Europe and the European settler colonies in the Americas of the late eighteenth century. Political philosophers of the time argued that self-determination—the freedom to rule one's own country—was a fundamental right of all peoples. The ideal of self-determination spread, and over the course of the nineteenth century, a globalized system of nearly 100 nation-states emerged. Many of these nation-states, such as Portugal, Holland, and France, ruled overseas territories as colonies of the mother country. Under the European Empire, self-determination did not apply to colonies. However, the United Nations incorporated the principle of national self-determination into its charter in 1945. This

principle was adopted by independence movements in European colonies around the globe and led ultimately to decolonization.

The Condition of Transnationality

Transnationalism and *transnationality* are terms that are increasingly being heard in the halls of academia as well as in popular media. Among other things, these terms suggest that the processes of globalization are raising new questions about cultural identities. Nationality and ethnicity may continue to be important in an era of globalization, but what should one make of the increasing number of people who live, work, and play in a way that is neither rooted in a tightly knit ethnic community nor territorially grounded in the traditional nation-state? What sorts of cultural identities do these people create and defend? Is a culture of transnationalism emerging?

Geographer Katharyne Mitchell (see Practicing Geography, page 206) suggests that we might think about some of the cultural implications of globalization in terms of a "condition of transnationality." She points out that the restructuring of the world economy caused by globalization has produced a great increase in the movement of people across borders. One of the key ways in which this movement differs from other historical migrations is that it tends not to be unidirectional and permanent. With the aid of new transportation and communications technologies, people are able to maintain social networks and physically move about in ways that transcend political boundaries. The experience of cultural "in-betweenness"—of living in and being linked to multiple places around the globe, while being rooted in no single place—has become fundamental to the identity of a growing group of people.

Mitchell's study of Hong Kong Chinese immigrants in Vancouver, British Columbia, illustrates the idea of transnationalism. In the 1980s, the Canadian government created a new category of immigration designed to attract business investment. Immigrants entering the country under this law had to establish businesses based in Canada. The main immigrants taking advantage of this law were Hong Kong Chinese businesspeople leaving the colony in anticipation of its handover to the People's Republic of China.

As it happened, these immigrants maintained business and social ties in both Hong Kong and Canada, moving freely and frequently between them. In the process, a whole set of cultural conflicts were ignited between long-time Vancouver residents and the transnational Chinese. Neighborhoods were transformed as the transnationals attempted to establish themselves economically and culturally. They rapidly bought up real estate, demolished old houses, built houses in uncharacteristic architectural styles, and redesigned residential landscaping. Their mobility—a culturally defining characteristic of transnationals—made them appear transient and rootless in the eyes of longer-term residents and weakened the legitimacy of their claims and practices. This case shows us that the cultural aspects of transnationalism are bound up with the economic and political processes of globalization.

Ethnic Separatism

Many independent countries—the large majority, in fact—are not nation-states but instead contain multiple national, ethnic, and religious groups within their boundaries. India, Spain, and South Africa provide examples of older multiethnic countries. **Figure 6.17** illustrates this point by dividing up South Africa linguistically. Many, though not all, multiethnic countries came into being in the second half of the twentieth

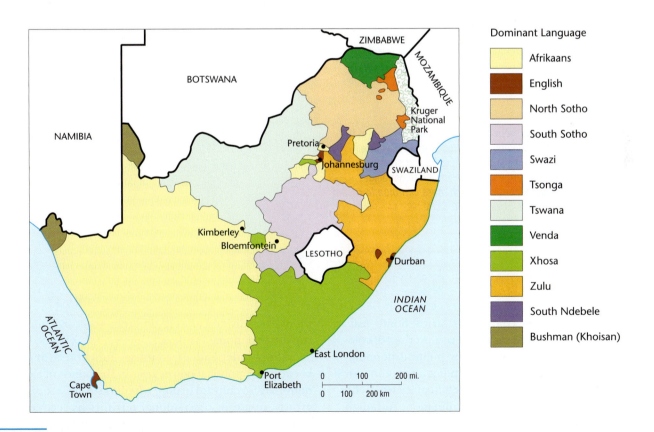

Dominant Language

- Afrikaans
- English
- North Sotho
- South Sotho
- Swazi
- Tsonga
- Tswana
- Venda
- Xhosa
- Zulu
- South Ndebele
- Bushman (Khoisan)

FIGURE 6.17 Languages of the Republic of South Africa, a multinational state. The mixture includes ten native tribal tongues and two languages introduced by settlers from Europe.

Ethnic Separation

- • Attempted genocide
- × Ethnic cleansing
- ○ Successful ethnic secession
- + Ongoing ethnic violence or strife
- − Largely nonviolent ethnic unrest
- ◉ Ethnic autonomy achieved

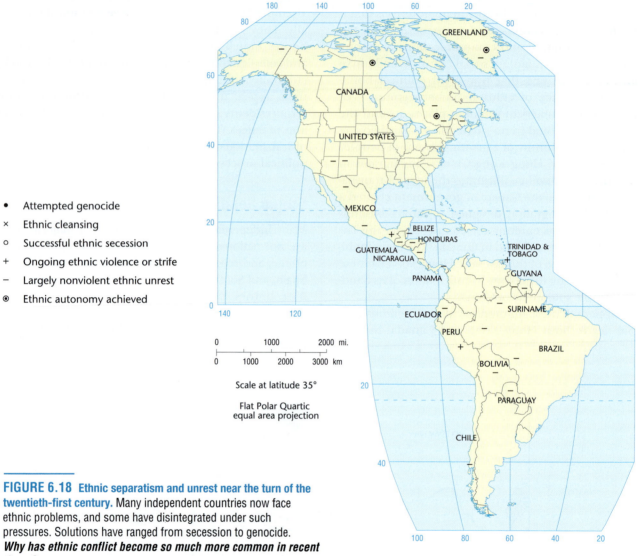

FIGURE 6.18 Ethnic separatism and unrest near the turn of the twentieth-first century. Many independent countries now face ethnic problems, and some have disintegrated under such pressures. Solutions have ranged from secession to genocide. *Why has ethnic conflict become so much more common in recent times?* (Sources: Lean, 1990: 50–51; Oldale, 1990; Smith, 1992: 140–142.)

century. These were former colonies in Africa, South Asia, and Southeast Asia whose boundaries were a product of colonialism. European colonial powers drew political boundaries without regard to the territories of indigenous ethnic or tribal groups. These boundaries remained in place when the countries gained the right of self-determination. Although these states are often culturally diverse, they are sometimes plagued by internal ethnic conflict. What's more, members of a single, territorially homogeneous ethnic group may find themselves divided among different states by culturally arbitrary international borders.

Many independent countries function as nation-states because the political power rests in the hands of a dominant, nationalistic cultural group, whereas sizable ethnic minority groups reside in the national territory as second-class citizens. This creates a centrifugal force, disrupting the country's unity. Many of the newest nation-states carved out of the former Soviet Union and Yugoslavia, such as Estonia,

Armenia, and Serbia, are relatively linguistically and ethnically homogeneous. These states' cultural homogeneity represents a centripetal force that supports national unity.

Globalization, particularly the establishment of rapid global communications, has encouraged once-isolated and voiceless minority populations to appeal their rights of autonomy and self-determination to the global community beyond their nation-state borders. Ethnic groups and indigenous peoples are cultural minorities living in a state dominated by a culturally distinct majority. Those who inhabit ethnic homelands (see Chapter 5) often seek greater autonomy or even full independence as nation-states (**Figure 6.18**). Even some old and traditionally stable multinational countries have felt the effects of separatist movements, including Canada and the United Kingdom. Certain other countries discarded the unitary form of government and adopted an ethnic-based federalism, in hopes of preserving the territorial boundaries of the state. The expression of ethnic nationalism ranges from public displays of cultural identity to organized protests and armed insurgencies. Occasionally, successful secessions occur, resulting in the birth of a new nation-state.

Francophones in Canada represent a cultural-linguistic minority group seeking secession. Approximately 7 million French-Canadians, concentrated in the province of Québec, form a large part of that country's total population of 31.6 million. Descended from French colonists who immigrated in the 1600s and 1700s, these Canadians lived under English or Anglo-Canadian rule and domination from 1760 until well into the twentieth century. Even the provincial government of Québec long remained in the hands of the English. A political awakening eventually allowed the French to gain control of their own homeland province, and as a result Québec differs in many respects from the rest of Canada. The laws of Québec retain a predominantly French influence, whereas the remainder of Canada adheres to English common law. French is the primary language of Québec and is heavily favored over English in provincial law, education, and government. In several elections, a sizable minority among the French-speaking population favored independence for Québec, and many Anglo-Canadians emigrated from the province. In 1995, more than half of the French-speaking electorate voted for independence, but the non-French minority in the province tipped the vote narrowly in favor of continued union with Canada. More recently, however, the campaign for independence seems to have weakened.

Reflecting on Geography

> Should Canada split into two independent countries? What would be the advantages and disadvantages for an independent Québec if this split occurred?

On a more general level, one result of unrest and separatist desires is that the international political map reflects a strong linguistic-religious character. Nevertheless, the distribution of cultural groups is so confoundingly complicated and peoples are so thoroughly mixed in many regions that ethnographic political boundaries can rarely be drawn to everybody's satisfaction.

cleavage model
A political-geographic model suggesting that persistent regional patterns in voting behavior, sometimes leading to separatism, can usually be explained in terms of tensions pitting urban against rural, core against periphery, capitalists against workers, and power group against minority culture.

The Cleavage Model

Why do so many cultural minorities seek political autonomy or independence? The **cleavage model,** originally developed by Seymour Martin Lipset and Stein Rokkan to explain voting patterns in electoral geography, sheds light on this phenomenon. It proposes to explain persistent regional patterns in voting behavior (which, in extreme cases, can presage separatism) in terms of tensions pitting the national core area against peripheral districts, urban against rural, capitalists against workers, and the dominant culture against minority ethnic cultures. Frequently, these tensions coincide geographically: an urban core area monopolizes wealth and cultural and political power while ethnic minorities, excluded from the power structure, reside in peripheral, largely rural, and less affluent areas.

The great majority of ethnic separatist movements shown in Figure 6.18 (pages 210–211), particularly those that have moved beyond unrest to violence or secession, are comprised of peoples living in national peripheries, away from the core area of the country. Every republic that seceded from the defunct, Russian-dominated Soviet Union lay on the borders of that former country. Similarly, the Slovenes and Croats, who withdrew from the former Yugoslavia, occupied border territories peripheral to Serbia, which contained the former national capital of Belgrade. Kurdistan is made up of the peripheral areas of Iraq, Iran, Syria, and Turkey—the countries that currently rule the Kurdish lands (**Figure 6.19**). Slovakia, long poorer and more rural than Czechia (The Czech Republic) and remote from the center of power at Prague, became another secessionist ethnic periphery. In a few cases, the secessionist peripheries were actually more prosperous than the political core area, and the separatists resented the confiscation of their taxes to support the less affluent core. Slovenia and Croatia both occupied such a position in the former Yugoslavia.

By distributing power, a federalist government reduces such core-periphery tensions and decreases the appeal of separatist movements. Switzerland, which epitomizes such a country, has been able to join Germans, French, Italians, and speakers of Raeto-Romansh into a single, stable independent country. Canada developed under Francophone pressure toward a Swiss-type system, extending considerable self-rule privileges even to the Inuit and Native American groups of the north. Russia, too, has adopted a more federalist structure to accommodate the demands of ethnic minorities, and 31 ethnic republics within Russia have achieved considerable autonomy. One of these, Chechnya (called Ichkeria by its inhabitants), has been fighting for independence.

An Example: The Sakha Republic

The Sakha Republic (also called Yakutia), in the huge Russian province of Siberia, provides a useful example of rising ethnic demands (**Figure 6.20**, page 214). The peripheral republic is enormous, forming one-fifth of Russia's land area, and is two and a half times the size of Alaska. Roughly 35 percent of its population of 1 million consists of ethnic

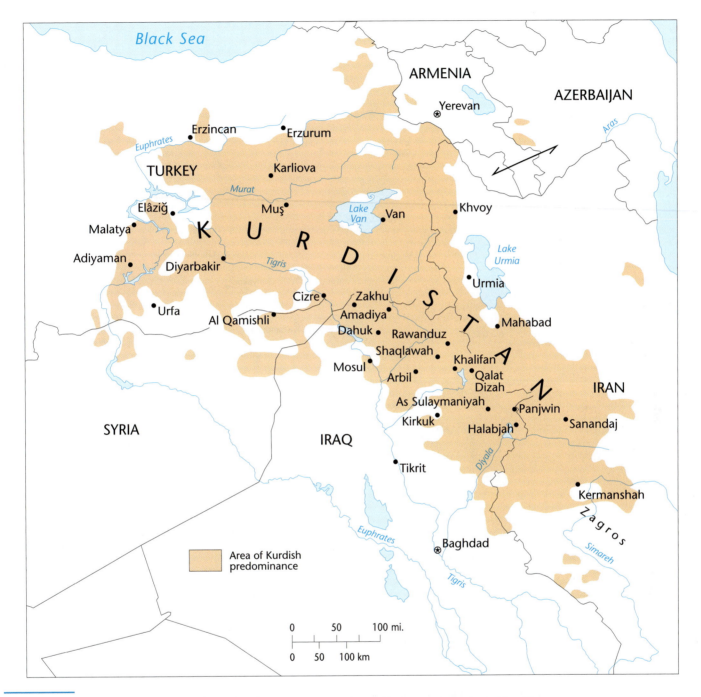

FIGURE 6.19 Kurdistan. This mountainous homeland of the Kurds now lies divided among several countries. The Kurds, numbering 25 million, have lived in this region for millennia. They seek independence and have waged guerrilla war against Iran, Iraq, and Turkey, but, so far, independence has eluded them. *What might cause so large and populous a nation to fail to achieve independence?* (Source: Office of the Geographer, U.S. Department of State.)

Sakha, or Yakuts, a people of Turkic origin (see Chapter 4). Russians, who outnumber the Sakha in the republic, are concentrated in 10 urban areas, whereas the Sakha are dominant in the rural/small-town core of the republic.

The demands of the Sakha led to a declaration of "state sovereignty" in 1990. The republic now has its own elected president and parliament, a flag and coat of arms, and a constitution (**Figure 6.21**, page 215). It has attained some measure of economic independence from Russia, especially with regard to authority over mineral rights. The republic can also prohibit nuclear testing on its territory. A survey in 1995 revealed that 72 percent of all ethnic Yakuts

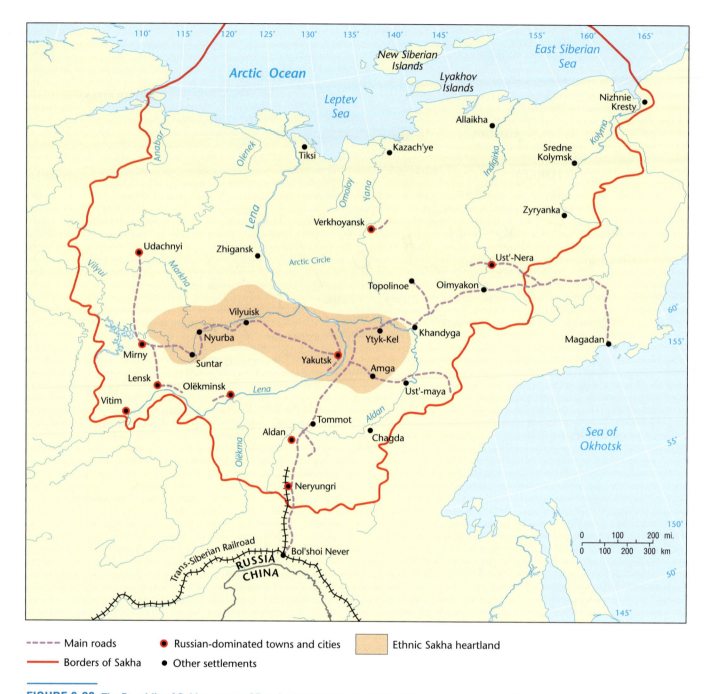

Main roads
Borders of Sakha
● Russian-dominated towns and cities
● Other settlements
Ethnic Sakha heartland

FIGURE 6.20 **The Republic of Sakha, a part of Russia, has achieved considerable autonomy as a result of ethnic considerations.** For its location in Russia, see Figure 6.18 (page 211). ***What would hinder the republic if it sought full independence?*** (Source: Jordan and Jordan-Bychkov, 2001: 4.)

felt more loyalty to Sakha than to Russia. Surprisingly, a third of all Russians living in the republic expressed this same loyalty.

The Sakha Republic does not seek independence from Russia. Still, its autonomy represents the embryo of a country. Ongoing Russian emigration from Sakha further complicates the matter. It is increasingly difficult, here as elsewhere in the modern world, to determine exactly what

an independent country is. *Sovereignty* is a political term whose meaning has become blurred.

Political Imprint on Economic Geography

The core-periphery economic differences implicit in the cleavage model reveal that the internal spatial arrangement of an independent country influences economic

For example, the U.S.–Canadian border in the Great Plains crosses an area of environmental and cultural similarity. Yet the presence of the border, representing the limits of jurisdiction of two different bodies of law and regulations, fostered differences in agricultural practices. In the United States, an act passed in the 1950s encouraged sheep raising by guaranteeing an incentive price for wool. No such law was passed in Canada. Consequently, sheep became far more numerous on the American side of the border, whereas Canadian farmers devoted more attention to swine (Figure 6.22).

Nature-Culture

What is the relationship between politics and the environment? How people use the land and natural resources is profoundly influenced by politics. Whether a particular habitat is conserved or degraded often has much to do with the structures of a country's land laws, tax codes, and agricultural policies. However, increasingly, politics is being defined by changing environmental conditions. National policies about environmental protection, guerrillas seeking a secure base for their operations, and the natural defense provided for an independent country by a surrounding sea all reveal an intertwining of environment and politics. How governments respond to ecological crises like the loss of biodiversity, pollution, and climate change has become an important political issue. Let's examine this complex two-way interaction between politics and the environment.

FIGURE 6.21 Coat of arms of the ethnic Republic of Sakha in Russia. The inscription is bilingual—Sakhan and Russian—and the picture of the horse rider is taken from ancient primitive rock art. *Why might this ancient image have been used in the coat of arms of a republic seeking increased autonomy?* (Source: Courtesy of the government of the Republic of Sakha.)

patterns, presenting a cultural interaction of politics and the economy. Moreover, laws differ from one country to another, which affects economic land use. As a result, political boundaries can take on an economic character as well.

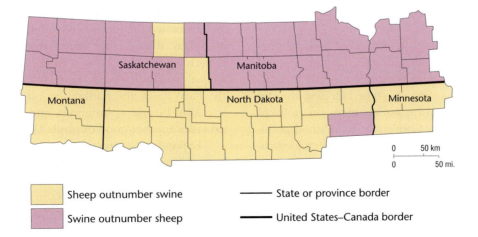

Sheep outnumber swine
Swine outnumber sheep
State or province border
United States–Canada border

FIGURE 6.22 The political impact on the economy. Government intervention can be seen in the choice of livestock in the border area between the United States and Canada. Sheep are more numerous than swine on the U.S. side of the boundary, partly because of government-backed price incentives for wool. *The map reflects conditions in the 1960s. Since then, the contrast has essentially disappeared. Why might that have happened?* (Source: Reitsma, 1971: 220–221. See also Reitsma, 1988.)

Chain of Explanation

When geographers Piers Blaikie and Harold Brookfield used the term *political ecology,* they were interested in trying to understand how political and economic forces affect people's relationships to the land. They suggested that focusing on proximate or immediate causes—for example, the farmer dumping pesticides in a river or the poor peasant cutting a patch of tropical forest—provided an inadequate and misleading explanation of human-environment relations. As an alternative, they developed the idea of a "chain of explanation" as a method for identifying ultimate causes. The chain of explanation begins with the individual "land managers," the people with direct responsibility for land-use decisions—the farmers, timber cutters, firewood gatherers, or livestock keepers. The chain of explanation then moves up in spatial scale, tracing the land managers' economic, cultural, and political relationships from the local to the national and, ultimately, to the global scale.

One of the primary areas addressed by the chain-of-explanation approach is the character of the state, particularly the way in which national land laws, natural resource policies, tax codes, and credit policies influence land-use decision making. For example, if a state assesses high taxes on land improvements, such as terracing and channeling, its tax policies actually create disincentives for land managers to implement soil conservation measures. Conversely, if a state provides cheap loans to land managers to build such structures, its credit policies encourage soil conservation. There are many examples of state influences on individual land-use decisions, leading Blaikie and Brookfield to argue that one cannot fully explain the causes of environmental problems without analyzing the role of the state.

Geopolitics

Spatial variations in politics and the spread of political phenomena are often linked to terrain, soils, climate, natural resources, and other aspects of the physical environment. The term **geopolitics** was originally coined to describe the influence of geography and the environment on political entities. Conversely, established political authority can be a powerful instrument of environmental modification, providing the framework for organized alteration of the landscape and for environmental protection.

geopolitics
The influence of the habitat on political entities.

folk fortress
A stronghold area with natural defensive qualities, useful in the defense of a country against invaders.

Before modern air and missile warfare, a country's survival was often aided by some sort of natural protection, such as surrounding mountain ranges, deserts, or seas; bordering marshes or dense forests; or outward-facing escarpments. Political geographers named such natural strongholds **folk fortresses.** The

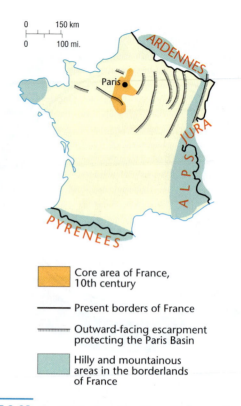

Core area of France, 10th century

Present borders of France

Outward-facing escarpment protecting the Paris Basin

Hilly and mountainous areas in the borderlands of France

FIGURE 6.23 **The distribution of landforms in France.** Terrain features such as ridges, hills, and mountains offer protection. Outward-facing escarpments form a folk fortress that protected the core area and capital of France until as recently as World War I. Hill districts and mountain ranges lend stability to French boundaries in the south and southeast.

folk fortress might shield an entire country or only its core area. In either case, it is a valuable asset. Surrounding seas have helped protect the British Isles from invasion for the past 900 years. In Egypt, desert wastelands to the east and west insulated the fertile, well-watered Nile Valley core. In the same way, Russia's core area was shielded by dense forests, expansive marshes, bitter winters, and vast expanses of sparsely inhabited lands. France—centered on the plains of the Paris Basin and flanked by mountains and hills such as the Alps, Pyrenees, Ardennes, and Jura along its borders—provides another good example (**Figure 6.23**).

Expanding countries often regard coastlines as the logical limits to their territorial growth, even if those areas belong to other peoples, as the drive across the United States from the East Coast to the Pacific Ocean in the first half of the nineteenth century has made clear. U.S. expansion was justified by the doctrine of manifest destiny, which was based on the belief that the Pacific shoreline offered the logical and predestined western border for the country. Underlying the doctrine was a racist ideology of Anglo-Saxon superiority, which held that Native Americans were savages blocking the progress of civilization and the productive use of the western lands. A similar doctrine led Russia to expand in the directions of the Mediterranean and Baltic seas and the Pacific and Indian oceans.

The Heartland Theory

Discussions of environmental influence, manifest destiny, and Russian expansionism lead naturally to the **heartland**

> **heartland theory**
> A 1904 proposal by Halford Mackinder that the key to world conquest lay in control of the interior of Eurasia.

theory of Halford Mackinder. Propounded in the early twentieth century and based on environmental determinism, the heartland theory addresses the balance of power in the world and, in particular, the possibility of world conquest based on natural habitat advantage. It held

that the Eurasian continent was the most likely base from which to launch a successful campaign for world conquest.

In examining this huge landmass, Mackinder discerned two environmental regions: the **heartland,** which lies remote from the ice-free seas, and the **rimland,** the densely populated coastal fringes of Eurasia in the east, south, and west (**Figure 6.24**). Far from the sea, the heartland was invulnerable to the naval power of rimland empires, but the cavalry and infantry of the heartland could spill out through diverse natural gateways and invade the rimland region. Mackinder thus reasoned that a

> **heartland**
> The interior of a sizable landmass, removed from maritime connections; in particular, the interior of the Eurasian continent.

> **rimland**
> The maritime fringe of a country or continent; in particular, the western, southern, and eastern edges of the Eurasian continent.

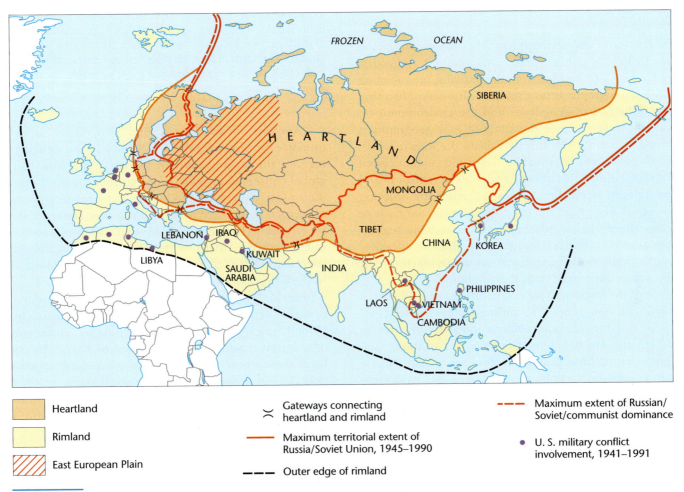

▨ Heartland	⌣ Gateways connecting heartland and rimland	--- Maximum extent of Russian/Soviet/communist dominance
▨ Rimland	— Maximum territorial extent of Russia/Soviet Union, 1945–1990	• U. S. military conflict involvement, 1941–1991
▨ East European Plain	--- Outer edge of rimland	

FIGURE 6.24 Heartland versus rimland in Eurasia. For most of the twentieth century, the heartland, epitomized by the Soviet Union and communism, was seen as a threat to America and the rest of the world, a notion based originally on the environmental deterministic theory of the political geographer Halford Mackinder. Control of the East European Plain would permit rule of the entire heartland, which, in turn, would be the territorial base for world conquest. During the cold war, 1945 to 1990, the United States and its rimland allies sought to counter this perceived menace by a policy of containment—resisting every expansionist attempt by the heartland powers. *(After Mackinder, 1904; Spykman, 1944.)*

unified heartland power could conquer the maritime countries with relative ease. He believed that the East European Plain would be the likely base for unification. As Russia had already unified this region at the time, Mackinder, in effect, predicted that the Russians would pursue world conquest.

Following Russia's communist revolution in 1917, the leaders of rimland empires and the United States employed a policy of containment. This policy, in no small measure, found its origin in Mackinder's theory and resulted in numerous wars to contain what was then considered a Russian-inspired conspiracy of communist expansion. Overlooked all the while were the fallacies of the heartland theory, particularly its reliance on the discredited doctrine of environmental determinism. In the end, Russia proved unable to hold together its own heartland empire, much less conquer the rimland and the world.

Geopolitics Today

In the post–cold war period, geopolitics has reemerged as a dynamic field of political-geographic thought. As geographers Gerard Toal and John Agnew explain, the meaning of political geography is now reversed. Instead of focusing on the influence of geography and the environment on politics, "it now becomes the study of how geography is informed by politics." By this they mean the ways in which political goals and ideologies, based on preconceived notions of cultural identities, regional stereotypes, and regional development hierarchies, influence the ordering of the world. How does the geopolitical outlook of a state structure the world into places of crisis or stability, regions of opportunity or danger, and states of allies or enemies? Many geographers distinguish this new focus on culture by labeling their approach "critical geopolitics."

One of the important aspects of critical geopolitics is a concern with how geopolitics influences our understanding of human-environment relations and affects the way we transform the environment. Consider, for example, the current scientific and policy debates over global climate change caused by greenhouse gas emissions. Worldwide debates must be placed in the historical geopolitical context of the global North's political and economic domination of the global South. From the South's perspective, according to Simon Dalby, "the North got rich by using fossil fuels for generations. Why should those in the South forgo the same possibilities just because they come on the development scene a little later?" According to advocates of the South, the North's ideas about restricting future emissions of greenhouse gases globally will have a disadvantageous effect on the South's economic development. Thus, questions of global environmental management are not restricted to the realm of science and technology; they also fall squarely in the realm of geopolitics.

FIGURE 6.25 Two visions of the landscape. Johnson Holy Rock looks out over Bear Butte State Park in the Black Hills near Sturgis, South Dakota. The Oglala Sioux Tribe of Pine Ridge Indian Reservation, of which Holy Rock is a member, considers the park area to be sacred ground. *What kinds of conflicts arise between conservation goals and indigenous peoples' rights, and how might we resolve them?* (Ann Heisenfelt/AP Photo.)

Another illustration of how geopolitics influences human-environment relations is seen in the debates concerning the conservation of global biodiversity. Current global conservation policy suggests that the most effective way to save the world's biodiversity is by creating protected areas such as national parks and reserves. However, Roderick Neumann has demonstrated that these protected areas, whether in the North or the South, were created in the historical context of European conquest and colonization. In the British Empire, protected areas were part of a grand economic development strategy to spatially reorder colonies into separate spheres of nature and culture. In the United States, park creation was conducted in the context of manifest destiny and the removal of Native Americans from their homelands and their placement on reservations. In both cases, the environmental management strategies of native cultures were denigrated as immoral and irrational, providing the justification for discarding local land and resource claims and practices. Today, those who have lost their land to protected areas argue bitterly that they bear the main costs of conservation (**Figure 6.25**). Thus, as with the case of global climate change, the North-South debates over strategies for the conservation of global biodiversity fall under the domain of geopolitics.

Warfare and Environmental Destruction

Of course, many political actions have an ecological impact, but perhaps none are as devastating as warfare.

FIGURE 6.26 A Kuwaiti oil field ablaze during the Persian Gulf War, 1991, gives new meaning to the expression "scorched earth." Severe ecological damage almost invariably accompanies warfare. ***Could modern war be waged without such damage?*** *(Noel Quidu/Gamma Liaison.)*

"Scorched earth," the systematic destruction of resources, has been a favored practice of retreating armies for millennia. Even military exercises and tests can be devastating. Certain islands in the Pacific were rendered uninhabitable, perhaps forever, by American hydrogen bomb testing in the 1950s. Actions during the Persian Gulf War of 1991 included an oil spill of 294 million gallons (1.1 billion liters) covering 400 square miles (1000 square kilometers) in the gulf waters, with the attendant loss of flora and fauna; the burning of oil fields; the mass bulldozing of sand by the Iraqis to make defensive berms, with consequent wind erosion and loss of vegetation; and the solid-waste pollution produced by 500,000 coalition forces in the Arabian Desert, including 6 million plastic bags discarded weekly by American forces alone (**Figure 6.26**).

Clearly, warfare—especially modern high-tech warfare—is environmentally catastrophic. From an ecological standpoint, it does not matter who started or won a war. Everyone loses when such destruction occurs, given the global interconnectedness of the life-supporting ecosystem.

CULTURAL LANDSCAPE

How does politics influence landscape? How are landscapes politicized? The world over, national politics is literally written on the landscape. State-driven initiatives for frontier settlement, economic development, and territorial control have profound effects on the landscape. Conversely, political writers and politicians look to landscape as a source of imagery to support or discredit political ideologies.

Imprint of the Legal Code

Many laws affect the cultural landscape. Among the most noticeable are those that regulate the land-survey system because they often require that land be divided into specific geometric patterns. As a result, political boundaries can become highly visible (**Figure 6.27**, page 220). In Canada, for example, the laws of the French-speaking province of Québec encourage land survey in long, narrow parcels, but most English-speaking provinces, such as Ontario, use a rectangular system. Thus, the political border between Québec and Ontario can be spotted easily from the air.

Legal imprints can also be seen in the cultural landscape of urban areas. In Rio de Janeiro, height restrictions on buildings have been enforced for a long time. The result is a waterfront lined with buildings of uniform height (**Figure 6.28**, page 220). By contrast, most American cities have no height restrictions, allowing skyscrapers to dominate the central city. The consequence is a jagged skyline for cities such as San Francisco or New York City. Many

FIGURE 6.27 Can you find the United States–Mexico border in this picture? The scene is near Mexicali, the capital of Baja California Norte. *Why does the cultural landscape so vividly reveal the political border?* (Courtesy of Terry G. Jordan-Bychkov.)

other cities around the world lack height restrictions, such as Malaysia's Kuala Lumpur, which has the world's tallest skyscrapers.

Perhaps the best example of how political philosophy and the legal code are written on the landscape is the so-called township and range system of the United States. The system is the brainchild of Thomas Jefferson—an early U.S.

president and one of the authors of the U.S. Constitution —who chaired a national committee on land surveying that resulted in the U.S. Land Ordinance of 1785. Jefferson's ideas for surveying, distributing, and settling the western frontier lands as they were cleared of Native Americans were based on a political philosophy of "agrarian democracy." Jefferson believed that political democracy had to be founded on economic democracy, which in turn required a national pattern of equitable land-ownership by small-scale independent farmers. In order to achieve this agrarian democracy, the western lands would need to be surveyed into parcels that could then be sold at prices within reach of family farmers of modest means.

Jefferson's solution was the township and range system, which established a grid of square-shaped "townships" with 6-mile (9.6-kilometer) sides across the Midwest and West. Each of these was then divided into 36 sections of 1 square mile (2.6 square kilometers), which were in turn divided into quarter-sections, and so on. Sections were to be the basic landholding unit for a class of independent farmers. Townships were to provide the structure for self-governing communities responsible for public schools, policing, and tax collection. With the exception of the 13 original colonies and a few other states or portions of states, a gridlike landscape was imposed on the entire country as a result of Jefferson's political philosophy and accompanying land-survey system (**Figure 6.29**).

Physical Properties of Boundaries

Demarcated political boundaries can also be strikingly visible, forming border landscapes. Political borders are

FIGURE 6.28 Legal height restrictions, or their absence, can greatly influence urban landscapes. Singapore (*left*), the city-state at the tip of the Malay Peninsula, lacks such restrictions, and its skyline is punctuated by spectacular skyscrapers. In Rio de Janeiro (*right*), by contrast, height restrictions allow the natural environment to provide the "high-rises." (Left: S. Thinakaran/AP Photo; Right: Martin Wendler/Peter Arnold, Inc.)

FIGURE 6.29 Imposing order on the land. This aerial view of Canyon County, Idaho, farmland reveals a landscape grid pattern. This pattern was imposed on the landscape across much of the United States following the passage of the Land Ordinance of 1785. *(David Frazier/Image Works.)*

usually most visible where restrictions limit the movement of people between neighboring countries. Sometimes such boundaries are even lined with cleared strips, barriers, pillboxes, tank traps, and other obvious defensive installations. At the opposite end of the spectrum are international borders, such as that between Tanzania and Kenya in East Africa, that are unfortified, thinly policed, and in many places very nearly invisible. Even so, undefended borders of this type are usually marked by regularly spaced boundary pillars or cairns, customhouses, and guardhouses at crossing points (**Figure 6.30**).

The visible aspect of international borders is surprisingly durable, sometimes persisting centuries or even millennia after the boundary becomes obsolete. Ruins of boundary defenses, some dating from ancient times, are common in certain areas. The Great Wall of China is probably the best-known reminder of past boundaries (**Figure 6.31**, page 222). Hadrian's Wall in England, which marks the northern border during one stage of Roman occupation and parallels the modern border between England and Scotland, is a similar reminder.

A quite different type of boundary, marking the territorial limits of urban street gangs, is evident in the central areas of many American cities. The principal device used by these gangs to mark their turf is spray-paint graffiti. Geographers David Ley and Roman Cybriwsky studied this phenomenon in Philadelphia. They found borders marked by externally directed, aggressive epithets, taunts, and obscenities placed there to warn neighboring gangs. A street gang of white youths, for example, plastered its border with an African-American gang's neighborhood with slogans such as "White Power" and "Do Not Enter [District] 21-W, _____." The gang's core area, its "home corner," contains internally supportive graffiti, such as "Fairmount Rules" or a roster of gang members. Thus, a perceptive observer can map gang territories on the basis of these political landscape features.

FIGURE 6.30 Even peaceful, unpoliced international borders often appear vividly in the landscape. Sweden and Norway insist on cutting a swath through the forest to mark their common boundary. *Why would they insist on making the boundary line so visible? (Courtesy of Terry G. Jordan-Bychkov.)*

FIGURE 6.31 The Great Wall of China is probably the most spectacular political landscape feature ever created and one of the few made by humans that is visible from outer space. The wall, which is 1500 miles (2400 kilometers) long, was constructed over many centuries by the Chinese in an ultimately unsuccessful attempt to protect their northern boundary from adjacent tribes of nomadic herders. **Can you think of comparable modern structures?** (Courtesy of Terry G. Jordan-Bychkov.)

outward like the spokes of a wheel to reach the hinterlands of the country, provide good indicators of central authority. In Germany, the rail network developed largely before unification of the country in 1871. As a result, no focal point stands out. However, the superhighway system of autobahns, encouraged by Hitler as a symbol of national unity and power, tied the various parts of the Reich to such focal points as Berlin or the Ruhr industrial district.

The visibility of provincial borders within a country also reflects the central government's strength and stability. Stable, secure countries, such as the United States, often permit considerable display of provincial borders. Displays aside, such borders are easily crossed. Most state boundaries in the United States are marked with signboards or other features announcing the crossing. By contrast, unstable countries, where separatism threatens national unity, often suppress such visible signs of provincial borders. Also in contrast, such "invisible" borders may be exceedingly difficult to cross when a separatist effort involves armed conflict.

Reflecting on Geography

What visible imprints of the Washington, D.C.–based central government can be seen in the political landscape of the United States?

The Impress of Central Authority

Attempts to impose centralized government appear in many facets of the landscape. Railroad and highway patterns focused on the national core area, and radiating

National Iconography on the Landscape

The cultural landscape is rich in symbolism and visual metaphor, and political messages are often conveyed through such means. Statues of national heroes or heroines and

FIGURE 6.32 The Statue of Liberty and the Rising Sun flag are widely recognized national symbols of the United States and Japan, respectively. **Can you think of other iconic landscapes that symbolize a nation?** (Left: Eyewire/Getty Images; Right: Kimimasa Mayama/EPA/Landov.)

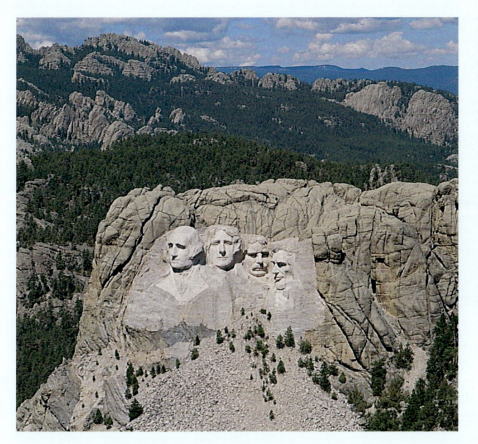

FIGURE 6.33 **Mount Rushmore,** in the Black Hills of South Dakota, presents a highly visible expression of American nationalism, an element of the political landscape. *If political landscapes are created by an elite, in an effort to legitimize and justify their control over territory, then who might disapprove of this monument?* (Brownie Harris/The Stock Market.)

of symbolic figures such as the goddess Liberty or Mother Russia form important parts of the political landscape, as do assorted monuments. The elaborate use of national colors can be visually very powerful as well. Landscape symbols such as the Rising Sun flag of Japan, the Statue of Liberty in New York Harbor, and the Latvian independence pillar in Riga (which stood untouched throughout half a century of Russian-Soviet rule) evoke deep patriotic emotions (**Figures 6.32** and **6.33**). The sites of heroic (if often futile) resistance against invaders, as at Masada in Israel, prompt similar feelings of nationalism.

Some geographers theorize that the political iconography of landscape derives from an elite, dominant group in a country's population and that its purpose is to legitimize or justify its power and control over an area. The dominant group seeks both to rally emotional support and to arouse fear in potential or real enemies. As a result, the iconographic political landscape is often controversial or contested, representing only one side of an issue. Look again at Figure 6.33. The area in which Mount Rushmore stands, the Black Hills, is sacred to the Native Americans who controlled the land before whites seized it. How might these Native Americans, the Lakota Sioux, perceive this monument? Are any other political biases contained in it? Cultural landscapes are

always complicated and subject to differing interpretations and meanings, and political landscapes are no exception.

Cultural-geographic studies of the political iconography of landscape have greatly enriched our understanding of political geography. Stephen Daniels, for example, demonstrated the importance of woodland landscapes for "naturalizing" ideas about the political and social order in late-eighteenth- and early-nineteenth-century England. Oak woodlands were imagined to embody the traditional conservative values associated with the economic and political dominance of the landed aristocracy. Newly planted coniferous woodlands of pine and fir were seen as a threat to the traditional order and a sign of the disruptive influence of the emerging class of industrial capitalists. Thus, debates over competing political ideologies were blended with debates over competing landscape aesthetics.

In a more contemporary study, geographer Gail Hollander has demonstrated the important symbolic role that the landscape and environment of the Florida Everglades have played in U.S. presidential elections (**Figure 6.34**, page 224). The symbolic role of the Everglades changed over time, from the 1928 to the 2004 presidential campaigns. In 1928, it was presented as worthless swampland. As such, it played a key role in the election of Herbert Hoover, who

FIGURE 6.34 Everglades National Park in Florida has long played a role in U.S. presidential politics. Visits to iconic national parks and protected landscapes are common in U.S. political campaigns and are meant to symbolize a candidate's commitment to protecting the environment. *(Courtesy of Roderick Neumann.)*

promised to drain it for agricultural development. By the 1970s, it was seen as an endangered wetland in need of protection and ecological restoration. Thus, in the 1996, 2000, and 2004 presidential campaigns, the candidates' positions on the Everglades were seen as indicators of their commitment to the environment. For example, when George W. Bush posed with pruning shears in the Everglades in May 2004, it offered him a chance to symbolically link his presidential campaign to the ecological restoration of what has come to be viewed as a national treasure.

CONCLUSION

Political spatial variations—from local voting patterns to the spatial arrangement of international power blocs—add yet another dimension to the complex human mosaic. In particular, nation-states operate as vital functional regions, which help shape many aspects of culture. Regions constantly change as political innovations ebb and flow across their surfaces. Political phenomena as varied as the nation-state, separatist movements, women's suffrage, and the territorial expansion of countries move along the paths of diffusion. Globalization interacts with political geography in complex and even contradictory ways. The forces of globalization have strengthened some features of the nation-state and weakened others.

Nature-culture relations are important to political geography, and the tools of political ecology help us understand the links between systems of power and the physical environment. Countries do not exist in an environmental vacuum. The spatial patterns of landforms often find reflection in boundaries, core areas, folk fortresses, and geopolitical strategies. Likewise, political culture very much influences our ideas and judgments about landscape and environment. Finally, politics leaves diverse imprints on the cultural land-

scape, and landscapes often provide the symbolism and visual metaphors to support or refute political ideologies. Political geography is clearly important to understanding the human mosaic

DOING GEOGRAPHY

The Complex Geography of Congressional Redistricting

Congressional redistricting normally happens every 10 years in the United States, following each national census. In some cases, such as Texas in 2002, redistricting occurs between censuses. Although soon proved wrong, newspaper accounts at the time predicted that Texas's midcensus redistricting would protect the Republican Party's majority in the U.S. Congress for the foreseeable future. Does this midcensus redistricting fall under the category of Republican gerrymandering, as some claim, or is it, as the Texas Republican Party argues, a case of necessary adjustments in response to population shifts? Either way, the case of Texas demonstrates how critically important the drawing of congressional district boundaries is to democratic governance.

This exercise requires you to identify cases of possible gerrymandering in your home state or an adjacent state. To do so, follow the steps below.

Steps to Identifying Gerrymandering

Step 1: Obtain a map of congressional district boundaries in your chosen state (http://nationalatlas.gov is one possible source). Once you have done so, see if you can visually identify districts that may have been gerrymandered. Figure 6.7 (page 200) and the discussion on page 199 should be helpful to you in identifying such districts.

Step 2: Having identified your candidate(s) for gerrymandering, address the following questions.

- What was it about the configuration of the boundaries that made you think the district(s) may have been gerrymandered?
- When were the boundaries drawn?
- Can you identify which of the major political parties was in power when the boundaries were drawn?
- Which party do these boundaries favor and why? That is, what are the racial, economic, religious, and ethnic characteristics of the district(s) that may suggest a particular party affiliation?

Step 3: Look at the proportion of major party registration in nearby districts to see if you can determine whether the boundary lines were drawn in order to dilute or to concentrate opposition votes.

The complex, convoluted shapes of some voting district boundaries, such as U.S. Congressional District 4 in Illinois, are evidence of gerrymandering. *(Courtesy of National Atlas.gov.)*

Key Terms

buffer state	p. 193
centrifugal force	p. 195
centripetal force	p. 195
cleavage model	p. 212
colonialism	p. 190
core area	p. 203
electoral geography	p. 199
enclave	p. 191
ethnographic boundary	p. 194
exclave	p. 192
federal state	p. 194
folk fortress	p. 216
geometric boundary	p. 194
geopolitics	p. 216
gerrymandering	p. 199
heartland	p. 217
heartland theory	p. 217
marchland	p. 193
nationalism	p. 195
nation-state	p. 206
natural boundary	p. 194
political geography	p. 189
regional trading blocs	p. 197
relic boundary	p. 194
rimland	p. 217
satellite state	p. 194
sovereignty	p. 189
state	p. 189
supranationalism	p. 197
supranational organization	p. 197
territoriality	p. 189
unitary state	p. 194

Political Geography on the Internet

You can learn more about political geography on the Internet at the following web sites:

European Union
http://europa.eu/index_en.htm
Here you can find information about the 27-member supranational organization that is, increasingly, reshaping the internal political geography of Europe.

Global Geopolitics Net
http://globalgeopolitics.net/
Sponsored by the Eurasia Research Center, this site presents information, analysis, and opinion on global politics, problems of intelligence gathering and analysis, counterterrorism, human rights, globalization, and other world issues.

International Boundary News Archive
http://www-ibru.dur.ac.uk/resources/newsarchive.html
This database contains more than 10,000 boundary-related reports from a wide range of news sources around the world dating from 1991 to March 2001, with additional reports from 2006 onward.

International Geographical Union (IGU): Commission on Political Geography
http://www.cas.muohio.edu/igu-cpg/
The objective of the commission is to study the main theoretical issues of political geography, including questions of the rise and fall of empires, the emergence of

SEEING GEOGRAPHY Post-9/11 Security Fences

Are these border fences or walls?

Post-9/11 border security: the boundary between Israeli and Arab-Palestinian lands (*left*) and the U.S.–Mexico border (*right*).

The question is more politically charged than it may seem at first. It is vehemently debated, and the side of the border from which one is observing greatly influences the answer. "Fence" suggests neighborliness; "wall" suggests isolation and exclusion. The two governments responsible for the structures, Israel and the United States, both argue that these are fences, which increase security for citizens on both sides. The U.S. Congress passed the Secure Fence Act following the 2001 terrorist attacks on the World Trade Center and the Pentagon. The act authorized the expenditure of more than $1 billion to build a 700-mile (1126-kilometer) fence on the U.S. border with Mexico to improve "homeland security." Critics argue that it is really a wall meant to stop the across-the-border flow of undocumented workers from Mexico and Central America and that there is little hope of its stopping terrorists from entering the United States. They also note that the three coastlines and the border with Canada remain without similar barriers.

The case of the barrier between Israeli and Arab-Palestinian lands is even more hotly disputed. In 2004, the Israeli government began building a physical barrier between its territory and that controlled by the Palestinian Authority. The Israel Ministry of Foreign Affairs claims that the "antiterrorist fence" is an act of self-defense against terrorists entering from Palestinian-ruled lands, such as the Gaza Strip. They also note that only 3 percent of the barrier is a concrete wall, whereas the remainder is a chain-link fence. The Palestinian Authority counters that the "wall" violates civil and human rights by cutting off communities' access to schools, workplaces, and families. The International Court of Justice seems to support the Palestinian position, ruling that construction of "the wall" is contrary to international law. Whichever side of the fence you come down on, these barriers are powerful reminders of the continuing relevance of international borders.

new geopolitical models, and contemporary challenges to the state. The site features the commission's newsletters.

Political Geography Specialty Group, Association of American Geographers

http://www.politicalgeography.org/

This site provides details about the activities and meetings of specialists in political geography and includes useful links to other sites featuring political geography and geopolitics.

United Nations

http://www.un.org

Search the worldwide organization with a membership that includes the large majority of independent countries. The site contains politically diverse information about such ventures as peacekeeping and conflict resolution.

United States "Where We Are" Project

http://nationalatlas.gov

An extremely useful site that features a wide range of interactive and dynamic maps "that capture and depict the patterns, conditions, and trends in American life."

Sources

Agnew, John. 1998. *Geopolitics: Re-Visioning World Politics.* London: Routledge.

Agnew, John. 2002. *Making Political Geography.* London: Arnold.

Blaikie, Piers, and Harold Brookfield. 1987. *Land Degradation and Society.* London: Methuen.

Blouet, Brian W. 1987. *Halford Mackinder: A Biography.* College Station: Texas A&M University Press.

Dalby, Simon. 2002. "Environmental Governance," in R. Johnston, P. Taylor, and M. Watts (eds.), *Geographies of Global Change: Remapping the World.* London: Routledge, 427–440.

Daniels, Stephen. 1988. "The Political Iconography of Woodland in Later Georgian England," in D. Cosgrove and S. Daniels (eds.), *The Iconography of Landscape.* Cambridge: Cambridge University Press, 43–82.

Elazar, Daniel J. 1994. *The American Mosaic: The Impact of Space, Time, and Culture on American Politics.* Boulder, Colo: Westview Press.

Gastner, M., C. Shalizi, and M. Newman. 2004. "Maps and Cartograms of the 2004 U.S. Presidential Election Results," University of Michigan Department of Physics and Center for the Study of Complex Systems.

Hollander, Gail. 2005. "The Material and Symbolic Role of the Everglades in National Politics." *Political Geography* 24(4): 449–475.

Jones, Martin, and Rhys Jones. 2004. "Nation States, Ideological Power and Globalization: Can Geographers Catch the Boat? *Geoforum* 35: 409–424.

Jordan, Bella Bychkova, and Terry G. Jordan-Bychkov. 2001. *Siberian Village: Land and Life in the Sakha Republic.* Minneapolis: University of Minnesota Press.

Jordan-Bychkov, Terry G., and Bella Bychkova Jordan. 2002. *The European Culture Area: A Systematic Geography,* 4th ed. Lanham, Md.: Rowman & Littlefield, Chapter 7.

Lean, Geoffrey, et al. 1990. *Atlas of the Environment.* New York: Prentice Hall.

Ley, David, and Roman Cybriwsky. 1974. "Urban Graffiti as Territorial Markers." *Annals of the Association of American Geographers* 64: 491–505.

Lipset, Seymour Martin, and Stein Rokkan (eds.). 1967. *Party Systems and Voter Alignments: Cross-National Perspectives.* New York: Free Press.

Mackinder, Halford J. 1904. "The Geographical Pivot of History." *Geographical Journal* 23: 421–437.

Mitchell, Katharyne. 1998. "Fast Capital, Race, and the Monster House," in R. George (ed.), *Burning Down the House: Recycling Domesticity.* Boulder, Colo.: Westview Press, 187–212.

Mitchell, Katharyne. 2002. "Cultural Geographies of Transnationality," in K. Anderson, M. Domosh, S. Pile, and N. Thrift (eds.), *Handbook of Cultural Geography.* London: Sage, 74–87.

Morrill, Richard L. 1981. *Political Redistricting and Geographic Theory.* Washington, D.C.: Association of American Geographers, Resource Publications.

Neumann, Roderick P. 1995. "Local Challenges to Global Agendas: Conservation, Economic Liberalization, and the Pastoralists' Rights Movement in Tanzania." *Antipode* 27(4): 363–382.

Neumann, Roderick P. 2002. "The Postwar Conservation Boom in British Colonial Africa." *Environmental History* 7(1): 22–47.

Neumann, Roderick P. 2004. "Nature-State-Territory: Toward a Critical Theorization of Conservation Enclosures," in R. Peet and M. Watts (eds.), *Liberation Ecologies,* 2nd ed. London: Routledge, 195–217.

Oldale, John. 1990. "Government-Sanctioned Murder." *Geographical Magazine* 62: 20–21.

O'Reilly, Kathleen, and Gerald R. Webster. 1998. "A Sociodemographic and Partisan Analysis of Voting in Three Anti-Gay Rights Referenda in Oregon." *Professional Geographer* 50: 498–515.

Ó Tuathail, Gearóid. 1996. *Critical Geopolitics.* Minneapolis: University of Minnesota Press.

Paulin, C., and John K. Wright. 1932. *Atlas of the Historical Geography of the United States.* New York: American Geographical Society and the Carnegie Institute.

Reitsma, Hendrik J. 1971. "Crop and Livestock Production in the Vicinity of the United States–Canada Border." *Professional Geographer* 23: 216–223.

Reitsma, Hendrik J. 1988. "Agricultural Changes in the American–Canadian Border Zone, 1954–1978." *Political Geography Quarterly* 7: 23–38.

Rumley, Dennis, and Julian V. Minghi (eds.). 1991. *The Geography of Border Landscapes.* London: Routledge.

Sack, Robert D. 1986. *Human Territoriality: Its Theory and History.* Studies in Historical Geography, No. 7. Cambridge: Cambridge University Press.

Smith, Dan. 1992. "The Sixth Boomerang: Conflict and War," in Susan George (ed.), *The Debt Boomerang*. London: Pluto Press.

Smith, Dan, et al. 1997. *The State of War and Peace Atlas,* 3rd ed. New York: Penguin.

Smith, Graham. 1999. "Russia, Geopolitical Shifts and the New Eurasianism." *Transactions of the Institute of British Geographers* 24: 481–500.

Spykman, Nicholas J. 1944. *The Geography of the Peace*. New York: Harcourt Brace.

Toal, Gerard, and John Agnew. 2002. "Introduction: Political Geographies, Geopolitics and Culture," in K. Anderson, M. Domosh, S. Pile, and N. Thrift (eds.), *Handbook of Cultural Geography*. London: Sage, 455–461.

Vanderbei, R. J. 2004. "Election 2004 Results." Available online at www.princeton.edu/~rvdb/JAVA/election2004/.

Ten Recommended Books on Political Geography

(For additional suggested readings, see *The Human Mosaic* web site: www.whfreeman.com/domosh12e)

Agnew, John. 1998. *Geopolitics: Re-Visioning World Politics*. London: Routledge. A leading political geographer critically examines the historical European perspective on world politics and shows how that vision of world order continues to influence geopolitical thinking.

Agnew, John. 2002. *Making Political Geography*. London: Arnold. This book provides an excellent overview of the field of political geography, highlighting the contributions of key thinkers from the nineteenth century to the present.

Dalby, Simon, and Gearóid Ó Tuathail (eds.). 1998. *Rethinking Geopolitics*. London: Routledge. Fifteen contributors to this postmodernist collection address questions of political identity and popular culture, state violence and genocide, militarism, gender and resistance, cyberwar, and the mass media. They suggest that political geography needs to be reconceptualized for the twenty-first century.

Herb, Guntram H., and David H. Kaplan (eds.). 1999. *Nested Identities: Nationalism, Territory, and Scale*. Lanham, Md.: Rowman & Littlefield. This collection of essays by 14 leading political geographers focuses on the geographical issue of territoriality using case studies of troubled countries and regions at different scales.

Hooson, David (ed.). 1994. *Geography and National Identity*. Oxford: Blackwell. Essays examine the connection between identity and homeland in a wide variety of settings and argue that the globalization of culture has strengthened the bonds between place and identity.

O'Loughlin, John (ed.). 1994. *Dictionary of Geopolitics*. Westport, Conn.: Greenwood Press. A basic reference book on political geography.

Olwig, Kenneth. 2002. *Landscape, Nature, and the Body Politic: From Britain's Renaissance to America's New World*. Madison: University of Wisconsin Press. This is an impressively researched historical study of the importance of landscape in shaping the ideas of nation and national identity in England and the United States.

Shelley, Fred M., J. Clark Archer, Fiona M. Davidson, and Stanley D. Brunn. 1996. *Political Geography of the United States*. New York: Guilford. A historical survey of the role of U.S. regionalism in shaping the American political system.

Wallerstein, Immanuel. 1991. *Geopolitics and Geoculture: Essays on the Changing World-System*. Cambridge: Cambridge University Press. A collection of Wallerstein's essays that link the collapse of the Soviet Union to the end of U.S. hegemony around the world.

Williams, Colin H. (ed.). 1993. *The Political Geography of the New World Order*. London: Belhaven. A collection of essays that explore the geopolitical consequences of the collapse of the Soviet Union and the rising importance of Europe and Japan.

Political-Geographical Journals

Geopolitics. This journal explores contemporary geopolitics and geopolitical change with particular reference to territorial problems and issues of state sovereignty. Published by Frank Cass. Volume 1 appeared in 1996.

Political Geography. This is a journal devoted exclusively to political geography. Formerly titled *Political Geography Quarterly*, the journal changed its name in 1992. Published by Elsevier. Volume 1 appeared in 1982.

Space and Polity. This journal is dedicated to understanding the changing relationships between the state and regional/local forms of governance. It highlights the work of scholars whose research interests lie in studying the relationships among space, place, and politics. Published by Carfax Publishing. Volume 1 appeared in 1997.

Answers

Figure 6.3 A, Armenia; B, Azerbaijan; C, Iran; a, Nagorno-Karabakh; b, the Nakhichevan Autonomous Republic; c, the Okhair Eskipara enclave; d, Sofulu enclave; e, Kyarki enclave; f, Bashkend enclave.

How can an ordinary landscape, such as a parking lot, become a sacred space?

Go to "Seeing Geography" on page 271 to learn more about this image.

7 THE GEOGRAPHY OF RELIGION
Spaces and Places of Sacredness

Religion is a core component of culture, lending vivid hues to the human mosaic. For many, religion is the most profoundly felt dimension of their identities. For this reason, it is important to clearly state what is meant by the term and provide a sense of the many ways in which religion can be manifest in people's lives. **Religion** can be defined as a more or less structured set of beliefs and practices through which people seek mental and physical harmony with the powers of the universe. The rituals of religion mark the events in our lives—birth, puberty, marriage, having children, and death—that are observed and celebrated. Religions often attempt not only to accommodate but also to influence the awesome forces of nature, life, and death. Religions help people make sense of their place in the world. In literal terms, the word *religion*—derived from the Latin *religare*—means "to fasten loose parts into a coherent whole."

But religion goes beyond a merely pragmatic set of rules for dealing with life's joys and sorrows. Most religions incorporate a sense of the supernatural that can be manifest in the concept of a God or gods who play a role in shaping human existence, in the notion of an afterlife that may involve a place of rest (or torment) for those who have died, or ideas of a soul that exists apart from our physical bodies and that may be released, or even born again, once we have died. This sense of the otherworldly is often spatially demarcated through the designation of sacred spaces, such as cemeteries, religious buildings, and sites of encounters with the supernatural.

In addition, religion is often at the heart of how people with very different worldviews can come to understand one another. So, on the one hand, the conquest of the Americas by the Iberians (people of modern-day Spain and Portugal) was often a violent affair, whereby the religious conversion of indigenous peoples to Christianity accompanied the political takeover. Temples were destroyed and coercion was often used to convert the natives to Christianity. On the other hand, the Virgin of Guadalupe, said to have appeared to the indigenous Mexican convert Juan Diego in 1531—only 10 years after Cortes's conquest of Mexico—is believed to be one of the most powerfully healing figures in the Americas to this day. Her portrait is a mixture of European and indigenous American symbols (**Figure 7.1**, page 232). In her kind manner of speaking to Juan Diego in Nahuatl (an indigenous language spoken by central Mexicans) and her resemblance to the earth goddess Tonantzín, she made sense to native Mexicans. For the Spaniards, dark-skinned Virgins had long been part of their religious symbolism. In the midst of the violence of conquest, then, the Virgin of Guadalupe provided a mother figure that was readily acceptable to native Mexicans, was familiar to Spaniards, and acted as a bridge by which people from these two very different cultures could understand one another.

religion
A social system involving a set of beliefs and practices through which people seek harmony with the universe and attempt to influence the forces of nature, life, and death.

FIGURE 7.1 The Virgin of Guadalupe. This is the image believed to have appeared to Juan Diego. As he opened his cape in the presence of Bishop Zumárraga of Mexico City, roses of Castile (a powerful symbol to Spaniards) fell onto the floor and this image was left behind on the fabric of the cape. The Virgin's downcast gaze and dark features spoke to Mexican Indians, as did the red belt about her waist, which indicates that she is pregnant. *(Mark Lennihan/ AP Photo.)*

Each of the world's major religions is organized according to more or less standardized practices and beliefs, and each is practiced in a similar fashion by millions, even billions, of adherents worldwide. Yet many people also express their religious faith in individual ways. Rituals and prayers can be adapted to fit particular circumstances or performed at home alone. Some religions, including the Taoic religions of East Asia, as well as Hinduism and Buddhism, are largely individual or family-oriented practices. Some people do not observe a widely recognized religion at all. They may be secular, holding no religious beliefs, or express skepticism—even hostility—toward organized religion. They may consider themselves to be faithful but not follow an organized expression of their beliefs. Or they may practice an unconventional belief system, or cult. The term *cult* is often used in a pejorative sense because it conjures images of mind control, mass suicide, and extreme veneration of a human leader. It is important to keep in mind that the practitioners of belief systems falling outside of the mainstream—for example, Mormonism, Scientol-ogy, and even Alcoholics Anonymous—strongly object to being labeled cult members.

Different types of religion exist in the world. One way to classify them is to distinguish between proselytic and ethnic faiths. **Proselytic religions,** such as Christianity and Islam, actively seek new members and aim to convert all humankind. For this reason, they are sometimes also referred to as **universalizing religions.** They instruct their faithful to spread the Word to all the Earth using persuasion and sometimes violence to convert the "heathen." The colonization of peoples and their lands is sometimes a result of the desire to convert them to the conqueror's religion. By contrast, each **ethnic religion** is identified with a particular ethnic or tribal group and does not seek converts. Judaism provides an

proselytic religions
Religions that actively seek new members and aim to convert all humankind.

universalizing religions
Also called proselytic religions, they expand through active conversion of new members and aim to encompass all of humankind.

ethnic religion
A religion identified with a particular ethnic or tribal group; does not seek converts.

example. In the most basic sense, a Jew is anyone born of a Jewish mother. Though a person can convert to Judaism, it is a complex process that has traditionally been discouraged. Proselytic religions sometimes grow out of ethnic religions—the evolution of Christianity from its parent Judaism is a good example.

Another distinction among religions is the number of gods worshipped. **Monotheistic religions,** such as Islam and Christianity, believe in only one God and may expressly forbid the worship of other gods or spirits. **Polytheistic religions** believe there are many gods. For example, Vodun (also spelled Voudou in Haiti or Voodoo in the southern United States) is a West African religious tradition with adaptations in the Americas wherever the enslavement of Africans was once practiced. Although, as with most major religions, there is one supreme God, it is the hundreds of spirits, or *iwa,* that Vodun adherents turn to in times of need. Some of the better-known spirits in the Haitian Voudou tradition are Danbala Wedo, the peaceful snake-god

monotheistic religion
The worship of only one god.

polytheistic religion
The worship of many gods.

who brings rain and fertility; Legba, the keeper of crossroads and doorways who is invoked at the beginning of all rituals; and Ezili Danto, the protective mother figure portrayed as a dark-skinned country woman.

Finally, the distinction between syncretism and orthodoxy is important. **Syncretic religions** combine elements of two or more different belief systems. Umbanda, a religion practiced in parts of Brazil, blends elements of Catholicism with a reverence for the souls of Indians, wise men, and historical Brazilian figures, along with a dash of nineteenth-century European spiritism, which is a set of beliefs about contacting spirits through mediums. Caribbean and Latin American religious practices often combine elements of European, African, and indigenous American religions. Sometimes, in order to continue practicing their religions, people in this region would hide statues of Afrocentric deities within images of Catholic saints. Or they would determine which Catholic figures were most like their own deities. Note in **Figure 7.2** the equation of Danbala, the

syncretic religions
Religions, or strands within religions, that combine elements of two or more belief systems.

FIGURE 7.2 Danbala and Saint Patrick. Danbala in Haitian Voudou is parallel to the Catholic Saint Patrick; both are associated with snakes. *(Left: Pierre Fugère Cherismè, Fonde-des-Negres, Haiti/Courtesy of Aid to Artisans, Inc.; Right: Mary Evans Picture Library.)*

orthodox religions
Strands within most major religions that emphasize purity of faith and are not open to blending with other religions.

snake-god of Haitian Voudou, with Saint Patrick, who is also associated with snakes. **Orthodox religions,** by contrast, emphasize purity of faith and are generally not open to blending with elements of other belief systems. The word *orthodoxy* comes from Greek and literally means "right" (*ortho*) "teaching" (*doxy*). Many religions, including Christianity, Judaism, Hinduism, and Islam, have orthodox strains. So, for instance, although some orthodox Jews closely follow a strict interpretation of the Oral Torah (a specific version of the Jewish holy book), moderate but committed Jews may observe only some or perhaps none of the dietary, marriage, and worship proscriptions observed by orthodox Jews. Intolerance of other religions, or of those fellow believers not seeming to follow the "proper" ways, is associated with **fundamentalism**

rather than orthodoxy. Many who consider themselves orthodox are in fact quite tolerant of other beliefs (see Subject to Debate).

The importance of religion to the contemporary study of cultural geography will be explored through our five themes. First, religious beliefs differ from one place to another, producing spatial variations that can be mapped as culture regions. Second, a high degree of mobility is characteristic of religions as they have spread historically through conquest, trade, and missionary activity. Violent conquests, diasporas, and the geographic expansion of religions through conversion have all played a role in shaping the contemporary world map of religious belief. Pilgrimages, too, illustrate how people can literally be set in

fundamentalism
A movement to return to the founding principles of a religion, which can include literal interpretation of sacred texts, or the attempt to follow the ways of a religious founder as closely as possible.

Subject to Debate
Religious Fundamentalism

Throughout history, religions have been one of the main ways in which people have attempted to make sense of the changing world around them. Although we may associate globalization with the fast pace and seemingly shrinking world of the past 30 years or so, people from diverse cultures have in fact been in contact across the globe for thousands of years. So how have religions helped people to cope with the changes brought by new ideas, new ways of doing things, and new belief systems?

One response is acceptance. The Muslim conquest of Iberia in the eighth century brought a centuries-long flourishing of culture to modern-day Spain and Portugal. Islamic rule here was noted for its humane and enlightened nature, whereas the rest of western Europe languished in the Dark Ages. The Muslims who ruled Iberia were renowned for their religious tolerance, and they included Jews and Christians as valued members of their governmental, scientific, and artistic communities.

Another response is intolerance. When Buddhism branched off from Hinduism, its parent religion, not all Hindus were pleased at the way Buddhism replaced Hinduism in some areas. Angkor Wat (see Figure 12.19, page 452), a temple complex in Cambodia, is replete with bas-relief carvings of Buddha figures with their faces either entirely chipped off or recarved to resemble Hindu deities. Hindus intolerant of the Khmer King

Jayavarman VII's Buddhist beliefs were responsible for defacing these sacred images on the king's death in A.D. 1220.

Today many religions, including Christianity, Judaism, Hinduism, and Islam, are experiencing intense fundamentalist movements. Fundamentalism means a return to the founding principles of a religion, which may include a literal interpretation of sacred texts and an attempt to follow the ways of a religious founder as closely as possible. Fundamentalists draw a sharp distinction between themselves and other practitioners of their religion, whom they do not believe to be following the proper religious principles, and between themselves and adherents of other faiths. Fundamentalism can be seen as an attempt to purify religious belief and practice in the face of modern influences that are thought to debase the religion. These tendencies have led to fundamentalists being regarded as antimodern and intolerant, although this view is strongly disputed by fundamentalists themselves.

Fundamentalism is an emotionally charged term because it is often used to portray its followers derogatorily as radical extremists. The tendency of the U.S. media to use the term *Islamic fundamentalists* as a synonym for *terrorists* is an unfortunate example of this. Yet there are connections between politics and religious fundamentalism. The political agenda of the U.S. government on matters of abortion, adoption, marriage, foreign policy, domestic security, gay rights, and the curriculum in public schools has been motion through religion. Third, the global spread of religious beliefs has led to the reach of some religions, such as Hinduism, beyond their traditional hearths, whereas other religions, such as Judaism, have become diluted in part through migration-related diffusion. Religions, like all elements of cultural geography, must either adapt, or not, to the globalizing world. But religions influence globalization as well; here we will consider the faith-based dimensions of the Internet. Fourth, the natural environment frequently plays an important part in faith-based belief structures, either because the forces of nature are viewed as potentially negotiable or because features of the natural landscape, such as rivers and mountains, are thought to exert a powerful influence over human destinies. Thus, the nature-culture dimension of religion is a key aspect of its cultural geography. Fifth, and finally, religious beliefs are often visible on the cultural landscape. For example, religious architecture, such as mosques, temples, and shrines, literally marks the landscape with the imprint of particular religious beliefs. How the spiritual shaping of some spaces as sacred comes about is an important statement about what, and who, matters, culturally speaking.

Region

What is the spatial patterning of religious faiths? Because religion, like all of culture, has a strong territorial tie, religious culture regions abound. The most basic kind of formal religious culture region depicts the spatial distribution of organized religions

notably influenced by politically conservative religious groups. Although not all of these groups are entirely fundamentalist in nature, many embrace fundamentalist Christian beliefs that espouse creationism and the sinfulness of homosexuality as well as question the separation of church and state. Islamism, a political ideology based in conservative Muslim fundamentalism, holds that Islam provides the political basis for running the state. Similar to the influence of Christian fundamentalism, Islamist influences in several Muslim-majority countries have set a conservative social agenda and have strongly questioned the separation of church and state.

Continuing the Debate

We often hear about religion in connection with violence. The next time you watch the nightly news broadcast or read a newspaper, keep these questions in mind:

- Do all religions have a dark, violent side to them, even those that profess a peaceful worldview?

- Is violence a necessary corollary to religious fundamentalism?

- What role does the media play in creating the perception—or inciting the practice—of religious antagonism?

Repent America, a fundamentalist Christian group, protests gay marriage outside San Francisco City Hall. *(© Dwayne Newton/PhotoEdit.)*

Major Religions

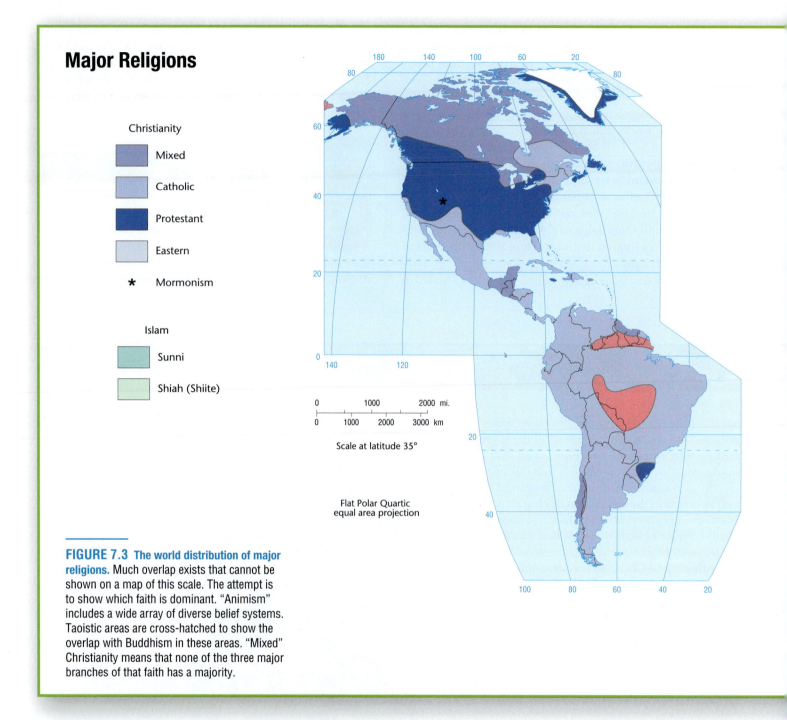

Christianity

- Mixed
- Catholic
- Protestant
- Eastern
- * Mormonism

Islam

- Sunni
- Shiah (Shiite)

0   1000   2000 mi.

0   1000   2000   3000 km

Scale at latitude 35°

Flat Polar Quartic
equal area projection

FIGURE 7.3 The world distribution of major religions. Much overlap exists that cannot be shown on a map of this scale. The attempt is to show which faith is dominant. "Animism" includes a wide array of diverse belief systems. Taoistic areas are cross-hatched to show the overlap with Buddhism in these areas. "Mixed" Christianity means that none of the three major branches of that faith has a majority.

(**Figure 7.3**). Some parts of the world exhibit an exceedingly complicated pattern of religious adherence, and the boundaries of formal religious culture regions, like most cultural borders, are rarely sharp (**Figure 7.4**, page 238). Persons of different faiths often live in the same province or town.

Judaism

Judaism is a 4000-year-old religion and the first major monotheistic faith to arise in southwestern Asia. It is the parent religion of Christianity and is also closely related to Islam. Jews believe in one God who created humankind for the purpose of bestowing kindness on them. As with Islam and Christianity, people are rewarded for their faith, are punished for violating God's commandments, and can atone for their sins. The Jewish holy book, or *Torah*, comprises the first five books of the Hebrew Bible. In contrast to the other monotheistic faiths, Judaism is not proselytic and has remained an ethnic religion through most of its existence.

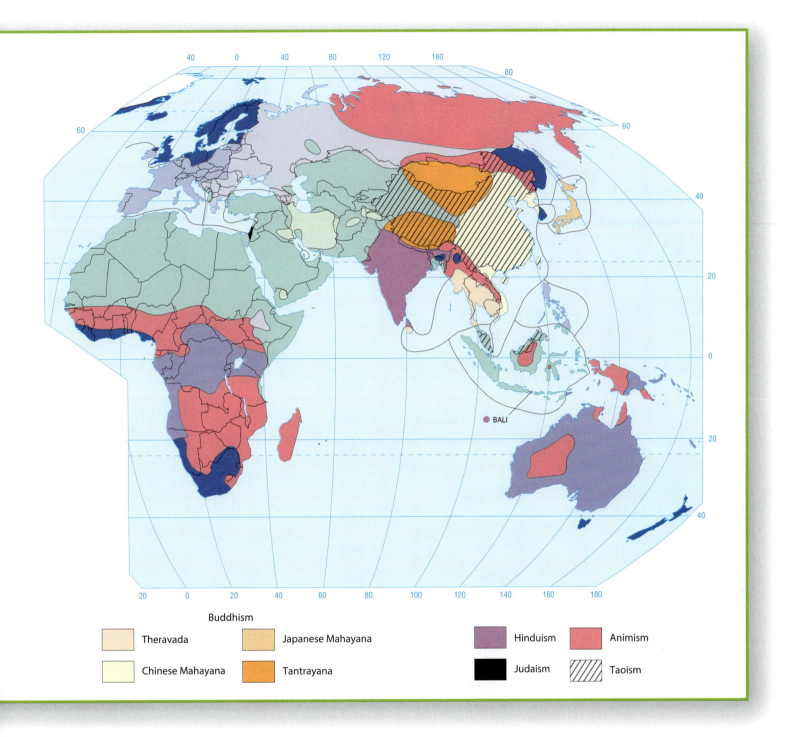

Buddhism

Theravada		Japanese Mahayana		Hinduism		Animism
Chinese Mahayana		Tantrayana		Judaism		Taoism

Judaism has split into a variety of subgroups, partly as a result of the Diaspora, a term that refers to the forced dispersal of the Jews from Palestine in Roman times and the subsequent loss of contact among the various colonies. Jews, scattered to many parts of the Roman Empire, became a minority group wherever they went. In later times, they spread throughout much of Europe, North Africa, and Arabia. Those Jews who lived in Germany and France before migrating to central and eastern Europe are known as the Ashkenazim; those who never left the Middle East and North Africa are called Mizrachim; and those from Spain and Portugal are known as Sephardim. Spain expelled its Sephardic Jews in 1492, the same year that Christopher Columbus set sail for the New World. It was not until the quincentennial of both events, in 1992, that the Spanish government issued an official apology for the expulsion.

The late nineteenth and early twentieth centuries witnessed large-scale Ashkenazic migration from Europe to America. The Holocaust that befell European Judaism during the Nazi years involved the systematic murder of perhaps

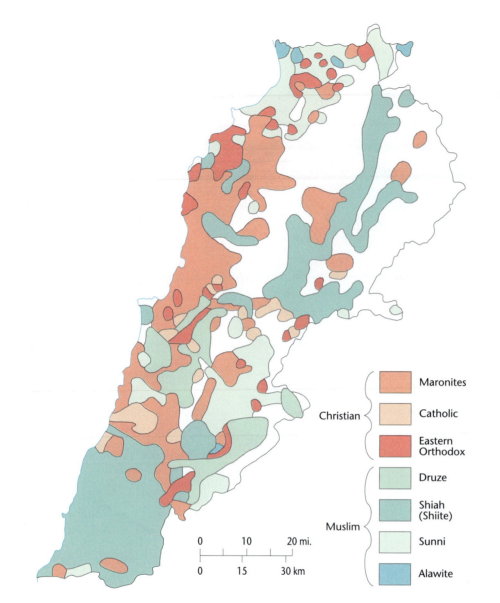

FIGURE 7.4 Distribution of religious groups in Lebanon. A land torn by sectarian warfare in recent times, Lebanon is one of the most religiously diverse parts of the world. Overall, Lebanon is today 37 percent Christian, 34 percent Shiite Muslim, 21 percent Sunni Muslim, and 7 percent Druze. Unshaded areas are largely unpopulated. *(Derived, with changes, from Klaer, 1996.)*

a third of the world's Jewish population, mainly Ashkenazim. Europe ceased to be the primary homeland of Judaism, and many of the survivors fled overseas, mainly to the Americas and later to the newly created state of Israel. Today, Judaism has about 13 million adherents throughout the world. At present, roughly 6 million, or a little less than half, live in North America, and slightly more than 5 million live in Israel.

Christianity

Christianity, a proselytic faith, is the world's largest religion, both in area covered and in number of adherents, claiming about a third of the global population (**Figure 7.5**). Christians are monotheistic, believing that God is a Trinity consisting of three persons: the Father, the Son, and the Holy Spirit. Jesus Christ is believed to be the incarnate Son of God who was given to humankind for the sake of redemption some

2000 years ago. Through Jesus' death and resurrection, all of humanity is offered redemption from sin and provided eternal life in heaven and a relationship with God.

Christianity, Islam, and Judaism are the world's three great monotheisms and share a common culture hearth in southwestern Asia (see Figure 7.14, page 245). All three religions venerate the patriarch Abraham and thus can be called Abrahamic religions. Because Judaism is the parent religion of Christianity, the two faiths share many elements, including the Torah, the first five books of what Christians call the Old Testament (which forms part of the Christian Bible or holy book); prayer; and a clergy. This is the basis for the term *Judeo-Christian,* which is used to describe beliefs and practices shared by the two faiths. Because Jesus was born into the Jewish faith, many Christians still accord Jews a special status as a chosen people and see Christianity as the natural continuation or fulfillment of Judaism.

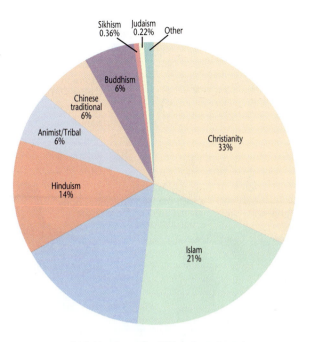

Total adds up to more than 100% due to rounding and
because upper-bound estimates were used for each group.

FIGURE 7.5 Major religions of the world. The world's major religions, by numbers of adherents expressed as a percentage of world population. Note that although Judaism is a prominent religion in the United States, it does not rank among the large religions of the world in terms of total number of adherents. Note that because of rounding and because upper-bound estimates were used for each group, percentages do not add up to 100. *(Source: adherents.com.)*

Christianity has long been fragmented into separate branches (see Figure 7.3, pages 236–237). The major division is threefold, made up of Roman Catholics, Protestants, and Eastern Christians. Western Christianity, which now includes Catholics and Protestants, was initially identified with Rome and the Latin-speaking areas that are largely congruent with modern western Europe, whereas the Eastern Church dominated the Greek world from Constantinople (now the city of Istanbul, Turkey). Belonging to the Eastern group are the Armenian Church, reputedly the oldest in the Christian faith and today centered among a people of the Caucasus region; the Coptic Church, originally the religion of Christian Egyptians and still today a minority faith there, as well as being the dominant church among the highland people of Ethiopia; the Maronites, Semitic descendants of seventh-century heretics who disagreed with some of early Christianity's beliefs and retreated to a mountain refuge in Lebanon (see Figure 7.4); the Nestorians, who live in the mountains of the Middle East and in India's Kerala state; and Eastern Orthodoxy, originally centered in Greek-speaking areas. Having converted many Slavic groups, Eastern Orthodoxy is today comprised of a variety of national churches, such as Russian, Greek,

Ukrainian, and Serbian Orthodoxy, with a collective membership of some 214 million (**Figure 7.6**).

Western Christianity splintered, most notably with the emergence of Protestantism in the Reformation of the 1400s and 1500s. As with all religious reformation movements, Protestantism sought to overcome what were viewed to be wrong practices of Roman Catholicism, such as the need for priestly or saintly mediation between humans and God, while still staying within the general framework of Christianity. Since then, the Roman Catholic Church, which alone includes 1.1 billion people, or more than one-sixth of humanity, has remained unified; but Protestantism, from its beginnings, tended to divide into a rich array of sects, which together have a total membership today of about 470 million worldwide.

In the United States and Canada, the denominational map vividly reflects the fragmentation of Western Christianity and the resulting complex pattern of religious culture regions (**Figure 7.7**, page 240). Numerous denominations imported from Europe were later augmented by Christian denominations that originated in America. The American frontier was a breeding ground for new religious groups, as individualistic pioneer sentiment found expression in splinter Protestant denominations. Today, in many parts of the country, even a relatively small community may contain the churches of half a dozen religious groups. As a result, we find today about 2000 different religious denominations and cults in the United States alone. In a broad Bible Belt across the South, Baptist and other conservative fundamentalist denominations dominate, and Utah is at the

FIGURE 7.6 Greek Orthodox Monastery. The bells of St. John's Monastery in Patmos, Greece. *(Courtesy of Patricia L. Price.)*

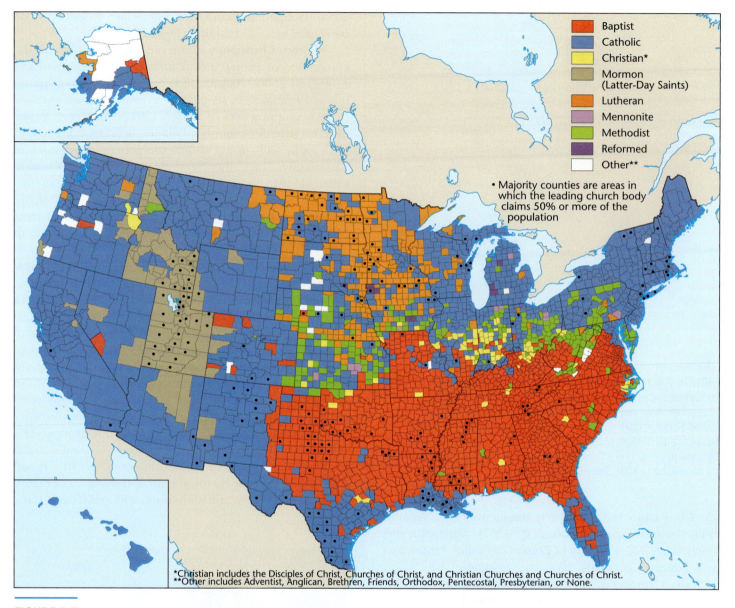

FIGURE 7.7 Leading Christian denominations in the United States, shown by counties. In the majority counties, the church or denomination indicated claimed 50 percent or more of the total church membership. The most striking features of the map are the Baptist dominance through the South, a Lutheran zone in the upper Midwest, Mormon (Latter-Day Saints) dominance in the interior West, and the zone of mixing in the American heartland. *Why do you think that so much of the western United States shows an adherence to Catholicism?* (Source: Jon T. Kilpinen, Valparaiso University; http://www.valpo. edu/geomet/geo/courses/geo200/religion.html. Data from Glenmary Research Center and U.S. Census.)

core of the Mormon realm. A Lutheran belt stretches from Wisconsin westward through Minnesota and the Dakotas, and Roman Catholicism is dominant in southern Louisiana, the southwestern borderland, and the heavily industrialized areas of the Northeast. The Midwest is a thoroughly mixed zone, although Methodism is the largest single denomination.

Today, Christianity is geographically widespread and highly diverse in its local interpretation. Although it is not as fast-growing as Islam, intense missionary efforts strive to increase the number of adherents. Because of this proselytic work, Africa and Asia are the fastest-growing regions for Christianity.

Islam

Islam, another great proselytic faith, claims more than a billion followers, largely in the desert belt of Asia and northern Africa and in the humid tropics as far east as Indonesia and the southern Philippines (see Figures 7.3, pages 236–237,

FIGURE 7.8 Muslims at prayer in Mecca, Saudi Arabia. *(Kazuyoshi Nomachi/ Pacific Press Service.)*

and 7.5, page 239). Adherents of Islam, known as Muslims (literally, "those who submit to the will of God"), are monotheists and worship one absolute God known as Allah. Islam was founded by Muhammad, considered to be the last and most important in a long line of prophets. The word of Allah is believed to have been revealed to Muhammad by the angel Gabriel (Jibrail) beginning in A.D. 610 of the Christian calendar in the Arabian city of Mecca (**Figure 7.8**). The Qur'an, Islam's holy book, is the text of these revelations and also serves as the basis of Islamic law, or sharia. Most Muslims consider both Jews and Christians to be, like themselves, "people of the Book," and all three religions share beliefs in heaven, hell, and the resurrection of the dead. Many biblical figures familiar to Jews and Christians, such as Moses, Abraham, Mary, and Jesus, are also venerated as prophets in Islam. Adherents to Islam are expected to profess belief in Allah, the one God whose prophet was Muhammad; pray five times daily at established times; give alms, or *zakat*, to the poor; fast from dawn to sunset in the holy month of Ramadan; and make at least one pilgrimage, if possible, to the sacred city of Mecca in Saudi Arabia. These duties are known as the Five Pillars of the faith.

Although not as severely fragmented as Christianity, Islam, too, has split into separate groups. Two major divisions prevail. Shiite Muslims, 16 percent of the Islamic total in diverse subgroups, form the majority in Iran and Iraq. Shiites believe that Ali, who was Muhammad's son-in-law, should have succeeded Muhammad. Sunni Muslims, who represent the Islamic orthodoxy (the word *Sunni* comes from *sunnah*, meaning "tradition"), form the large majority worldwide (see Figure 7.3). Islam has not undergone a

reformation parallel to that undergone by Christianity with Protestantism. However, liberal movements within Sunni and Shiite Islam do have reformation as their goal.

Islam's strength is greatest in the Arabic-speaking lands in Southwest Asia and North Africa, although the world's largest Islamic population is found in Indonesia. Other large clusters live in Pakistan, India, Bangladesh, and western China. Because of successful conversion efforts in non-Muslim areas and high birthrates in predominantly Muslim areas, Islam is the fastest-growing world religion. In the United States, more people convert to Islam than to any other religion, and Islam will soon surpass Judaism to become America's second-largest faith.

Hinduism

Hinduism, a religion closely tied to India and its ancient culture, claims about 800 million adherents (see Figures 7.3 and 7.5). Hinduism is a decidedly polytheistic religion. Although Hindus recognize one supreme god, Brahman, it is his many manifestations that are worshipped directly. Some principal Hindu deities include Vishnu, Shiva, and the mother goddess Devi. Ganesha, the elephant-headed god depicted in **Figure 7.9** (page 242), is often revered by Hindu university students because Ganesha is the god of wisdom, intelligence, and education. Believing that no one faith has a monopoly on the truth, most Hindus are notably tolerant of other religions (see Subject to Debate, page 234).

Hindus strive to locate the harmonious and eternal truth, *dharma*, which is within each human being. Social divisions, or castes, separate Hindu society into four major cate-

FIGURE 7.9 Ganesha. The elephant-headed god of wisdom, intelligence, and education is revered by many Hindu university students. *(Punit Puranjpe/ Reuters/Landov.)*

gories, or *varna,* based on occupational categories: priests (Brahmins), warriors (Kshatriyas), merchants and artisans (Vaishyas), and workers (Shudras). Castes are related to dharma inasmuch as dharma implies a set of rules for each varna that regulate their behaviors with regard to eating, marriage, and use of space. All Hindus also share a belief in reincarnation, the idea that, although the physical body may die, the soul lives on and is reborn in another body. Related to this is the notion of *karma.* Karma can be viewed almost as a causal law, which holds that what an individual experiences in this life is a direct result of that individual's thoughts and deeds in a past life. Likewise, all thoughts and deeds, both good and bad, affect an individual's future lives. Ultimately

moksha, or liberation of the soul from the cycle of death and rebirth, will occur when the worldly bonds of the material self fall away and one's pure essence is freed. *Ahimsa,* or the principle of nonviolence, involves veneration of all forms of life. This implies a principle of noninjury to all sentient creatures, which is why many Hindus are vegetarians.

Hinduism has splintered into diverse groups, some of which are so distinctive as to be regarded as separate religions. Jainism, for instance, is an ancient outgrowth of Hinduism, claiming perhaps 4 million adherents, almost all of whom live in India, and traces its roots back more than 25 centuries. Although they reject Hindu scriptures, rituals, and priesthood, the Jains share the Hindu belief in ahimsa and reincarnation. Jains adhere to a strict asceticism, a practice involving self-denial and austerity. For example, they practice veganism, a form of vegetarianism that prohibits the consumption of all animal-based products, including milk and eggs. Sikhism, by contrast, arose much later, in the 1500s, in an attempt to unify Hinduism and Islam. Centered in the Punjab state of northwestern India, where the Golden Temple at Amritsar serves as the principal shrine, Sikhism has about 23 million followers. Sikhs are monotheistic and have their own holy book, the Adi Granth.

No standard set of beliefs prevails in Hinduism, and the faith takes many local forms. Hinduism includes very diverse peoples. This is partly a result of its former status as a proselytic religion. A Hindu majority on the Indonesian island of Bali suggests the religion's former missionary activity (**Figure 7.10**). Today, most Hindus consider their religion an ethnic one, whereby one is acculturated into the Hindu community by birth; however, conversion to Hinduism is also allowed.

Buddhism

Hinduism is the parent religion of Buddhism, which began 25 centuries ago as a reform movement based on the teachings of Prince Siddhartha Gautama, "the awakened one" (**Figure 7.11**). He promoted the four "noble truths": life is full of suffering, desire is the cause of this suffering, cessation of suffering comes with the quelling of desire, and an Eightfold Path of proper personal conduct and meditation permits the individual to overcome desire. The resultant state of enlightenment is known as *nirvana.* Those few individuals who achieve nirvana are known as Buddhas. The status of Buddha is open to anyone regardless of social status, gender, or age. Because Buddhism derives from Hinduism, the two religions share many beliefs, such as dharma, reincarnation, and ahimsa. Buddhism and its parent religion, Hinduism, as well as the related faiths Jainism and Sikhism, are known as dharmic religions.

Today, Buddhism is the most widespread religion in South and East Asia, dominating a culture region stretch-

FIGURE 7.10 **Hindu temple in Bali, Indonesia.** Besakih, known as the Mother Temple of Bali, is the largest Hindu temple complex on the island, occupying the slopes of Mount Agung. *(Courtesy of Patricia L. Price.)*

ing from Sri Lanka to Japan and from Mongolia to Vietnam. In the process of its proselytic spread, particularly in China and Japan, Buddhism fused with native ethnic religions such as Confucianism, Taoism, and Shintoism to form syncretic faiths that fall into the Mahayana division of Buddhism. Southern, or Theravada, Buddhism, dominant in Sri Lanka and mainland Southeast Asia, retains the greatest similarity to the religion's original form, whereas a variation known as Tantrayana, or Lamaism, prevails in Tibet and Mongolia (see Figure 7.3, pages 236–237). Buddhism's tendency to merge with native religions, particularly in China, makes it difficult to determine the number of its adherents. Estimates range from 350 million to more than 500 million people (see Figure 7.5, page 239). Although Buddhism in China has become mingled with local faiths to become part of a composite ethnic religion, elsewhere it remains one of the three great proselytic religions in the world, along with Christianity and Islam.

FIGURE 7.11 **Buddhism is one of the religious faiths of South Korea.** Here an image of the Buddha is carved from a rocky bluff to create sacred space and a local pilgrimage shrine. *(Courtesy of Terry G. Jordan-Bychkov.)*

Taoic Religions

Confucianism, Shinto, and Taoism together make up the faiths that center on Tao, or the force that balances and orders the universe. Derived from the teachings of philosopher Kung Fu-tzu (551–479 B.C.), Confucianism was later promoted by China's Han dynasty (206 B.C.–A.D. 220) as the official state philosophy. Thus, it has bureaucratic, ethical, and hierarchical overtones that have led some to question whether it is more a way of life than a religion in the proper sense. In China, Confucianism's formalism is balanced by Taoism's romanticism. Taoism, which is both an established religion and a philosophy, emphasizes the dynamic balance depicted by the Chinese yin/yang symbol. The "three jewels of Tao" are humility, compassion, and moderation. Shinto, once the state religion of Japan, has long been blended with Buddhism, as well as infused with a Confucian-derived legal system, but at its core Shinto is an animistic religion. Because Tao is found in nature, there is some overlap with the animistic faiths discussed next. In addition, because of their fluid nature, Taoic religions tend toward syncretism and have blended with other faiths, particularly Buddhism. People may simultaneously practice elements of all of these faiths. For these reasons, it is difficult to provide an exact number of adherents, although estimates hold there to be about half a billion. Taoic religions center on East Asia (see Figure 7.3).

Animism/Shamanism

Peoples in diverse parts of the world often retain indigenous religions and are usually referred to collectively as animists (**Figure 7.12**). For the most part, animists do not form organized or recognized religious groups but instead practice as ethnic religions common to clans or tribes. In addition, most animists follow oral, rather than written, traditions and thus do not have holy books. A tribal religious figure, called by some a shaman, usually serves as an intermediary between the people and the spirits.

Currently numbering perhaps 240 million, animists believe that nonhuman beings and inanimate objects possess spirits or souls. These spirits are believed to live in rocks and rivers, on mountain peaks and celestial bodies, in forests and swamps, and even in everyday objects. As such, objects are considered to be alive, and, depending on the local beliefs, objects and animals may assume human forms or engage in human activities such as speech. For some animists, the objects in question do not actually possess spirits but rather are valued because they have a particular potency to serve as a link between a person and the omnipresent god. Followers of Wicca, a contemporary neopagan religion derived from pre-Christian European practices of reverence for the mother goddess and the horned god, worship the god and goddess who are thought to inhabit everything. Ritual tools used by Wiccans include the *athame* (dagger), *boline* (knife), and *besom* (broom), which are used in ceremonies to direct energy as practitioners communicate with the god and goddess.

Animistic elements can also pervade established religions. Japanese Shinto adherents worship *kami*, or spirits,

animists
Adherents of animism, the idea that souls or spirits exist not only in humans but also in animals, plants, rocks, natural phenomena such as thunder, geographic features such as mountains or rivers, or other entities of the natural environment.

FIGURE 7.12 Druids from the Mistletoe Foundation bless mistletoe in Worcestershire, England. Celtic Druids have their roots in the ancient, pre-Christian reverence for natural elements such as streams, hills, and plants. Mistletoe is sacred to the Druidic faith. *(Andrew Fox/Corbis.)*

FIGURE 7.13 Pet grave marker. Pumpkin the cat was apparently baptized into the Christian faith and now lies at rest in Pet Haven Cemetery in Miami, Florida. *(Courtesy of Patricia L. Price.)*

that inhabit natural objects such as waterfalls and mountains. We should not classify such systems of belief as primitive or simple because they can be extraordinarily complex. Even in places such as the United States, where the majority of the inhabitants would not see themselves as animists, such beliefs are pervasive. For example, many people in the United States believe that their pets possess souls that ascend to heaven on death (**Figure 7.13**).

Mobility

How did the geographical distribution of religions and denominations, of regions and places, come about? Why do religions mandate or encourage mobility through pilgrimage? The spatial patterning of religions, denominations, and secularism is the product of innovation and cultural diffusion. To a remarkable degree, the origin of the major religions was concentrated spatially in three principal culture hearth areas (**Figure 7.14**).

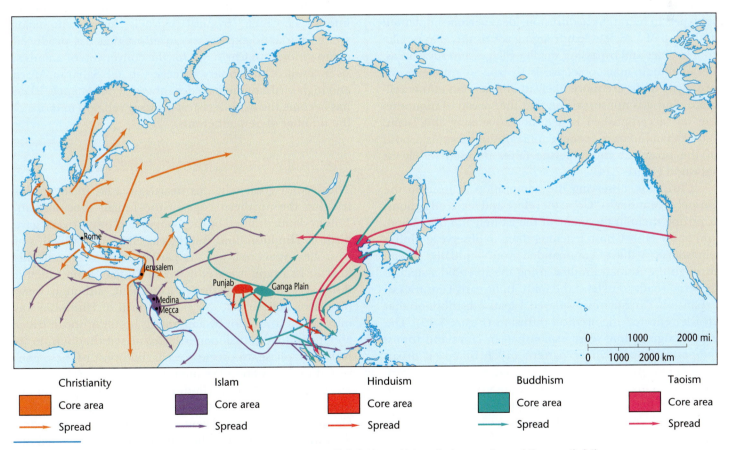

FIGURE 7.14 The origin and diffusion of five major world religions. Christianity and Islam, the two great proselytic monotheistic faiths, arose in Semitic southwestern Asia and spread widely through the Old World. Hinduism and Buddhism both originated in the northern reaches of the Indian subcontinent and spread throughout southeastern Eurasia. Taoic religions originated in East Asia, relocating with Chinese and Japanese migration regionally and, more recently, to North and South America.

culture hearth
A focused geographic area
where important innovations
are born and from which they
spread.

A **culture hearth** is a focused geographic area where important innovations are born and from which they spread. Many religions mandate periodic return of the faithful to these culture hearths in order to confirm or renew their faith.

The Semitic Religious Hearth

All three of the great monotheistic faiths—Judaism, Christianity, and Islam—arose among Semitic peoples who lived in or on the margins of the deserts of southwestern Asia, in the Middle East (see Figure 7.14, page 245). Judaism, the oldest of the three, originated some 4000 years ago. Only gradually did its followers acquire dominion over the lands between the Mediterranean and the Jordan River—the territorial core of modern Israel. Christianity, whose parent religion is Judaism, originated here about 2000 years ago. Seven centuries later, the Semitic culture hearth once again gave birth to a major faith when Islam arose in western Arabia, partly from Jewish and Christian roots.

Religions spread by both relocation and expansion diffusion. Recall from Figure 1.8 (page 10) that expansion diffusion can be divided into hierarchical and contagious subtypes. In hierarchical diffusion, ideas become implanted at the top of a society, leapfrogging across the map to take root in cities and bypassing smaller villages and rural areas. Because their main objective is to convert nonbelievers, proselytic faiths are more likely to diffuse than ethnic religions, and it is not surprising that the spread of monotheism was accomplished largely by Christianity and Islam, rather than Judaism. From Semitic southwestern Asia, both of the proselytic monotheistic faiths diffused widely, as shown in Figure 7.14.

Christians, observing the admonition of Jesus in the Gospel of Matthew—"Go ye therefore and teach all nations, baptizing them in the name of the Father, and of the Son, and of the Holy Ghost, teaching them to observe all things whatsoever I have commanded you"—initially spread through the Roman Empire, using the splendid system of imperial roads to extend the faith. In its early centuries of expansion, Christianity displayed a spatial distribution that clearly reflected hierarchical diffusion (**Figure 7.15**). The early congregations were established in cities and towns, temporarily producing a pattern of Christianized urban centers and pagan rural areas. Indeed, traces of this process remain in our language. The Latin word *pagus*, "countryside," is the root of both *pagan* and *peasant*, suggesting the ancient connection between non-Christians and the countryside.

The scattered urban clusters of early Christianity were created by such missionaries as the apostle Paul, one of Jesus' disciples who moved from town to town bearing the news of the emerging faith. In later centuries, Christian missionaries often used the strategy of converting kings or tribal leaders, setting in motion additional hierarchical diffusion. The Russians and Poles were converted in this manner. Some Christian expansion was militaristic, as in the reconquest of Iberia from the Muslims and the invasion of Latin America. Once implanted in this manner, Christianity spread farther by means of contagious diffusion. When applied to religion, this method of spread is called **contact conversion** and is the result of everyday associations between believers and nonbelievers.

contact conversion
The spread of religious
beliefs by personal
contact.

The Islamic faith spread from its Semitic hearth area in a predominately militaristic manner. Obeying the command in the Qur'an that they "do battle against them until there be no more seduction from the truth and the only worship be that of Allah," the Arabs expanded westward across North Africa in a wave of religious conquest. The Turks, once converted by the Arabs, carried out similar Islamic conquests. In a different sort of diffusion, Muslim missionaries followed trade routes eastward to implant Islam hierarchically in the Philippines, Indonesia, and the interior of China. Tropical Africa is the current major scene of Islamic expansion, an effort that has produced competition with Christians for the conversion of animists. As a result of missionary successes in sub-Saharan Africa and high birthrates in its older sphere of dominance, Islam has become the world's fastest-growing religion in terms of the number of new adherents.

The Indus-Ganges Religious Hearth

The second great religious hearth area lay in the plains fringing the northern edge of the Indian subcontinent. This lowland, drained by the Ganges and Indus rivers, gave birth to Hinduism and Buddhism. Hinduism, which is at least 4000 years old, was the earliest faith to arise in this hearth. Its origin apparently lay in Punjab, from which it diffused to dominate the subcontinent, although some historians believe that the earliest form of Hinduism was introduced from Iran by emigrating Indo-European tribes about 1500 B.C. Missionaries later carried the faith, in its proselytic phase, to overseas areas, but most of these converted regions were subsequently lost to other religions.

Branching off from Hinduism, Buddhism began in the foothills bordering the Ganges Plain about 500 B.C. (see Figure 7.14). For several centuries it remained confined to the Indian subcontinent, but missionaries later carried the religion to China (100 B.C. to A.D. 200), Korea and Japan (A.D. 300 to 500), Southeast Asia (A.D. 400 to 600), Tibet

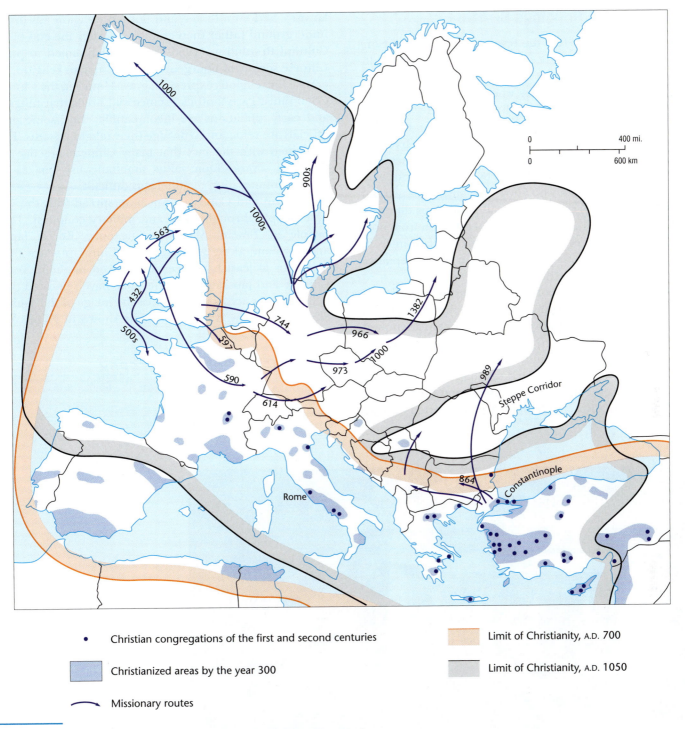

- •   Christian congregations of the first and second centuries
-    Christianized areas by the year 300
-    Missionary routes

Limit of Christianity, A.D. 700

Limit of Christianity, A.D. 1050

FIGURE 7.15 The diffusion of Christianity in Europe, first to eleventh centuries.

(A.D. 700), and Mongolia (A.D. 1500). Buddhism developed many regional forms throughout Asia, except in India, where it was ultimately accommodated within Hinduism.

The diffusion of Buddhism, like that of Christianity and Islam, continues to the present day. Some 2 million Buddhists live in the United States, about the same number as Episcopalians. Mostly, their presence is the result of relocation diffusion by Asian immigrants to the United States, where immigrant Buddhists outnumber Buddhist converts by three to one.

The East Asian Religious Hearth

Kong Fu-tzu and Lao Tzu, the respective founders of Confucianism and Taoism, were contemporaries who reputedly once met with each other. Both religions were adopted widely throughout China only when the ruling elite promoted them. In the case of Confucianism, this was several centuries after the master's death. During his life, Kong Fu-tzu wandered about with a small band of disciples trying to convince rulers to put his ideas on good governance into practice. But he was shunned even by lowly peasants, who criticized him as "a man who knows he cannot succeed but keeps trying." Thus, early attempts at contagious diffusion were unsuccessful, whereas hierarchical diffusion from politicians and schools eventually spread Confucianism from the top down. Taoism, as well, did not gain wide acceptance until it was promoted by the ruling Chinese elite.

After 1949, China's communist government officially repressed organized religious expression, dismissing it as a relic of the past. In other words, the government attempted to erect an absorbing barrier that would not only halt the spread of religion but would also erase it from Chinese public and private life. As noted in Chapter 1, absorbing barriers are rarely completely successful, and this example is no exception. Although temples were converted to secular uses, and even looted and burned, religion was driven underground rather than eradicated. After the end of the Cultural Revolution (1966–1976), which aimed to purify Chinese society of bourgeois excesses such as religion, tolerance for religious expression grew. However, the Chinese Communist Party's official stance still holds that religious and party affiliations are incompatible; thus, some party officials are reluctant to divulge their religious status. This, combined with the fact that many Chinese practice elements of several Taoic faiths simultaneously, makes the precise enumeration of adherents difficult.

Taoism and Confucianism have spread with the Chinese people through trade and military conquest. Thus, people in Taiwan, Malaysia, Singapore, Korea, Japan, and Vietnam, along with mainland China, practice these beliefs or at least have been influenced by them. Today, Chinese and Japanese migrants alike have relocated their belief systems across the globe, as the image of the Tsubaki Shinto Shrine in Washington State shown in **Figure 7.16** confirms.

Reflecting on Geography

Why, in the entire world, did only three culture hearths produce so many major religious faiths?

FIGURE 7.16 Shinto shrine in the United States. With the migration of Japanese people in the nineteenth and twentieth centuries, particularly to the West Coast of the United States, Shinto traditions and religious structures followed. This image depicts the Tsubaki Shrine in Granite Falls, Washington. *(© Alexander Marten Zhang.)*

FIGURE 7.17 Religious pilgrims. Visitors number in the millions annually and can provide a holy site's main source of revenue. *Upper left:* the grotto where the Virgin Mary appeared to a girl in 1858 in Lourdes, France. *Lower left:* the Temple of the Emerald Buddha in Bangkok, Thailand. *Above:* the Great Mosque in Touba, Senegal. *(Upper left: Franz-Peter Tschauner/dpa/Corbis; Lower left: Kevin R. Morris/Corbis; Above: Richard List/Corbis.)*

Religious Pilgrimage

pilgrimages
Journeys to places of religious importance.

For many religious groups, journeys to sacred places, or **pilgrimages,** are an important aspect of faith-based mobility (**Figure 7.17**). Pilgrimages are typical of both ethnic and proselytic religions. They are particularly significant for followers of Islam, Hinduism, Shintoism, and Roman Catholicism. Along with missionaries, pilgrims constitute one of the largest groups of people voluntarily on the move for religious reasons. Pilgrims do not aim to convert people to their faith through their journeys. Rather, they enact in their travels a connection with the sacred spaces of their faith. Indeed, some religions man-

date pilgrimage, as is the case with Islam, where pilgrimage to Mecca forms one of the Five Pillars.

The sacred places visited by pilgrims vary in character. Some have been the setting for miracles; a few are the regions where religions originated or areas where the founders of the faith lived and worked; others contain sacred physical features such as rivers, caves, springs, and mountain peaks; and still others are believed to house gods or are religious administrative centers where leaders of the religion reside. Examples include the Arabian cities of Mecca and Medina in Islam, which are cities where Muhammed resided and thus form Islam's culture hearth; Rome, the home of the Roman Catholic pope; the French town of Lourdes, where the Virgin Mary is said to have appeared to a girl in 1858; the Indian city of temples on the holy Ganges River, Varanasi, a destination for Hindu, Buddhist, and Jain pilgrims; and Ise, a shrine complex located in the culture hearth of Shintoism in Japan. Places of pilgrimage might be regarded as places of spatial convergence, or nodes, of functional culture regions.

Religion provides the stimulus for pilgrimage by offering those who participate the purification of their souls or the attainment of some desired objective in their lives, by mandating pilgrimage as part of devotion, or by allowing the faithful to connect with important historic sites of their faith. For this reason, pilgrims often journey great distances to visit major shrines. Other sites of lesser significance draw pilgrims only from local districts or provinces. Pilgrimages can have tremendous economic impact because the influx of pilgrims is a form of tourism.

In some localities, the pilgrim trade provides the only significant source of revenue for the community. Lourdes, a town of about 16,000, attracts between 4 million and 5 million pilgrims each year, many seeking miraculous cures at the famous grotto where the Virgin Mary reportedly appeared. Not surprisingly, among French cities, Lourdes ranks second only to Paris in number of hotels, although most of these are small. Mecca, a small city, annually attracts millions of Muslim pilgrims from every corner of the Islamic culture region as they perform the *hajj*, or Fifth Pillar of Islam, which all able-bodied Muslims who can afford the journey are encouraged to perform at least once in their lifetime. By land, by sea, and (mainly) by air, the faithful come to this hearth of Islam, a city closed to all non-Muslims (**Figure 7.18**).

FIGURE 7.18 Religious segregation. Access to Mecca is restricted, allowing Muslims only. The Saudi Arabian government believes that permitting non-Muslim tourists to visit Mecca and Medina would disturb the sanctity of these sacred places. *(vario images GmbH & Co. KG./Alamy.)*

Such mass pilgrimages obviously have a major impact on the development of transportation routes and carriers, as well as other provisions such as inns, food, water, and sanitation facilities.

Globalization

How has religion adapted to globalization? To what degree is religion relevant in today's global world? Religion is seen by many people as providing a stable anchor, one that is particularly necessary amidst the flux and turmoil that characterizes globalization. Religious customs, meetings, and obligations serve to form a strong basis of community. Typically, these communities are experienced locally; however, through long-distance pilgrimage, the faithful from across the globe may gather together in one place. Today's increased mobility and communication make possible religious communities at a truly global scale.

The Rise of Evangelical Protestantism in Latin America

As economic, social, and political changes occur in places, cultural forces such as religion must adapt. For instance, in Chapter 4 we saw how languages change over time in response to immigration, urbanization, and trade-based exchanges. At other times, cultural forces drive changes in politics, economics, and societies. In Chapter 5, we saw how Hispanic immigration to the United States has resulted in major shifts in consumption, political behavior, and social attitudes.

A good example of how the cultural, economic, social, and political arenas work together is provided by the contemporary religious landscape of Latin America and the Caribbean. In this region, Roman Catholicism has dominated the religious landscape since the time of the Iberian conquerors in the late 1400s. More Catholics live in this region than in any other on Earth, with three out of every five Catholics residing here. Brazil is the largest Catholic country in the world in terms of population, with more than 120 million adherents. However, the Catholic Church has been on the decline in Brazil and throughout the region in recent decades. Less than 80 percent of Brazilians are Catholics today, whereas 50 years ago that figure was more than 95 percent. What has happened to the region's Catholics?

Briefly, more and more Latin Americans believe that the Catholic Church has failed to keep in touch with the needs and concerns of contemporary urban societies. Birth

FIGURE 7.19 **Protestant storefront church.** This Pentecostal church, Igreja o Brasil para Cristo, is one of the fastest-growing evangelical Protestant churches in Brazil. The garage-front arrangement shown here is located in the mining town of Lencois. Some of the 25 worshippers at this service went into a trance. *(Ponkawonka.)*

control, divorce, and persistent poverty are issues that are simply not addressed by Roman Catholicism to the satisfaction of many Latin Americans. As a result, more and more are turning to evangelical Protestant faiths, such as the Seventh-Day Adventist and Pentecostal churches (**Figure 7.19**). The focus of these churches on thrift, sobriety, and resolving problems directly rather than through the mediation of priests is appealing to many. Others find their needs are best addressed by an array of African-based spiritist religions, which include Umbanda, Candomblé, and Santería. Whether Catholicism will ultimately reform from within to become more current, or whether it will continue to lose out to rival denominations, is a key question in regard to the Latin American and Caribbean religious landscape.

Religion on the Internet

In the mid-fifteenth century, Gutenberg's printing press made the Bible available to a mass audience. The Internet, proponents argue, represents a similar technological revolution, one that will inevitably attract new adherents to religious faiths. Gathering together regularly in a holy place—a mosque, temple, church, or shrine—is at the heart of most organized religions. Traditionally, people have come together to receive sacraments, sing, pray, and celebrate major life events in houses of worship. With the tens of thousands of religious web sites now in existence, it is no longer necessary to physically go to a place of worship. Rather, you can sit in front of a computer, at any time of the day or night, and read a holy book, submit and read online prayers, engage in theological debate, or watch broadcasts of religious services. A survey conducted by the Pew Foundation found that the number of Internet users seeking religious or spiritual information online grew 94% from 2000 to 2002, more than any other category of Internet use except for online banking.

For cultural geographers, the most interesting dimension of online worship concerns what it does to the role of place in a global society. Clearly, the Internet has made it possible to practice religion in a way that is not linked to a specific place of worship, but is this a good thing? Supporters of practicing religion online contend that it is because people can become members of virtual communities that would not otherwise have been available to them in times of spiritual need. Now illness, invalidism, or the pressures of a busy life need not present a barrier to worship. Detractors argue that solitary worship online erodes the place-based communities

FIGURE 7.20 Yoga class in session. Many people in the United States have incorporated Hindu and Buddhist spiritual practices into their lives. *(Ryan McVay/Getty Images.)*

that are at the heart of many religions. Recall from Chapter 1 that debates center on whether globalization dilutes the role of place and of place-based differences that make up the human mosaic. As people pick and choose elements from the different faiths available online, will religions converge into a watered-down, homogeneous "McFaith"?

Reflecting on Geography

> Are virtual religious communities created through the Internet an acceptable substitute for traditional place-based religious communities?

Religion's Relevance in a Global World

Geographer Roger Stump points to a twentieth-century trend toward fragmentation and dominance of particular religions in regions of the United States. Baptists in the South, Lutherans in the upper Midwest, Catholics in the Northeast and Southwest, and Mormons in the West each dominate their respective regions more thoroughly today than at the turn of the twentieth century. Each of the four traditions is socially conservative and has a strong, long-standing infrastructure. Other experts, however, believe that American culture is becoming more religiously mixed, with weakening regional borders around religious core areas. Newer religious influences, too, are making an entrance onto the

American religious stage. For example, if you have ever practiced yoga or meditation, you are engaging in Hindu and Buddhist practices (**Figure 7.20**). Many Americans do yoga or meditate to enhance physical well-being and/or spiritual growth, to relieve stress, or even to keep up with the latest trends, rather than as part of a religious practice. Yet they are among the growing number of Americans who have found a blend of Eastern and Western religious practices to be compatible with their beliefs and lifestyles (**Figure 7.21**).

The number of nonreligious, secular persons in the world is estimated at 913 million at present (see Figure 7.5). Across the United States, the level of religious faith varies, as depicted in **Figure 7.22**. The U.S. Religious Landscape Survey, released in 2008, found that 16.1 percent of Americans are not affiliated with any particular religion, while 5 percent of Americans do not believe in God or a universal spirit. In some parts of the world, especially in much of Europe, religious affiliation has declined, giving way to *secularization* (**Figure 7.23**, page 254). In some instances, the retreat from organized religion has resulted from a government's active hostility toward a particular faith or toward religion in general. The French government has long—since the French Revolution, in fact—discouraged public displays of religious faith. (France's long-standing secularism is evident on the map in Figure 7.23.) This extends not only to France's traditional Roman Catholicism but also to Jews and the rising

Supernatural Experiences and Beliefs

	Total %	Christians %
% who have . . .		
Been in touch with dead	29	29
Had ghostly experience	18	17
Consulted psychic	15	14
% believe in . . .		
Spiritual energy in trees, etc.	26	23
Astrology	25	23
Reincarnation	24	22
Yoga as a spiritual practice	23	21
Evil eye (i.e., casting of curses)	16	17
Sample size	2003	1589

FIGURE 7.21 **Blending East and West.** This survey demonstrates that a fair number of Christians hold beliefs and engage in practices that might be thought of as Eastern, such as reincarnation and yoga. Furthermore, some Christians said they had experienced supernatural phenomena, such as being in touch with the dead or having a ghostly experience. For many people, a blend of beliefs and practices characterizes their everyday lived experience of religious faith. *(Source: www. pewforum.org.)*

Muslim population. In 2004, French law banned the wearing of skullcaps, headscarves, or large crosses in public schools, and in 2011 burqas—veils worn by some Muslim women—were outlawed in all public areas. In other cases, we can attribute the decline in religious affiliation to the failure of religions to meet the needs of rural folk cultures or to adapt to the contemporary urban scene, as we saw with Roman Catholicism in Latin America. Such patterns once again reveal the inherent spatial variety of humankind and invite analysis by the cultural geographer.

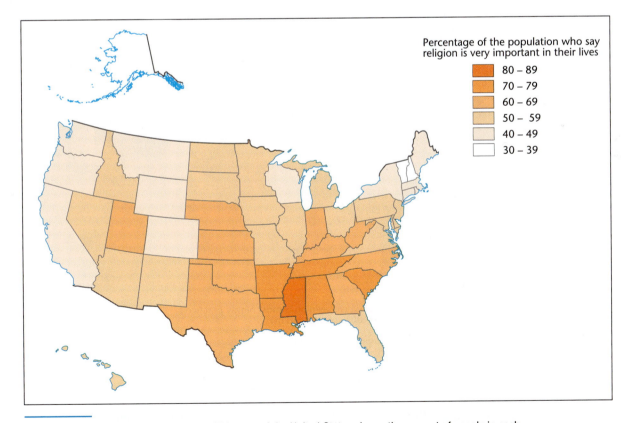

Percentage of the population who say religion is very important in their lives

- 80 – 89
- 70 – 79
- 60 – 69
- 50 – 59
- 40 – 49
- 30 – 39

FIGURE 7.22 **Importance of Religion.** This map of the United States shows the percent of people in each state who said that religion is very important in their lives. *What factors might explain why this map shows these particular regional patterns?* *(Source: www.pewforum.org.)*

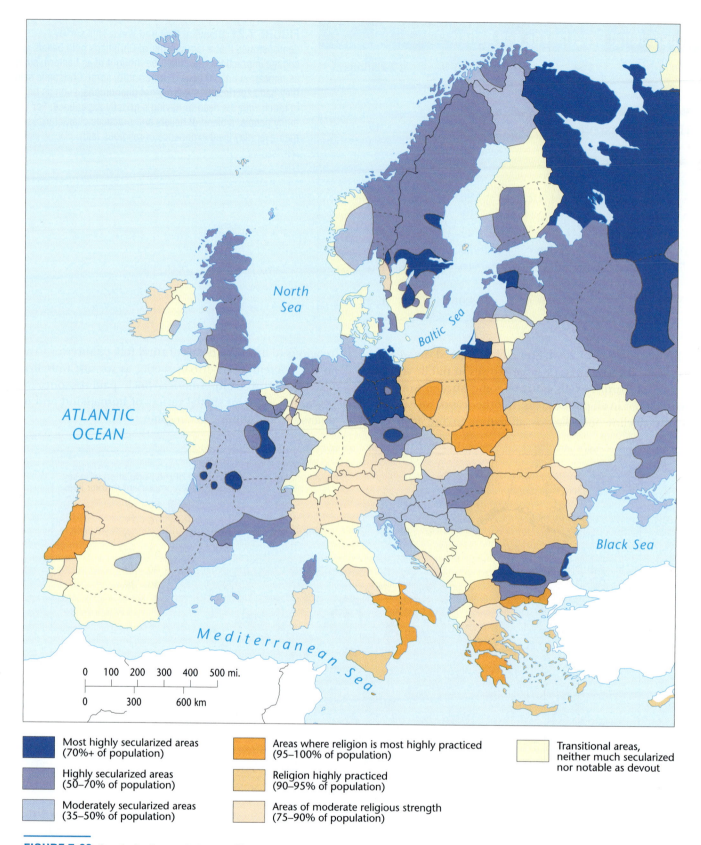

Most highly secularized areas
(70%+ of population)

Areas where religion is most highly practiced
(95–100% of population)

Transitional areas,
neither much secularized
nor notable as devout

Highly secularized areas
(50–70% of population)

Religion highly practiced
(90–95% of population)

Moderately secularized areas
(35–50% of population)

Areas of moderate religious strength
(75–90% of population)

FIGURE 7.23 Secularized areas in Europe. These areas, in which Christianity has ceased to be of much importance, occur in a complicated pattern. ***What causal forces might have been at work?*** In all of Europe, some 190 million people report no religious faith, amounting to 27 percent of the population. (*Source: Jordan-Bychkov and Jordan, 2002: 104.*)

Nature-Culture

What is the relationship between religion and the natural habitat? How does religion impact our modification of the environment and shape our perception of nature? Does the habitat influence religion? All these questions, and more, fit into the theme of cultural ecology.

Appeasing the Forces of Nature

One of the main functions of many religions is the maintenance of harmony between a people and their physical environment. Thus, religion is perceived by its adherents to be part of the **adaptive strategy** (one of the cultural tools needed to survive in a given environment); for that reason, physical environmental factors, particularly natural hazards and disasters, exert a powerful influence on the development of religions.

adaptive strategy
The unique way in which each culture uses its particular physical environment; those aspects of culture that serve to provide the necessities of life—food, clothing, shelter, and defense.

Environmental influence is most readily apparent in animistic faiths. In fact, an animistic religion's principal goal is to mediate between its people and the spirit-filled forces of nature. Animistic ceremonies are often intended to bring rain, quiet earthquakes, end plagues, or in some other way manipulate environmental forces by placating the spirits believed responsible for these events. Sometimes the link between religion and natural hazard is visual. The great pre-Columbian temple pyramid at Cholula, near Puebla in central Mexico, strikingly mimics the shape of the awesome nearby volcano Popocatépetl, which towers to the menacing height of nearly 18,000 feet (5500 meters).

Rivers, mountains, trees, forests, and rocks often achieve the status of sacred space, even in the great religions. The Ganges River and certain lesser streams such as the Bagmati in Nepal are holy to the Hindus, and the Jordan River has special meaning for Christians, who often transport its waters in containers to other continents for use in baptism (**Figure 7.24**). Most holy rivers are believed to possess soul-cleansing abilities. Hindu geographer Rana Singh speaks of the "liquid divine energy" of the Ganges "nourishing the inhabitants and purifying them."

Mountains and other high places likewise often achieve sacred status among both animists and adherents of the great religions (**Figure 7.25**, page 256). Mount Fuji is sacred in Japanese Shintoism, and many high places are revered in Christianity, including Mount Sinai. Some mountains tower so impressively as to inspire cults devoted exclusively to them. Mount Shasta, a massive snowcapped

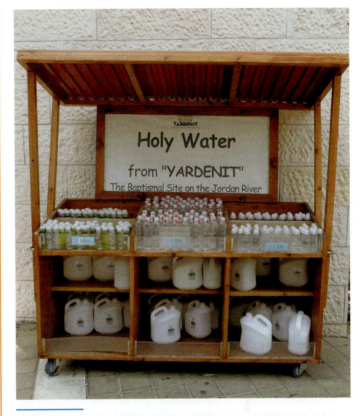

FIGURE 7.24 Holy water from the Jordan River. This water is taken from the Jordan River, at the site where Jesus is thought to have been baptized by John the Baptist. It can be purchased and used in rituals. *(Courtesy of Patricia L. Price.)*

volcano in northern California, near the Oregon border, serves as the focus of no fewer than 30 New Age cults, the largest of which is the "I Am" religion, founded in the 1930s (see Figure 7.25, right). These cults posit that the Lemurians, denizens of a lost continent, established a secret city inside Mount Shasta. Geographer Claude Curran, who studied the Shasta cults, found that, although few of the adherents live near the mountain, pilgrimages and festivals held during the summer swell the population of nearby towns and contribute to the local economy.

Animistic nature-spirits lie behind certain practices found in the great religions. *Feng shui*, which literally means "wind and water," refers to the practice of harmoniously balancing the opposing forces of nature in the built environment. A feature of Asian religions that emphasizes Tao, the dynamic balance found in nature, feng shui involves choosing environmentally auspicious sites for locating houses, villages, temples, and graves. The homes of the living and the resting places of the dead must be aligned with the cosmic forces of the world in order to assure good luck, health, and prosperity. Although the practice of feng shui dates back some 7000 years, contemporary people practice

FIGURE 7.25 **Two high places that have evolved into sacred space.** *Left:* The reddish sandstone Uluru, or Ayers Rock, in central Australia is sacred in Aboriginal animism. *Right:* Snowy Mount Shasta in California is venerated by some 30 New Age cults (see Huntsinger and Fernández-Giménez, 2000.) ***Why do mountains so often inspire such worship?*** *(Left: Michael Fogden/Bruce Coleman; Right: Joel Sartore/ National Geographic.)*

its principles. **Figure 7.26** depicts a high-rise condominium in Hong Kong's Repulse Bay neighborhood that incorporates feng shui principles in its design. The square opening in the building's center is said to provide passage to the bay for the dragon that dwells in the hill behind the building, allowing the dragon to drink from the waters of the bay and return to its abode unencumbered. Some Westerners have also adopted the principles of feng shui. Office spaces as well as homes are arranged according to its basic ideas. For example, artificial plants, broken articles, and paintings depicting war are thought to bring negative energy into living spaces and so should be avoided.

Although the physical environment's influence on the major Western religions is less pronounced, it is still evident. Some contemporary adherents to the Judeo-Christian tradition believe that God uses plagues to punish sinners, as in the biblical account of the 10 plagues inflicted on Egypt, which forced the Israelites into the desert. Modern-day droughts, earthquakes, and hurricanes are interpreted by some as God's punishment for wrongdoing, whereas others argue that this is not so. Environmental stress can, however, evoke a religious response not so different from that of animistic faiths. Local ministers and priests often attempt to alter unfavorable weather conditions with

FIGURE 7.26 *Condominium in Hong Kong.* The square opening in this building is supposed to allow for the passage of the dragon who resides in the hill behind. *(Courtesy of Ari Dorfsman.)*

special services, and there are few churchgoing people in the Great Plains of the United States who have not prayed for rain in dry years.

The Impacts of Belief Systems on Plants and Animals

Every known religion expresses itself in food choices, to one degree or another. In some faiths, certain plants and livestock, as well as the products derived from them, are in great demand because of their roles in religious ceremonies and traditions. When this is the case, the plants or animals tend to spread or relocate with the faith.

For example, in some Christian denominations in Europe and the United States, celebrants drink from a cup of wine that they believe is the blood of Christ during the sacrament of Holy Communion. The demand for wine created by this ritual aided the diffusion of grape growing from the sunny lands of the Mediterranean to newly Chris-

tianized districts beyond the Alps in late Roman and early medieval times. The vineyards of the German Rhine were the creation of monks who arrived from the south between the sixth and ninth centuries. For the same reason, Catholic missionaries introduced the cultivated grape to California, in an example of relocation diffusion. In fact, wine was associated with religious worship even before Christianity arose. Vineyard-keeping and winemaking spread westward across the Mediterranean lands in ancient times in association with worship of the god Dionysus.

Religion also can often explain the absence of individual crops or domestic animals in an area. The environmentally similar lands of Spain and Morocco, separated only by the Strait of Gibraltar, show the agricultural impact of food taboos. On the Spanish, Roman Catholic side of the strait, pigs are common and pork a delicacy, but in Muslim Morocco on the African side, only about 12,000 swine can be found throughout the entire country. Islam's prohibition of pork explains this contrast. **Figure 7.27** maps the

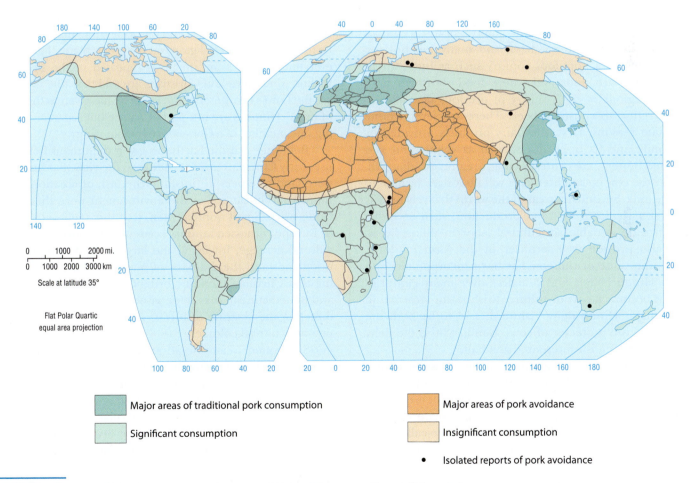

FIGURE 7.27 Consumption and avoidance of pork are influenced by religion. Some religions and churches—such as Islam, Judaism, and Seventh-Day Adventism—prohibit the eating of pork. Cultural groups with a traditional fondness for pork include central Europeans and the Chinese. *(Based in part on Simoons, 1994.)*

pork taboo. Judaism also imposes restrictions against pork and other meats, as is stated in the following passage from the Book of Leviticus:

> *These shall ye not eat, of them that chew the cud, or of them that divide the hoof: . . . the swine, though he divide the hoof, and be cloven-footed, yet he cheweth not the cud; he is unclean unto you.*

Religious taboos can even function as absorbing barriers, preventing diffusion of foods, drinks, and practices that violate the taboo. Mormons, who are encouraged to avoid caffeine, have not taken part in the American fascination with coffee. Sometimes these barriers are permeable. Certain Pennsylvania Dutch churches, for example, prohibit cigarette smoking but do not object to member farmers raising tobacco for sale in the commercial market.

The case of India's sacred cows provides an intriguing example of how religious beliefs shape the role of animals in society, in this instance avoiding the use of cows as food while emphasizing the animal's sacred as well as practical functions. Although only one out of every three of India's Hindus practices vegetarianism, almost none of India's Hindu population will eat beef, and many will not use leather. Why, in a populous country such as India where many people are poor and where food shortages have plagued regions of the country in previous decades, do people refuse to consume beef? There are several quite legitimate reasons. First, the dairy products provided by cow's milk and its by-products, such as yogurt and *ghee* (clarified butter), are central to many regional Indian cuisines. If the cow is slaughtered, it will no longer be able to provide milk. Second, in areas of the world that rely heavily on local agriculture for food production, such as India, cows provide valuable agricultural labor in tilling fields. Cows also provide free fertilizer in the form of dung. Finally, the value of the cow has been incorporated into Indian Hindu beliefs and practices over many centuries and has become a part of culture. Krishna, an incarnation of the major Hindu deity Vishnu, is said to be both the herder and the protector of cows. Nandi, who is the deity Shiva's attendant, is represented as a bull.

Think for a moment about what you consider appropriate to eat. Perhaps beef is part of your diet, but would you eat horsemeat? What is so different about a cow and a horse? Has the scare about mad cow disease affected your consumption of beef? Has the threat of so-called bird flu curtailed your poultry intake? How about dog or cat meat? What makes certain animals pets in some cultures and dinner in others? Through this sort of questioning, you may come to realize that practices that seem second nature—such as what you will and will not eat—are in fact

the result of long histories that are very different from place to place.

Ecotheology

Ecotheology is the name given to a rich and abundant body of literature studying the role of religion in habitat modification. More exactly, ecotheo-

> **ecotheology**
> The study of the influence of religious belief on habitat modification.

logians ask how the teachings and worldviews of religion are related to our attitudes about modifying the physical environment. In some faiths, human power over natural forces is assumed. The Maori people of New Zealand, for example, believe that humans represent one of six aspects of creation, the others being forest/animals, crops, wild food, sea/fish, and winds/storms. In the Maori worldview, people rule over all of these except winds/storms.

The Judeo-Christian tradition also teaches that humans have dominion over nature, but it goes further, promoting the view that the Earth was created especially for human beings, who are separate from and superior to the natural world. This view is implicit in God's message to Noah after the Flood, promising that "every moving thing that lives shall be food for you, and as I gave you the green plants, I give you everything." The same theme is repeated in the Book of Psalms, where Jews and Christians are told that "the heavens are the Lord's heavens, but the Earth he has given to the sons of men." Humans are not part of nature but separate, forming one member of a God-nature-human trinity.

Believing that the Earth was given to humans for their use, early Christian thinkers adopted the view that humans were God's helpers in finishing the task of creation, that human modifications of the environment were therefore God's work. Small wonder, then, that the medieval period in Europe witnessed an unprecedented expansion of agricultural acreage, involving the large-scale destruction of woodlands and the drainage of marshes. Nor is it surprising that Christian monastic orders, such as the Cistercian and Benedictine monks, supervised many of these projects, directing the clearing of forests and the establishment of new agricultural colonies.

Subsequent scientific advances permitted the Judeo-Christian West to modify the environment at an unprecedented rate and on a massive scale. This marriage of technology and theology is one cause of our modern ecological crisis. The Judeo-Christian religious heritage, in short, has for millennia promoted an instrumentalist view of nature that is potentially far more damaging to the habitat than an organic view of nature in which humans and nature exist in balance. Yet there is considerable evidence to the contrary. For example, the Orthodox Church in Russia is working to create wildlife preserves on monastery lands. The Patriarch of Constantinople, leader of Eastern Orthodox Christianity, has made the fight against pollution a church policy, declaring that damaging the natural habitat constitutes a sin against God. The Church of England has declared that abuse of nature is blasphemous, and throughout the monotheistic religions, the green teachings of long-dead saints, such as Christianity's St. Francis of Assisi, who treasured birds and other wildlife, now receive heightened attention.

Indeed, the Judeo-Christian tradition is not lacking in concern for environmental protection. In the Book of Leviticus, for example, farmers are instructed by God to let the land lie fallow one year in seven and not to gather food from wild plants in that "Sabbath of the land." Robin Doughty, a humanistic cultural geographer, suggests that "Western Christian thought is too rich and complex to be characterized as hostile toward nature," although he feels that Protestantism, "in which worldly success symbolizes individual predestination," may be more conducive to "ecological intemperance." Beyond that, according to geographer Janel Curry-Roper, some fundamentalist Protestant sects herald ecological crisis and environmental deterioration as a sign of the coming Apocalypse, Christ's return, and the end of the present age. Thus, they welcome ecological collapse and, obviously, are unlikely to be of much help in solving the problem of environmental degradation.

Some conservative, fundamentalist Protestants, however, have adopted conservationist views, citing biblical admonitions. The Flood story from the Old Testament, in which Noah saves diverse animals by bringing them onto the ark, is now viewed by many fundamentalists as a call to protect endangered species. An ecotheological focus underlies the multidenominational National Religious Partnership for the Environment, which includes many evangelical Protestant members. The hope is to mobilize the Christian Right against wanton environmental destruction in the same way in which they oppose abortion.

The great religions of southern and eastern Asia, and many animistic faiths, also highlight teachings and beliefs that protect nature. In Hinduism, for example, geographer Deryck Lodrick found that the doctrine of ahimsa had resulted in the establishment of many animal homes, refuges, and hospitals, particularly in the northwestern part of India. The hospitals, or *pinjrapoles,* are maintained by the Jains. In this view of the world, people are part of and at harmony with nature. Such religions would presumably not threaten the ecological balance.

However, real-world practices do not always reflect the stated ideals of religions. Both Buddhism and Hinduism protect temple trees but demand huge quantities of wood for cremations (**Figure 7.28**). Traditional Hindu cremations, for example, place the corpse on a pile of wood, cover it with more wood, and burn it during an open-air rite that can last up to six hours. The construction of funeral pyres is estimated to strip some 50 million trees from India's countryside annually. In addition, the ashes

FIGURE 7.28 Wood gathered for Hindu cremations at Pashupatinath, on the sacred river Bagmati in Nepal. These cremations contribute significantly to the ongoing deforestation of Nepal and reveal the underlying internal contradiction in Hinduism between conservation as reflected in the doctrine of ahimsa and sanctioned ecologically destructive practices. *(Courtesy of Terry G. Jordan-Bychkov.)*

are later swept into rivers, and the burning itself releases carbon dioxide into the atmosphere, contributing to the pollution of waterways and the atmosphere. The use of wood, and its placement on the ground, provides an important symbolic connection between the body and the earth. It does not, however, lead to efficient burning; on average, 880 pounds (400 kilograms) of wood is required to cremate a single corpse. A "green cremation system" is currently under development by a New Delhi–based nonprofit organization. By placing the first layer of wood on a grate, and placing a chimney over the pyre, wood use can be reduced by 75 percent. Although traditional Hindus balk at the notion of breaking with conventional practice, severe wood shortages and the escalating cost of wood will likely make green cremations an increasingly popular option in the future. As this example illustrates, the idea of a link between godliness and greenness is a global one. In the years following a conference in Italy in the mid-1980s— which brought together environmentalists and religious leaders representing Buddhism, Christianity, Hinduism, Islam, and Judaism—some 130,000 projects have arisen linking the green teachings some see as inherent in these faiths to the ecology movement.

Ecofeminists have also entered this debate. They point out that the rise of the all-powerful male sky-deity of Semitic monotheism came at the expense of earth goddesses of fertility and sustainability. In their view, because the Judeo-Christian tradition elevated a sky-god remote from the Earth,

FIGURE 7.29 Bumper sticker. The "mother" in question is Mother Earth. *(Courtesy of Jeffrey Gordon.)*

the harmonious relationship between people and the habitat was disrupted. The ancient holiness of ecosystems perished, endangering huge ecoregions. The **Gaia hypothesis** possesses an ecofeminist spirit, wherein the Earth is seen as a mother figure who reacts to humankind's environmental depredations through a variety of self-regulating mechanisms. The popular "love your mother" bumper sticker displays an image of the planet Earth and is a humorous example of this notion (**Figure 7.29**).

> **Gaia hypothesis**
> The theory that there is one interacting planetary ecosystem, Gaia, that includes all living things and the land, waters, and atmosphere in which they live; further, that Gaia functions almost as a living organism, acting to control deviations in climate and to correct chemical imbalances, so as to preserve Earth as a living planet.

CULTURAL LANDSCAPE

In what forms does religion appear in the cultural landscape? How does the visibility of religion differ from one faith or denomination to another? Religion is a vital aspect of culture; its visible presence can be quite striking, reflecting the role played by religious motives in the human transformation of the landscape. In some regions, the religious aspect is the dominant visible evidence of culture, producing sacred landscapes. At the opposite extreme are areas almost purely secular in appearance. Religions differ greatly in visibility, but even those least apparent to the eye usually leave some subtle mark on the countryside.

Religious Structures

The most obvious religious contributions to the landscape are the buildings erected to house divinities or to shelter worshippers. These structures vary greatly in size, function, architectural style, construction material, and degree of ornateness (**Figure 7.30**, page 262). To Roman Catholics, for example, the church building is literally the house of God, and the altar is the focus of key rituals. Partly for these reasons, Catholic churches are typically large, elaborately decorated, and visually imposing. In many towns and villages, the Catholic house of worship is the focal point of the settlement, exceeding all other structures in size and grandeur. In medieval European towns, Christian cathedrals were the tallest buildings, representing the supremacy of religion over all other aspects of life.

For many Protestants—particularly the traditional Calvinistic chapel-goers of British background, including Presbyterians and Baptists—the church building is, by contrast, simply a place to assemble for worship. The result is an unsanctified, small, simple structure that appeals less to the senses and more to personal faith. For this reason, traditional Protestant houses of worship typically are not designed for comfort, beauty, or high visibility but instead

Patricia's Notebook

My Travels to Jerusalem

Patricia Price

Panoramic shot of Jerusalem, Israel, site of some of the world's most contested sacred spaces. In the foreground is the iconic Dome of the Rock, sacred to Muslims, which is located on the Temple Mount, site of the destroyed Second Temple and thus also sacred to Jews. *(Courtesy of Patricia L. Price.)*

One of the best things about being a geographer is that I always have a good excuse to travel the world! Among the most interesting places I have visited thus far is Israel. It's hard to think of any place on Earth where so much globally significant history, passion, and conflict are packed into such a small area. Because I'm a cultural geographer whose work emphasizes the humanistic theme of how it is that societies establish meaningful ties to places—in other words, how we dwell on Earth—I always have my eyes open for traces of meaning-making on the landscape.

In Jerusalem, for instance, you can walk the Via Dolorosa route, visiting all 14 Stations of the Cross walked by Jesus from his condemnation by the Romans to the site of his crucifixion and death. You can see the iconic golden Dome of the Rock on Jerusalem's Temple Mount, which shelters the rock (hence the name) from which Muhammad is said to have ascended to heaven. Jews also believe the site to be sacred because it is said to be the place where Abraham prepared to sacrifice his son Isaac and is the site of the Western Wall, the remnant of the ancient Jewish temple. Non-Muslims are forbidden from entering the Dome of the Rock for reasons of political conflict and control as well as pollution taboos (which are held by most religions, including Islam). Some Jews voluntarily decline to set foot on the Temple Mount as well, believing it too holy for mere mortal presence.

So much of what we hear and see about this place on our nightly news depicts only the political tensions existing between Israelis and Palestinians. So it is eye-opening to wander the streets of Jerusalem and witness Jews, Christians, and Muslims, orthodox and secular alike, living their daily lives in this city. Yes, there is tension. Mobility and access are often thwarted or at least slowed down, thanks mostly to security concerns. But, as with anyplace else on Earth, most people in Jerusalem are merely going about their business.

I always tell my students that, if given the chance to travel anywhere in the world, they should seriously consider visiting the Middle East. Directly experiencing peoples and places goes a long way toward dispelling the myths and stereotypes about those peoples and places, which are particularly prevalent in this part of the world.

Posted by Patricia Price

---

appear deliberately humble (see Figure 7.30, bottom left). Similarly, the religious landscape of the Amish and Mennonites in rural North America, the "plain folk," is very subdued because they reject ostentation in any form. Some of their adherents meet in houses or barns, and the churches that do exist are very modest in appearance, much like those of the southern Calvinists.

Islamic mosques are usually the most imposing structures in the landscape, whereas the visibility of Jewish synagogues varies greatly. Hinduism has produced large

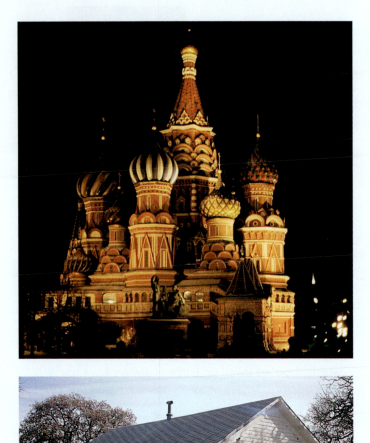

FIGURE 7.30 Traditional religious architecture takes varied forms. St. Basil's Cathedral on Red Square in Moscow (*upper left*) reflects a highly ornate Russian landscape presence, whereas the plain board chapel in the American South (*lower left*) demonstrates the opposite tendency of favoring visual simplicity by British-derived Protestants. The ornate Hindu temple in Varanasi, India (*above*), offers still another sacred landscape. (*Source: Terry G. Jordan-Bychkov.*)

numbers of visually striking temples for its multiplicity of gods, but much worship is practiced in private households with their own personal shrines (**Figure 7.31**).

Nature religions such as animism generally place only a subtle mark on the landscape. Nature itself is sacred, and imposing features already present in the landscape—mountains, waterfalls, and grottos, for example—mean that few human-made shrines are needed. There are, not surprisingly, exceptions. In Korea, for example, where animism merged with Buddhism and the Chinese composite religion, animistic shrines survive in the landscape (**Figure 7.32**).

Houses of worship can also reveal subtler content and messages. In Polynesian Maori communities of New

Zealand, for example, the *marae,* a structure linked to the pagan gods of the past, generally stands alongside the Christian chapel, which reflects the Maoris' conversion (**Figure 7.33**, page 264). The landscape thus reflects the blending of two faiths.

Paralleling this contrast in church styles are attitudes toward roadside shrines and similar manifestations of faith. Catholic culture regions typically abound with shrines, crucifixes, and other visual reminders of religion, as do some Eastern Orthodox Christian areas. Protestant areas, by contrast, are bare of such symbolism. Their landscapes, instead, display such features as signboards advising the traveler to "get right with God," a common sight in the southern United States (**Figure 7.34**, page 264).

FIGURE 7.31 Temples dedicated to ancestors in Bali, Indonesia. Bali's population practices a blend of Hinduism, animism, and ancestor worship. These temples to ancestors are a feature of every Balinese home, and offerings of incense, food, and flowers are made three times per day to show reverence. *(Courtesy of Ari Dorfsman.)*

Faithful Details

Not only buildings but also other architectural choices, such as color and landscaping elements, can convey spiritual meaning through the landscape. Consider the color green. The image of a Muslim mosque in northern Nigeria in **Figure 7.35** (page 265) depicts how green—sacred in Islam—appears often on mosque domes and minarets, the tall circular towers from which Muslims are called to prayer. The prophet Muhammad declared his favorite color to be green, and his cloak and turban were said to be green. In Christianity, green also has positive connotations, symbolizing fertility, freedom, hope, and renewal. It is used in decorations, such as banners, and for clergy vestments, especially those associated with the Trinity and with the Epiphany season.

Water, often used as a decorative element in fountains, baths, and pools, has religious meaning. In fact, all three of the great desert monotheisms—Islam, Christianity, and

FIGURE 7.32 Stacked stones at a pilgrimage site in South Korea perpetuate an ancient animistic/shamanistic practice. These stones had meaning in ancient times, meaning that has since been lost. *(Courtesy of Terry G. Jordan-Bychkov.)*

FIGURE 7.33 In the Maori community of Tikitiki, on North Island, New Zealand, a traditional Polynesian *marae,* or shrine (on the left), stands beside a Christian chapel (on the right). The religious landscape, in this odd juxtaposition, tells us much about the composite faith held by the modern Maori. *(Courtesy of Terry G. Jordan-Bychkov.)*

Judaism—consider water to have a special status in religious rituals. In general, followers of all three religions believe water has the ability to purify and cleanse the body as well as to purify and cleanse the soul of sins. Muslims use water in ablutions, or washing, of the hands, face, or ears before daily prayers and other rituals. Footbaths, or ablution fountains, are important features located outside of mosques, and decorating with water, particularly pools and fountains, is a common practice in Islamic architecture. Central to Christianity is baptism, where the newborn infant or adult convert is sprinkled with or fully immersed in holy water, symbolizing

the cleansing of original sin for infants and repentance or cleansing from sin for adults. Jews may engage in a ritual bath, known as a *mikveh,* before important events, such as weddings (for women), or on Fridays, the evening of the Jewish Sabbath (for men). The story of the Great Flood, found in the Book of Genesis (common to both Jews and Christians), depicts the washing away of the sins of the world so that it could be born anew.

Landscapes of the Dead

Religions differ greatly in the type of tribute each awards to its dead. These variations appear in the cultural landscape. The few remaining Zoroastrians, called Parsees, who preserve a once-widespread Middle Eastern faith now confined to parts of India, have traditionally left their dead exposed to be devoured by vultures. Thus, the Parsee dead leave no permanent mark on the landscape. Hindus cremate their dead. Having no cemeteries, their dead, as with the Parsees, leave no obvious mark on the land. Yet on the island of Bali in Indonesia, Hinduism has blended with animism such that temples to family ancestors occupy prominent places outside the houses of the Balinese, a landscape feature that does not exist in India itself (see Figure 7.31, page 263).

In Egypt, spectacular pyramids and other tombs were built to house dead leaders. These monuments, as well as the modern graves and tombs of the rural Islamic folk of Egypt, lie on desert land not suitable for farming (**Figure 7.36**). Muslim cemeteries are usually modest in appearance, but spectacular tombs are sometimes erected for aristocratic persons, giving us such sacred structures as

FIGURE 7.34 **Billboard in Three Forks, Montana.** Protestant billboards often use American-style advertising slogans to promote church attendance. *(Nancy H. Belcher.)*

the Taj Mahal in India (**Figure 7.37**, page 266), one of the architectural wonders of the world.

Taoic Chinese typically bury their dead, setting aside land for that purpose and erecting monuments to their deceased kin. In parts of China—reflecting its pre-communist history—cemeteries and ancestral shrines take up as much as 10 percent of the land in some districts, greatly reducing the acreage available for agriculture.

Christians also typically bury their dead in sacred places set aside for that purpose. These sacred places vary significantly from one Christian denomination to another. Some graveyards, particularly those of Mennonites and southern Protestants, are very modest in appearance, reflecting the reluctance of these groups to use any symbolism that might be construed as idolatrous. Among certain other Christian groups, such as Roman Catholics and members of the various Orthodox Churches, cemeteries are places of color and elaborate decoration (**Figure 7.38**, page 266).

Cemeteries often preserve truly ancient cultural traits because people as a rule are reluctant to change their practices relating to the dead. The traditional rural cemetery of the southern United States provides a case in point. Freshwater mussel shells are placed atop many of the elongated grave mounds, and rose bushes and cedars are planted throughout the cemetery. Recent research suggests that the use of roses may derive from the worship of an ancient, pre-Christian mother goddess of the Mediterranean lands. The rose was a symbol of this great goddess, who could restore life to the dead. Similarly, the cedar evergreen is an age-old pagan symbol of death and eternal life, and the use of shell decoration derives from an animistic custom in West Africa, the geographic origin of slaves in the

FIGURE 7.35 Muslim mosque in northern Nigeria. Areas with Muslim populations enjoy a landscape where mosque minarets and domes are frequently green in color. *(Courtesy of Diane Rawson/Photo Researcher.)*

FIGURE 7.36 Landscape of the dead, landscape of the living. The Beni Hassan Islamic necropolis in central Egypt lies in the desert, just beyond the irrigable land; nearby, the living make intensive use of every parcel of land watered by the Nile River. The mud brick structures are all tombs. Thus, the doubly dead landscape is sacred, whereas the realm of living plants and people is profane. *(Courtesy of Terry G. Jordan-Bychkov.)*

Built as a Muslim tomb, the Taj Mahal is perhaps the most impressive religious structure in the world. (*Pallava Bagla/Corbis.*)

American South. Although the present Christian population of the South is unaware of the origins of their cemetery symbolism, it seems likely that their landscape of the dead contains animistic elements thousands of years old, revealing truly ancient beliefs and cultural diffusions.

Reflecting on Geography

What is the most visible element of the religious landscape where you live?

Sacred Space

Sacred spaces consist of natural and/or human-made sites that possess special religious meaning that is recognized as worthy of devotion, loyalty, fear, or esteem. By virtue of their sacredness, these special places might be avoided by the faithful, sought out by pilgrims, or barred to members of other religions (see Figure 7.18, page 250). Often, sacred

sacred spaces
Areas recognized by a religious group as worthy of devotion, loyalty, esteem, or fear to the extent that they become sought out, avoided, inaccessible to nonbelievers, and/or removed from economic use.

FIGURE 7.38   Two different Christian landscapes of the dead. In the Yucatán Peninsula of Mexico, the dead rest in colorful, aboveground crypts, whereas in Amana, Iowa, communalistic Germans prefer tidiness, order, and equality. (*Left: Macduff Everton/Corbis; Right: Courtesy of Terry G. Jordan-Bychkov.*)

space includes the site of supposed supernatural events or is viewed as the abode of gods. Cemeteries, too, are also generally regarded as a type of sacred space. So is Mount Sinai, described by geographer Joseph Hobbs as endowed "with special grace," where God instructed the Hebrews to "mark out the limits of the mountain and declare it sacred." Conflict can result if two religions venerate the same space. In Jerusalem, for example, the Muslim Dome of the Rock, on the site where Muhammad is believed to have ascended to heaven, stands above the Western Wall, the remnant of the ancient Jewish temple (**Figure 7.39**; see Patricia's Notebook on page 261). These two religions literally claim the same space as their holy site, which has led to conflict.

Sacred space is receiving increased attention in the world. In the mid-1990s, the internationally funded Sacred Land Project began to identify and protect such sites, 5000 of which have been cataloged in the United Kingdom alone. Included are such places as ancient stone circles, pilgrim routes, holy springs, and sites that convey mystery or great natural beauty. This last type falls into the category of mystical places: locations unconnected with established religion where, for whatever reason, some people believe that extraordinary, supernatural things can happen. The Bermuda Triangle in the western Atlantic, where airplanes and ships supposedly disappear, is a mystical place and, in effect, a vernacular culture region. Some people find the expanses of the American Great Plains mystical, and writer Jonathan Raban spoke of them as "a landscape ideally suited to the staging of the millennium, open to the gaze of the Almighty." Sometimes the sacred space of vanished ancient religions never loses—or

FIGURE 7.39 Sacred spaces. The three so-called Abrahamic faiths—Judaism, Christianity, and Islam—arose in the same place and thus their adherents revere some of the same holy sites. *Left:* Mount Sinai, in North Africa, is where Moses is said to have received the Ten Commandments as the Israelites continued their journey out of Egypt and is also mentioned several times in the Qur'an. *Right:* Jews pray at the Western Wall in Jerusalem, in Israel. Standing above the wall, on the site of the destroyed Jewish temple, is one of the holiest sites for Muslims: the golden-capped mosque called the Dome of the Rock. (See also the photograph in Patricia's Notebook, page 261.) *(Left: Konrad Wothe/age fotostock; Right: Gary Cralle/Gettyone.)*

FIGURE 7.40 Our Lady of Clearwater. This 60-foot (18-meter) image appeared in December 1996 in the window of a bank building in Clearwater, Florida. Because of its resemblance to Our Lady of Guadalupe, Catholics—including a sizable Mexican immigrant population as well as Catholic pilgrims from around the region—gather in the parking lot every day for the celebration of Mass. *(Courtesy of Patricia L. Price.)*

later regains—the functional status of mystical place, which is what has happened at Stonehenge in England.

In his 1957 book *The Sacred and the Profane,* the renowned religious historian Mircea Eliade suggested that all societies, past and present, have sacred spaces. According to Eliade, sacred spaces are so important because they establish a geographic center on which society can be anchored. Through what he termed *hierophany,* a sacred space emerges from the profane, ordinary spaces surrounding it. A contemporary example of hierophany is provided by the appearance of an image of the Virgin of Guadalupe on a bank building in Clearwater, Florida, transforming the profane space of finance into a Catholic pilgrimage site (**Figure 7.40**). For more information on another type of sacred space—sites originally marked by tragedy and violence—see Practicing Geography.)

PRACTICING GEOGRAPHY

Kenneth Foote came to geography the way many people do: he simply found it more interesting than anything else he'd studied before. He began college knowing he was interested in cities and thought he would become an architect or an urban planner. However, at the University of Wisconsin, where he went to school, most of the classes on cities were offered by the Geography Department. As Professor Foote says, "I soon found geography—particularly landscape studies and cultural and historical geography—more interesting than urban planning."

Professor Foote has focused on places and what happens to them in the aftermath of violent or tragic events. He has studied the sites of the Oklahoma City bombing and the September 11, 2001, terrorist attacks in New York City, among others. "For me the question is why some of these places are transformed into powerful and evocative monuments whereas others disappear, or are actually effaced, from view entirely." His book *Shadowed Ground* (see Doing Geography) focuses on the United States, while his recent work concentrates on Europe. "Once I began to focus on the transformative effect of violence on people and place," he writes, "I realized there were many ways I could extend my work. For me, a good research question almost always suggests another and another and another." His current research examines the very different ways in which European countries have interpreted places associated with World Wars I and II; the Holocaust; civil wars; and events of social, political, and economic violence in the twentieth and twenty-first centuries. During the past few years, he has become very interested in how national shrines and memorials are created, managed, and sometimes manipulated by certain groups. "I am becoming more and more interested in the way that the legacies of the cold war will be remembered and commemorated throughout Europe."

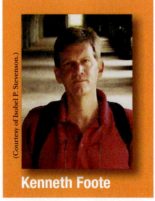

(Courtesy of Isobel P. Stevenson.)

Kenneth Foote

Detailed case studies of places and events form the building blocks of Professor Foote's research. He uses photographs and maps to document what has happened at a place. He supplements this visual evidence by interviewing people about what happened as well as by searching through newspapers and other documents to see what the written record says happened. Fieldwork—actually going to the place in question and consulting human, visual, and written sources in those places—is usually at the heart of how cultural geographers do their work, and Professor Foote is no exception. Because he enjoys traveling, meeting new people, and visiting libraries and archives, Professor Foote enjoys research more than any other part of his job. "I love to be on the road investigating interesting places, talking with people, or trying to locate good sources of local history. I try to be on the road at least one month a year, sometimes more.

CONCLUSION

Religion is firmly interwoven in the fabric of culture, a bright hue in the human mosaic; religions vary greatly from one region to another, creating diversity so profound as to give special significance to James Griffith's admonition that we should all "learn, respect, and walk softly." Yet religions tend to share some common elements, the most basic being that they provide a structure of faith that anchors communities. This religious spatial variation leads us to ask how these distributions came to be, a question best answered by examining methods of cultural diffusion. Some religions actively encourage their own diffusion through missionary activity or conquest. Other religions erect barriers to expansion diffusion by restricting membership to one particular ethnic group. People, as well, are set in motion thanks to spiritual practices, most notably pilgrimages.

In today's globalizing world, it is worth exploring how—or even if—the world's great religions will become significantly modified in response to changing contexts. The common practice of Internet-based religious practice brings into focus the debates over the role of place in a global world.

The theme of nature-culture reveals some fundamental ties between religion and the physical environment. One major function of many religious systems, particularly the animistic faiths, is to appease and placate the forces of nature in order to achieve and maintain harmony between humans and the physical environment. Plants and animals spread—or not—with the spread of religions. The followers of various religions differ in their outlooks on environmental modification by humans.

The cultural landscape abounds with expressions of religious belief. Places of worship—mosques, temples, churches, and shrines—differ in appearance, distinctiveness, prominence, and frequency of occurrence from one religious culture region to another. These buildings provide a visual index to the various faiths, as do their related architectural elements. Cemeteries and sacred spaces also add a special effect to the landscape that tells us about the religious character of the population. In all these ways, and more, the five themes of cultural geography prove relevant to the study of the world's religions.

DOING GEOGRAPHY

The Making of Sacred Spaces

For this exercise, you will use Professor Foote's ideas about how spaces become sacred. In his book *Shadowed Ground: America's Landscapes*

of Violence and Tragedy, Professor Foote argues that what a society chooses to forget about its past is at least as important as what it remembers in shaping an image of itself that it can present to the world. Professor Foote discusses how certain places in the United States have, or have not, become sacred sites. He notes, for example, that the battlefields of the Revolutionary and Civil wars are well marked and visited by thousands every year. Yet the 1692 execution site of the supposed witches of Salem, Massachusetts, is unmarked. How could the exact site of such an infamous event in U.S. history be impossible to locate? Professor Foote realized that this was not limited to the United States, writing, "Indeed, soon after my first trip to Salem, I found myself in Berlin before the reunification of Germany. There I came across similar places—Nazi sites such as the Gestapo headquarters and Reich chancellery—that have lain vacant since just after World War II and seem to be scarred permanently by shame."

Professor Foote suggests that sites of important events can be sanctified, obliterated, purposely ignored, or fall somewhere in between. Sanctified sites are set apart from other sites, are carefully maintained, are often publicly owned, and attract annual visitors and ceremonies. At the other end of the spectrum, the sites of particularly shameful, stigmatized, or violent events simply may be obliterated from the landscape in an attempt to forget about them.

Steps to Understanding Sacred Spaces

Step 1: Think about sites near where you now live and where a violent or tragic event occurred, for example, a murder, a freak accident, a natural disaster, an assassination, an act of terrorism, or a heinous crime. Virtually all places experience some such event. If nothing readily comes to mind, inquire, and inevitably you will discover something for this exercise, even if you live in a rural area or a college town.

Step 2: Once you have identified a site, consider its current status. Is it sanctified, is it obliterated, or does it fall somewhere in between? Elements of the landscape, such as signs or monuments (or their lack) will indicate that status of the site.

Step 3: Take note of any debates that have occurred about what to do with this site. Has the status of the site changed over time? How?

When people are unsure of how an event fits into their history, the site where the event occurred may inhabit a sort of limbo, remaining unmarked until the society in question comes to terms with the event. This is the case with many sites of racially motivated, violent, or shameful acts in the United States, such as the Memphis, Tennessee, motel where Martin Luther King, Jr., was assassinated in 1968 or the Manzanar, California, internment camp where Japanese-Americans were confined during World War II. Only recently have both sites become prominent on the landscape after decades of neglect. Consider the following questions:

- Are there any newsworthy events happening now that may mark a place as a site of tragedy?
- What sites, whether close by to you or far away, do you expect to become more sacred over time?
- Can you think of an event that is so horrific that the place where it occurred will be forever forgotten and erased from the map?

Cemetery at Manzanar National Historic Site. The Japanese inscription reads "Soul Consoling Tower." Fifteen of the 146 Japanese-Americans who died at this Relocation Center are buried here. A total of 110,000 Japanese-Americans were interned during World War II in camps like Manzanar. After the 1941 Japanese attack on Pearl Harbor, Hawaii, Japanese-Americans were deemed "enemy aliens." It was not until 1992—nearly 50 years after the end of the war—that Manzanar was declared a National Historic Site. *(George Ostertag/age fotostock.)*

Key Terms

Geography of Religion on the Internet

You can learn more about the geography of religion on the Internet at the following web sites:

Get Religion
http://www.getreligion.org/

Founded because (quoting journalist William Schneider) "the press . . . just doesn't get religion," this blog attempts to ferret out the ghosts lurking in news coverage of religion.

History of Religion
http://www.mapsofwar.com/ind/history-of-religion.html

"See 5000 years of religion in 90 seconds" by launching this dynamic map. The timeline shows major developments in the history of world religions.

Our Lady of Guadalupe
http://www.sancta.org/opro1.html

On the "ora pro nobis" ("pray for us") section of this web site, you can post your prayer or request to the Virgin of Guadalupe and read the heartfelt posts of other devotees.

Pew Forum on Religion & Public Life
http://pewforum.org

Seeking to promote a greater understanding of the intersection between religion and public affairs, the Pew Forum conducts polls, surveys, and demographic analysis in the United States and around the world.

World Council of Churches
http://www.oikoumene.org/

An interdenominational group headquartered in Geneva, Switzerland, the World Council of Churches works for greater cooperation and understanding among different faiths.

Sources

Curran, Claude W. 1991. "Mt. Shasta, California, and the I Am Religion," in *Abstracts, The Association of American Geographers 1991 Annual Meeting, April 13–17, Miami, Florida.* Washington, D.C.: Association of American Geographers, 42.

Curry-Roper, Janel M. 1990. "Contemporary Christian Eschatologies and Their Relation to Environmental Stewardship." *Professional Geographer* 42: 157–169.

Doughty, Robin W. 1981. "Environmental Theology: Trends and Prospects in Christian Thought." *Progress in Human Geography* 5: 234–248.

Eliade, Mircea. 1987 [1957]. *The Sacred and the Profane: The Nature of Religion.* Willard R. Trask (trans.). San Diego: Harcourt, Inc.

Foote, Kenneth E. 1997. *Shadowed Ground: America's Landscapes of Violence and Tragedy.* Austin: University of Texas Press.

SEEING GEOGRAPHY Parking Lot Shrine

How can an ordinary landscape, such as a parking lot, become a sacred space?

Parking lot shrine to the Virgin of Guadalupe, Self-Help Graphics and Art, East Los Angeles.

This large statue of the Virgin of Guadalupe inhabits the corner of the parking lot of Self-Help Graphics and Art, an artists' cooperative in East Los Angeles. According to Michael Amezcua, a metalwork artist at Self-Help, neighborhood residents actively use this shrine, leaving offerings and petitions to the Virgin in this outdoor grotto. Every December 12, neighborhood residents gather here to celebrate the feast day of the Virgin of Guadalupe.

We could interpret this unique parking lot shrine to be a variation on the yard shrine, a common feature of the residential landscape of the southwestern United States. Throughout the Mexican-American homeland region of the Southwest, it is common to see a small shrine to the Virgin of Guadalupe in people's front yards. This, in turn, is a variation on the Mexican custom of maintaining elaborate shrines and altars to important religious figures. These are sometimes located inside the house in a space reserved especially for altars (called a *nicho*). Or they can be located outside the house, near the front door, because Mexican homes usually do not have yard space in front of the house. On December 12, altars all over Mexico and Mexican-American areas in the United States are elaborately decorated with candles, ribbons, and flowers. Devotees of the Virgin of Guadalupe celebrate with family and neighbors throughout the night. Since turquoise blue is considered to be the favorite color of the Virgin of Guadalupe, this color abounds. Because the Virgin of Guadalupe is also the official patron saint of Mexico, the red, white, and green of the Mexican flag is also prevalent in altar decorations.

In Miami, another Latino population center in the United States, yard shrines are built to La Virgen de la Caridad de Cobre. Like the Virgin of Guadalupe, La Virgen de la Caridad is an American manifestation of the Virgin Mary. This Virgin hails from the copper mining village of Cobre, in Cuba. She is thought to protect seafarers and has been adopted by Cuban rafters as their patron saint. Yemayá, who is an *orisha,* or spirit, in the Afrocentric Cuban religious tradition of Santería, is often equated with La Virgen de la Caridad de Cobre, as both are associated with the sea.

What does the fact that this particular shrine is built in a parking lot rather than a yard say about what, and how, spaces become sacred? Can you think of a more mundane place than a parking lot for such a hallowed figure as the Virgin of Guadalupe? Perhaps the drive-through architecture of strip malls, highways, and parking lots that is so closely associated with the urban landscape of Los Angeles has something to do with it. Where else, other than in Los Angeles, can you find a sacred parking lot?

Griffith, James S. 1992. *Beliefs and Holy Places: A Spiritual Geography of the Pimería Alta.* Tucson: University of Arizona Press.

Hobbs, Joseph J. 1995. *Mount Sinai.* Austin: University of Texas Press.

Huntsinger, Lynn, and María Fernández-Giménez. 2000. "Spiritual Pilgrims at Mount Shasta, California." *Geographical Review* 90: 536–558.

Jordan-Bychkov, Terry G., and Bella Bychkova Jordan. 2002. *The European Culture Area: A Systematic Geography,* 4th ed. Lanham, Md.: Rowman & Littlefield.

Klaer, Wendelin. 1996. "Survey: Lebanon." *The Economist.* February 24, l996: 333.

Lodrick, Deryck O. 1981. *Sacred Cows, Sacred Places: Origins and Survivals of Animal Homes in India.* Berkeley: University of California Press.

Madden, Mary, and Lee Rainie. 2003. "America's Online Pursuits." Washington, D.C.: Pew Research Center. Available online at http://www.pewinternet.org/Reports/2003/Americas-Online-Pursuits/1-Summary-of-Findings.aspx

Pew Forum on Religion & Public Life. 2008. U.S. Religious Landscape Survey. Washington, D.C.: Pew Research Center. Available online at http://religions.pewforum.org/pdf/report-religious-landscape-study-full.pdf.

Raban, Jonathan. 1996. *Bad Land: An American Romance.* New York: Pantheon.

Simoons, Frederick J. 1994. *Eat Not This Flesh: Food Avoidances in the Old World,* 2nd ed. Madison: University of Wisconsin Press.

Singh, Rana P. B. 1994. "Water Symbolism and Sacred Landscape in Hinduism." *Erdkunde* 48: 210–227.

Stump, Roger W. (ed.). 1986. "The Geography of Religion." Special issue, *Journal of Cultural Geography* 7: 1–140.

Ten Recommended Books on the Geography of Religion

(For additional suggested readings, see *The Human Mosaic* web site: www.whfreeman.com/domosh12e)

Esposito, John, Susan Tyler Hitchcock, Desmond Tutu, and Mpho Tutu. 2004. *Geography of Religion: Where God Lives, Where Pilgrims Walk.* Washington, D.C.: National Geographic. In the tradition of the *National Geographic,* this comprehensive reference book is beautifully illustrated.

Gottlieb, Roger S. (ed.). 1995. *This Sacred Earth: Religion, Nature and Environment.* London: Routledge. A good introduction to ecotheology.

Halvorson, Peter L., and William M. Newman. 1994. *Atlas of Religious Change in America, 1952–1990*. Atlanta: Glenmary Research Center. This atlas shows the changing pattern of denominational membership in the United States in the second half of the twentieth century, revealing the strengthening of some religions and the weakening of others.

Harpur, James. 1994. *The Atlas of Sacred Places*. New York: Henry Holt. This volume is a useful depiction of the location of sacred places, or "sacred space," as we labeled it; you will be amazed by their number and diversity.

Jordan, Terry G. 1982. *Texas Graveyards: A Cultural Legacy*. Austin: University of Texas Press. The only book by a geographer on this aspect of the religious landscape, this volume is very revealing about a state possessing a wide array of ethnic groups.

O'Brien, Joanne, and Martin Palmer. 2007. *The Atlas of Religion: Mapping Contemporary Challenges and Beliefs*. Berkeley and Los Angeles: University of California Press. Maps the nature, extent, and influence of the world's major religions. Depicts how religions spread across space, their role in global issues such as hunger as well as conflict, and the locations of sacred places.

Park, Chris. 1994. *Sacred Worlds: An Introduction to Geography and Religion*. London: Routledge. An introductory text on the geography of religion.

Pui-lan, Kwok (ed.). 1994. *Ecotheology: Voices from South and North*. New York: World Council of Churches Publications. A very readable sampling of recent ecotheological thought.

Stoddard, Robert H., and Alan Morinis (eds.). 1997. *Sacred Places, Sacred Spaces: The Geography of Pilgrimages*. Baton Rouge: Geoscience Publications. Published by the Department of Geography and Anthropology at Louisiana State University, these 14 essays, each by a different author, draw attention to Christian, Muslim, Hindu, and lesser Asian pilgrimage traditions, as seen from the perspective of cultural geography.

Stump, Roger W. 2000. *Boundaries of Faith: Geographical Perspectives on Religious Fundamentalism*. Lanham, Md.: Rowman & Littlefield. After describing the background of fundamentalist movements within various world religions occurring in several countries, the author explains commonalities and clarifies political implications. One of the first examinations of the spatial strategies inherent in fundamentalism.

What differences can you "read" in these landscapes? Can you determine their locations?

Go to "Seeing Geography" on page 313 to learn more about these images.

AGRICULTURE
The Geography of the Global Food System

Every one of us depends, either directly or indirectly, on agriculture for our survival. It is easy to forget that urban-industrial society relies on the food surplus generated by farmers and herders and that, without agriculture, there would be no cities, universities, factories, or offices.

Agriculture, the tilling of crops and rearing of domesticated animals to produce food, feed, drink, and fiber, has been the principal enterprise of humankind throughout recorded history. Even today, agriculture remains by far the most important economic activity in the world, using more land than any other activity and employing about 40 percent of the working population. In some parts of Asia and Africa, more than 75 percent of the labor force is devoted to agriculture. North Americans, on the other hand, live in an urban society in which less than 2 percent of the population work as agriculturists. Likewise, Europe's labor force is as thoroughly nonagricultural as North America's. Nearly half of the world's population, however, continues to live in farm villages.

Region

How is the theme of region relevant to agriculture? For thousands of years, farmers have found ways to cope with a range of environmental conditions, creating in the process an array of different types of food-producing systems. Collectively, these farming practices have constructed formal **agricultural regions.** During the past 500 years, colonialism, industrialization, and globalization have greatly altered existing practices and created new types of agricultural regions, such as plantations. Increasingly, large-scale, energy-intensive agriculture overcomes local environmental constraints through irrigation; land reclamation; and the use of synthetic fertilizers, chemical pesticides, herbicides, and genetic engineering. Consequently, in order to understand agricultural regions, we need to consider the role of the environment, cultural factors, and the political and economic forces that influence a region.

Swidden Cultivation

Many of the peoples of tropical lowlands and hills in the Americas, Africa, and Southeast Asia practice a land-rotation agricultural system known as **swidden cultivation.** The term *swidden* is derived from an old English term meaning "burned clearing." Using machetes, axes, and chainsaws, swidden cultivators chop away the undergrowth from small patches of land and kill the trees by removing a strip of bark completely around the trunk. After the dead vegetation dries out, the farmers set it on fire to clear the land. Because of these clearing techniques, swidden cultivation is also called slash-and-burn agriculture. Working with digging sticks or hoes, the farmers then plant a variety of crops in the ash-covered clearings, varying from the maize (corn), beans, bananas, and manioc of Native Americans to the yams and nonirrigated rice grown by hill tribes in Southeast Asia (**Figure 8.1**, page 276). Different crops typically share the same clearing, a practice called **intercropping.** This technique allows taller, stronger crops to shelter lower,

FIGURE 8.1 **A farmer in Tojo, central Sulawesi, surveys the results of clearing and burning a patch of forest for planting.** What looks like destruction is actually part of a cycle that begins and ends in forest when swidden cultivation is practiced sustainably. *(Reuters/Yusuf Ahmad/Landov.)*

more fragile ones; reduces the chance of total crop losses from disease or pests; and provides the farmer with a varied diet. The complexity of many intercropping systems reveals the depth of knowledge acquired by swidden cultivators over many centuries. Relatively little tending of the plants is necessary until harvest time, and no fertilizer is applied to the fields because the ashes from the fire are a sufficient source of nutrients.

The planting and harvesting cycle is repeated in the same clearings for perhaps three to five years, until soil fertility begins to decline as nutrients are taken up by crops and not replaced. Subsequently, crop yields decline. These fields then are temporarily left fallow, and new clearings are prepared to replace them. Because the farmers periodically shift their cultivation plots, another commonly used term for the system is *shifting cultivation*. The abandoned cropland lies fallow for 10 to 20 years before farmers return to clear it and start the cycle again. Swidden cultivation represents one form of **subsistence agriculture:** food production mainly for the family and local community rather than for market.

subsistence agriculture
Farming to supply the minimum food and materials necessary to survive.

Although the technology of swidden may be simple, it has proved to be an efficient and adaptive strategy. Swidden farming, unlike some modern systems, is ecologically sustainable and has endured for millennia. Indeed, contemporary studies suggest that some swidden systems have actually enhanced biodiversity. Furthermore, swidden returns more calories of food for the calories spent on cultivation than does modern mechanized agriculture. In many tropical forest regions, swidden cultivation, unlike Western plantations, has left most of the forest intact over centuries of continuous use.

Nonetheless, swidden cultivation can be environmentally destructive under certain conditions. In poor countries with large landless populations, one often finds a front of pioneer swidden farmers advancing on the forests. In such situations, a range of institutional, economic, political, and demographic factors restrict poor farmers' abilities to employ the methods of swidden agriculture in a sustainable way. In many tropical countries, for example, a small proportion of the population owns most of the best agricultural land, forcing the majority of farmers to clear forests to gain access to land. Another condition that may diminish the sustainability of swidden cultivation occurs when a population experiences a sudden increase in its rate of growth and political or social conditions restrict its mobility. Swidden cultivation, still widely practiced throughout the tropics, is thus a highly variable system, occurring in both sustainable and unsustainable forms.

Paddy Rice Farming

Peasant farmers in the humid tropical and subtropical parts of Asia practice a highly distinctive type of agriculture called **paddy rice farming.** Rice, the dominant paddy crop,

paddy rice farming
The cultivation of rice on a paddy, or small flooded field enclosed by mud dikes, practiced in the humid areas of the Far East.

FIGURE 8.2 Cultivation of rice on the island of Bali, Indonesia. Paddy rice farming traditionally entails enormous amounts of human labor and yields very high productivity per unit of land. *What are the disadvantages of such a system?*
(Denis Waugh/Tony Stone Images.)

forms the basis of civilizations in which almost all the caloric intake is of plant origin. From the monsoon coasts of India through the hills of southeastern China and on to the warmer parts of Korea and Japan stretches a broad region of diked, flooded rice fields, or paddies, many of which are perched on terraced hillsides (**Figure 8.2**). The terraced paddy fields form a striking cultural landscape (see Figure 1.16, page 23).

A paddy rice farm of only 3 acres (1 hectare = 2.47 acres) is usually adequate to support a family because irrigated rice provides a very large output of food per unit of land. Still, paddy farmers must till their small patches intensively to harvest enough food. A system of irrigation that can deliver water when and where it is needed is key to success. In addition, large amounts of fertilizer must be applied to the land. Paddy farmers often plant and harvest the same parcel of land twice per year, a practice known as **double-cropping.** These systems are extremely productive, yielding more food per acre than many forms of industrialized agriculture in the United States.

double-cropping
Harvesting twice a year from the same parcel of land.

The modern era has witnessed a restructuring of paddy rice farming in more developed countries, such as Japan, Korea, and Taiwan. In some cases, the terrace structure has been reengineered to produce larger fields that can be worked with machines. In addition, dams, electric pumps, and reservoirs now provide a more reliable water supply, and high-yielding seeds, pesticides, and synthetic fertilizers boost production further. Most paddy rice farmers now produce mainly for urban markets.

Peasant Grain, Root, and Livestock Farming

In colder, drier Asian farming regions that are climatically unsuited to paddy rice farming—as well as in the river valleys of the Middle East, in parts of Europe, in Africa, and in the mountain highlands of Latin America and New Guinea—farmers practice a diverse system of agriculture based on bread grains, root crops, and herd livestock (**Figures 8.3 and 8.4**, page 278). Many geographers refer to these farmers as **peasants,** recognizing that they often represent a distinctive **folk culture** strongly rooted in the land. Peasants generally are small-scale farmers who own their fields and produce both for their own subsistence and for sale in the market. The dominant grain crops in these regions are wheat, barley, sorghum, millet, oats, and maize. Common cash crops—some of them raised for export—include cotton, flax, hemp, coffee, and tobacco.

peasant
A farmer belonging to a folk culture and practicing a traditional system of agriculture.

folk culture
A small, cohesive, stable, isolated, nearly self-sufficient group that is homogenous in custom and race; characterized by a strong family or clan structure, order maintained through sanctions based in the religion or family, little division of labor other than between the sexes, frequent and strong interpersonal relationships, and a material culture consisting mainly of handmade goods.

These farmers also raise herds of cattle, pigs, sheep, and, in South America, llamas and alpacas. The livestock pull the plow; provide milk, meat, and wool; serve as beasts of burden; and produce manure for the fields. They also consume a portion of the grain harvest. In some areas, such as the Middle Eastern river valleys, the use of irrigation helps support this peasant system. In general,

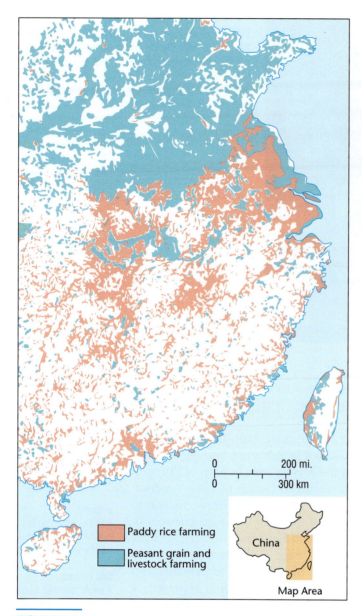

FIGURE 8.3 Two agricultural regions in China. The intricacies of culture region boundaries are suggested by the distribution of two types of agriculture in Taiwan and the eastern part of China. All such cultural-geographical borders are difficult to draw. *What might account for the more fragmented distribution of paddy rice farming, as contrasted with peasant grain, root, and livestock farming? Where would you draw the cultural boundary between the two types of agriculture?* (Source: Chuan-jun, 1979.)

however, most modern agricultural technologies are beyond the financial reach of most peasants.

Plantation Agriculture

In certain tropical and subtropical areas, Europeans and Americans introduced a commercial agricultural system called **plantation agriculture**. A **plantation** is a landholding

FIGURE 8.4 Peasant grain, root, and livestock agriculture in highland New Guinea. Distinctive "raised fields" with sweet potato mounds are found here among the farmers of highland New Guinea. These people raise diverse crops but give the greatest importance to sweet potatoes, together with pigs. *Why might they go to the trouble of creating these small mounds?* (Courtesy of Terry G. Jordan-Bychkov.)

devoted to capital-intensive, large-scale, specialized production of one tropical or subtropical crop for the global market. Each plantation district in the tropical and subtropical zones tends to specialize in one crop. Plantation agriculture has long relied on large amounts of manual labor, initially in the form of slave labor and later as wage labor. The plantation system originated in the 1400s on Portuguese-owned sugarcane-producing islands off the coast of tropical West Africa—São Tomé and Principe—but the greatest concentrations are now in the American tropics, Southeast Asia, and tropical South Asia. Most plantations lie near the seacoast, close to the shipping lanes that carry their produce to nontropical lands such as Europe, the United States, and Japan.

Workers usually live right on the plantation, where a rigid social and economic segregation of labor and management produces a two-class society of the wealthy and the poor. As a result of the concentration of ownership and production, a handful of multinational corporations, such as Chiquita and Dole, control the largest share of plantations globally. Tension between labor and management is not uncommon, and the societal ills of the plantation system remain far from cured (**Figure 8.5**).

Plantations provided the base for European and American economic expansion into tropical Asia, Africa, and Latin America. They maximize the production of luxury crops for

plantation agriculture
A system of monoculture for producing export crops requiring relatively large amounts of land and capital; originally dependent on slave labor.

plantation
A large landholding devoted to specialized production of a tropical cash crop.

FIGURE 8.5 Plantation agriculture. This sign was erected by the management at the entrance to a banana plantation in Costa Rica. "Welcome to Freehold Plantation, a workplace where labor harmony reigns; in mutual respect and understanding, we united workers produce and export quality goods in peace and harmony." *How does this message suggest that, in fact, not all is harmonious here and that the tension of the two-class plantation system might simmer below the surface?* (Courtesy of Terry G. Jordan-Bychkov.)

Europeans and Americans, such as sugarcane, bananas, coffee, coconuts, spices, tea, cacao, pineapples, rubber, and tobacco (**Figure 8.6**). Similarly, textile factories require cotton, sisal, jute, hemp, and other fiber crops from the plantation areas. Much of the profit from these plantations is exported, along with the crops themselves, to Europe and North America, another source of political friction between countries of the global North and South.

Market Gardening

The growth of urban markets in the last few centuries also gave rise to other commercial forms of agriculture, including **market gardening,** also known as truck farming. Unlike plantations, truck farms are located in developed countries and specialize in intensively cultivated nontropical fruits, vegetables, and vines. They raise no livestock. Many districts concentrate on a single product, such as wine, table grapes, raisins, olives, oranges, apples, lettuce, or potatoes, and the entire farm output is raised for sale rather than for consumption on the farm. Many truck farmers participate in cooperative marketing arrangements and depend on migratory seasonal farm laborers to harvest their crops. Market garden districts appear in most industrialized countries. In the United States, a broken belt of market gardens extends from California eastward through the Gulf and Atlantic coast states, with scattered districts in other parts of the country. The lands around the Mediterranean Sea are dominated by market gardens. In regions with mild climates, such as California, Florida, and the Mediterranean region, winter vegetables are a common market crop raised for sale in colder regions of the higher latitudes.

> **market gardening**
> Farming devoted to specialized fruit, vegetable, or vine crops for sale rather than consumption.

Livestock Fattening

In **livestock fattening,** farmers raise and fatten cattle and hogs for slaughter. One of the most highly developed fattening areas is the famous Corn Belt of the U.S. Midwest, where farmers raise corn and soybeans

> **livestock fattening**
> A commercial type of agriculture that produces fattened cattle and hogs for meat.

FIGURE 8.6 Tea plantation in the highlands of Papua New Guinea. Although profitable for the owners and providing employment for a small labor force, the plantation recently displaced a much larger population of peasant grain, root, and livestock farmers. This is one result of globalization. *Should the government have prevented such a displacement?* (Courtesy of Terry G. Jordan-Bychkov.)

FIGURE 8.7 Cattle feedlot for beef production. This feedlot, in Colorado, is reputedly the world's largest. *What ecological problems might such an enterprise cause?* (William Strode/Woodfin Camp.)

to feed cattle and hogs. Typically, slaughterhouses are located close to feedlots, creating a new meat-producing region, which is often dependent on mobile populations of cheap immigrant labor. A similar system prevails over much of western and central Europe, though the feed crops there more commonly are oats and potatoes. Other zones of commercial livestock fattening appear in overseas European settlement zones such as southern Brazil and South Africa.

One of the central traditional characteristics of livestock fattening is the combination of crops and animal husbandry. Farmers breed many of the animals they fatten, especially hogs. In the last half of the twentieth century, livestock fatteners began to specialize their activities; some concentrated on breeding animals, others on preparing them for market.

feedlot
A factorylike farm devoted to either livestock fattening or dairying; all feed is imported and no crops are grown on the farm.

In the factorylike **feedlot,** farmers raise imported cattle and hogs on purchased feed (**Figure 8.7**). Increases in the amount of land and crop harvest dedicated to beef production have accompanied the growth in feedlot size and number. In the United States, across Europe, and in European settlement zones around the globe, 51 to 75 percent of all grain raised goes to livestock fattening.

The livestock fattening and slaughtering industry has become increasingly concentrated, on both the national and global scales. In 1980 in the United States, the top four companies accounted for 41 percent of all slaughtered cattle. By 2000, the top four companies were slaughtering 81 percent of all feedlot cattle. One company alone accounted for 35 percent. Corporate conglomerates such as ConAgra and Cargill control much of the beef supply through their domination of the grain market and ownership of feedlots and slaughterhouses. The concentration of the industry has extended north and south across the border, primarily spurred by the 1994 North American Free Trade Agreement (NAFTA). By 1999, two U.S.-based companies accounted for 50 percent of Canada's slaughtered cattle. Cargill owns beef operations in more than 60 countries. One industry observer concludes that three multinational companies control the entire global beef industry.

Grain Farming

Grain farming is a type of specialized agriculture in which farmers grow primarily wheat, rice, or corn for commercial markets. The United States is the world's leading wheat and corn exporter. The United States, Canada, Australia, the European Union (EU), and Argentina together account for more than 85 percent of all wheat exports, while the United States alone accounts for about 70 percent of world corn exports. Wheat belts stretch through Australia, the Great Plains of interior North America, the steppes of Russia and Ukraine, and the pampas of Argentina. Farms in these areas generally are very large, ranging from family-run wheat farms to giant corporate operations (**Figure 8.8**).

Widespread use of machinery, synthetic fertilizers, pesticides, and genetically engineered seed varieties enables grain farmers to operate on this large scale. The

FIGURE 8.8 A "wheat landscape" in the Palouse, a grain farming region on the borders of Washington and Oregon. Grain elevators are a typical part of such agricultural landscapes. The raising of one crop such as wheat over entire regions is called *monoculture*. **What problems might be linked to monoculture?** *(Courtesy of Terry G. Jordan-Bychkov.)*

planting and harvesting of grain is more completely mechanized than any other form of agriculture. Commercial rice farmers employ such techniques as sowing grain from airplanes. Harvesting is usually done by hired migratory crews operating corporation-owned machines (**Figure 8.9**). Perhaps grain farming's ultimate development is the **suitcase farm,** which is found in the Wheat Belt of the northern Great Plains of the United States. The people who own and operate these farms do not live on the land. Most of them own several suitcase farms, lined up in a south-to-north row through the Plains states. They keep fleets of farm machinery, which they send north with crews of laborers along the string of suitcase farms to plant, fertilize, and harvest the wheat. The progressively later ripening of the grain as one moves north allows these farmers to maintain and harvest crops

suitcase farm
In American commercial grain agriculture, a farm on which no one lives; planting and harvesting are done by hired migratory crews.

FIGURE 8.9 Mechanized wheat harvest on the Great Plains of the United States. North American grain farmers operate in a capital-intensive manner, investing in machines, chemical fertilizers, and pesticides. **What long-term problems might such methods cause? What benefits are realized in such a system?** *(Roger Du Buisson/The Stock Market.)*

on all their farms with the same crew and the same machinery. Except for visits by migratory crews, the suitcase farms are uninhabited.

Such highly mechanized, absentee-owned, large-scale operations, or *agribusinesses,* have mostly replaced the small, husband- and wife-operated American family farm, an important part of the U.S. rural heritage. Geographer Ingolf Vogeler documented the decline of small family farms in the American countryside and argued that U.S. governmental policies, prompted by the forces of globalization, have consistently favored the interests of agribusiness, thereby hastening the decline. Family-owned farms continue to play an important role, but they now operate mostly as large agribusinesses that own or lease many far-flung grain fields.

Dairying

In many ways, the specialized production of dairy goods closely resembles livestock fattening. In the large dairy belts of the northern United States from New England to the upper Midwest, western and northern Europe, southeastern Australia, and northern New Zealand, the keeping of dairy cows depends on the large-scale use of pastures. In colder areas, some acreage must be devoted to winter feed crops, especially hay. Dairy products vary from region to region, depending in part on how close the farmers are to their markets. Dairy belts near large urban centers usually produce milk, which is more perishable, while those farther away specialize in butter, cheese, or processed milk. An extreme case is New Zealand, which, because of its remote location from world markets, produces much butter.

Reflecting on Geography

Why is dairying confined to northern Europe and the overseas lands settled by northern Europeans?

As with livestock fattening, in recent decades a rapidly increasing number of dairy farmers have adopted the feedlot system and now raise their cattle on feed purchased from other sources. Often situated on the suburban fringes of large cities for quick access to market, the dairy feedlots operate like factories. Like industrial factory owners, feedlot dairy owners rely on hired laborers to help maintain their herds. Dairy feedlots are another indicator of the rise of globalization-induced agribusiness and the decline of the family farm. By easing trade barriers, globalization compels U.S. dairy farmers to compete with producers in other parts of the world. Huge feedlots, a factory-style organization of production, automation, the concentration of ownership, and the increasing size of dairy farms are responses to this intense competition.

Nomadic Herding

In the dry or cold lands of the Eastern Hemisphere, particularly in the deserts, prairies, and savannas of Africa, the Arabian Peninsula, and the interior of Eurasia, **nomadic livestock herders** graze cattle, sheep, goats, and camels. North of the tree line in Eurasia, the cold tundra forms a zone of nomadic herders who raise reindeer. The common characteristic of all nomadic herding is mobility. Herders move with their livestock in search of forage for the animals as seasons and range conditions change. Some nomads migrate from lowlands in winter to mountains in summer; others shift from desert areas during the rainy season to adjacent semiarid plains in the dry season or from tundra in summer to nearby forests in winter. Some nomads herd while mounted on horses, such as the Mongols of East Asia, or on camels, such as the Bedouin of the Arabian Peninsula. Others, such as the Rendile of East Africa, herd cattle, goats, and sheep on foot.

> **nomadic livestock herder**
> A member of a group that continually moves with its livestock in search of forage for its animals.

Their need for mobility dictates that the nomads' few material possessions be portable, including the tents used for housing (**Figure 8.10**). Their mobile lifestyle also affects how wealth is measured. Typically, in nomadic cultures wealth is based on the size of livestock holdings rather than on the accumulation of property and personal possessions. Usually, the nomads obtain nearly all of life's necessities from livestock products or by bartering with the farmers of adjacent river valleys and oases.

For a number of different reasons, nomadic herding has been in decline since the early twentieth century. Some national governments established policies encouraging nomads to practice **sedentary cultivation** of the land. This practice was begun in the nineteenth century by British and French colonial administrators in North Africa because it allowed greater control of the people by the central governments. Today, many nomads are voluntarily abandoning their traditional life to seek jobs in urban areas or in the Middle Eastern oil fields. Recent severe drought in sub-Saharan Africa's Sahel region, which decimated nomadic livestock herds, was a further impetus to abandon nomadic life.

> **sedentary cultivation**
> Farming in fixed and permanent fields.

In recent decades, research conducted by geographers and anthropologists in Africa's semiarid environments has revealed the sound logic of nomadic herding practices. These studies demonstrate that nomadic cultures' pasture and livestock management strategies are rational responses to an erratic and unpredictable environment. Rainfall is highly irregular in time and space, and herding practices must adjust. The most important nomadic strategy is mobility, which allows herders to take fullest advantage of the resulting variations in range productivity. These findings

FIGURE 8.10 **Nomadic pastoralists in West Africa.** This Fulbe-Waila pastoralist household in Chad is moving their livestock to new pastures. Mobility is key to their successful use of the variable and unpredictable environment of this region of Africa. They are taking with them all of their possessions, including their shelter—a hut, which they have packed on top of one of their cows. *(Frank Kroenke/ Peter Arnold, Inc.)*

have led to a new appreciation of nomadic herding cultures and may cause governments to reconsider sedentarization programs, thus postponing the demise of herding cultures.

Livestock Ranching

ranching
The commercial raising of herd livestock on a large landholding.

Superficially, **ranching** might seem similar to nomadic herding. It is, however, a fundamentally different livestock-raising system. Although both nomadic herders and livestock ranchers specialize in animal husbandry to the exclusion of crop raising and both live in arid or semiarid regions, livestock ranchers have fixed places of residence and operate as individuals rather than within a communal or tribal organization. In addition, ranchers raise livestock on a large scale for market, not for their own subsistence.

Livestock ranchers are found worldwide in areas with environmental conditions that are too harsh for crop production. They raise only two kinds of animals in large numbers: cattle and sheep. Ranchers in the United States, Canada, tropical and subtropical Latin America, and the warmer parts of Australia specialize in cattle raising. Midlatitude ranchers in cooler and wetter climates specialize in sheep. Sheep production is geographically concentrated to such a degree that only three countries—Australia, China, and New Zealand—account for 56 percent of the world's export wool.

Urban Agriculture

The United Nations (UN) calculated that in 2008 the human species passed a milestone. For the first time in history, more people live in cities than in the countryside. As this global-scale rural-to-urban migration gained momentum, a distinct form of agriculture rose in significance. We might best call this **urban agriculture.** Millions of city dwellers, especially in Third World countries, now produce enough vegetables, fruit, meat,

urban agriculture
The raising of food, including fruit, vegetables, meat, and milk, inside cities, especially common in the Third World.

and milk from tiny urban or suburban plots to provide most of their food, often with a surplus to sell. In China, urban agriculture now provides 90 percent or more of all the vegetables consumed in the cities. In the African metropolises of Kampala and Dar es Salaam, 70 percent of the poultry and 90 percent of the leafy vegetables consumed in the cities, respectively, come from urban lands. In 2010, the UN Food and Agriculture Organization convened the first symposium on urban agriculture in Africa in recognition of the important role it will play in feeding the continent's twenty-first-century urban residents.

Geographer Susanne Freidberg has conducted research demonstrating the importance of urban agriculture to family income and food security in West Africa. Focusing on the city of Bobo-Dioulasso in Burkina Faso, Freidberg showed that though plots were small, urban agriculture offered residents "a culturally meaningful way to fulfill their roles as food producers and family providers." In its heyday in the 1970s and 1980s, urban farming provided substantial incomes from vegetable sales in both the domestic and export markets. Since then, collapsing demand and the deterioration of environmental conditions have threatened the enterprise and undermined cooperation and trust within Bobo-Dioulasso's urban agricultural communities.

In the developed world, urban agriculture is becoming more prevalent despite, or perhaps in reaction to, the fact that food sourcing is increasingly globalized. The reasons for the boom in urban agriculture in cities such as Vancouver, Canada, and London are complex. However, almost everywhere it is seen as a partial solution to the problems of food insecurity and lack of access to nutritious foods in cities. The work of geographer Nik Heynen has shown that solving food insecurity is integral to improving urban social justice in the United States. It tends to be poorer residents, often members of marginalized minority groups, who lack access to a healthy diet. Similarly in Europe, the World Health Organization (WHO) has identified the poor availability of and inequitable access to nutritious vegetables and fruits in cities, especially for vulnerable groups, as a major twenty-first-century problem for the region. According to the WHO, the promotion of local food production in European cities will aid in reducing urban poverty and inequalities.

Reflecting on Geography

What types of agriculture occur in the three most densely populated areas of the world? Are these two characteristics—type of agriculture and population density—linked?

Farming the Waters

Most of us don't think of the ocean when discussing agriculture. In fact, however, every year more and more of our

aquaculture
The cultivation, under controlled conditions, of aquatic organisms, primarily for food but also for scientific and aquarium uses.

mariculture
A branch of aquaculture specific to the cultivation of marine organisms, often involving the transformation of coastal environments and the production of distinctive new landscapes.

animal protein is produced through **aquaculture:** the cultivation and harvesting of aquatic organisms under controlled conditions. Aquaculture includes **mariculture,** shrimp farming, oyster farming, fish farming, pearl cultivation, and more. In the heart of Brooklyn, New York, urban aquaculture thrives in a laboratory, where fish bound for New York City restaurants are harvested from indoor tanks. Aquaculture has its own distinctive cultural landscapes of containment ponds, rafts, nets, tanks, and buoys (**Figure 8.11**). Like terrestrial agriculture, there is a marked regional character to aquacultural production.

Aquaculture is an ancient practice, dating back at least 4500 years. Older local practices, sometimes referred to as traditional aquaculture, involve simple techniques such as constructing retention ponds to trap fish or "seeding" flooded rice fields with shrimp. Commercial aquaculture, the industrialized, large-scale protein factories that are driving today's production growth, is a contemporary phenomenon. The greatest leaps in technology and production have occurred mostly since the 1970s.

FIGURE 8.11 Emerging mariculture landscapes. As mariculture expands, coastal landscapes and ecosystems are transformed, as in the case of marine fish farming off Langkawi Island, Malaysia. (© age fotostock/SuperStock.)

At the turn of the twenty-first century, aquaculture was experiencing phenomenal growth worldwide, growing nearly four times faster than all terrestrial animal food-producing sectors combined. Aquaculture is a key reason that per capita protein availability has mostly kept pace with population growth in recent decades. For example, China's per capita protein supply derived from aquaculture went from 1 kilogram in 1970 to nearly 24 kilograms in 2004. By 2008, aquaculture was producing nearly one-half of all fish and seafood consumed worldwide, compared with only 4 percent in 1970. Aquaculture's share of production is projected to continue growing as demand for seafood increases and wild fish stocks decline.

The extraordinary expansion of food production by aquaculture has come with high costs to the environment and human health. As with industrialized agriculture, most commercial aquaculture relies on large energy and chemical inputs, including antibiotics and artificial feeds made from the wastes of poultry and hog processing. Such production practices tend to concentrate toxins in farmed fish, creating a potential health threat to consumers. The discharge from fish farms, which can be equivalent to the sewage from a small city, can pollute nearby natural aquatic ecosystems. Around the tropics, especially tropical Asia, the expansion of commercial shrimp farms is contributing to the loss of highly biodiverse coastal mangrove forests.

Although aquaculture can take place just about anywhere that water is found, strong regional patterns do exist. Mariculture is prevalent along tropical coasts, particularly in the mangrove forest zone. Marine coastal zones in general, especially in protected gulfs and estuaries, have high

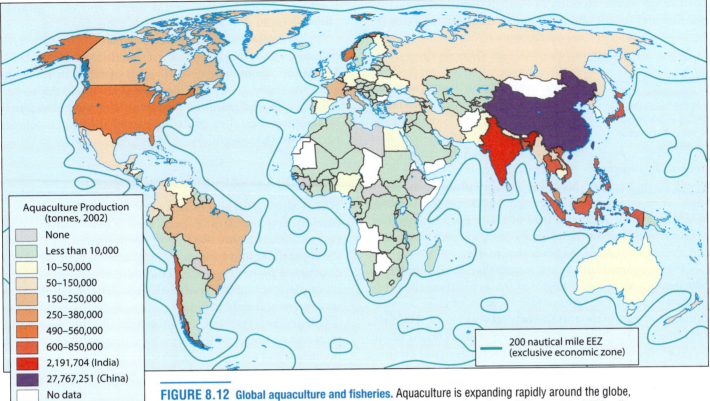

FIGURE 8.12 Global aquaculture and fisheries. Aquaculture is expanding rapidly around the globe, led by China, the world's top producer. Virtually all mariculture takes place within countries' 200-mile coastal territory known as the EEZ (exclusive economic zone). The EEZ is also the site of most commercial fishing, which has greatly depleted wild fish stocks and raised the need for increased mariculture production. *(Source: The Global Education Project, Fishing and Aquaculture.)*

Legend for figure:

Aquaculture Production (tonnes, 2002)
- None
- Less than 10,000
- 10–50,000
- 50–150,000
- 150–250,000
- 250–380,000
- 490–560,000
- 600–850,000
- 2,191,704 (India)
- 27,767,251 (China)
- No data

— 200 nautical mile EEZ (exclusive economic zone)

concentrations of mariculture. On a global scale, China dwarfs all other regions, producing over two-thirds of the world's farmed seafood (**Figure 8.12**). Asia and Pacific regions combined produce over 90 percent of the world's total.

If you order shrimp, trout, or salmon for dinner at any of the myriad restaurant chains, you will almost certainly be eating farmed protein. As the populations of wild species decline and the price of capture fish increases, farmed seafood will increasingly move into the breach. If trends continue, we may be among the last generations to enjoy affordable fresh-caught wild fish.

Nonagricultural Areas

Areas of extreme climate, particularly deserts and subarctic forests, do not support any form of agriculture. Such lands are found predominantly in much of Canada and Siberia. Often these areas are inhabited by **hunting-and-gathering** groups of native peoples, such as the Inuit, who gain a livelihood by hunting game, fishing where possible, and

hunting and gathering
The killing of wild game and the harvesting of wild plants to provide food in traditional cultures.

gathering edible and medicinal wild plants. At one time, all humans lived as hunter-gatherers. Today, fewer than 1 percent of humans do. Given the various inroads of the modern world, even these people rarely depend entirely on hunting and gathering. In most hunting-and-gathering societies, a division of labor by gender occurs. Males perform most of the hunting and fishing, whereas females carry out the equally important task of gathering harvests from wild plants. Hunter-gatherers generally rely on a great variety of animals and plants for their food.

Mobility

How does the theme of mobility help us understand the spatial and cultural patterns of agricultural production and food consumption? Some of the variation among the agriculture regions we've discussed results from **cultural diffusion.** Agriculture and its many components are inventions; they arose as innovations in certain source areas and diffused to other

cultural diffusion
The spread of elements of culture from the point of origin over an area.

parts of the world. Mobility, as we shall see, is key in the expansion and functioning of the modern global food system.

Origins and Diffusion of Plant Domestication

domesticated plant
A plant deliberately planted and tended by humans that is genetically distinct from its wild ancestors as a result of selective breeding.

Agriculture probably began with the domestication of plants. A **domesticated plant** is one that is deliberately planted, protected, cared for, and used by humans. Such plants are genetically distinct from their wild ancestors because they result from selective breeding by agriculturists. Accordingly, they tend to be bigger than wild species, bearing larger and more abundant fruit or grain. For example, the original wild "Indian maize" grew on a cob only one-tenth the size of the cobs of domesticated maize.

Plant domestication and improvement constituted a process, not an event. It began as the gradual culmination of hundreds, or even thousands, of years of close association between humans and the natural vegetation. The first step in domestication was perceiving that a certain plant was useful, which led initially to its protection and eventually to deliberate planting.

Cultural geographer Carl Johannessen suggests that the domestication process can still be observed. He believes that by studying current techniques used by native subsistence farmers in places such as Central America, we can gain insight into the methods of the first farmers of prehistoric antiquity. Johannessen's study of the present-day cultivation of the *pejibaye* palm tree in Costa Rica revealed that native cultivators actively engage in seed selection. All choose the seed of fresh fruit from superior trees, ones that bear particularly desirable fruit, as determined by size, flavor, texture, and color. Superior seed stocks are built up gradually over the years, with the result that elderly farmers generally have the best selections.

Seeds are shared freely within family and clan groups, allowing rapid diffusion of desirable traits.

The widespread association of female deities with agriculture suggests that it was women who first worked the land. Recall the almost universal division of labor in hunting-gathering-fishing societies. Because women had day-to-day contact with wild plants and their mobility was constrained by childbearing, they probably played the larger role in early plant domestication.

Locating Centers of Domestication

When, where, and how did these processes of plant domestication develop? Most experts now believe that the process of domestication was independently invented at many different times and locations. Geographer Carl Sauer, who conducted pioneering research on the origins and dispersal of plant and animal domestication, was one of the first to propose this explanation.

Sauer believed that domestication did not develop in response to hunger. He maintained that necessity was not the mother of agricultural invention, because starving people must spend every waking hour searching for food and have no time to devote to the leisurely experimentation required to domesticate plants. Instead, he suggested this invention was accomplished by peoples who had enough food to remain settled in one place and devote considerable time to plant care. The first farmers were probably sedentary folk rather than migratory hunter-gatherers. He reasoned that domestication did not occur in grasslands or large river floodplains, because primitive cultures would have had difficulty coping with the thick sod and periodic floodwaters. Sauer also believed that the hearth areas of domestication must have been in regions of great biodiversity where many different kinds of wild plants grew, thus providing abundant vegetative raw material for experimen-

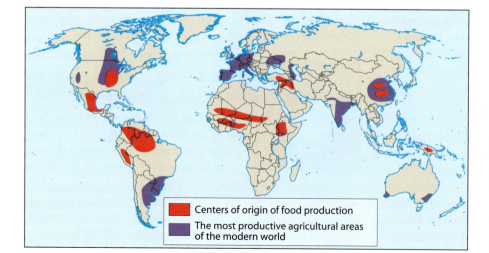

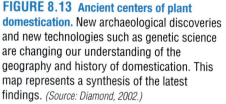

FIGURE 8.13 Ancient centers of plant domestication. New archaeological discoveries and new technologies such as genetic science are changing our understanding of the geography and history of domestication. This map represents a synthesis of the latest findings. *(Source: Diamond, 2002.)*

■ Centers of origin of food production
■ The most productive agricultural areas of the modern world

tation and crossbreeding. Such areas typically occur in hilly districts, where climates change with differing sun exposure and elevation above sea level.

Geographers, archaeologists, and, increasingly, genetic scientists continue to investigate the geographic origins of domestication. Because the conditions conducive to domestication are relatively rare, most agree that agriculture arose independently in at most nine regions (**Figure 8.13**). All of these have made significant contributions to the modern global food system. For example, the Fertile Crescent in the Middle East is the origin of the great bread grains of wheat, barley, rye, and oats that are so key to our modern diets. This region is also home to the first domesticated grapes, apples, and olives. China and New Guinea provided rice, bananas, and sugarcane, while the African centers gave us peanuts, yams, and coffee. Native Americans in Mesoamerica created another important center of domestication, from which came crops such as maize (corn), tomatoes, and beans. Farmers in the Andes domesticated the potato. While crop diffusions out of these nine regions have occurred over the millennia, the forces of globalization have now made even the rarest of local domesticates available around the world.

The dates of earliest domestication are continually being updated by new research findings. Until recently, archaeological evidence suggested that the oldest center is the Fertile Crescent, where crops were first domesticated roughly 10,000 years ago. However, domestication dates for other regions are constantly being pushed back by new discoveries. Most dramatically, in the Peruvian Andes archaeologists recently excavated domesticated seeds of squash and other crops that they dated to 9240 years before the present. These seeds were associated with permanent dwellings, irrigation canals, and storage structures, suggesting that farming societies were established in the Americas 10,000 years ago, similar to the Fertile Crescent date.

Pets or Meat? Tracing Animal Domestication

domesticated animal
An animal kept for some utilitarian purpose whose breeding is controlled by humans and whose survival is dependent on humans; domesticated animals differ genetically and behaviorally from wild animals.

A **domesticated animal** is one that depends on people for food and shelter and that differs from wild species in physical appearance and behavior as a result of controlled breeding and frequent contact with humans. Animal domestication apparently occurred later in prehistory than did the first planting of crops—with the probable exception of the dog, whose companionship with humans appears to be much more ancient. Typically, people value domesticated animals and take care of them for some utilitarian purpose. Certain domesticated animals, such as the pig and the dog, probably attached themselves voluntarily to human settlements to feast on garbage. At first, perhaps, humans merely tolerated these animals, later adopting them as pets or as sources of meat.

The early farmers in the Fertile Crescent deserve credit for the first great animal domestications, most notably that of herd animals. The wild ancestors of major herd animals—such as cattle, pigs, horses, sheep, and goats—lived primarily in a belt running from Syria and southeastern Turkey eastward across Iraq and Iran to central Asia. Farmers in the Middle East were also the first to combine domesticated plants and animals in an integrated system, the antecedent of the peasant grain, root, and livestock farming described earlier. These people began using cattle to pull the plow, a revolutionary invention that greatly increased the acreage under cultivation. In other regions, such as southern Asia and the Americas, far fewer domestications took place, in part because suitable wild animals were less numerous. The llama, alpaca, guinea pig, Muscovy duck, and turkey were among the few American domesticates.

Modern Mobilities

Over the past 500 years, European exploration and colonialism were instrumental in redistributing a wide variety of crops on a global scale: maize and potatoes from North America to Eurasia and Africa, wheat and grapes from the Fertile Crescent to the Americas, and West African rice to the Carolinas and Brazil.

The diffusion of specific crops continues, extending the process begun many millennia ago. The introduction of the lemon, orange, grape, and date palm by Spanish missionaries in eighteenth-century California, where no agriculture existed in the Native American era, is a recent example of **relocation diffusion.** This was part of a larger process of multidirectional diffusion. Eastern Hemisphere crops were introduced to the Americas, Australia, New Zealand, and South Africa through the mass emigrations from Europe over the past 500 years. Crops from the Americas diffused in the opposite direction. For example, chili peppers and maize, carried by the Portuguese to their colonies in South Asia, became staples of diets all across that region (**Figure 8.14**, page 288).

relocation diffusion
The spread of an innovation or other element of culture that occurs with the bodily relocation (migration) of the individual or group responsible for the innovation.

indigenous technical knowledge
Highly localized knowledge about environmental conditions and sustainable land-use practices.

In cultural geography, our understanding of agricultural diffusion focuses on more than just the crops; it also includes an analysis of the cultures and **indigenous technical knowledge** systems in which they are embedded. For example, geographer Judith Carney's study of the diffusion of African rice (*Oryza glaberrima*), which was domesticated independently in the inland delta area of West Africa's Niger River, shows the

FIGURE 8.14 **Chili peppers in Nepal and Korea.** A Tharu tribal woman of lowland Nepal prepares a condiment made of chili peppers from her garden, and in South Korea chili peppers dry under a plastic-roofed shed. This crop comes from the Indians of Mexico. *How might it have diffused so far and become so important in Asia?* For the answer, see Andrews, 1993. *(Courtesy of Terry G. Jordan-Bychkov.)*

importance of indigenous knowledge. European planters and slave owners carried more than seeds across the Atlantic from Africa to cultivate in the Americas. The Africans taken into slavery, particularly women from the Gambia River region, had the knowledge and skill to cultivate rice. Slave owners actively sought slaves from specific ethnic groups and geographic locations in the West African rice-producing zone, suggesting that they knew about and needed Africans' skills and knowledge. Carney argues that the "association of agricultural skills with certain African ethnicities within a specific geographic region" means that research on agricultural diffusion must address the relation of culture to technology and the environment.

green revolution
The recent introduction of high-yield hybrid crops and chemical fertilizers and pesticides into traditional Asian agricultural systems, most notably paddy rice farming, with attendant increases in production and ecological damage.

Not all innovations involve expansion diffusion and spread wavelike across the land; less orderly patterns are more typical. The **green revolution** in Asia provides an example. The green revolution is a product of modern agricultural science that involves the development of high-yield hybrid varieties of crops, increasingly genetically engineered, coupled with extensive use of chemical fertilizers. The high-yield crops of the green revolution tend to be less resistant to insects and diseases, necessitating the widespread use of

pesticides. The green revolution, then, promises larger harvests but ties the farmer to greatly increased expenditures for seed, fertilizer, and pesticides. It enmeshes the farmer in the global corporate economy. In some countries, most notably India, the green revolution diffused rapidly in the latter half of the twentieth century. By contrast, countries such as Myanmar resisted the revolution, favoring traditional methods. An uneven pattern of acceptance still characterizes the paddy rice areas today.

The green revolution illustrates how cultural and economic factors influence patterns of diffusion. In India, for example, new hybrid rice and wheat seeds first appeared in 1966. These crops required chemical fertilizers and protection by pesticides, but with the new hybrids India's 1970 grain production output was double its 1950 level. However, poorer farmers—the great majority of India's agriculturists—could not afford the capital expenditures for chemical fertilizer and pesticides, and the gap between rich and poor farmers widened. Many of the poor became displaced from the land and flocked to the overcrowded cities of India, aggravating urban problems. To make matters worse, the use of chemicals and poisons on the land heightened environmental damage.

The widespread adoption of hybrid seeds has created another problem: the loss of plant diversity or genetic variety. Before hybrid seeds diffused around the world, each farm developed its own distinctive seed types through the annual harvest-time practice of saving seeds from the better plants for the next season's sowing. Enormous genetic diversity vanished almost instantly when farmers began purchasing hybrids rather than saving seed from the last harvest. "Gene banks" have belatedly been set up to preserve what remains of domesticated plant variety, not just in the areas affected by the green revolution but also in the American Corn Belt and many other agricultural regions where hybrids are now dominant. In sum, the green revolution has been a mixed blessing.

Labor Mobility

Agriculture, more than any other modern economic endeavor, is constrained by the rhythms of nature. Biological cycles associated with planting and harvesting are reflected in cycles of labor demand. Seeds must be planted at the right moment to take advantage of seasonal conditions. When crops ripen, they must be harvested, often in a matter of days or weeks. In between, labor demand is minimal, so farmers face a dilemma. They need to mobilize a large labor force for harvesting crops, but keeping a year-round work force raises farm production costs. One solution has been to rely on migratory labor.

> **migrant workers**
> Most broadly, this term refers to people working outside of their home country. Migrant workers are particularly critical to large-scale commercial agriculture.

In the United States, the use of **migrant workers** has been central to the growth and profitability of farming. In his study of labor and landscape in California, geographer Don Mitchell found that the mechanization and intensification of farming created ever greater extremes in seasonal labor demand, lessening the demand for labor during nonharvest periods while still requiring manual labor for the increased harvests. Farmers thus developed an agricultural industry based on migratory labor, employing cultural and racial stereotypes to depress farm wages and tighten employers' control over farmworkers. Mitchell noted that through much of the twentieth century, California growers surmised that "hispanic, black, or Asian workers . . . were 'naturally' better suited to agricultural tasks," and prevailing racist attitudes allowed them to "pay nonwhite workers a lower wage than white workers." Ultimately, the federal government provided the legal mechanism, called the *Bracero* Program, by which contract workers were brought from Mexico to California during periods of peak labor demand. Migrant workers lived in substandard housing, were paid less than a living wage, and were deported back to Mexico if they complained.

The case of California's migrant workers is not unusual. Geographer Gail Hollander has documented the use of migratory labor in the development of Florida's sugarcane region. Florida produces 20 percent of the U.S. sugar supply, which until the 1990s was harvested entirely by hand by migrant workers imported seasonally from "former slave plantation economies of the Caribbean." Like their California counterparts, growers relied on racial stereotypes to argue that only blacks were suitable for cutting cane in Florida. The federal government also established a special federal immigration program like *Bracero* to import Caribbean migrant workers for the Florida harvest and repatriate them afterwards. Migrant farmworkers remain ubiquitous today in other regions, too, such as the EU. A recent agricultural boom in Mediterranean coastal Spain relies heavily on North African migrant workers, many of them undocumented immigrants, who are subject to persistent racial prejudices (**Figure 8.15**).

FIGURE 8.15 Migrant farmworkers: a global phenomenon. International migrant farmworkers, such as these African immigrants harvesting lettuce in southern Spain, are critical to production in the global food system. *(Mark Eveleigh/Alamy.)*

Globalization

How have the processes of globalization altered the geography of agriculture? How does globalization affect the availability and variety of food in specific places?

For most of human history, people obtained their provisions locally and had locally distinct dietary cultures. The development of global markets over the past 500 years has shifted cultural food preferences and altered the ecology of vast areas of the planet.

Local-Global Food Provisioning

As European maritime explorers brought far-flung cultures into contact, a multitude of crops were diffused around the globe. The processes of exploration, colonization, and globalization created new regional cuisines (imagine Italian cuisine without tomatoes from the Americas) and at the same time simplified the global diet to a disproportionate reliance on only three grains: wheat, rice, and maize.

The expansion of European empires in the seventeenth and eighteenth centuries was inseparable from the expansion of tropical plantation agriculture. Plantations in warm climates produced what were then luxury foods for markets in the global North, which had developed seemingly insatiable appetites for sugar, tea, coffee, and other tropical crops. The expansion of plantation agriculture had profound effects on local ecology, all but obliterating,

for example, the forests of the Caribbean and the tropical coasts of the Americas (**Figure 8.16**).

In short, we have witnessed the development of a global food system that, for better or worse, has freed consumers in the affluent regions of the world from the constraints of local ecologies. Fresh strawberries, bananas, pears, avocados, pineapples, and many, many other types of temperate, subtropical, and tropical produce are available in our urban supermarkets any day, any time of the year. On the other hand, the emphasis on a relatively small number of staple crops desired by northern consumers can mean the abandonment of local crop varieties and a decline in the associated biological diversity. Imported refined wheat from the global North enters poor tropical countries by the shipload, altering local dietary cultures and undercutting the ability of local farmers to sell their crops at a profitable price.

These are the general patterns, but the globalization of food and agriculture has complex effects on culture and ecology that vary by location and spatial scale. These complexities are best illustrated by a case study from the Peruvian Andes. Geographer Karl Zimmerer (see Practicing Geography; also see Rod's Notebook, page 292, for a different look at the globalization of food and agriculture) conducted extensive fieldwork among Quichua peasant farmers in the Paucartambo Andes to determine the effects of economic change on indigenous agricultural practices and the genetic diversity of local crops. He wanted to test the general hypothesis that as globalization and national economic policies integrate indigenous farmers into market production,

FIGURE 8.16 An oil palm plantation in Malaysia. Plantation agriculture continues to expand in many Third World countries. While plantation-grown export commodities can be important to national economies, the accompanying destruction of tropical rain forests is a high ecological price to pay. *(Stuart Franklin/Magnum.)*

PRACTICING GEOGRAPHY

Penn State University geography professor Karl Zimmerer discovered the discipline of geography through his interests in agriculture. As an undergraduate, he was a student intern at the Land Institute in Salina, Kansas, where research was focused on developing alternatives to conventional agriculture and on the food supply. As Zimmerer recalls, "Its director, Wes Jackson, suggested that I consider the field of geography for graduate school. I read books, visited departments, talked to students and listened to faculty lectures, and recognized that this was the discipline for me." Geography offered him the ideal disciplinary setting for pursuing his broad interests in agriculture and food, which range from environmental to political issues.

Professor Zimmerer finds three aspects of practicing geography particularly compelling and exciting. "First is the challenge of understanding the multifaceted changes of agriculture in Latin America and the United States. . . . Part of my interest is the fusion of the human world and the natural world that takes place in agriculture." In particular, he is interested in the new changes in the areas of biotechnology and corporate agribusiness.

"Second is the excitement and intellectual engagement of fieldwork [and] working closely with the diverse people in farming sites. When in the Andean countries, most often Bolivia and Peru, I speak Quechua as well as Spanish. Using these languages is not only exciting and engaging, but it also continually reinforces for me the importance of languages in addition to English."

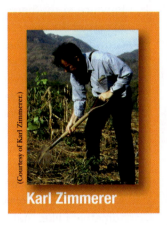

(Courtesy of Karl Zimmerer.)

Karl Zimmerer

The third area of practicing geography that appeals to Professor Zimmerer is communicating his knowledge through teaching, speaking, and writing. "Often I find myself faced with new and welcome challenges in understanding, interpreting, and telling the stories of the changes in agriculture." As an example, through the feedback that he has received from his teaching and public speaking, Zimmerer "became increasingly aware of this diverse, expanding, and possibly hopeful (although potentially explosive) interface of agriculture and conservation." The agriculture-conservation interface subsequently has become an important part of his research and writing.

His work in the field typically involves a variety of methods and techniques, including a lot of interviewing and related qualitative ethnographic methods. "I also do a fair amount of archival research," he explains, "since many of my projects on agriculture have a historical dimension and since they are often located in places where archival research is the main way of gaining information and awareness about the past." Professor Zimmerer is no stranger to quantitative methods, which are important for studying the environmental aspects of agriculture, such as biodiversity and soil properties. "In the latter sort of research I often use biogeography and ecology field methods and have also used laboratory techniques." Most recently his research has taken a historical turn, focusing on ideas of sustainable land use in the Americas over a 500-year period.

---

the diversity of crops declines, ultimately resulting in genetic erosion (i.e., a decline in the genetic diversity of cultivars).

Zimmerer's study produced surprising findings on the complex relationships among culture, economy, and the environment. On the question of whether farmers must abandon crop diversity in order to adopt new, commercially oriented high-yielding varieties, he found there was no simple answer. In fact, it was the more well-off peasants, heavily involved in commercial farming, who had the resources and land to cultivate diverse crops and "enjoy their agronomic, culinary, cultural, and ritual values." Among these values was the use of diverse, noncommercial potato varieties in local bartering. The ability to use noncommercial varieties in this way is valued in the local culture because it is a traditional

way to cement interpersonal bonds. Such uses emphasize the cultural importance of crop diversity. Zimmerer discovered that the cultural relevance of crops was a strong motivation for planting by well-off farmers. At least 90 percent of the genetically diverse crops had been conserved, even as the Quichua were further integrated into commercial production for the market.

A number of lessons can be drawn from Zimmerer's study, chief among them the need to carefully examine the effects of globalization on culture and environment, rather than simply assuming that local agricultural practices will give way to the demands of the marketplace. The Quichua farmers who benefited most from their participation in the market were those best able to cultivate traditional

Rod's Notebook

The Importance of Place in the Global Food System

Roderick Neumann

Olive growers large and small deliver their harvests for milling into oil at the Sierra de Cazorla Olive Growers' Cooperative in Andalusia, Spain. Such local co-ops are critical in place-based marketing of global agricultural commodities such as olive oil. *(Courtesy of Roderick Neumann.)*

December is olive harvest season in southern Spain, when farmers large and small bring their crops to nearby mills for processing. On a recent research trip to Jaén Province in southern Spain, my colleague Gail Hollander and I observed the process at the mill of the Sierra de Cazorla Olive Growers' Cooperative. We watched as families with 5-foot tow trailers lined up next to huge corporate-owned dumptrucks to unload their harvests onto 150-foot conveyer belts leading to the oil presses. We were witnessing a centuries-old harvest ritual refitted to agroindustrial mass production.

Olives and olive oil are staple foods in the Mediterranean diet. Thus, people in the region are sensitized to varying tastes, colors, and consistencies of olive oils to which most North Americans are oblivious. Approach most North Americans about olive oil and they are likely to respond with something like, "Italian food." In one sense this is correct, since Italy is indeed the largest exporter of processed olive oil. But Spain is the world's largest producer of olives, the greatest share coming from Jaén Province. Most of Spain's harvest, however, is sent to Italy for repackaging and export under Italian brand names that are more familiar to North American shoppers.

This system of mass production of olive oil for Italian export firms worked well for Spanish farmers until recent years. The forces of globalization are now challenging Spain's dominance of the market. North African countries, such as Tunisia, are expanding olive production and competing with Spain for access to the global mass market of undifferentiated oil. In interviews with farmers in Jaén, we discovered that they are looking for ways to differentiate their product from mass-produced oils. Similar to the way fine wines are marketed, they seek to emphasize the unique qualities of oils that are produced in specific places with particular climate and soil characteristics. The idea is to find a niche market of high-end consumers who are appreciative of the distinct qualities and characteristics of local olive varieties.

Jaén farmers have responded to the challenges of globalization by creating new place-based products. Olive oil is being bottled with the "*denominación de origen*" labeling, a third-party certification system that guarantees consumers that the product originates from a specific place. We discovered that the number of marketing cooperatives using the *denominación de origen* designation has exploded across Jaén. Each co-op offers an array of olive-based products. We are still enjoying, for example, our bottle of extra-virgin Royal Aniversario 10 olive oil from the Sierre de Cazorla *denominación de origen,* a product that many Spaniards consider to be of the highest quality anywhere. Currently *denominación de origen* olive oil is sold mostly to an appreciative domestic Spanish market. Farmers are eyeing North America and East Asia as future export markets as affluent consumers learn of the complexities of olive oil quality and seek new culinary experiences. We were reminded that place matters in the global food system.

Posted by Rod Neumann

varieties, which functioned as an expression of cultural identity and their sense of place. Cultural values, not merely a strict economic or ecological calculus, critically influenced farming decisions.

The von Thünen Model

Geographers and others have long tried to understand the distribution and intensity of agriculture based on transportation costs to market. Long before globalization took hold, the nineteenth-century German scholar-farmer Johann Heinrich von Thünen developed a **core-periphery** model to address the problem. In his model, von Thünen proposed an "isolated state" that had no trade connections with the outside world; possessed only one market, located centrally in the state; had uniform soil and climate; and had level terrain throughout. He further assumed that all farmers located the same distance from the market had equal access to it and that all farmers sought to maximize their profits and produced solely for market. Von Thünen created this model to study the influence of distance from market and the concurrent transport costs on the type and intensity of agriculture.

> **core-periphery**
> A concept based on the tendency of both formal and functional culture regions to consist of a core or node, in which defining traits are purest or functions are headquartered, and a periphery that is tributary and displays fewer of the defining traits.

Figure 8.17 presents a modified version of von Thünen's isolated-state model, which reflects the effects of improvements in transportation since the 1820s, when von Thünen proposed his theory. The model's fundamental feature is a series of concentric zones, each occupied by a different type of agriculture, located at progressively greater distances from the central market.

Reflecting on Geography

> Why should we study spatial models, such as von Thünen's, when they do not depict reality?

For any given crop, the intensity of cultivation declines with increasing distance from the market. Farmers near the market have minimal transportation costs and can invest most of their resources in labor, equipment, and supplies to

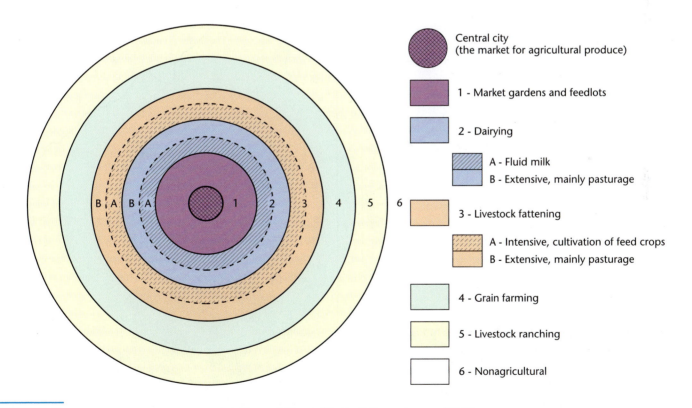

Central city (the market for agricultural produce)

1 - Market gardens and feedlots

2 - Dairying

  A - Fluid milk
  B - Extensive, mainly pasturage

3 - Livestock fattening

  A - Intensive, cultivation of feed crops
  B - Extensive, mainly pasturage

4 - Grain farming

5 - Livestock ranching

6 - Nonagricultural

FIGURE 8.17 Von Thünen's isolated-state model. The model is modified to fit the modern world better, showing the hypothetical distribution of types of commercial agriculture. Other causal factors are held constant to illustrate the effect of transportation costs and differing distances from the market. The more intensive forms of agriculture, such as market gardening, are located nearest the market, whereas the least intensive form (livestock ranching) is most remote. Compare this model to the real-world pattern of agricultural types in Uruguay, South America, shown in Figure 8.18, page 294. *Why does the model have the configuration of concentric circles?*

augment production. Indeed, because their land is more valuable and subject to higher taxes, they have to farm intensively to make a bigger profit. With increasing distance from the market, farmers invest progressively less in production per unit of land because they have to spend progressively more on transporting produce to market. The effect of distance means that highly perishable products such as milk, fresh fruit, and garden vegetables need to be produced near the market, whereas peripheral farmers have to produce nonperishable products or convert perishable items into a more durable form, such as cheese or dried fruit.

The concentric-zone model describes a situation in which highly capital-intensive forms of commercial agriculture, such as market gardening and feedlots, lie nearest to market. The increasingly distant, successive concentric belts are occupied by progressively less intensive types of agriculture, represented by dairying, livestock fattening, grain farming, and ranching.

How well does this modified model describe reality? As we would expect, the real world is far more complicated. For example, the emergence of **cool chains** for agricultural commodities—the refrigeration and transport technologies that bring fresh produce from fields around the globe to our dinner tables—have collapsed distance. Still, on a world scale, we can see that intensive commercial types of agriculture tend to occur most commonly near the huge urban markets of northwestern Europe and the eastern United States. An even closer match can be observed in smaller areas, such as in the South American nation of Uruguay (**Figure 8.18**).

cool chain
The refrigeration and transport technologies that allow for the distribution of perishables.

The value of von Thünen's model can also be seen in the underdeveloped countries of the world. Geographer Ronald Horvath made a detailed study of the African region centering on the Ethiopian capital city of Addis Ababa. Although noting disruptions caused by ethnic and environmental differences, Horvath found "remarkable

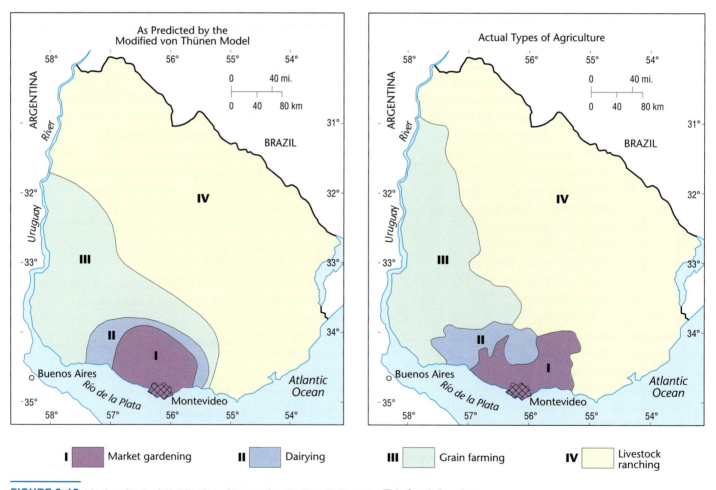

FIGURE 8.18 Ideal and actual distribution of types of agriculture in Uruguay. This South American country possesses some attributes of von Thünen's isolated state, in that it is largely a plains area dominated by one city. *In what ways does the spatial pattern of Uruguayan agriculture conform to von Thünen's model? How is it different? What might cause the anomalies?* (For the answers, see Griffin, 1973.)

parallels between von Thünen's crop theory and the agriculture around Addis Ababa." Similarly, German geographer Ursula Ewald applied the model to the farming patterns of colonial Mexico during the period of Spanish rule, concluding that even this culturally and environmentally diverse land provided "an excellent illustration of von Thünen's principles on spatial zonation in agriculture."

Will the World Be Fed?

Are famine and starvation inevitable as the world's population grows, as Thomas Malthus predicted (see Chapter 3)? Or will our agricultural systems successfully feed nearly 7 billion people? In trying to answer these questions, we face a paradox. Today, nearly 1 billion people are malnourished, some to the point of starvation. Almost every year, we read of food shortages occurring somewhere in the world. In 2008, there were food riots in 30 countries around the world when food prices shot beyond the reach of hundreds of thousands of the urban poor. Between 1990 and 2010, the number of hungry people in western and southern Asia and sub-Saharan Africa increased by tens of millions (**Figure 8.19**, pages 296–297).

Yet—and this would astound Malthus—food production has grown more rapidly than the world population over the past 40 or 50 years. Per capita, more food is available today than in 1950, when only about half as many people lived on Earth. Production continues to increase. From 1996 to 2006, world food production increased at an annual rate of 2.2 percent, and hunger was reduced by 30 percent in more than 30 countries. Thus, paradoxically, on a global scale there is enough food produced to feed everyone, while famines and malnutrition prevail. If the world food supply is sufficient to feed everyone and yet hunger afflicts one of every six or seven persons, then Malthus was wrong about the limits on population growth but right about the persistence of deprivation.

What explains this paradox of dearth amidst plenty? As geographer Thomas Bassett and economist Alex Winter-Nelson show in their *Atlas of World Hunger,* one must ask both where and why people are hungry. We see in Figure 8.19 where people are going hungry on a global scale. Why they are hungry is complex and varies geographically and historically. For example, Bassett and Winter-Nelson explain that the crisis of HIV/AIDS in southern Africa affects mostly 15- to 49-year-olds, the most productive segment of the population. The epidemic thus negatively affects food production and raises the level of food insecurity, especially in rural areas.

To a great extent international political economics, not global food shortages, causes hunger and starvation. International trade favors the farmers of wealthier countries through systems of government subsidies that make the prices of their agricultural exports artificially low. Third World farmers find it difficult to compete. Many Third World countries do not grow enough food to feed their populations, and they cannot afford to purchase enough imported food to make up the difference. As a result, famines can occur even when plenty of food is available. Millions of Irish people starved in the 1840s while adjacent Britain possessed enough surplus food to have prevented this catastrophe. Bangladesh suffered a major famine in 1974, a year of record agricultural surpluses in the world.

Internal government policies are also an important cause of famine. The roots of the largest famine of the twentieth century lie in the agricultural policies of the Chinese government's 1958–1961 Great Leap Forward. The Chinese government required peasants to abandon their individual fields and work collectively on large, state-run farms. This policy of collectivization succeeded in boosting food production in some cases but failed in most. Thirty million rural Chinese died of starvation during the Great Leap Forward. Misguided government policies triggered one of the first famines of the twenty-first century as well. In the early 2000s, Zimbabwe's President Robert Mugabe clung to power by demonizing white commercial farmers. In 2002, he threatened Zimbabwe's commercial farmers with imprisonment if they continued to farm. Other government policies discouraged planting and cultivation, thus producing another human-caused famine.

Even when major efforts are made to send food from wealthy countries to famine-stricken areas, the poor transportation infrastructure of Third World countries often prevents effective distribution. Political instability can disrupt food shipments, and the donated food often falls into the hands of corrupt local officials. Such was the case in Somalia in the 1990s, where warring factions in the capital city of Mogadishu prevented food aid from getting to starving populations. So while the trigger for famine may be environmental, there are deep-seated political and economic problems that conspire to block famine relief.

The Growth of Agribusiness

Globalization and its impact on agriculture have been referred to throughout this chapter. Globalization, you will recall, involves the restructuring of the world economy by multinational corporations thriving in an era of free-trade capitalism, rapid communications, improved transport, and computer-based information systems. When applied to agriculture, globalization tends to produce **agribusiness:** a modern farming system that is totally commercial, large-scale, mechanized, and dependent upon chemicals,

globalization
The binding together of all the lands and peoples of the world into an integrated system driven by capitalistic free markets, in which cultural diffusion is rapid, independent states are weakened, and cultural homogenization is encouraged.

agribusiness
Highly mechanized, large-scale farming, usually under corporate ownership.

Mapping Hunger

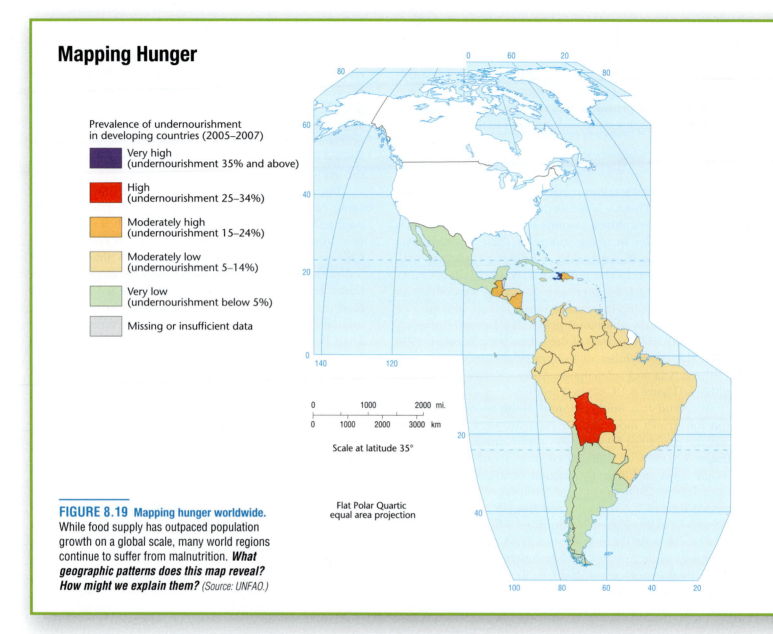

Prevalence of undernourishment
in developing countries (2005–2007)

- Very high
  (undernourishment 35% and above)
- High
  (undernourishment 25–34%)
- Moderately high
  (undernourishment 15–24%)
- Moderately low
  (undernourishment 5–14%)
- Very low
  (undernourishment below 5%)
- Missing or insufficient data

Scale at latitude 35°

0   1000   2000 mi.
0 1000 2000 3000 km

Flat Polar Quartic
equal area projection

FIGURE 8.19 Mapping hunger worldwide.
While food supply has outpaced population
growth on a global scale, many world regions
continue to suffer from malnutrition. ***What
geographic patterns does this map reveal?
How might we explain them?*** (*Source: UNFAO.*)

monoculture
The raising of only one crop
on a huge tract of land in
agribusiness.

hybrid seeds, genetic engineering, and the practice of **monoculture** (raising a single specialty crop on vast tracts of land). Furthermore, agribusinesses are often vertically integrated; that is, corporations own the land as well as the processing and marketing facilities. Vertical integration takes a variety of forms depending on the nature of production processes and markets. The case of the "global chicken" provides a useful illustration of vertically integrated agribusiness as well as the interconnections of changing cultural values and global food production.

The origins of the "global chicken" can be found in the shift from beef to poultry as the preferred protein source in American dietary culture, which underwent fundamental changes during the post–World War II period.

From 1945 to 1995, per capita consumption of chicken in the United States rose from 5 to 70 pounds (2.25 to 31.5 kilograms) and by 1990 surpassed that of beef. This was a startling development in American culture, where the myth of the cowboy herding cattle on the open range has been so central to an imagined national identity. Since the 1990s, the per capita consumption of chicken has continued to grow as that of beef continues to decline, especially since the publicity about mad cow disease.

Advances in U.S. agrotechnologies for the breeding, nutrition, housing, and processing of chickens largely account for increases in production efficiency (**Figure 8.20**, page 298). This has allowed the U.S. poultry industry, largely centered in the South, to become the world's single largest supplier of broilers. As the taste for chicken spreads world-

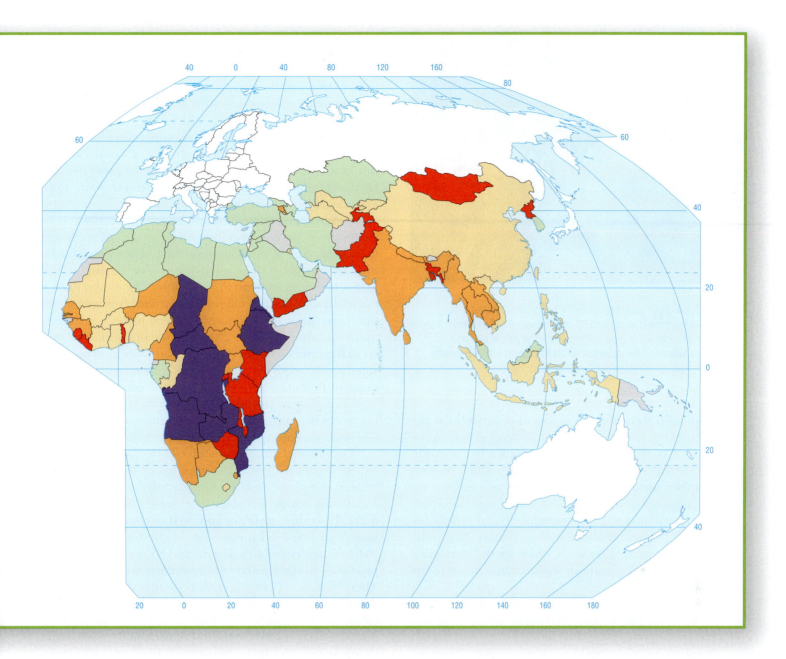

wide, U.S. producers have been positioned to gain increasing shares of an expanding market. From 1980 to 2002, world trade in broilers grew nearly 500 percent, while the U.S. share of that trade rose from 22.2 to 46.1 percent. China has been the hottest import market, because rising affluence there has led to increasing per capita consumption. At the same time, China is increasing its production and its exports of poultry. It is likely to become a major competitor with the United States for access to other Asian markets.

Farmers who produce a single commodity, such as poultry, must produce far more than the local market can consume in order to be profitable. Thus, they must sell in both national and global markets, access to which requires a dependence on multinational agribusinesses. So pervasive is the reach of agribusiness that many poultry farmers no longer own the chickens they produce; multinational corporations do. Farmers contract with multinationals to receive farm inputs such as chicks and feed and to cover other expenses related to marketing and transport. When the chickens mature, they are trucked to the contracting corporation's processing plant where they are weighed and corporate expenditures are deducted from the farmer's shares. The farmers take their earnings to pay the mortgages on their lands and buildings, and the chickens are processed for the global food system.

The Ongoing Green Revolution

The green revolution generally refers to the transfer of agro-industrial technological packages to Third World countries.

FIGURE 8.20 The "global chicken." Poultry consumption has skyrocketed worldwide, propelled by new industrialized systems of meat production and changing food preferences. *How might cultural differences influence the structure of international trade in poultry products?* (Robert Nickelsburg/Time Life Pictures/Getty Images.)

It can also been seen as one component of agricultural globalization, along with countless "rural development" projects in Third World countries, usually funded by the World Bank or the International Monetary Fund. These projects often displace small-scale peasant farmers to make way for larger enterprises and even multinational agribusinesses. The green revolution is thus a continuation and geographic expansion of the industrialization of agriculture. Key to the industrialization process are seeds, which are the foundation of cultivation. Whoever controls seeds controls access to the next crop harvest.

Control of seeds has been consolidated in fewer and fewer companies. The five biggest hybrid vegetable seed suppliers control 75 percent of the global market, and the ten largest agrochemical manufacturers command 85 percent of the world supply. Four corporations supply more than two-thirds of the U.S. consumption of hybrid seed maize. Sometimes single companies—Monsanto, for example—both supply the seeds and manufacture the pesticides. What's more, the genetic engineering of seeds is often done in-house. This arrangement allows Monsanto to genetically engineer "Roundup Ready" seed varieties. Roundup is an herbicide manufactured by Monsanto, and its Roundup Ready gene builds in greater tolerance to higher doses. The seeds essentially became vehicles to sell more herbicide.

Genetically modified (GM) crops, the products of biotechnology, are seen by many as another aspect of globalization. Genetic engineering produces new organisms through gene splicing. Pieces of DNA can be recombined with the DNA of other organisms to produce new properties, such as pesticide tolerance or disease resistance. DNA can be transferred not only between species but also between plants and animals, which makes this technology truly revolutionary and unlike any other development since the beginning of domestication. Agribusinesses are often able to patent the processes and resulting genetically engineered organisms and, thus, claim legal ownership of new life-forms.

Political, economic, and environmental problems resulting from the concentration of ownership of seeds and the production of pesticide-resistant GM seeds are beginning to emerge in the United States. The Department of Justice began an antitrust investigation of Monsanto's activities in the seed market. One reason for the investigation was the sharp rise in corn and soybean seed prices, which have more than doubled since 2001. The use of Roundup Ready seeds has resulted in the emergence and spread of so-called superweeds, weeds that have evolved resistance to increasing doses of Roundup herbicide. Thus, the gains in yield that were initially produced by Roundup Ready seeds are starting to decline, suggesting the environmental limitations of GM seeds that promote the use of pesticides.

Commercial production of GM crops began in the United States in 1996. The technology has now spread around the globe, but the United States still dominates, accounting for two-thirds of the world's acreage (**Figure 8.21**). Two crops, soybeans and corn, account for the rapid growth of GM food production in the United States. By 2010, 93 percent of all soybeans and 86 percent of all corn produced in the United States were genetically modified. Their genetically engineered resistance to disease and drought are an important reason for the spread of GM crops, but that's only part of the story. For example, all of the GM soybean seeds in the United States are engineered to tolerate greater doses of synthetic herbicides produced by agrochemical companies.

If you provision your household from a U.S. supermarket, you have undoubtedly ingested GM foods. Whether or not one finds this troubling is closely related to the strength of certain cultural norms and values that vary from region to region and country to country. In England and western Europe, where national identities are strongly linked to the countryside, agrarian culture, and regional cuisines, there has been a lot of opposition to biotechnology in agriculture. In response to public pressure, major supermarket chains, such as Sainsbury's in England, have refused since 1998 to sell GM foods. In the United States, the response has been far more muted, so much so that the expansion of GM crop planting has proceeded virtually without public debate.

genetically modified (GM) crops
Plants whose genetic characteristics have been altered through recombinant DNA technology.

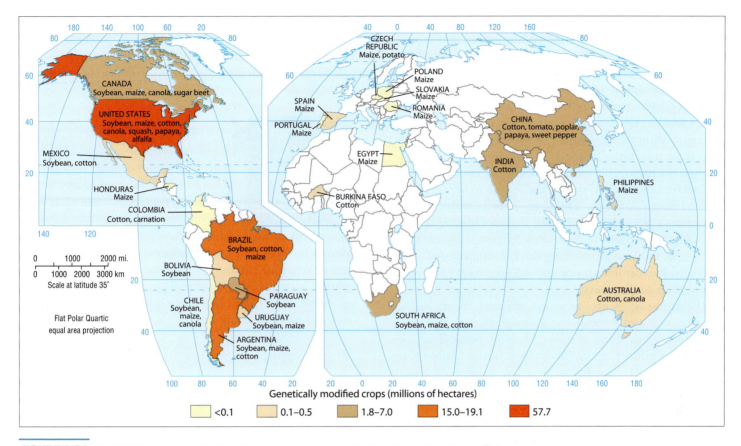

FIGURE 8.21 **Worldwide use of genetically altered crop plants, especially maize and soybeans.** This diffusion has occurred despite the concerns of health scientists, environmentalists, and consumers. *What problems might arise?* *(Source: ISAAA.)*

These cultural differences are now coming to the fore of globalization debates as the EU challenges the United States over international trade in GM seeds and foods.

Food Fears

A globalized food system is vulnerable to events that threaten food safety. Recent events, such as outbreaks of mad cow disease in Great Britain and the contamination of imported produce by *Salmonella* or the virus that causes hepatitis A in the United States, have raised anxiety levels among consumers about eating food transported from distant lands. Responses to such crises in the global food system vary culturally and individually and may be based more on the perception of risk than on the probability of illness or death. The way in which national governments and consumers respond to a particular food safety problem can fundamentally reshape geographic patterns of agricultural production on a global scale. Conversely, the idea that a culturally symbolic food may be tainted and life-threatening can shake the strongest of cultural identities.

The case of mad cow disease in Great Britain—where beef and dairy consumption has long been associated with British cultural identity—provides a good illustration. Mad cow disease is the vernacular term for bovine spongiform encephalopathy (BSE), a disease that attacks the central nervous system of cattle. The disease has an incubation period of several years, and there is no cure or vaccine. It was first discovered in Great Britain in 1986. While 95 percent of the cases have been in the United Kingdom (U.K.), new cases were documented in Austria, Finland, Slovenia, and Canada in 2001 and in the United States in 2003.

In 1996, the British government announced a link between BSE and a variant of Creutzfeldt-Jakob disease (vCJD), which causes a progressive deterioration of brain tissue in humans. The precise link between BSE and vCJD is unknown, but scientists think that vCJD is caused when people ingest contaminated brain, spinal cord, and other organs from BSE-infected cattle. Such organ matter is typically contained in ground beef. By 2002, doctors had attributed 115 deaths in the United Kingdom to vCJD, and deaths were also reported in France, Ireland, and elsewhere.

The fallout from the BSE and vCJD scares altered the geography of beef production, consumption, and trade in complex ways. The cattle production processes that led to the BSE crisis resulted from the growing industrialization and intensification of livestock fattening. High-protein cattle feed was produced from infected sheep and cow bones and organs. Cows that ate the feed contracted BSE, thus spreading the epidemic. These beef production practices have now been banned. In addition, most countries banned the import of cattle and beef from countries where BSE was discovered and shifted their trade to other regions, such as South America. The U.K. industry suffered three BSE crises (1988, 1996, and 2000) that had the net result of reducing its exports of live cows to zero (Figure 8.22). Since U.K. cattle were entering the EU export market, EU beef and live cattle exports also suffered and have yet to regain their old markets.

Consumers' fears of contaminated beef initially reduced demand for cattle worldwide, though the downward trend has been reversed, mostly because of new demand in middle-income countries. In the United Kingdom and the EU, the BSE scares have reinforced a long-term pattern of declining demand for beef. Similarly, Japanese consumers continue to be wary of imported beef; demand has yet to recover. In 2003, when BSE was discovered in a single cow in the United States, all exports were temporarily curtailed and the industry's global competitiveness suffered long-term damage.

FIGURE 8.22 Vulnerabilities in the global food system. The industrialization and globalization of agriculture have heightened the possibilities for and consequences of food contamination, most sensationally illustrated by the case of mad cow disease. The mad cow crisis led to the destruction of tens of thousands of cattle in the United Kingdom, demoralizing many farming communities and crippling a major export industry. (© P. Ashton/South West News Service.)

The case of mad cow disease demonstrates the global interconnectivity of food-producing and food-consuming regions, the vulnerability of the global food system, and the role of cultural norms and values surrounding questions of food safety.

Nature-Culture

How are nature-culture relations expressed through the production and consumption of food? Agriculture has been the fundamental encounter between nature and culture for more than 10,000 years, as human labor is mixed with nature's bounty to produce our sustenance. What we eat and how we eat it is a basic source of cultural identity. At the same time, thousands of years of agricultural use of the land have led to massive alterations in our natural environment.

Technology over Nature?

Historically, climate and the physical environment have exerted the greatest influence on shaping agriculture. People have had to adjust their subsistence strategies and techniques to the prevailing regional climate conditions. In addition, soils have played an influential role in both agricultural practices and food provisioning. Swidden cultivation, in part, reflects an adaptation to poor tropical soils, which rapidly lose their fertility when farmed. Peasant agriculture, by contrast, often owes its high productivity to the long-lived fertility of local volcanic soils. Terrain has also influenced agriculture, as farmers tended to cultivate relatively level areas (Figure 8.23). In sum, the constraints of climate, soil, and terrain have historically limited the types of crops that could be grown and the cultivation methods that could be practiced.

In recent centuries, markets, technology, and capital investment have greatly altered the spatial patterns of agriculture that climate and soils had historically shaped. Expanding global-scale markets for agricultural commodities such as sugar, coffee, and edible oils have reduced millions of acres of biologically diverse tropical forests to monocrop plantations. Synthetic fertilizers and petroleum-based insecticides and herbicides, widely available in developed countries after World War II, helped boost agricultural productivity to unimagined levels. Massive dams and large-scale irrigation systems have caused the desert to bloom from central Asia to the Americas, converting, for example, the semiarid Central Valley of California into the world's most productive agricultural region.

The ecological price of such technological miracles is high. Drainage and land reclamation destroy wetlands and associated biodiversity. The application of synthetic fertil-

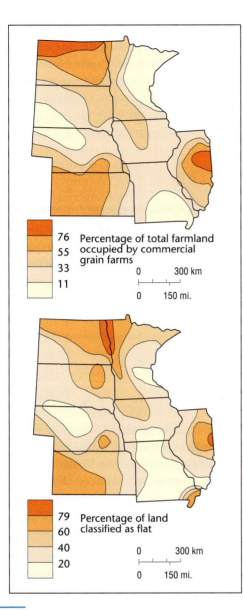

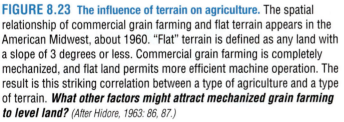

FIGURE 8.23 The influence of terrain on agriculture. The spatial relationship of commercial grain farming and flat terrain appears in the American Midwest, about 1960. "Flat" terrain is defined as any land with a slope of 3 degrees or less. Commercial grain farming is completely mechanized, and flat land permits more efficient machine operation. The result is this striking correlation between a type of agriculture and a type of terrain. *What other factors might attract mechanized grain farming to level land?* (After Hidore, 1963: 86, 87.)

izers results in nutrient-rich runoff from farms that enters freshwater systems as pollution, lowering water quality, destroying aquatic habitat, and reducing biodiversity. Agrochemicals also enter the environment as runoff, as residue on food crops, and in the tissues of livestock. These chemicals ultimately reduce biodiversity, pollute water systems, and cause increases in the rates of cancer and birth defects in humans and animals.

In the context of global climate change, great concern exists about the long-term sustainability of modern agricultural practices. In many parts of the world, groundwater is being pumped to the surface faster than it can be replenished and reservoirs are approaching the end of their life spans. Well-and-pump irrigation has drastically lowered the water table in parts of the American Great Plains, particularly Texas, causing ancient springs to go dry. Climate change data from the distant past and computer models of future climate conditions suggest that the U.S. Southwest is likely entering a drier climate regime. Lake Mead, the giant reservoir in Arizona and Nevada that helps supply water to California's cities and farms, was only 49 percent full in 2006 and is unlikely ever to be full again. This region, the most agriculturally productive in the world, will face increasingly difficult questions regarding the viability of current farming and land-use practices.

Another area where arid land irrigation has had severe ecological consequences lies on the borderland between Kazakhstan and Uzbekistan in central Asia. The once-huge Aral Sea has become so diminished by the diversion of irrigation water from the rivers flowing into it that large areas of dry lakebed now lie exposed. Not only was the local fishing industry destroyed, but noxious, chemical-laden dust storms now blow from the barren lakebed onto nearby settlements, causing assorted health problems. Irrigation water diverted to huge cotton fields, then, destroyed an ecosystem and produced another desert.

Sustainable Agriculture

As cultural geography studies by Zimmerer and many others have shown, local and indigenous knowledge about ecological conditions can be a foundation for sustainable agriculture (refer to Practicing Geography on page 291). **Sustainability**—the survival of a land-use system for centuries or millennia without destruction of the environmental base—is the central ecological issue confronting agriculture today. The case of the Quichua peasants offers an optimistic assessment of indigenous knowledge as the basis for long-term sustainability. Their response to contemporary market pressures suggests that development and conservation can be compatible. Their knowledge of complex and variable ecological conditions in the Andes has allowed them to farm highly diverse crop varieties, a practice that in some cases has been strengthened by economic development.

Another example of sustainable indigenous agriculture is the paddy rice farming that occurs near the margins of the Asian wet-rice region, where unreliable rainfall causes harvests to vary greatly from one year to the next. Farmers have developed complex cultivation strategies to

sustainability
The survival of a land-use system for centuries or millennia without destruction of the environmental base, allowing generation after generation to continue to live there.

avert periodic famine, including growing many varieties of rice. These farmers, including those in parts of Thailand, almost universally rejected the green revolution. The simplistic advice given to them by agricultural experts working for the Thai government was inappropriate for their marginal lands. Based on generations of experimentation, the local farmers knew that their traditional adaptive strategy of diversification was superior. In West Africa, peasant grain, root, and livestock farmers have also developed adaptations to local environmental influences. They raise a multiplicity of crops on the more humid lands near the coast. Moving inland toward the drier interior, farmers plant fewer kinds of crops but grow more drought-resistant varieties. Having observed many cases like these in which local practices have proved effective and sustainable, most geographers now agree that agricultural experts need to consider indigenous knowledge when devising development plans.

Intensity of Land Use

intensive agriculture
The expenditure of much labor and capital on a piece of land to increase its productivity. In contrast, *extensive* agriculture involves less labor and capital.

Great spatial variation exists in the intensity of rural land use. In **intensive agriculture,** a large amount of human labor or investment capital, or both, is put into each acre or hectare of land, with the goal of obtaining the greatest output. Intensity can be calculated by measuring either energy input or level of productivity. In much of the world, especially the paddy rice areas of Asia, high intensity is achieved through prodigious use of human labor, which results in a rice output per unit of land that is the highest in the world. In Western countries, high intensity is achieved

through the use of massive amounts of investment capital for machines, fertilizers, and pesticides, resulting in the highest agricultural productivity per capita found anywhere.

Many geographers support the theory that increased land-use intensity is a common response to population growth. As demographic pressure mounts, farmers systematically discard the more geographically extensive adaptive strategies to focus on those that provide greater yield per unit of land. In this manner, the population increase is accommodated. The resultant farming system may be riskier, because it offers fewer options and possesses greater potential for environmental modification, but it does yield more food—at least in the short run. Other geographers reject this theory, arguing instead that increases in population density follow innovations, such as the introduction of new high-calorie crops, which lead to greater land-use intensity.

The Desertification Debate

Over the millennia, as dependence on agriculture grew and as population increased, humans made ever larger demands on the forests. With the rise of urban civilization and conquering empires, the human transformation of forests to fields accelerated and expanded. In many parts of China, India, and the Mediterranean lands, forests virtually vanished. In transalpine Europe, the United States, and some other areas, they were greatly reduced (**Figure 8.24**).

Grasslands suffered similar modifications. Farmers occasionally plowed grasslands that were too dry for sustainable crop production, and herders sometimes damaged semiarid pastures through overgrazing. The result

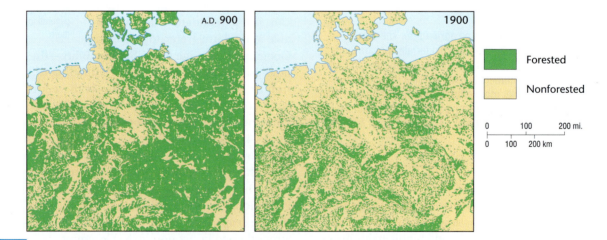

FIGURE 8.24 **The agricultural impact on the forest cover of central Europe from A.D. 900 to 1900.** Extensive clearing of the forests, mostly before 1350, was tied largely to expansion of farmland. The distribution of forests in 1900 closely resembles that of hills and mountain ranges. ***Why might this pattern have developed?*** *(Redrawn from Darby, 1956: 202–203.)*

desertification

A process whereby human actions unintentionally turn productive lands into deserts through agricultural and pastoral misuse, destroying vegetation and soil to the point where they cannot regenerate.

could be **desertification,** a process first studied half a century ago by geographer Rhoads Murphey. He argued that farmers caused substantial parts of North Africa to be added to the margins of the Sahara Desert. He noted the catastrophic decline of countries such as Libya and Tunisia in the 1500 years since the time of Roman rule, when North Africa served as the "granary of the Empire," yielding huge wheat harvests.

More recent research on desertification has centered on the Sahel, a semiarid tropical savanna region just south of the Sahara Desert in Africa (**Figure 8.25**). A series of droughts in the 1970s and 1980s raised concern that the Sahel was becoming desertified. This was a focal point of discussion at the 1977 UN Conference on Desertification. One theory advanced at the conference was that farmers and pastoralists were overusing the land, destroying vegeta-

tion to such an extent that the plant life could not regenerate. Lands that had been covered with pastures and fields could become permanently joined with the adjacent Sahara. In sum, researchers theorized that Africans were overgrazing rangelands and using poor cultivation practices, both of which practices were causing the desert to spread southward.

In the intervening years, substantial evidence, much of it obtained through satellite images, has raised questions about this theorized link between land use and the advancing Sahara Desert. Satellite imagery suggests to many researchers that the semiarid lands possess more resiliency than was once thought. They claim that, since 1960, the Sahara-Sahel boundary has not migrated steadily south but fluctuated as it always has, responding to wetter and drier years. New research by geographers and anthropologists also challenges the generalized claim that Africans were misusing the land. These findings suggest that African

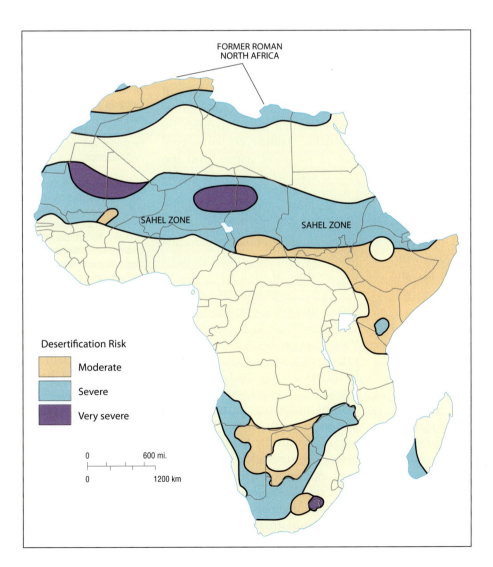

Desertification Risk

Moderate

Severe

Very severe

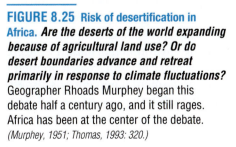

FIGURE 8.25 **Risk of desertification in Africa.** *Are the deserts of the world expanding because of agricultural land use? Or do desert boundaries advance and retreat primarily in response to climate fluctuations?* Geographer Rhoads Murphey began this debate half a century ago, and it still rages. Africa has been at the center of the debate.

(Murphey, 1951; Thomas, 1993: 320.)

land-use practices in the region are highly adapted to the variable and unpredictable environments of the Sahel. A great deal of careful research is needed to distinguish the fluctuations caused by climate variability from permanent ecological damage caused by human misuse. For now, the debate continues over the extent and sources of desertification in the Sahel.

Reflecting on Geography

> What might be some ecological consequences of expanding cropland to meet the world's rising food needs?

Environmental Perception by Agriculturists

People perceive the physical environment through the lenses of their culture. Each person's agricultural heritage can be influential in shaping these perceptions. This is not surprising, because human survival depends on how successfully people can adjust their ways of making a living to environmental conditions.

The Great Plains of the United States provides the setting for how an agricultural experience in one environment influenced farmers' environmental perceptions and subsequent behavior when they began farming in the Great Plains. The Plains farmers who came from the humid eastern United States consistently underestimated the problem of drought in their new home. In the 1960s, geographer Thomas Saarinen pointed out that almost every Great Plains farmer, even the oldest and most experienced of the farmers, underestimated the frequency of the dry periods. By contrast, **culturally preadapted** German-speaking immigrants from the steppes of Russia and Ukraine, an area very much like the American Great Plains, accurately perceived the new land and experienced fewer problems due to drought.

culturally preadapted
A complex of adaptive traits and skills possessed in advance of migration by a group, giving it survival ability and competitive advantage in occupying the new environment.

Farmers rely on climatic stability. A sudden spell of unusual weather events can change agriculturists' environmental perceptions. Geographer John Cross studied Wisconsin agriculture following a series of floods, droughts, and other anomalies. He found that two-thirds of all Wisconsin dairy farmers now believe the climate is changing for the worse, and fully one-third told him that continued climatic variability threatened their operations. Perhaps they perceive the environmental hazard to be greater than it really is, but they make decisions based on their perceptions.

Environmental perceptions also operate at the level of society. Such is the case in the U.S. Southwest, where early-twentieth-century dam building on the Colorado River allowed for the expansion of irrigated agriculture and an urban population boom that is still going strong. The widespread perception that water is abundant, however, has been based on a misreading of the climate. The early 1920s, when the Colorado River flow was divided among western states, marked an unusually wet period in the region's climate. Officials greatly overestimated average river flow when allocating water rights, thus creating a perception of water abundance. The region is now entering a drier phase, partly associated with global warming, which is bound to put agriculture in increasing conflict with urban demands. Since migration into the Southwest—where urban growth outpaces that of the rest of the country—continues, we can conclude that perceptions have yet to catch up with the reality of water scarcity.

Don't Panic, It's Organic

Alarmed by the ecological and health hazards of chemical-dependent, industrialized agriculture, a small counterculture movement emerged in the United States and Europe in the 1960s and 1970s. Geographer Julie Guthman labels this the organic farming movement. For Guthman, the movement in the United States saw in **organic agriculture** a solution to a range of social, cultural, political, and environmental ills. These included the loss of small family-owned and -operated farms, environmental pollution from industrial agriculture, corporate control of the food system, and nutritional deficiencies of highly processed foods.

organic agriculture
A form of farming that relies on manuring, mulching, and biological pest control and rejects the use of synthetic fertilizers, insecticides, herbicides, and genetically modified crops.

When the organic food movement was in its infancy, there was no way to differentiate organically produced animals and crops from the product of what has come to be called **conventional agriculture.** Movement advocates, many of them based in California, invented new certification systems in the 1970s that focused on the technical aspects of organic agriculture, particularly the absence of artificial fertilizers and petrochemical-based pesticides and herbicides. Certification systems created uniform standards and definitions for organic production but lacked enforcement powers to control fraud. In 1979, California passed the first organic law in the United States, but it took 11 more years before the state added enforcement powers. At the federal level, opposition from corporate agribusiness interests delayed regulatory legislation on organic farming until 1990 and full implementation for another decade.

conventional agriculture
The widely adopted commercial, industrialized form of farming that uses a range of synthetic fertilizers, insecticides, and herbicides to control pests and maximize productivity; a term that emerged following the creation of alternative forms, such as organic farming.

By legislating regulatory standards for organic agriculture, the state and federal governments provided organic farmers the basis for differentiating their product in the marketplace. This in turn allowed producers and retailers to charge consumers a premium. Premium pricing has encouraged many producers to switch to organic agriculture. The organic food market is now the most rapidly growing (and most profitable) agricultural sector. Affluent consumer demand for organics in First World countries, while only 1 to 3 percent of national retail sales, is growing at the phenomenal pace of 20 to 25 percent annually. Land-use practices reflect this growing demand. In the United States, organic crop acreage increased 11 percent between 2001 and 2003, with larger increases in vegetables and fruits, which now account for 4 and 2 percent, respectively, of total U.S. farm acreage.

In Guthman's assessment, the organic farming movement, because it focused solely on the technical aspects of organic production, fell short of addressing many of the social and political ills it sought to correct. Her study of California organic agriculture demonstrated that organic farming is readily incorporated into large-scale agribusiness enterprises. Indeed, some existing agribusinesses merely purchased existing organic farms as a means of diversifying their operations and tapping the profits from premium pricing. While organic agriculture has produced environmental and health benefits (for farm laborers and consumers), its effect on the historic decline of family farms and their rural communities has been minimal.

Green Fuels from Agriculture

biofuel
Broadly, this term refers to any form of energy derived from biological matter, increasingly used in reference to replacements for fossil fuels in internal combustion engines, industrial processes, and the heating and cooling of buildings.

Henry Ford fueled his first car with alcohol, and Rudolf Diesel ran his engine on fuel made from peanut oil. They soon abandoned these **biofuels,** however, for nonrenewable fossil fuels derived from "rock oil," and the rest is history. The modern global economy is dependent on fossil fuels to produce everything that we eat, wear, listen to, read, and live in. Now fossil fuels, particularly oil, have become scarce, causing prices and political uncertainties to increase. In addition, fossil-fuel combustion is a source of atmospheric greenhouse gases, which are partly responsible for global warming. Thus, an urgent search is under way to find alternative, renewable fuel supplies, and agriculture has become one of the main sources.

How can agriculture be a source of energy for industry? By tapping the simple process of fermentation, many plant materials can be converted to a combustible alcohol, ethanol. In the United States, corn is the main source of ethanol, and new ethanol plants are springing up across the Corn Belt in the midwestern United States. The U.S. Department of Agriculture planned for ethanol production to double from 2005 to 2009 and then continue to expand. This shift in corn production from food to fuel affects nearly every aspect of the field crop sector, as well as livestock production, habitat protection, and the global grain trade. For example, corn acreage in the midwestern United States is projected to increase nearly 15 percent, replacing crops such as soybeans and taking over pastures and land set aside for conservation.

Brazil, which is the closest international rival of the United States in ethanol production, grows and ferments sugarcane, which yields twice as many gallons of ethanol per acre as does corn. The policies of the Brazilian government have encouraged the creation of a delivery infrastructure (plants, tanks, and pumps) and the manufacture and sale of "flex" cars that can run on either gasoline or ethanol. By 2006 the country had freed itself from a dependence on imported oil, an achievement that many countries dream of emulating.

Will biofuels from agriculture produce a sustainable, environmentally friendly alternative to fossil fuels? Biofuels appear to offer tremendous promise, but many problems remain, such as shifting crops from food to fuel production (**Figure 8.26**). The environmental benefits of biofuels also may be reduced by some of the associated costs of increasing ethanol and biodiesel production. Biodiesel, fuel made from vegetable oils, takes less energy to produce than crop-based ethanol, but it is typically more expensive than petroleum-derived diesel. These issues are explored further in Subject to Debate (page 306).

FIGURE 8.26 Corn Belt ethanol plant. Plants such as this one in Colorado have been springing up in cornfields across the midwestern United States as demand for alternative fuel sources increases. Corn is the main ingredient in making ethanol as a gasoline additive in the United States. *(Rick Wilking/Reuters/Corbis.)*

Subject to Debate

Can Biofuels Save the Planet?

In 2005, in response to diminishing oil reserves and global warming, the U.S. government mandated that biofuels be added to gasoline. Governments around the world have implemented similar initiatives to increase renewable fuel use. Globally, the most promising environmental outcome of increased biofuel use is a decrease in greenhouse gasses. Growing plants consume atmospheric carbon dioxide. Using them for fuel thus recycles an important greenhouse gas, in contrast with fossil fuels, which release stored carbon into the atmosphere when combusted.

Biofuel demand is transforming agriculture around the world, but the energy and environmental benefits are uncertain. Because U.S. corn cultivation is so thoroughly industrialized, ethanol production consumes as much fossil fuel as it replaces. By some estimates, corn ethanol production *uses more* energy than it supplies. Brazil's sugarcane ethanol industry has a far better energy balance of 1 unit of fossil fuel input to 8 units of biofuel output. These energy gains may be offset by other environmental costs. Sugarcane cultivation has created a monocultural desert that is expected to double in acreage by 2014. Many fear this will contribute to deforestation. Likewise, in the United States, 35 million acres of land now set aside for soil and wildlife conservation may be plowed to grow corn for ethanol.

The increase in biofuel production raises social and ethical questions. For example, the world grain trade is controlled by giant multinational corporations that stand to gain the most from government subsidies and rising fuel prices. On the other hand, farmers are receiving near-record prices for their corn. Using cropland to produce fuel is already raising food prices for consumers and threatening food security in the poorest countries. The emerging food-fuel battle could pit 800 million affluent motorists against the world's 2 billion poorest people.

Continuing the Debate

Think about the future role biofuels will play in addressing the linked crises of energy supply and global warming and consider these questions:

- What do you think can be done to make biofuels more promising environmentally?

- How can food security for the poor be assured as biofuel use increases?

- Who do you think will benefit the most from the expansion of biofuel use? Small farmers or agribusiness? High-income countries or low-income countries? Consumers or corporations?

The biofuel controversy. Many U.S. politicians and farm lobbyists promote ethanol from corn as an alternative fuel source for cars. Many scientists and environmentalists argue otherwise. *(Pat LaCroix/Photographer's Choice/Getty Images.)*

CULTURAL LANDSCAPE

What is the agricultural component of the cultural landscape? What might we learn about agriculture by examining its unique landscape? According to the UN Food and Agriculture Organization, 37.3 percent of the world's land area is cultivated or pastured. In this huge area, the visible imprint of humankind might best be called the **agricultural landscape.** The agricultural landscape often varies even over short distances, telling us much about local cultures and subcultures. Moreover, it remains in many respects a window on the past, and archaic features abound. For this reason, the traditional rural landscape can teach us a great deal about the cultural heritage of its occupants.

agricultural landscape
The cultural landscape of agricultural areas.

In Chapter 3, we discussed some aspects of the agricultural landscape, in particular the rural settlement forms. We saw the different ways in which farming people situate their dwellings in various cultures. In Chapter 2, we considered traditional rural architecture, another element in the agricultural landscape. In this chapter, we attend to a third aspect of the rural landscape: the patterns of fields and property ownership created as people occupy land for the purpose of farming.

Survey, Cadastral, and Field Patterns

A **cadastral pattern** is one that describes property ownership lines, whereas a field pattern reflects the way that a farmer subdivides land for agricultural use. Both can be greatly influenced by **survey patterns,** the lines laid out by surveyors prior to the settlement of an area. Major regional contrasts exist in survey, cadastral, and field patterns, for example, unit-block versus fragmented landholding and regular, geometric survey lines versus irregular or unsurveyed property lines.

cadastral pattern
The shapes formed by property borders; the pattern of land ownership.

survey pattern
A pattern of original land survey in an area.

hamlet
A small rural settlement, smaller than a village.

Fragmented farms are the rule in the Eastern Hemisphere. Under this system, farmers live in farm villages or smaller **hamlets.** Their landholdings lie splintered into many separate fields situated at varying distances and lying in various directions from the settlement. One farm can consist of 100 or more separate, tiny parcels of land (**Figure 8.27**). The individual plots may be roughly rectangular in shape, as in Asia and southern Europe, or they may lie in narrow strips.

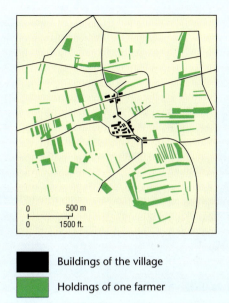

Buildings of the village

Holdings of one farmer

FIGURE 8.27 Fragmented landholdings surround a French farm village. The numerous fields and plots belonging to one individual farmer are shaded. Such fragmented farms remain common in many parts of Europe and Asia. *What are the advantages and disadvantages of this system?* (After Demangeon, 1946.)

The latter pattern is most common in Europe, where farmers traditionally worked with a bulky plow that was difficult to turn. The origins of the fragmented farm system go back to an early period of peasant communalism. One of its initial justifications was a desire for peasant equality. Each farmer in the village needed land of varying soil composition and terrain. Travel distance from the village was to be equalized. From the rice paddies of Japan and India to the fields of western Europe, the fragmented holding remains a prominent feature of the cultural landscape.

Unit-block farms, by contrast, are those in which all of the farmer's property is contained in a single, contiguous piece of land. Such forms are found mainly in the overseas area of European settlement, particularly the Americas, Australia, New Zealand, and South Africa. Most often, they reveal a regular, geometric land survey. The checkerboard of farm fields in the rectangular survey areas of the United States provides a good example of this cadastral pattern (**Figure 8.28**, page 308).

The American township and range system, discussed in Chapter 6, first appeared after the Revolutionary War as an orderly method for parceling out federally owned land for sale to pioneers. It imposed a rigid, square, graph-paper pattern on much of the American countryside; geometry triumphed over physical geography (**Figure 8.29**, page 308). Similarly, roads follow section and township lines, adding to the checkerboard character of the

FIGURE 8.28 American township and range survey creates a checkerboard, illustrated well by irrigated agriculture in the desert of California's Imperial Valley. *(Glowimages/Getty Images.)*

American agricultural landscape. Canada adopted an almost identical survey system, which is particularly evident in the Prairie Provinces.

Reflecting on Geography

What advantages does the checkerboarded North American rural landscape offer? What disadvantages?

Equally striking in appearance are long-lot farms, where the landholding consists of a long, narrow unit-block stretching back from a road, river, or canal (**Figure 8.30**). Long-lots lie grouped in rows, allowing this cadastral survey pattern to dominate entire districts. Long-lots occur widely in the hills and marshes of central and western Europe, in parts of Brazil and Argentina, along the rivers of French-settled Québec and southern Louisiana, and in parts of Texas and northern New Mexico. These unit-block farms are elongated because such a layout provides each farmer with fertile valley land, water, and access to transportation facilities, either roads or rivers. In French America, long-lots appear in rows along streams, because waterways provided the chief means of transport in colonial times. In the hill lands of central Europe, a road along the valley floor provides the focus, and long-lots reach back from the road to the adjacent ridge crests.

Some unit-block farms have irregular shapes rather than the rectangular or long-lot patterns. Most of these

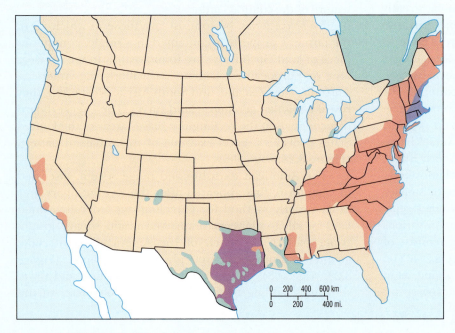

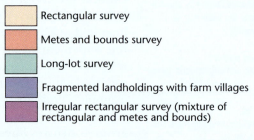

Rectangular survey

Metes and bounds survey

Long-lot survey

Fragmented landholdings with farm villages

Irregular rectangular survey (mixture of rectangular and metes and bounds)

0 200 400 600 km
0   200   400 mi.

FIGURE 8.29 Original land-survey patterns in the United States and southern Canada. The cadastral patterns still retain the imprint of the various original survey types. The map is necessarily generalized, and many local exceptions exist. ***What impact on rural life might the different patterns have?***

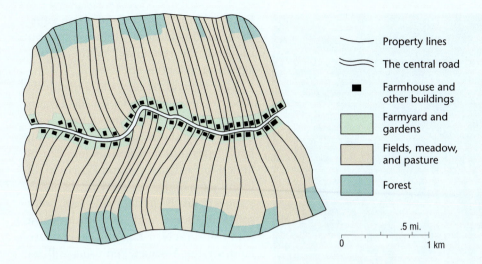

Property lines

The central road

■ **Farmhouse and other buildings**

Farmyard and gardens

Fields, meadow, and pasture

Forest

.5 mi.

0   1 km

FIGURE 8.30 **A long-lot settlement in the hills of central Germany.** Each property consists of an elongated unit-block of land stretching back from the road in the valley to an adjacent ridgecrest, part of which remains wooded.

result from metes and bounds surveying, which makes much use of natural features such as trees, boulders, and streams. Parts of the eastern United States were surveyed under the metes and bounds system, with the result that farms there are much less regular in outline than those where rectangular surveying was used (**Figure 8.31**).

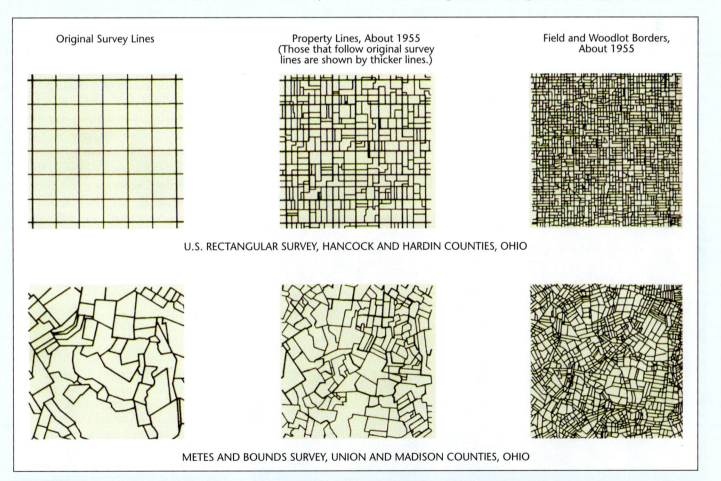

Original Survey Lines

Property Lines, About 1955
(Those that follow original survey lines are shown by thicker lines.)

Field and Woodlot Borders, About 1955

U.S. RECTANGULAR SURVEY, HANCOCK AND HARDIN COUNTIES, OHIO

METES AND BOUNDS SURVEY, UNION AND MADISON COUNTIES, OHIO

FIGURE 8.31 **Two contrasting land-survey patterns, rectangular and metes and bounds.** Both types were used in an area of west-central Ohio. Note the impact these survey patterns had on modern cadastral and field patterns. ***What other features of the cultural landscape might be influenced by these patterns?*** *(After Thrower, 1966: 40, 63, 84.)*

FIGURE 8.32 Traditional fence in the mountains of Papua New Guinea. The fence is designed to keep pigs out of sweet potato gardens. The modern age has had an impact, as revealed in the use of tin cans to decorate and stabilize the fence. Each culture has its own fence types, adding another distinctive element to the agricultural landscape. *(Courtesy of Terry G. Jordan-Bychkov.)*

Fencing and Hedging

Property and field borders are often marked by fences or hedges, heightening the visibility of these lines in the agricultural landscape. Open-field areas, where the dominance of crop raising and the careful tending of livestock make fences unnecessary, still prevail in much of western

Europe, India, Japan, and some other parts of the Eastern Hemisphere, but much of the remainder of the world's agricultural land is enclosed.

Fences and hedges add a distinctive touch to the cultural landscape (**Figure 8.32**). Because different cultures have their own methods and ways of enclosing land, types of fences and hedges can be linked to particular groups. Fences in different parts of the world are made of substances as diverse as steel wire, logs, poles, split rails, brush, rock, and earth. Those who visit rural New England, western Ireland, or the Yucatán Peninsula will see mile upon mile of stone fence that typifies those landscapes. Barbed-wire fences swept across the American countryside a century ago, but remnants of older styles can still be seen. In Appalachia, the traditional split-rail zigzag fence of pioneer times survives here and there. As do most visible features of culture, fence types can serve as indicators of cultural diffusion.

The hedge is a living fence. In Brittany and Normandy in France and in large areas of Great Britain and Ireland, hedgerows are a major aspect of the rural landscape (**Figure 8.33**). To walk or drive the roads of hedgerow country is to experience a unique feeling of confinement quite different from the openness of barbed wire or unenclosed landscapes. In recent decades, hedgerows have been disappearing as landholdings have been consolidated and grown larger. The removal of hedgerows means not only the loss of a defining feature of the rural landscape but also a decline in habitat for many rare plants, mammals, and birds. In response, the U.K. government passed regulations in 1997 to protect hedgerows in England and Wales; these regulations appear to have slowed their removal.

FIGURE 8.33 Hedgerows in England. Archetypical hedgerow landscapes, such as the one in Shropshire, England, historically common throughout western Europe, are now disappearing as landownership has become increasingly concentrated. *How might agricultural landscapes with hedgerows support greater biodiversity than those without?* *(David Paterson/Picade.)*

CONCLUSION

We have seen that the ancient human endeavor called agriculture varies markedly from region to region and is reflected in formal agricultural regions; we have also seen that we can better understand this complicated pattern through the themes of mobility, the global food system, nature-culture interactions, and agricultural landscape. Once again, we have seen the interwoven character of the five themes of cultural geography. In many fundamental ways, the agricultural revolution changed humankind. In equally dramatic fashion, the industrial revolution sparked further changes. We will use the five themes to guide an exploration of the industrial world in the next chapter.

DOING GEOGRAPHY

The Global Geography of Food

Until recently, people throughout history obtained the food they needed either by growing it themselves or procuring it directly from nearby farmers. The choice and availability of food were limited and changed seasonally. About 100 years ago, this situation began to change dramatically as the pace of urbanization and industrialization accelerated. Today, very few people in developed countries grow their own food or even know where the food they eat was produced or who produced it. Where does our food come from? Who produces and sells it? Your task for this exercise is to find out.

This exercise can be organized as a group or individual project. As a group project, people can be assigned to research particular categories of food, such as meat and poultry, cereals and grains, fruits and vegetables, and dairy products. As an individual project, you should begin with a typical day's meals and identify all the ingredients (don't forget the seasonings and cooking oils used in preparation).

Steps to Tracing the Global Geography of Food

Step 1: The project starts at the food markets where you usually shop. For much of the information you will need, you can refer to the labels on the food items. For some items, such as fish, poultry, and meat, you may need to speak to the butcher or store manager. Find out, as specifically as possible, where the food item was produced. Find out the name of the company that marketed the product and, if available, the name of the parent company.

Step 2: After collecting this basic information, you will need to head to the library and perhaps log on to the Internet to do further research (the web sites listed in this chapter should be helpful). Organize a list of companies and the food products

they market; then locate the geographic origins of each food product.

Step 3: Now look for patterns.

- Which and how many companies are involved and what proportion of the food supply does each control?
- What proportion and which kinds of food are produced in other countries? Do certain kinds of foods tend to be produced closer to the market than others?
- Do certain kinds of food tend to be marketed by large corporations more than others?
- Can you think of explanations for the patterns you identify?

You may wish to take this investigation to a greater depth.

- Can you determine from your research what the conditions are where the food is produced? For example, what landscape changes occur when regions begin producing for the global food system?
- How is production structured? Is it organized into large corporate plantations or small peasant farm plots?
- What are the ethical and social justice dimensions of food production?
- What are the conditions for workers?
- Have concerns over the treatment of animals been raised?
- Have environmental or human health concerns been raised?

A bountiful produce display, common in U.S. supermarkets. *Where does the bounty come from, who grows it, and how is it made available for our tables no matter what the season?* (Associated Press.)

Key Terms

Agricultural Geography on the Internet

You can learn more about agricultural geography on the Internet at the following web sites:

Agriculture, Food, and Human Values (AFHVS)

http://www.afhvs.org

Founded in 1987, AFHVS promotes interdisciplinary research and scholarship in the broad areas of agriculture and rural studies. The organization sponsors an annual meeting and publishes a journal by the same name.

Food First

http://www.foodfirst.org

Founded in 1975 by author-activist Francis Moore Lappé, Food First is a nonprofit, "people's" think tank and clearinghouse for information and political action. The organization highlights root causes and value-based solutions to hunger and poverty around the world, with a commitment to establishing food as a fundamental human right.

International Food Policy Research Institute, Washington, D.C.

http://www.ifpri.cgiar.org

Learn about strategies for more efficient planning for world food supplies and enhanced food production from a group concerned with hunger and malnutrition. Part of this site deals with domesticated plant biodiversity.

Resources for the Future

http://www.rff.org

This well-respected center for independent social science research was the first U.S. think tank on the environment and natural resources. This site contains a great deal of information related to the environmental aspects of global food and agriculture.

United Nations Food and Agriculture Organization (FAO), Rome, Italy

http://www.fao.org

Discover an agency that focuses on expanding world food production and spreading new techniques for improving agriculture as it strives to predict, avert, or minimize famines.

United States Department of Agriculture, Washington, D.C.

http://www.usda.gov

Look up a wealth of statistics about American farming from the principal federal regulatory and planning agency dealing with agriculture.

Urban Agriculture Notes

http://www.cityfarmer.org

This is the site of Canada's Office of Urban Agriculture. It concerns itself with all manner of subjects, from rooftop gardens to composting toilets to air pollution and community development. It encompasses mental and physical health, entertainment, building codes, rats, fruit trees, herbs, recipes, and much more.

World Bank Group, Washington, D.C.

http://www.worldbank.org

Read about an agency that provides development funds to countries, particularly in economically distressed regions. It is a driving force behind globalization and agribusiness.

SEEING GEOGRAPHY Reading Agricultural Landscapes

What differences can you "read" in these landscapes? Can you determine their locations?

Two types of contemporary agricultural landscapes.

Take a careful look at each photo and systematically identify the differences in each, beginning with the one on the left. The most striking aspect of this aerial landscape shot is the abrupt division between the cultivated land at the top and the noncultivated land at the bottom. Looking closely, we see that an irrigation channel forms the boundary between the two. A second prominent feature of the landscape is the checkerboard pattern of the fields and the straight roads forming their boundaries. Other details emerge as you look more closely. For example, the settlement pattern consists of isolated, sparsely arranged farmsteads separated by large expanses of cultivated fields. You might also note that the uncultivated land is brown and treeless and that trees in the cultivated portion are found only along the watercourses.

The landscape features in the photo on the right are nearly the opposite of those in the one on the left. Settlement is clustered in a densely populated village centered on a church and town square. The fields are of irregular sizes and shapes and form a band of cultivated land around the concentrations of houses, some of which are built of stone. There is no clear evidence of irrigation. Trees and shrubs are concentrated in the outermost band but also occur throughout the landscape, which overall appears verdant.

Putting all these visual clues together leads us to conclude that the landscape on the left must be somewhere in the western United States. We know this region was surveyed and settled under the township and range system, which explains the isolated farmsteads and checkerboard pattern. We also know that much of the western United States is arid or semiarid, which explains the need for irrigation and the general lack of trees and green vegetation in the bottom half of the photo. In fact, it is in Mack, Colorado, where irrigation meets the desert. The landscape on the right is probably located in Europe. The large church in the center and dense cluster of houses suggest the settlement pattern of a historical market town. The irregular fields and their close proximity to the town are explained by deep historical patterns of land-ownership and the reliance on foot travel in preindustrial agriculture. The verdant landscape and absence of irrigation suggest the temperate climate characteristic of western Europe. In fact, it is the vineyard region of Saône-et-Loire, France.

Worldwatch Institute

http://www.worldwatch.org

Learn about a privately financed organization focused on long-range trends, particularly food supply, population growth, and ecological deterioration.

Your Food Environment Atlas

http://maps.ers.usda.gov/FoodAtlas/

A powerful, interactive mapping site that currently includes 90 indicators of the food environment ranging from store/restaurant proximity to income and poverty measures.

Sources

Andrews, Jean. 1993. "Diffusion of Mesoamerican Food Complex to Southeastern Europe." *Geographical Review* 83: 194–204.

Bassett, Thomas, and Alex Winter-Nelson. 2010. *The Atlas of World Hunger.* Chicago: University of Chicago Press.

Binns, T. 1990. "Is Desertification a Myth?" *Geography* 75: 106–113.

Bourne, Joel, Jr. 2007. "Biofuels: Boon or Boondoggle." *National Geographic Magazine* 212(4): 38–59.

Boyd, William, and Michael Watts. 1997. "Agroindustrial Just-In-Time: The Chicken Industry and Postwar American Capitalism," in M. Watts and D. Goodman (eds.), *Globalizing Food:*

Agrarian Questions and Global Restructuring. London: Routledge, 192–225.

Carney, Judith. 2001. *Black Rice: The African Origins of Rice Cultivation in the Americas*. Cambridge, Mass.: Harvard University Press.

Chakravarti, A. K. 1973. "Green Revolution in India." *Annals of the Association of American Geographers* 63: 319–330.

Chuan-jun, Wu (ed.). 1979. "China Land Utilization." Map. Beijing: Institute of Geography of the Academica Sinica.

Cowan, C. Wesley, and Patty J. Watson (eds.). 1992. *The Origins of Agriculture: An International Perspective*. Washington, D.C.: Smithsonian Institution Press.

Cross, John A. 1994. "Agroclimatic Hazards and Fishing in Wisconsin." *Geographical Review* 84: 277–289.

Darby, H. Clifford. 1956. "The Clearing of the Woodland in Europe," in William L. Thomas, Jr. (ed.), *Man's Role in Changing the Face of the Earth*. Chicago: University of Chicago Press, 183–216.

Demangeon, Albert. 1946. *La France*. Paris: Armand Colin.

Diamond, Jared. 2002. "Evolution, Consequences and Future of Plant and Animal Domestication." *Nature* 418: 700–707.

Dillehay, T., J. Rossen, T. Andres, and D. Williams. 2007. "Preceramic Adoption of Peanut, Squash, and Cotton in Northern Peru." *Science* 316(5833): 1890–1893.

Ewald, Ursula. 1977. "The von Thünen Principle and Agricultural Zonation in Colonial Mexico." *Journal of Historical Geography* 3: 123–133.

Freidberg, Susanne. 2001. "Gardening on the Edge: The Conditions of Unsustainability on an African Urban Periphery." *Annals of the Association of American Geographers* 91(2): 349–369.

Griffin, Ernst. 1973. "Testing the von Thünen Theory in Uruguay." *Geographical Review* 63: 500–516.

Grigg, David B. 1969. "The Agricultural Regions of the World: Review and Reflections." *Economic Geography* 45: 95–132.

Griliches, Zvi. 1960. "Hybrid Corn and the Economics of Innovation." *Science* 132 (July 26): 275–280.

Guthman, Julie. 2004. *Agrarian Dreams: The Paradox of Organic Farming in California*. Berkeley: University of California Press.

Hewes, Lewlie. 1973. *The Suitcase Farming Frontier: A Study in the Historical Geography of the Central Great Plains*. Lincoln: University of Nebraska Press.

Heynen, Nik. 2010. "Cooking up Non-Violent Civil Disobedient Direct Action for the Hungry: Food Not Bombs and the Resurgence of Radical Democracy." *Urban Studies* 47(6): 1225–1240.

Hidore, John J. 1963. "Relationship Between Cash Grain Farming and Landforms." *Economic Geography* 39: 84–89.

Hollander, Gail. 2008. *Raising Cane in the 'Glades: The Global Sugar Trade and the Transformation of Florida*. Chicago: University of Chicago Press.

Horvath, Ronald J. 1969. "Von Thünen's Isolated State and the Area Around Addis Ababa, Ethiopia." *Annals of the Association of American Geographers* 59: 308–323.

ISAAA. 2007. International Service for the Acquisition of Agri-biotech Applications web site. http://www.isaaa.org.

Johannessen, Carl L. 1966. "The Domestication Processes in Trees Reproduced by Seed: The Pejibaye Palm in Costa Rica." *Geographical Review* 56: 363–376.

Mathews, Kenneth, Jason Bernstein, and Jean Buzby. 2003. "International Trade of Meat and Poultry Products and Food Safety Issues," in Jean Buzby (ed.), *International Trade and Food Safety: Economic Theory and Case Studies*. Washington, D.C.: U.S. Department of Agriculture Economic Research Service.

Millstone, Erik, and Tim Lang. 2003. *The Penguin Atlas of Food*. New York: Penguin.

Mitchell, Don. 1996. *The Lie of the Land: Migrant Workers and the California Landscape*. Minneapolis: University of Minnesota Press.

Murphey, Rhoads. 1951. "The Decline of North Africa Since the Roman Occupation: Climatic or Human?" *Annals of the Association of American Geographers* 41: 116–131.

Norberg-Hodge, Helena, Todd Merrifield, and Steven Gorelick. 2002. *Bringing the Food Economy Home: Local Alternatives to Global Agribusiness*. London: Zed.

Popper, Deborah E., and Frank Popper. 1987. "The Great Plains: From Dust to Dust." *Planning* 53(12): 12–18.

Saarinen, Thomas F. 1966. *Perception of Drought Hazard on the Great Plains*. University of Chicago, Department of Geography, Research Paper No. 106. Chicago: University of Chicago Press.

Sauer, Carl O. 1952. *Agricultural Origins and Dispersals*. New York: American Geographical Society.

Sauer, Jonathan D. 1993. *Historical Geography of Crop Plants*. Boca Raton, Fla.: CRC Press.

Thomas, David S. G. 1993. "Sandstorm in a Teacup? Understanding Desertification." *Geographical Journal* 159(3): 318–331.

Thomas, David S. G., and Nicholas J. Middleton. 1994. *Desertification: Exploding the Myth*. New York: Wiley.

Thrower, Norman J. W. 1966. *Original Survey and Land Subdivision*. Chicago: Rand McNally.

U.S. Department of Agriculture. 2004. Economic Research Service web site. www.ers.usda.gov.

Vogeler, Ingolf. 1981. *The Myth of the Family Farm: Agribusiness Dominance of United States Agriculture*. Boulder, Colo.: Westview Press.

von Thünen, Johann Heinrich. 1966. *Von Thünen's Isolated State: An English Edition of Der Isolierte Staat*. Carla M. Wartenberg (trans.). Elmsford, N.Y.: Pergamon.

Westcott, Paul. 2007. *Ethanol Expansion in the United States: How Will the Agricultural Sector Adjust?* Washington, D.C.: U.S. Department of Agriculture Economic Research Service.

Whittlesey, Derwent S. 1936. "Major Agricultural Regions of the Earth." *Annals of the Association of American Geographers* 26: 199–240.

Wilken, Gene C. 1987. *Good Farmers: Traditional Agricultural and Resource Management in Mexico and Central America*. Berkeley: University of California Press.

Zimmerer, Karl. 1996. *Changing Fortunes: Biodiversity and Peasant Livelihood in the Peruvian Andes*. Berkeley: University of California Press.

Ten Recommended Books on Agricultural Geography

(For additional suggested readings, see *The Human Mosaic* web site: www.whfreeman.com/domosh12e)

Bassett, Thomas, and Alex Winter-Nelson. 2010. *The Atlas of World Hunger.* Chicago: University of Chicago Press. This book maps out the geography and causes of world hunger from a critical social science perspective.

Clay, Jason. 2004. *World Agriculture and the Environment: A Commodity-by-Commodity Guide to Impacts and Practices.* Washington, D.C., and Covelo, Calif.: Island Press. The book describes the environmental effects resulting from the production of 22 major crops; it is global in scope and encyclopedic in detail.

Denham, Tim, Jose Iriarte, and Luc Vrydaghs (eds.). 2007. *Rethinking Agriculture: Archeological and Ethnoarcheological Perspectives.* Walnut Creek, Calif.: Left Coast Press. An edited volume bringing together geographers, anthropologists, and archaeologists to present the latest research findings on the origins of early agriculture in non-Eurasian regions.

Galaty, John G., and Douglas L. Johnson (eds.). 1990. *The World of Pastoralism: Herding Systems in Comparative Perspective.* New York: Guilford. A multidisciplinary collection of essays spanning five continents that analyze the productivity of different animal-herding practices and their contributions to herding societies.

Middleton, Nick, and David S. G. Thomas (eds.). 1997. *World Atlas of Desertification,* 2nd ed. London: Arnold. Look at the cartographic evidence and decide for yourself whether the deserts of the world are enlarging at the expense of agricultural lands.

Millstone, Erik, and Tim Lang. 2003. *The Penguin Atlas of Food.* New York: Penguin. A book packed with information on global agriculture on a wide range of topics, including genetically modified crops, fast food, organic farming, and more, all presented in brilliantly detailed maps.

Sachs, Carolyn E. 1996. *Gendered Fields: Rural Women, Agriculture, and Environment.* Boulder, Colo.: Westview Press. An exploration of the commonalities and differences in rural women's experiences and their strategies for dealing with the challenges and opportunities of rural living.

Sauer, Carl O. 1969. *Seeds, Spades, Hearths, and Herds.* Cambridge, Mass.: MIT Press. The renowned American cultural geographer presents his theories on the origins of plant and animal domestication—the beginnings of agriculture.

Turner, B. L., II, and Stephen B. Brush (eds.). 1987. *Comparative Farming Systems.* New York: Guilford. An interdisciplinary collection of essays that integrate socioeconomic, political, environmental, and technical elements of farming systems in Latin America, Anglo America, Africa, Asia, and Europe.

Watts, M., and D. Goodman (eds). 1997. *Globalizing Food: Agrarian Questions and Global Restructuring.* London: Routledge. This edited volume, containing primarily the work of geographers, analyzes globalization and the biotechnological revolution in agriculture.

Journals in Agricultural Geography

Agriculture and Human Values. An interdisciplinary journal dedicated to the study of ethical questions surrounding agricultural practices and food. Published by Kluwer. Volume 1 appeared in 1984. Visit the homepage of the journal at http://www.kluweronline.com/issn/0889–048X/current.

Journal of Agrarian Change. A journal focusing on agrarian political economy, featuring both historical and contemporary studies of the dynamics of production, property, and power. Published by Blackwell. Volume 1 appeared in 2000. Formerly titled the *Journal of Peasant Studies,* which began publication in 1973. Visit the homepage of the journal at http://www.blackwellpublishing.com/journal.asp?ref=1471-0358&site=1.

Journal of Rural Studies. An international interdisciplinary journal ranked as the best of its kind. Published by Pergamon, an imprint of Elsevier Science, Amsterdam, the Netherlands. Volume 1 appeared in 1985. Visit the homepage of the journal at http://www.elsevier.nl/locate/jrurstud.

Buildings that house a set of shoe factories in Guangdong Province, China. *(Manufacturing #2, Shift Change, Yuyuan Shoe Factory, Gaobu Town, Guangdong Province, 2004. Copyright © Edward Burtynsky. Courtesy of Charles Cowles Gallery, New York/Nicholas Metivier Gallery, Toronto.)*

How is this Chinese landscape connected to the Walmart located in your neighborhood?

Go to "Seeing Geography" on page 346 to learn more about this image.

9 GEOGRAPHY OF ECONOMIES
Industries, Services, and Development

Almost every facet of our lives is affected in some way by economic activity. On a Friday night out, you might drive in a car to a single outlet in a nationwide chain of restaurants, where you order chicken raised indoors several states away on special enriched grain, brought by refrigerated truck to a deep freeze, and cooked in an electric deep fryer. The car you drive most likely was manufactured in several different locations around the world, assembled somewhere else, brought by container ship or truck-bed to your local automobile dealership, and purchased through financing provided by a bank or other financial service provider. Later, at a movie, you buy a candy bar manufactured halfway across the country. You then enjoy a series of machine-produced pictures that flash in front of your eyes so rapidly that they seem to be moving. Just about every object or event in your life is affected, if not actually created, by economics. What we mean by *economics* is how goods (from the food we eat to the Internet we use) are produced, distributed, financed, sold, and consumed by people. These diverse activities greatly shape culture and are in turn shaped by cultural preferences, ideas, and beliefs. Economies function differently in different parts of the world. In this chapter, we examine differences and similarities of geographies of economic activities from a variety of vantage points that range from our own neighborhoods to the entire globe.

Region

How can we understand why certain regions of the world are relatively wealthy while other regions of the world are considered poor? Many of you have had the experience of traveling to countries or regions where the everyday conditions of living seem so much more difficult than the conditions you are used to at home. If you haven't traveled yourself to these places, you certainly have seen enough television shows, movies, and Internet features to understand that the everyday lives of people in different parts of the world, including their access to housing, food, and health care, may differ greatly from yours. You may also notice that there are economic differences within your own neighborhood, town, or city. On a global scale, the differences between those parts of the world that are considered wealthy and those that are less wealthy often are explained with reference to stages of economic development.

We've discussed the idea of development throughout the book, particularly in Chapter 3, where we examined how development is often correlated with a region's demographic structure. In this chapter we focus on the economic aspects of development in order to help explain how and why people living in different parts of the world have differing access to resources such as food, health care, and housing. First, however, it is important to understand exactly what is meant by economic development and how this meaning was formulated, because the term is a powerful one and carries with it several assumptions that are often overlooked.

The term *economic development* emerged in the post–World War II era when scholars as well as government and policy officials, particularly those in the United States, became concerned that the standard of living in many

Gross Domestic Product

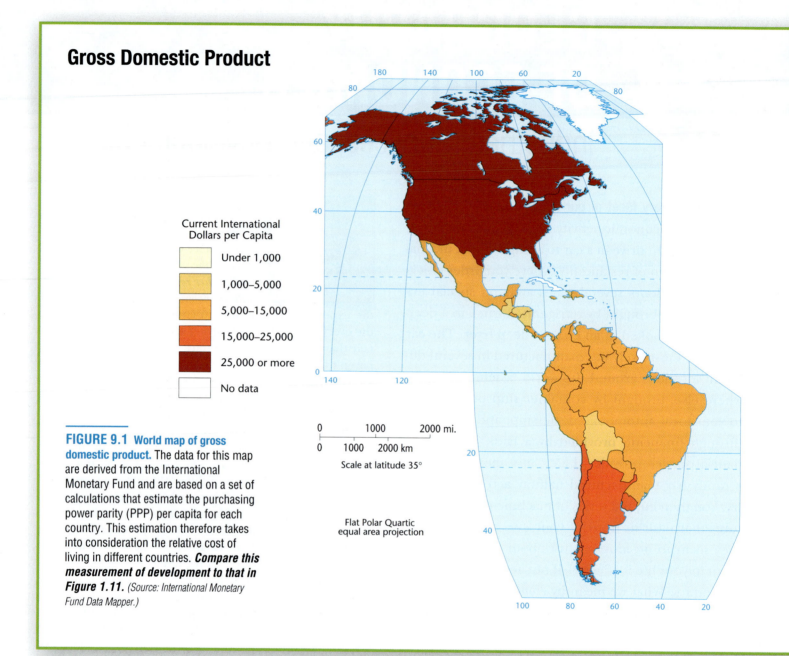

Current International
Dollars per Capita

- Under 1,000
- 1,000–5,000
- 5,000–15,000
- 15,000–25,000
- 25,000 or more
- No data

0   1000   2000 mi.

0   1000   2000 km

Scale at latitude 35°

Flat Polar Quartic
equal area projection

FIGURE 9.1 **World map of gross domestic product.** The data for this map are derived from the International Monetary Fund and are based on a set of calculations that estimate the purchasing power parity (PPP) per capita for each country. This estimation therefore takes into consideration the relative cost of living in different countries. *Compare this measurement of development to that in Figure 1.11.* (Source: International Monetary Fund Data Mapper.)

regions of the world was quite low. By standard of living, they were referring to such things as literacy rates, infant mortality, life expectancy, and poverty levels. They realized that, in general, a low standard of living was correlated with economies that were based primarily on subsistence agriculture. These officials and scholars believed that introducing new technologies and skills would enable these regions to develop more productive forms of agriculture and that industrialization—the transformation of raw materials into commodities—would follow. These new economic initiatives then would lead to a higher standard of living. This

viewpoint—categorizing regions of the world using the criterion of economic structure and then assuming that one could raise the standard of living in certain regions by transforming that economic structure—is what came to be known as economic development.

Today, economic development as a concept refers to two related but somewhat different ideas. The first is a way of categorizing the regions of the world. Regions are categorized according to particular measures of economic growth, such as GDP (gross domestic product). GDP is a measurement of the total value of all goods and services produced

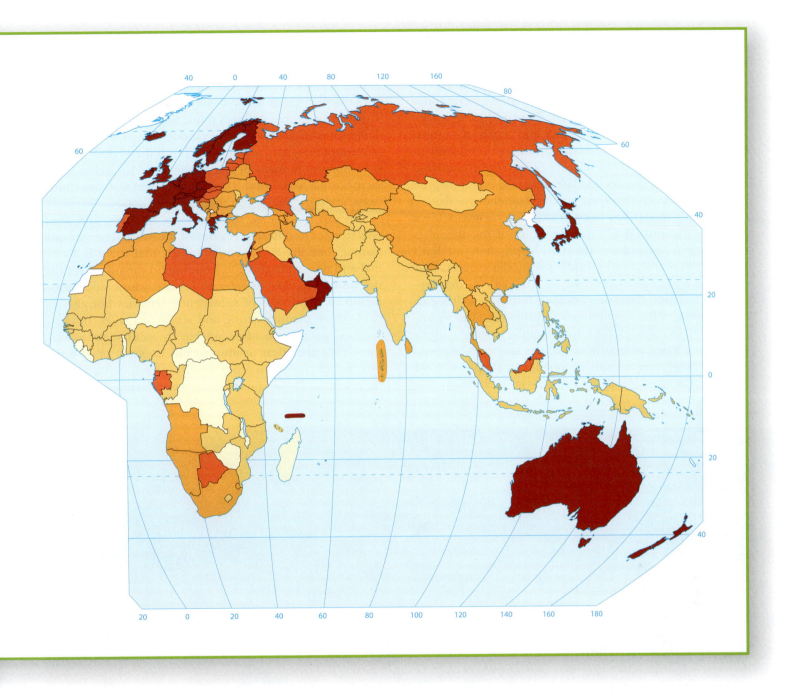

within a particular region or country within a set time period. Those regions with relatively high GDPs are considered developed regions; those with relatively low GDPs are considered developing regions (**Figure 9.1**). It is important to keep in mind, however, as we learned in Chapter 1, that GDP is only one indicator of development—*economic* development—while other indicators look toward different standards of living such as literacy rates and infant mortality. The second meaning refers to the actual process of economic growth. It can refer to the shift of a region's economy from one based on subsistence agriculture to one that relies

on industrialization. Thus, economic development refers both to a process and to a way of categorizing the different regions of the world.

One of the earliest and most influential models of the process of economic development was suggested by the economist W. W. Rostow in 1962. Rostow posited that economic development was a process that all regions of the world would go through and experience in similar ways. His model assumed that regions would progress in a linear fashion through economic stages. The first stage, which he called a "traditional" economy, is one that is based on agriculture and has limited access to or knowledge of advanced technologies. This first stage would be followed by a series of other stages characterized by increasing technological sophistication and the introduction of industrialization. Rostow's final stage, called the "age of high mass consumption," is the stage that he said characterized such economies as those of the United States and much of western Europe. In this stage, industrialization has developed to such a point that goods and services are readily available and most people can accumulate wealth to such a degree that they don't need to worry about subsistence needs.

Rostow's model has been criticized by scholars for a variety of reasons. While the simplicity of the model is appealing, its assumptions about the world often just don't stand up to reality. For example, the model suggests that countries and regions proceed on this path to development in isolation from one another. But as we will see in this chapter, different countries and their economies are interlinked in complex ways; for example, if one country's economy is based primarily on services and consumption, it needs other regions and economies to supply its food and manufactured goods, and vice versa. In addition, the model assumes that all economies will develop without obstacles from other countries, and we know that this is rarely true, since, for example, some countries may deliberately try to stop economic development in other countries that they think will become their competitors. We also know that different cultures think, feel, and act differently. The proposed goal of Rostow's model—high levels of mass consumption—may not be considered the highest level of development to many people in the world for whom other factors, such as political freedom or decent working conditions, are more important than material goods.

In this chapter we use the terms *developing* and *developed* to refer to different regions of the world, but we don't assume, like the Rostow model does, that there is a normal and linear progression from one to the other. We use the term *developing* to refer to regions whose economies include a good deal of subsistence activities but also include some manufacturing and service activities. In these regions, people are often unable to accumulate wealth, since they are often producing just enough, or at times not enough, food and other resources for their own immediate needs. These countries therefore have a lower GDP

FIGURE 9.2 Oil field at sunset. Primary industries extract natural resources from the Earth. *(Bill Ross/Corbis.)*

and are considered relatively poor. We use the term *developed* to refer to those regions of the world whose economies are based more on manufacturing and services. These regions have a higher GDP and are considered wealthier because people there are better able to accumulate resources.

In order to fully understand the differences and similarities between developing and developed economic regions, it is helpful to distinguish the types of economic activities that characterize them. In general, scholars divide types of economic activities into three broad categories. **Primary industries** refer to activities that involve extracting natural resources from the earth. Fishing, farming, hunting, lumbering, oil extraction, and mining are examples of primary industries (**Figure 9.2**). Although primary industries are located throughout the world, it is in the developing world that these economic activities dominate. **Secondary industries** process the raw materials extracted by primary industries, transforming them into more usable forms. Ore is converted into steel; logs are milled into lumber; and fish are processed and canned. *Manufacturing* is a more common way of referring to secondary industries. Manufacturing activities are found throughout the world, though, as we will discuss later in this chapter, they tend to cluster in particular areas because of favorable circumstances such as cheap power sources or available labor. In many parts of the developed world, where people import the bulk of their manufactured products, economic activities are dominated by **services.** Services refer to all the different types of work necessary to move goods and resources around and deliver them to people. So wide is the range of services that some geographers find it useful to distinguish three different types: transportation/communication services, producer services, and consumer services. We will discuss these services in more detail later in the chapter. In general, countries that are dominated by primary industries tend to be in the developing category, while countries that are dominated by services are considered to be in the developed category.

But, of course, the world and its economies are complex, and there are many exceptions to the general association of different types of economic activities and stages of economic development as outlined above. Countries in the Persian Gulf region, such as Saudi Arabia, are considered fairly developed, yet their economies are dominated by a primary industry, oil extraction. On the other hand, countries like the Russian Federation that were industrialized long ago are often placed in the category of developing regions. So stages of economic activity don't necessarily explain why some countries are rich and some countries are poor; other factors need to be considered, too, such as political stability, cultural values, and so on. Wealth and poverty are not always indicators of social development, because money does not always correlate to a higher standard of living. For example, Figure 9.1 (pages 318–319) groups countries in the world by a strict economic indicator: GDP. This map, however, does not totally correspond to the Human Development Index (see Figure 1.11, pages 14–15), the measurement compiled by the United Nations that includes other variables in addition to income, such as literacy, life expectancy, and education.

Understanding why some regions of the world are richer than other regions, and why some groups of people have better access to such things as education and health care, is a difficult affair. We can see that it is related to a region's economic structure—whether its economy is dominated by primary industry, secondary industry, or services—but we know that economic structure is not the only factor to consider. In the next section, we take a closer look at the changing geography of industrialization itself: where it began and how it diffused; its differing impacts on the globe; and its relationship to contemporary movements of people, money, and things around the world. This will help us understand the uneven distribution of wealth around the globe.

primary industry
An industry engaged in the extraction of natural resources, such as agriculture, lumbering, and mining.

secondary industry
An industry engaged in processing raw materials into finished products; manufacturing.

services
The range of economic activities that provide services to industry.

Mobility

How can we understand the changing locations of industrial activities? As we have seen, economic development is related to industrialization. In this section we consider the factors needed for industrialization and how those factors have shifted over time and through space. We start by considering the beginnings of industrialization and its impacts around the world.

Origins of the Industrial Revolution

Until the industrial revolution, society and culture remained overwhelmingly rural and agricultural. Cities certainly existed—as centers of political power, education, and innovation—but the majority of people lived in the countryside, working to procure food through agriculture. To be sure, secondary industry already existed in this setting. For as long as *Homo sapiens* have existed, we have fashioned tools, weapons, utensils, clothing, and other objects, but traditionally these items were made by hand, laboriously and slowly. Before about 1700, most such manufacturing was carried out in two rather distinct systems: cottage industry and guild industry.

cottage industry
A traditional type of manufacturing in the pre–industrial revolution era, practiced on a small scale in individual rural households as a part-time occupation and designed to produce handmade goods for local consumption.

guild industry
A traditional type of manufacturing in the pre–industrial revolution era, involving handmade goods of high quality manufactured by highly skilled artisans who resided in towns and cities.

Cottage industry, by far the more common system, was practiced in farm homes and rural villages, usually as a sideline to agriculture. Objects for family use were made in each household, usually by women. Additionally, most villages had a cobbler, miller, weaver, and smith, all of whom worked part-time at these trades in their homes. Skills passed from parents to children with little formality.

By contrast, the **guild industry** consisted of professional organizations of highly skilled, specialized artisans engaged full-time in their trades and based in towns and cities. Membership in a guild came after a long apprenticeship, during which the apprentice learned the skills of the profession from a master. Although the cottage and guild systems differed in many respects, both depended on hand labor and human power.

The industrial revolution began in England in the early 1700s. First, machines replaced human hands in the fashioning of finished products, rendering the word *manufacturing* (meaning "made by hand") technically obsolete. No longer would the weaver sit at a hand loom and painstakingly produce each piece of cloth. Instead, large mechanical looms were invented to do the job faster and cheaper. Second, human power gave way to various other forms of power: water power, the burning of fossil fuels, and later electricity and the energy of the atom fueled the machines. Men and women, once the producers of handmade goods, became tenders of machines.

The initial breakthrough came in the secondary, or manufacturing, sector. More specifically, it occurred in the British cotton textile cottage industry, centered at that time in the district of Lancashire in northwestern England. At first the changes were on a small scale. Mechanical spinners and looms were invented, and flowing water was harnessed to drive the looms. During this stage, manufacturing industries were still largely rural and dispersed, because they were tethered to their individual power sources. Sites where rushing streams could be found, especially those with waterfalls and rapids, were ideal locations. Later in the eighteenth century, the invention of the steam engine provided a better source of power, and a shift away from water-powered machines occurred. The steam engine also allowed textile producers and other manufacturers to move away from their power source—water—and to relocate instead in cities that better suited their labor and transportation needs. Manufacturing enterprises generally began to cluster together in or near established cities.

Traditionally, metal industries had been small-scale, rural enterprises, carried out in small forges situated near ore deposits. Forests provided charcoal for the smelting process. The chemical changes that occurred in the making of steel remained mysterious even to the craftspeople whose job it was, and much ritual, superstition, and ceremony were associated with steel making. Techniques had changed little since the beginning of the Iron Age, 2500 years earlier. The industrial revolution radically altered all this. In the eighteenth century, a series of inventions by iron makers living in Coalbrookdale, in the English Midlands, allowed the old traditions, techniques, and rituals of steel making to be swept away and replaced by a scientific, large-scale industry. Coke, nearly pure carbon derived from high-grade coal, was substituted for charcoal in the smelting process. Large blast furnaces replaced the forge, and efficient rolling mills took the place of hammer and anvil. Mass production of steel resulted, and that steel was used to make the machines that in turn created more industry.

Primary industries also were revolutionized. Coal mining was the first to feel the effects of the new technology. The adoption of the steam engine required huge amounts of coal to fire the boilers, and the conversion to coke in the smelting process further increased the demand for coal. New mining techniques and tools were invented, and coal mining became a large-scale, mechanized industry. However, coal, which is heavy and bulky, was difficult to transport. As a result, manufacturing industries that relied heavily on coal began flocking to the coalfields to be near the supply. Similar modernization occurred in the mining of iron ore, copper, and other metals needed by rapidly growing industries.

The industrial revolution also affected the service industries, most notably in the development of new forms of transportation. Traditional wooden sailing ships gave way to steel vessels driven by steam engines, and later railroads became more prevalent. The need to move raw materials and finished products from one place to another both cheaply and quickly was the main stimulus that led to these transportation breakthroughs. Without them, the impact of the industrial revolution would have been minimized.

Once in place, the railroads and other innovative modes of transport associated with the industrial revolution fostered additional cultural diffusion. Ideas spread more rapidly and easily because of this efficient transportation network. In fact, the industrial revolution itself was diffused through these new transportation technologies. We take up this idea in the next section.

Diffusion of the Industrial Revolution

Great Britain maintained a virtual monopoly on the industrial revolution well into the 1800s. Indeed, the British government actively tried to prevent the diffusion of the various inventions and innovations that made up the industrial revolution. After all, they gave Britain an enormous economic advantage and contributed greatly to the growth and strength of the British Empire. Nevertheless, this technology finally diffused beyond the bounds of the British Isles (**Figure 9.3**), with continental Europe feeling the impact first. In the last half of the nineteenth century, the industrial revolution took firm root in the coalfields of Belgium, Germany, and other nations of northwestern and central Europe. The diffusion of railroads in Europe provides a good index of the spread of the industrial revolution there. The United States began rapid adoption of this new technology in about 1850, followed half a century later by Japan, the first major non-Western nation to undergo full industrialization. In the first third of the twentieth century, the diffusion of industry and modern transport spilled over into Russia and Ukraine.

Until fairly recently, most of the world's manufacturing plants were clustered together in pockets within several regions, particularly Anglo America, Europe, Russia, and Japan. In the United States, secondary industries once clustered mainly in the northeastern part of the country, a region referred to as the American Manufacturing Belt (**Figure 9.4**, page 324). Across the Atlantic, manufacturing occupied the central core of Europe, which was surrounded by a less industrialized periphery throughout the mid-twentieth century. Japan's industrial complex was located along the shore of the Inland Sea and throughout the southern part of the country.

The Locational Shifts of Secondary Industry

The diffusion of industrialization around the globe continues today, but not in the same manner in which the industrial revolution spread from England to Europe, the United States, and Japan. During the diffusion of industry in the nineteenth and early twentieth centuries, there were very few global transportation or communication networks. Therefore, for the most part, industries were predominantly national in scale. All of their activities were located within national boundaries, from their headquarters to their supply of raw materials to the labor for their plants. However, by the mid to late twentieth century, with

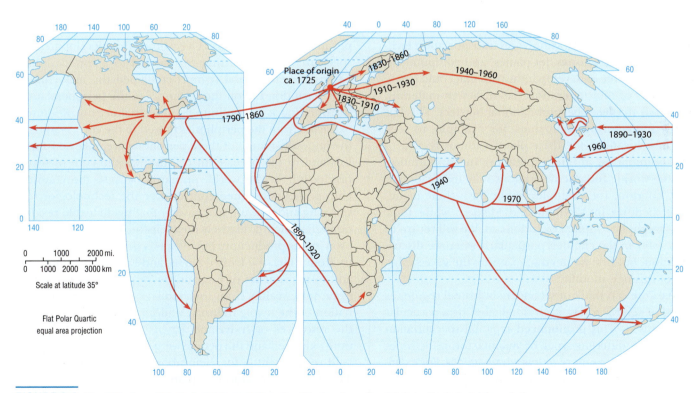

FIGURE 9.3 **The diffusion of the industrial revolution.** By diffusion from Great Britain, the industrial revolution has changed cultures in much of the world. *Why might the industrial revolution have originated in so small and peripheral a country?*

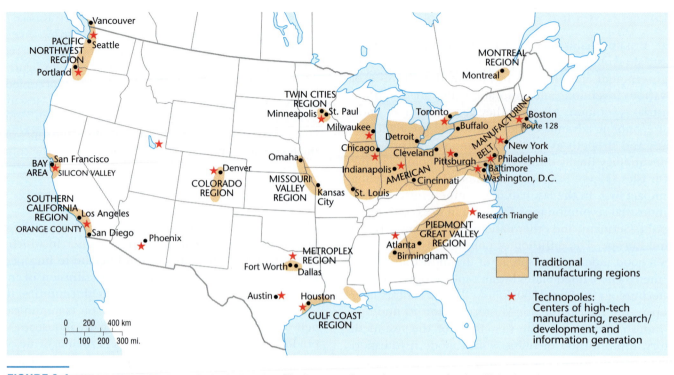

FIGURE 9.4 Major regions of industry in Anglo America. The largest and most important region is still the American Manufacturing Belt, the traditional industrial core of the United States. Dispersal of manufacturing to other regions occurred after World War II and now involves mainly high-tech and information-based enterprises, or technopoles. *Why do high-tech industries have a distribution different from that of more traditional manufacturing?*

the advent of new communication and transportation technologies such as telephones, jet planes, and computers, it had become possible to develop industries that could locate different parts of their operation (labor, power, materials, markets, etc.) in places beyond national borders. This has allowed manufacturing companies, for example, to seek out the best locations for their plants without regard to national boundaries, in effect spreading industrialization around the world. Today, corporations that are headquartered in London, for example, might have their manufacturing facilities in Vietnam and Mexico, while their materials might be supplied by Indonesia. We will return to this topic later in the chapter, particularly when we discuss globalization, but first we need to examine the geographic impacts of this new mobility.

To start, we can think of secondary industrial regions as consisting of several zones, each dominated by a particular kind of industry. This pronounced regional specialization occurs because different types of manufacturing activities find different locations more or less advantageous. For example, iron- and steel-processing plants tend to locate close to their mining source, given that transporting these primary materials is very expensive, while textile mills, which are heavily dependent on labor, tend to locate in areas that can sup-

ply inexpensive workers. Over time, this development of specialized manufacturing regions led to a core-periphery pattern, where several regions contained the major industries, each drawing on the resources of the peripheral areas surrounding it to continue its industrial activity. Resources extracted from the peripheries flowed to the core, leading to the impoverishment of these peripheral areas. For example, England in the nineteenth century was the core of textile production in the world, primarily because the country's entrepreneurs retained control over the technologies (the spinning and weaving machinery) that made textile production efficient and relatively inexpensive. The cotton that was used in this production, however, was shipped into the country from India, the United States, and Egypt. The merchants and entrepreneurs of England became very wealthy from this domination of the textile industry, while the resources of India and Egypt were extracted and the people were paid relatively little for their labor. The resultant geographical pattern —one of the fundamental realities of our age—is often referred to as **uneven development** or regional

uneven development
The tendency for industry to develop in a core-periphery pattern, enriching the industrialized countries of the core and impoverishing the less industrialized periphery. This term is also used to describe urban patterns in which suburban areas are enriched while the inner city is impoverished.

disparity. In the United States, that pattern of uneven development ended when local entrepreneurs developed the technologies to open their own factories, thus leading to the development of an American-based textile industry.

Although the manufacturing dominance of the developed countries persists, a major global geographical shift is currently under way. In virtually every core country, much of the secondary sector is in marked decline, especially traditional mass-production industries, such as steel making, that require a minimally skilled, blue-collar workforce. In such districts, factories are closing and blue-collar unemployment rates are at the highest level since the Great Depression of the 1930s. In the United States, for example, where manufacturing employment began a relative decline around 1950, nine out of every ten new jobs in recent years have been low-paying service positions. The manufacturing industries surviving and now booming in the core countries are mainly those requiring a highly skilled or artisanal workforce, such as high-tech firms and companies producing high-quality consumer goods. Because it is often difficult for the blue-collar workforce to acquire the new skills needed in such industries, many old manufacturing districts lapse into deep economic depression. Moreover, high-tech manufacturers employ far fewer workers than heavy industries and tend to be geographically concentrated in very small districts, sometimes called **technopoles** (see Figure 9.4).

technopole
A center of high-tech manufacturing and information-based industry.

deindustrialization
The decline of primary and secondary industry, accompanied by a rise in the service sectors of the industrial economy.

The word **deindustrialization** describes the decline and fall of once-prosperous factory and mining areas, such as the east and west Midlands of England (**Figure 9.5**, page 326). Manufacturing industries lost by the core countries relocate to newly industrializing lands that were once the periphery. South Korea, Taiwan, Singapore, Brazil, Mexico, coastal China, and parts of India, among others, have experienced a major expansion of manufacturing. Geographers and others often refer to these countries as newly industrializing countries (NICs). Companies move to these areas for many different reasons, including cheaper labor costs, lower environmental standards, and the relative proximity of these plants to their expanding markets outside the traditional core. This ongoing locational shift in manufacturing

transnational corporations
Companies that have international production, marketing, and management facilities.

regions is largely the work of **transnational corporations.** One can no longer think of decisions about market location, labor supply, or other aspects of industrial planning within the framework of a single plant controlled by a single owner. Instead, we now deal with a highly complex international corporate structure that is able to coordinate spatially diffuse production, marketing, and management facilities. In other words, transnational corporations, many of which are headquartered in places like New York or London, locate their manufacturing plants where labor and resources are cheapest and then ship their products to the places where they can be sold for the most profit. Yet even with this massive diffusion of industry, the geographic pattern of the world's manufacturing sites remains quite uneven (**Figure 9.6**, page 327). Only three countries in the world account for approximately 50 percent of the world's manufacturing output: the United States, China and Japan. (See Seeing Geography at the end of the chapter.)

Reflecting on Geography

Identify and discuss some of the reasons that the "old" manufacturing core still retains its dominance in the global manufacturing system.

A good example of corporations that are transnational are those that produce automobiles. According to geographer Peter Dicken, both Ford and General Motors produce almost two-thirds of their cars outside the United States—most often in Europe but also in Canada, Brazil, and Mexico. In comparison to the automobile industry, the manufacturing plants of which are scattered throughout the world, the manufacturing facilities of the textile and garment industries are disproportionately located in China, which has a large pool of relatively low-wage workers and a government eager to offer incentives. However, not all types of textile production are located in China. While the bulk of mass-produced items are made in China, much high-end, designer clothing tends to be produced in different places and under different conditions. The activity described in Doing Geography at the end of the chapter will help clarify and explain some of these distinctions within the complex global textile and garment industry. This is all part of the process of globalization.

The Locational Shifts of Service Industries

The decline of primary and secondary industries in the older developed core, or deindustrialization, has ushered in an era in these regions widely referred to as the **postindustrial phase.** Both the United States and Canada are in the postindustrial era, as are most countries of Europe as well as Japan. In this postindustrial phase, service industries are most prevalent. Scholars have identified three different types of service industries.

postindustrial phase
The phase of a society characterized by the dominance of the service sector of economic activity.

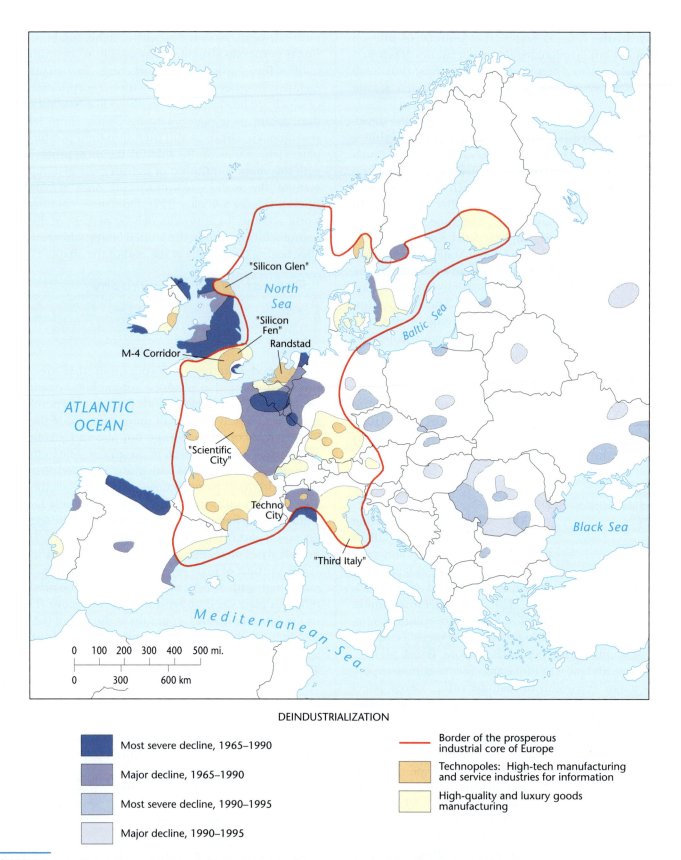

DEINDUSTRIALIZATION

Most severe decline, 1965–1990

Major decline, 1965–1990

Most severe decline, 1990–1995

Major decline, 1990–1995

Border of the prosperous industrial core of Europe

Technopoles: High-tech manufacturing and service industries for information

High-quality and luxury goods manufacturing

FIGURE 9.5 Industrial regions and deindustrialization in Europe. New, prosperous centers of industry specializing in high-quality goods, luxury items, and high-tech manufacture have surpassed older centers of heavy industry—both primary and secondary. The regions in decline were earlier centers of the industrial revolution. *Why might the industrial districts in decline not have shared the new prosperity? Why did eastern Europe fall so far behind?* (Source: Jordan-Bychkov and Jordan, 2002: 300.)

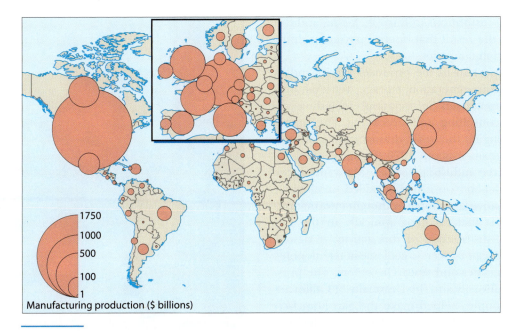

FIGURE 9.6 **Map of world manufacturing production.** Notice that the old industrial regions of the United States, Europe, and Japan still dominate much of the world's manufacturing, although countries like China and India are important new centers of manufacturing. *Identify several factors that have contributed to this uneven distribution of industrialization.* *(Adapted from Dicken, 2007.)*

**transportation/
communication services**
The range of economic activities that provide transport and communication to businesses.

Transportation/communication services, part of both the industrial and postindustrial phases, include transportation, communication, and utility (such as power companies) services. Highways, railroads, airlines, pipelines, telephones, radio, television, and the Internet are all referred to as transportation/communication services. All facilitate the distribution of goods, services, and information. Modern industries require well-developed transport systems, and every industrial district is served by a network of such facilities.

Major regional differences exist in the relative importance of the various modes of transport. In Russia and Ukraine, for example, highways are not very important to industrial development; instead, railroads—and, to a lesser extent, waterways—carry much of the transport load. Indeed, Russia still lacks a paved transcontinental highway. In the United States, by contrast, highways reign supreme, while the railroad system has declined. Western European nations rely heavily on a greater balance among rail, highway, and waterway transport. Meanwhile, electronic transfers of funds and telecommunications between computers continents apart add a new dimension and speed to the exchange of data and ideas.

producer services
The range of economic activities required by producers of goods.

Producer services are services required by the manufacturers of goods, including services such as insurance, legal services, banking, advertising, wholesaling, retailing, consulting, information generation, and real estate transactions. Such businesses represent one of the major growth sectors in postindustrial economies, and a geographical segregation has developed in which manufacturing is increasingly shunted to the peripheries while corporate headquarters and the producer-related service activities remain, for the most part, in the core. Some of these producer service activities, however, are now moving to such places as the Caribbean and India, to take advantage of educated but low-wage workers. Nevertheless, the main control centers for these large companies are still located in the major cities of the West.

An inherent problem with this spatial arrangement is that it leads to more uneven development: global corporations invest in secondary industry in the peripheries, but profits flow back to the core, where the corporate headquarters are located. As early as 1965, American-based corporations took, on average, about four-fifths of their net profits out of Latin America in this way. As a result, the industrialization of less developed countries actually increases the power of the world's established industrial nations. Consequently, although industrial technology has spread everywhere, the basic industrial power of the planet is more centralized today than ever before. As we said earlier, global corporations are based mainly in several large cities in the older industrial regions—places like New York, Tokyo, and London (see the section on globalization in Chapter 10). Similarly, loans for industrial development come from banking institutions in Europe, Japan, and the United States, with the result that interest payments drain away from poor to rich countries.

Increasingly important in the producer service industries are the collection, generation, storage, retrieval, and processing of computerized information, including research, publishing, consulting, and forecasting. The impact of computers is changing the world dramatically, a process that has accelerated since about 1970, drastically reshaping each of the three industrial sectors. Many producer service businesses depend on a highly skilled, intelligent, creative, and imaginative labor force. Although information-generating activity is focused geographically in the old industrial core, the distribution of this activity, if viewed on a more local scale, can be seen to coalesce in technopoles around major universities and research centers. The presence of Stanford University and the University of California at Berkeley, for example, helped make the San Francisco Bay Area a major center of such industry. Similar technopoles have developed near Harvard and MIT in New England and near the Raleigh–Durham–Chapel Hill Research Triangle of North Carolina (see Figure 9.4, page 324). These **high-tech corridors** (see the section on new urban landscapes in Chapter 11)—or "silicon landscapes," as some have dubbed them—occupy relatively little area. In other words, the information economy is highly focused geographically, contributing to and heightening uneven development spatially. In Europe, for example, the emerging core of producer service industries is even more confined geographically than the earlier concentration of manufacturing (see Figure 9.5, page 326).

Consumer services are those provided to the general public, and they

high-tech corridor
An area along a limited-access highway that houses offices and other services associated with high-tech industries.

consumer services
The range of economic activities that facilitate the consumption of goods.

FIGURE 9.8 **Tourism reaches even into remote areas.** Small charter buses now deliver climbers to a thatched tourist hut, which lacks running water, at the foot of Mount Wilhelm in highland Papua New Guinea—an area totally unknown to the outside world as late as 1930. Increasing numbers of tourists from Europe, North America, and Japan seek out such places. *How might these places change as a result of tourism?* (Courtesy of Terry G. Jordan-Bychkov.)

include education, government, recreation/tourism, and health/medicine. Many of these activities are also shifting their locations, since the Internet has made it possible to provide these services remotely. For example, some American executives find themselves so busy at work and with their families that they have hired personal assistants to schedule their travel, medical appointments, and so on. Some of these services are now located in India. Similarly, some educational activities can now be done through the Internet. A company called TutorVista employs 600 tutors in India to help American students with their homework. The company has over 10,000 subscribers.

One of the most rapidly expanding activities included under consumer services is tourism. By 1990, this industry already accounted for 5.5 percent of the world's economy, generated $2.5 trillion in income, and employed 112 million workers—more than any other single industrial activity, and amounting to 1 of every 15 workers in the world. Just a decade later, the total income generated had risen to $4.5 trillion, and tourism employed 1 of every 12 workers. This trend toward the increased importance of tourism has continued and spread through most regions of the world, often in spite of terrorist attacks directed against tourists, as in Bali in 2002 and Mumbai in 2008. The recent global recession, however, has greatly impacted travel worldwide, leading to an estimated 4 percent decline in the number of interna-

International Tourism 2011—Forecast

	2010	Forecast 2011
World	+6.7%	4% to 5%
Europe	+3.2%	2% to 4%
Asia and the Pacific	+12.6%	7% to 9%
Americas	+7.7%	4% to 6%
Africa	+6.4%	4% to 7%
Middle East	+13.9%	7% to 10%

FIGURE 9.7 Intensity of tourism. As you can see from the table, tourism increased in 2010 and is forecast to increase in 2011. *What variables do you think explain the fact that some regions of the world are experiencing a much larger increase in tourism than other parts of the world?* (Source: United Nations World Tourism Organization, 2011.)

tional tourist arrivals in 2009. Like all other forms of industry, tourism varies greatly in importance from one region to another (**Figure 9.7**), with some countries, particularly those located on tropical islands, depending principally on tourism to support their national economies (**Figure 9.8**).

Globalization

How has globalization affected industries, services, and development? As we have seen, the location of industries and services has been shifting over the past 50 years or so. This mobility is created by—and then reinforces—the interlinked economic networks that we refer to as globalization. Many companies now function globally in the sense that they can seek out the best locations around the world to locate their factories, obtain their services, and sell their products. Let's think through how this actually works.

Labor Supply

labor-intensive industry
An industry for which labor costs represent a large proportion of total production costs.

Labor-intensive industries are those industries in which labor costs form a large part of total production costs. Examples include industries that depend on skilled workers producing small objects of high value, such as computers, cameras, and watches, and industries that require large numbers of semiskilled workers, such as the textile and garment industries. Manufacturers consider several characteristics of labor in deciding where to locate factories: availability of workers, average wages, necessary skills, and worker productivity. Workers with certain skills tend to live and work in a small number of places, partly as a result of the need for higher education or for person-to-person training in handing down such skills. Consequently, manufacturers often seek locations where these skilled workers live.

In recent decades, with the increasing effects of globalization, two trends have been evident in terms of the relationship between the location of industry and labor supply. On the one hand, the increased mobility of people has in some ways lessened the locational influence of the labor force. Large numbers of workers around the globe have migrated to manufacturing regions. Today, labor migration is at an all-time high, lessening labor's influence on industrial location. On the other hand, particularly for industries that are reliant on large pools of labor and that are transnational in structure, the location of labor has become even more important. These industries are often referred to as footloose, in the sense that they shift the location of their facilities in search of cheap labor. We discussed this earlier in regard to the textile and garment industries (see the section on secondary industry), but here we can see how globaliza-

tion is enabling more industries to locate their production facilities close to labor supply.

Some companies are even locating their producer service activities in places where they can take advantage of cheaper labor costs. In the 1990s, many companies began to move what we typically called secretarial jobs—filing, typing, document formatting, and so on—out of their corporate headquarters and into places and spaces that were cheaper, both in terms of rental of office space and, especially, in terms of labor costs. More recently, some types of businesses, such as accounting firms, Wall Street investment houses, advertising agencies, and insurance companies, have found it cost effective to **outsource** other white-collar jobs that are generally considered more skilled, such as legal research, financial analysis, and accounting. Chennai (formerly known as Madras), India, for example, is a primary site of these more skilled outsourced activities (**Figure 9.9**, page 330). Here, educated and highly skilled workers are trained to do these jobs; they are paid the equivalent of $10,000 to $20,000 a year, compared to the average annual salary of $100,000 for a similar worker in New York City. Furthermore, although lower labor costs are the major motivation behind these locational shifts, Wall Street companies have also found that having their junior analysts and researchers located in India keeps them from the temptations of insider trading and leads to better decision making.

outsource
The physical separation of some economic activities from the main production facility, usually for the purpose of employing cheaper labor.

A new global division of labor, then, seems to have emerged. Behind these changes in the international labor market lies strategic thinking by directors of global corporations. According to a U.S. Department of Commerce study, as early as the mid-1970s, 298 American-based global corporations employed as many as 25 percent of their workers outside the United States. Since then, as we have seen, the practice has become even more common. Such factories and offices, despite relocation costs, quickly drive up corporate profit margins. In addition, the ability of these corporations to shift the production of a given product to faraway lands has a weakening effect on organized labor inside the United States.

Markets

Geographically, a **market** includes the area in which a product may be sold in a volume and at a price profitable to the manufacturer. The size and distribution of markets are generally the most important factors in determining the spatial distribution of industries. With globalization, markets have expanded to cover most of the regions in the world, but even so the costs of shipping may prohibit particular kinds of manufactured goods from being sold worldwide. Some

market
The geographical area in which a product may be sold in a volume and at a price profitable to the manufacturer.

FIGURE 9.9 Back-office workers in Chennai, India. This photo was taken inside the Office Tiger call center, a company that provides professional support services. *There are still many call centers located in the United States, but increasingly these services are being outsourced. Why?*
(James Pomerantz/Corbis.)

manufacturers have to situate their factories among their customers to minimize costs and maximize profits. Such industries include those that manufacture a weight-gaining finished product, such as bottled beverages, or a bulk-gaining finished product, such as metal containers or bottles. In other words, if weight or bulk is added to the raw materials in the manufacturing process, location near the market is economically desirable because of the high transportation costs. Similarly, if the finished product is more perishable or time-sensitive than the raw materials, as with baked goods and local newspapers, a location near the market is also required. In addition, if the product is more fragile than the raw materials that go into its manufacture, as with glass items, the industry will be attracted to locations near its market. In each of these cases—gain in weight or bulk, perishability, or fragility—transportation costs of the finished product are much higher than those of the raw materials. On the other hand, items that become easier to transport after manufacturing are often produced in locations close to the site of the raw materials.

As a rule, we can say that the greatest market potential exists where the largest numbers of people live. This is why many manufacturing firms today are looking to the large population centers in such countries as India, China, and Russia for what they consider their emerging markets. The term *emerging markets* refers to places within the global economy that have recently been opened to foreign trade and where populations are just beginning to accumulate capital that they can spend on goods and services. In China, for example, globalization has led to a huge increase in industrialization (as discussed earlier), and these industries cre-

ate jobs for people that allow them to purchase consumer goods. China's large population base (close to 1.5 billion people) serves as both a great source of labor and an expanding consumer market. Because China has more than 40 cities with populations exceeding a million (**Figure 9.10**) and also has a stable government that encourages foreign investment, transnational companies find the opportunities for low-cost production and expanding consumption there appealing.

Governments and Globalization

So far, we've considered the roles of labor, corporations, and consumers in globalizing processes, but governments are major players, too. Governmental policies shape economic activities on scales that range from urban areas, to regions, to nations, to the globe. Governments often intervene directly in decisions about industrial location. Such intervention typically results from a desire to encourage foreign investment; to create national self-sufficiency by diversifying industries; to bring industrial development and a higher standard of living to poverty-stricken provinces; to establish strategic, militarily important industries that otherwise would not develop; or to halt agglomeration in existing industrial areas. Such governmental influence becomes most pronounced in highly planned economic systems, particularly in certain socialist countries such as China, but it exists to some extent in almost every industrial nation.

Other types of government influence come in the form of tariffs, import-export quotas, political obstacles to the free movement of labor and capital, and various methods of

FIGURE 9.10 **Chinese cities with a population over 1 million.** As you can see from this map, there are more than 40 cities in China with populations that exceed 1 million. Each of these agglomerations represents a large consumer market and source of labor. ***Why do you think most of these cities are increasing in population?*** (Source: Thomas Brinkhoff: City Population, http://www.citypopulation.de.)

hindering transportation across borders. Tariffs, in effect, reduce the size of a market area proportionally to the amount of tariff imposed. A similar effect is produced when the number of border-crossing points is restricted. In some parts of the world, especially Europe, the impact of tariffs and borders on industrial location has been greatly reduced by the establishment of free-trade blocs—groups of nations that have banded together economically and abolished most tariffs. Of these associations, the European Union (EU) is perhaps the most famous. Composed of 27 nations, the EU has succeeded in abolishing tariffs within its area. The North American Free Trade Agreement (NAFTA) among the United States, Canada, and Mexico—with future expansion to include other countries—is a similar arrangement in the Western Hemisphere. At a much larger spatial scale, the World Trade Organization (WTO) administers trade agreements and settles trade disputes, including those over tariffs, throughout much of the world. With nearly 150 member countries, the WTO regulates approximately 97 percent of world trade. The WTO was formed in 1995, replacing the international organization known as GATT (General Agreement on Tariffs and Trade) that had been established after World War II. The degree to which free trade—trade between countries that is not regulated by governments—as promoted by these transnational organizations is actually beneficial to different groups of people is, however, subject to debate (see Subject to Debate, pages 332–333).

Transnational corporations, which scatter their holdings across international borders, would seem to suffer from such political regulations. In reality, however, multinational enterprises are well placed to take advantage of some government policies. Various countries act differently to encourage or discourage foreign investment, creating major spatial discontinuities in opportunities for the global corporations. Areas where foreign investment is encouraged are often called **export processing zones (EPZs),** although in China they are called special economic zones (SEZs). In general, these zones are designated areas of countries where governments create conditions conducive to export-oriented production, including trade

export processing zones (EPZs)
Designated areas of countries where governments create conditions conducive to export-oriented production.

Subject to Debate

Is Free Trade Fair Trade?

Around the world, demonstrations by groups such as the campus-based United Students Against Sweatshops and larger political protests, such as those in different cities where the World Trade Organization (WTO) has met, have raised public awareness of some of the problems that stem from free trade. Most of these protesters, and the scholars and thinkers whose work they build on, think that free trade benefits only the wealthy living in the more prosperous countries and regions of the world, at the expense of workers in less wealthy regions. On the other side of the debate—the side associated with such organizations as the World Bank, the WTO, and large transnational corporations—are scholars and thinkers who believe that free trade throughout the world is the most efficient way to equalize prosperity. Is there some middle ground? Can free trade also be fair trade?

Let's start with thinking about what free trade really means. As we've learned in this book, global trade has a fairly long history. But for most of that history, countries found ways to regulate how that trade would occur. For example, in the late nineteenth and early twentieth centuries, when industrialization was diffusing to different parts of the world, many countries realized that their newly forming industries needed to be protected from outside competition. Even countries like the United States, with developed industries, did not necessarily want competition from foreign companies. What these countries did, then, was to impose tariffs, or

taxes, on certain imports so that foreign products were very expensive and thus not readily purchased. In addition to tariffs, many countries today use other forms of state control to protect national economic interests—things like government subsidies to farmers or price controls that keep product prices low.

Advocates of free trade believe that these tariffs, controls, and subsidies should be eliminated, thereby creating a situation in which people who need or want goods (the demand side) would be supplied with those goods in the most efficient way. Most importantly for this debate, they believe that this efficiency will eventually lead to an equitable distribution of wealth. The regions of the world that produce textiles, for example, will see rising wages as the demand for these products increases. In contrast, antiglobalization activists believe that free trade benefits only the wealthy countries that are able to support their higher standard of living by exploiting workers living in less wealthy countries. The people who produce textiles, for example, often do so in sweatshop conditions— long working hours, little pay, no recourse to labor organizing, and so on—that will be maintained as long as the demand for these cheaper products continues. Free trade and globalization, they believe, translate into a race to the bottom, not the top.

Can free trade ever be fair trade? Free-trade advocates believe it can. Over time, economic globalization will lead to rising prosperity in less wealthy countries.

Antiglobalization advocates believe it can't. Free trade and globalization will exacerbate existing inequalities. Is there any middle ground here? Part of the problem in figuring this out is that we have no way to test free trade and see what happens. Most

concessions, exemptions from certain types of legislation, provision of physical infrastructure and services, and waivers of restrictions on foreign ownership. According to Peter Dicken, about 90 percent of these zones are located in Latin America, the Caribbean, Mexico, and Asia (**Figure 9.11**, page 334). Those located along the U.S.-Mexican border are populated by American-owned assembly plants called maquiladoras (**Figure 9.12**, page 334), most of which use low-wage Mexican labor, predominantly

women, to produce textiles, clothing, and electronics for the export market.

Economic Globalization and Cultural Change

Economic globalization brings with it enormous changes, and those changes have profound impacts on people's lives. As we have already seen in this chapter, the globalization of

countries and many regions (the European Union, for example) still regulate their trade in some manner, and organizations like the World Bank and the WTO also function as international regulatory bodies.

Continuing the Debate

Think about the pros and cons of free trade and consider these questions:

- Do you see examples in your own life or from your own state or region that show how global free trade is helping people attain a higher standard of living?

- Do you see examples of people's lives becoming harder because of free trade?

- How is it possible for free trade to be both helpful and harmful at the same time (and sometimes in the same place)?

Students (joined by others) in front of the Nike Town store in downtown Chicago, protesting working conditions in sweatshops. *Do you think protests like these are effective forms of social activism? What other activities could help promote better working conditions around the world?* (Russell Gordon/DanitaDelimont.com.)

industry can radically alter regional economies, while the increase in international trade that is a major consequence of globalization allows for the circulation of new and different types of consumer goods and services throughout much of the world. Some very fundamental changes in how people live their everyday lives often accompany these large economic transformations. Increased interregional trade often leads to intercultural contact and the introduction of different ways of living and working to formerly isolated regions.

The locational shifts of industries into new regions of the world can precipitate large movements of people from rural farming areas into cities to work in factories. In some places in the world, economic globalization has restructured families and reshaped gender roles. For example, many transnational companies seek out a female labor force to work in their factories, since they believe women are more reliable and pliant workers and also because in general they earn less than men.

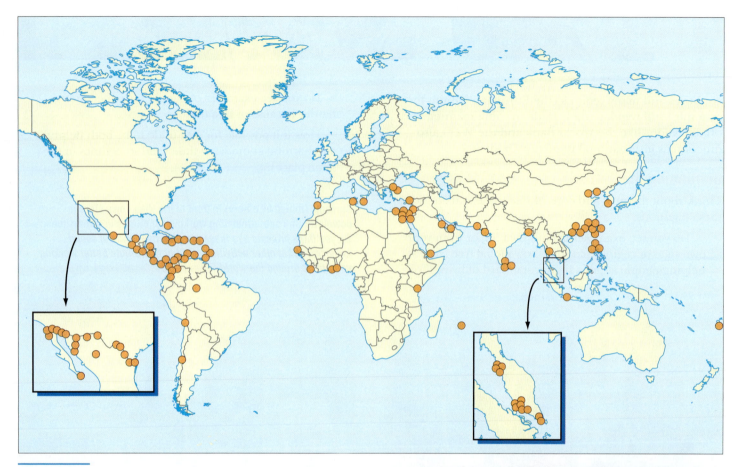

FIGURE 9.11 **Global sites of export processing zones.** Notice the concentration of these zones along the U.S.-Mexican border. **What explains this phenomenon?** (Adapted from Dicken, 2003.)

Even though globalization has spread industries and services throughout the world, not all groups of people have benefited equally. Some regions of the world have remained relatively underdeveloped, serving the global economy as sources of raw materials and low-wage labor, while other regions have benefited from the cheaper costs of production elsewhere that have raised their standards of living. This can be explained by a large range of factors: the particular histories and physical geographies of certain countries, their situation within the geopolitical world (see Chapter 6), international trade agreements (refer again to Subject to Debate on pages 332–333), and the cultures in different regions and the types of people who live there. As we can see, economic development is an uneven process that brings with it enormous changes that benefit some, but not all.

FIGURE 9.12 **Maquiladora in Nuevo Laredo, Mexico.** In this maquiladora, Mexican workers are completing telephone repairs for AT&T. **Why are many of these workers women?** (Bob Daemmrich/ The Image Works.)

Nature-Culture

How do economic activities affect the natural environment? Industrialization is a process that transforms natural resources into commodities for human use. The theme of nature-culture, then, is central to understanding industrialization and economic development. Some scholars would go so far as to say that we have tacitly agreed to the ongoing destruction of the planet in exchange for living comfortably today, for most economic activity creates serious ecological problems. In other words, examining the various impacts that economic activities have had on the natural environment provides an important example of how humans have modified the Earth.

Renewable Resource Crises

At first glance, it might seem that only finite resources are affected by industrialization, but even renewable resources such as forests and fisheries are endangered. The term *renewable resources* refers to those that can be replenished naturally at a rate sufficient to balance their depletion by human use. But that balance is often critically disrupted by the effects of industrialization. So while deforestation is an ongoing process that began at least 3000 years ago, the industrial revolution drastically increased the magnitude of the problem. In just the last half-century, a third of the world's forest cover has been lost. Lumber use tripled between 1950 and 2000, and the demand for paper increased fivefold. Today, we are witnessing the rapid destruction of one of the last surviving great woodland ecosystems, the tropical rain forest (**Figure 9.13**). The most intensive rain-forest clearing is occurring in the East Indies and Brazil, and commercial lumber interests are largely responsible (**Figure 9.14**, page 336). Although trees represent a renewable resource when properly managed, too many countries are in effect mining their forests. Canadians and Americans can only hypocritically chastise countries such as Brazil and Indonesia for not protecting their tropical rain forests, because their own midlatitude rain forests in the Pacific Northwest, British Columbia, and Alaska continue to suffer severe damage as a result of unwise lumbering practices. In any case, foreign rather than Brazilian interests now hold logging rights to nearly 30 million acres (12 million hectares) of Amazonian rain forest. Even when forests are converted into scientifically managed "tree farms," as is true in most of the developed world, ecosystems are often destroyed. Natural ecosystems have plant and animal diversity that cannot be sustained under the monoculture of commercial forestry.

Similarly, overfishing has brought a crisis to many ocean fisheries, a problem compounded by the pollution of many of the world's seas. The total fish catch of all countries combined rose from 84 million metric tons in 1984 to more than 100 million metric tons by 2003, causing some species to decline. As a result, in recent years there has been a significant drop in the levels of fish catch worldwide. For example, salmon in Pacific coastal North America and cod in the Maritime Provinces of Canada can be said to have reached a "marine biological crisis." Overfishing is a global problem, with some estimates suggesting that almost 30 percent of the fish species in the world are near a state of collapse because of overfishing or habitat loss caused by pollution. Some good news was reported in 2009 in the North Sea: for the first time in over a decade, the size of the spawning stock of cod reached a level that scientists agreed was sustainable. Experts credited an array of conservation techniques for the turnaround.

FIGURE 9.13 Destruction of tropical rain forest near Madang in Papua New Guinea by Japanese lumbering interests. The entire forest was leveled to extract a relatively small number of desired trees. *Why are such overtly destructive policies employed?* (Courtesy of Terry G. Jordan-Bychkov.)

FIGURE 9.14 The tropical rain forest of the Amazon Basin. In Brazil, the rain forest is under attack by settlers, ranchers, and commercial loggers. Its removal will intensify the impact of greenhouse gases, especially carbon dioxide, because the forest acts to convert those gases into benign forms. About 10,000 square miles (26,000 square kilometers) of Brazil's tropical rain forest are cleared each year. *(Source: Worldwatch Institute web site.)*

Acid Rain

Secondary and service industries pollute the air, water, and land with chemicals and other toxic substances. The burning of fossil fuels by power plants, factories, and automobiles releases acidic sulfur oxides and nitrogen oxides into the air; these chemicals are then flushed from the atmosphere by precipitation. The resultant rainfall, called **acid rain,** has a much higher acidity than normal rain. Overall, 84 percent of the world's human-produced energy is generated by burning fossil fuels, making acid rain a prevalent phenomenon.

Acid rain can poison fish, damage plants, and diminish soil fertility. This was a particularly acute problem in the post–World War II era, when scientists had not yet recognized the degree to which acidity was destroying ecosystems. For example, by 1980 more than 90 lakes in the seemingly pristine Adirondack Mountains of New York were "dead," meaning devoid of fish life, and 50,000 lakes in eastern Canada faced a similar fate. Since about 1990, the acid-rain problem has become less severe in many parts of the world as new and cleaner technologies for burning fossil fuels have been adopted by companies in the United States and globally.

acid rain
Rainfall with much higher acidity than normal, caused by sulfur and nitrogen oxides derived from the burning of fossil fuels being flushed from the atmosphere by precipitation, with lethal effects for many plants and animals

Global Climate Change

Most scientists now agree that we have entered a phase of **global warming** caused by industrial activity and, most particularly, by the greatly increased amount of carbon dioxide (CO_2) produced by burning fossil fuels (however, see Subject to Debate in Chapter 1, page 22, to examine more fully why some disagree about the extent to which industrial activities contribute to the increase in CO_2). Fossil fuels—coal, petroleum, and natural gas—are burned to create the energy that powers the world's factories, and as that

global warming
The pronounced climatic warming of the Earth that has occurred since about 1920 and particularly since the 1970s.

burning has increased, so has the amount of CO_2 released into the atmosphere. The eight hottest years on record all occurred in the period 1990–2001, based on records compiled at more than 14,000 locations. In 2002, a huge ice mass the size of Rhode Island broke off from Antarctica and fragmented into icebergs in the ocean, and since then scientists have noted a general shrinking of glaciers throughout much of the world.

greenhouse effect
A process in which the increased release of carbon dioxide and other gases into the atmosphere, caused by industrial activity and deforestation, permits solar short-wave heat radiation to reach the Earth's surface but blocks long-wave outgoing radiation, causing a thermal imbalance and global heating.

At issue is the so-called **greenhouse effect.** Every year automobiles and industry produce billions of tons of CO_2 worldwide by burning fossil fuel, at a level 75 percent greater than that in 1860. By some estimates, the atmospheric concentration of CO_2 has climbed to the highest level in 180,000 years (**Figure 9.15**). In addition, the ongoing destruction of the world's rain forests adds huge additional amounts of CO_2 to the atmosphere. Although CO_2 is a natural component of the Earth's atmosphere, the freeing of this huge additional amount is altering the chemical composition of the air. Carbon dioxide, only one of the absorbing gases involved in the greenhouse effect, permits solar short-wave heat radiation to reach the Earth's surface but acts to block or trap long-wave outgoing radiation, causing a thermal imbalance and global heating.

The greenhouse effect could warm the global climate enough to melt or partially melt the polar ice caps, causing the sea level to rise and inundate the world's coastlines. To begin to mitigate the effects of this looming crisis, 38 industrialized countries signed the Kyoto Protocol in 2001. This document, originally discussed in Kyoto in 1997, binds these countries to reducing their emissions of greenhouse gases so that their 2012 levels of emissions will be less than their 1990 levels. The United States agreed to this protocol in 1997 but has since changed its position and is no longer a part of the agreement, though the drafting of the Copenhagen Accord at the 2009 United Nations Climate Change Conference is a hopeful sign of a renewed commitment on the part of the United States to reducing emissions.

Ozone Depletion

Potentially even more serious is the depletion of the upper-atmosphere ozone layer, which acts to shield humans and all other forms of life from the most harmful types of solar radiation. Several chemicals, including the Freon used in refrigeration and air conditioning, are almost certainly the main culprits. These refrigeration systems are extensively used in large manufacturing plants and are critical for such things as the global trade in food commodities.

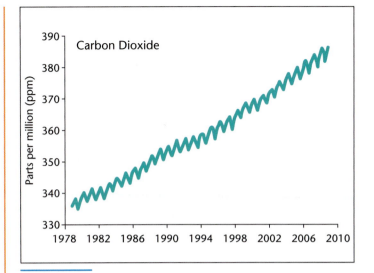

FIGURE 9.15 The global average of CO_2 concentrations. You can see the steady rise of CO_2 concentrations in the atmosphere from 1979 through 2009. *(Sources: The NOAA Annual greenhouse Gas Index and NOAA/ESRI Global Monitoring Division web site.)*

Because of particular stratospheric conditions, the ozone layer has been most depleted in the Arctic and Antarctic regions. The ozone decrease in the Arctic was first noticed in the early 1990s, with the observation of an ozone hole in that region comparable to the one first detected in the Antarctic during the 1980s. In 2006, the hole in the ozone layer over the Antarctic was the largest ever measured, in terms of both surface area and the actual loss of mass. Other areas of the world also are affected by the reduction in the ozone layer. In the middle latitudes, where most of the world's population lives, ozone levels have dropped an average of 5 percent per decade since 1979.

Most of the industrialized countries of the world contribute large amounts of these chemicals, and they signed the Montreal Protocol in 1989, which was aimed at reducing the ozone-damaging substances. The problem is that there is a significant time lag between when these substances are released into the atmosphere and when they are finally dissipated. The 2006 measurements show a significant reduction in the level of damaging substances within the lower atmosphere but not yet in the higher atmosphere. Scientific estimates now suggest that it will take at least another 50 to 75 years to see recovery of the ozone layer.

Reflecting on Geography

If we cannot be certain that global warming and upper-level ozone depletion are caused by industrial activity rather than being natural fluctuations or cycles, should we take action or simply wait and see what happens?

Environmental Sustainability

The key issue in all these industry-related ecological problems is sustainability. A sustainable environment is defined as one that would last indefinitely, providing adequate resources for the world's population. Can our present industry-based way of life continue without causing those resources to run out, leading to ecological collapse?

The Environmental Sustainability Index (ESI) has been devised to measure, country by country, the level of progress toward sustainability. The highest possible score on the ESI is 100 and the lowest, 0. No fewer than 21 "core indicators" involving 67 different variables are considered (Figure 9.16). These include air and water quality, biodiversity, population pressures, private business sector responsiveness, level of governmental intervention, and so on. In 2005, the highest-ranked country was Finland, at 75.1; the lowest was North Korea, at 29.2. Although far from being a perfect measure, the ESI clearly points to the regions of the world where ecosystems are most highly stressed. These lie mainly in the tropics and subtropics.

Developing new sources of energy is one of the keys to environmental sustainability. For example, notable progress is being made in the use of wind power to generate electricity. In the United States alone, wind power produced about 1 percent of the total electricity generated in

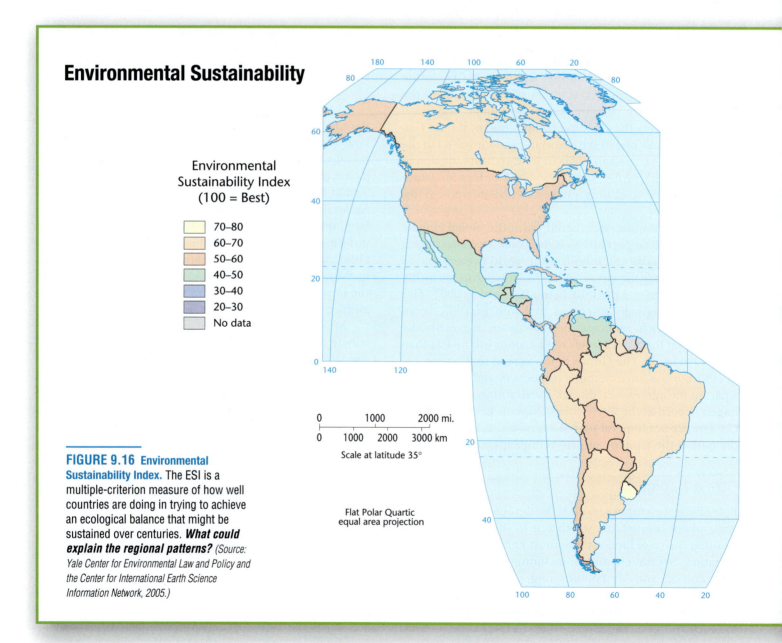

FIGURE 9.16 Environmental Sustainability Index. The ESI is a multiple-criterion measure of how well countries are doing in trying to achieve an ecological balance that might be sustained over centuries. *What could explain the regional patterns?* (Source: Yale Center for Environmental Law and Policy and the Center for International Earth Science Information Network, 2005.)

Environmental Sustainability

Environmental Sustainability Index (100 = Best)

- 70–80
- 60–70
- 50–60
- 40–50
- 30–40
- 20–30
- No data

0   1000   2000 mi.
0   1000   2000   3000 km
Scale at latitude 35°

Flat Polar Quartic
equal area projection

2008. This represents a significant increase over that of previous years, although the United States lags far behind such countries as Denmark, Spain, and Germany. Texas, California, Iowa, and Minnesota are the leading wind-power states.

Another development fostering sustainability is **ecotourism,** defined as responsible travel that does not harm ecosystems or the well-being of local people. Ecotourism arose when it was recognized that even seemingly benign industries such as tourism can create ecological problems. Ecotourists tend to visit out-of-the-way places with exotic, healthy ecosystems. They disdain the comforts of large hotels and resorts, preferring more spartan conditions. Revenue from ecotourism is helping to rescue Uganda's mountain gorillas from extinction, especially now that the government has realized that the wildlife serves as a valuable tourist attraction.

Also notable has been the rise of the **Greens,** political activists who advocate an emphasis on environmental issues. Many countries now have Green political parties. Also active in environmental advocacy are groups such as the Sierra Club, Greenpeace, and the Nature Conservancy.

ecotourism
Responsible travel that does not harm ecosystems or the well-being of local people.

Greens
Activists and organizations, including political parties, whose central concern is addressing environmental deterioration.

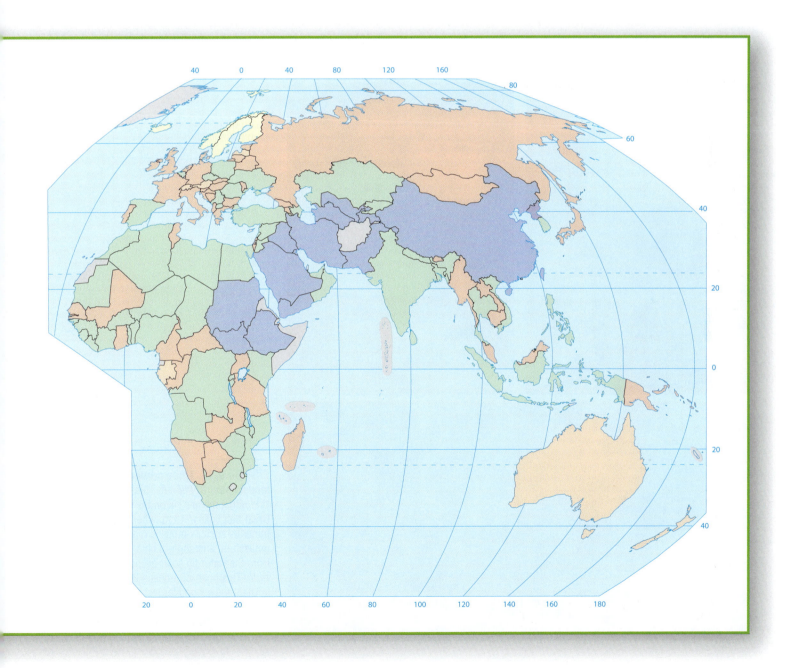

CULTURAL LANDSCAPE

In what ways have economic activities altered the cultural landscape? It is difficult to look at any particular landscape and not see some evidence of the production, circulation, and consumption of goods and services. Factory buildings of course provide an obvious example, but perhaps less obvious are the spaces where goods and services circulate (highways, trains, airports, etc.) and are consumed (malls, shops, offices, etc.). Economic landscapes, then, are prominent visible features of our surroundings and form part of daily life.

Each level of industrial activity produces its own distinctive landscape. Primary industries exert perhaps the most drastic impact on the land. The resulting landscapes contain slag heaps, clear-cut commercial forests, strip-mining scars, open-pit mines, and "forests" of oil derricks (**Figure 9.17**). The destruction of nature is less apparent in other types of primary industrial landscapes. The fishing villages of Portugal or Newfoundland even attract tourists (**Figure 9.18**). In still other cases, efforts are made to restore the preindustrial landscape. Examples include the establishment of artificial grasslands in old strip-mining areas of the American Midwest and the creation of recreational ponds in old mine pits along interstate highways.

Among the most obvious features of the landscapes of secondary industry, or manufacturing, are factory buildings. Early- to mid-nineteenth-century industrial landscapes are easy to identify, because the technologies that ran these factories were reliant on water power. Hence, the mill buildings that housed the machinery were designed in

FIGURE 9.17 Three primary industrial landscapes. *Upper left:* This bizarrely colored "industrial mosaic" is on the margins of the Great Salt Lake in Utah, where chemicals and minerals such as metallic magnesium, potassium, and sodium chloride are derived from the water by solar evaporation. *Upper right:* This open-pit mine is Bingham Canyon in Utah, the second largest in the world. *Bottom:* This mind-boggling artificial "Alp" is the result of potash mining near Kassel in Germany. *(Courtesy of Terry G. Jordan-Bychkov.)*

FIGURE 9.18 **A fishing village in Newfoundland.** Primary industrial landscapes can be pleasing to the eye. ***What problems might such a landscape also suggest?*** *(Courtesy of Terry G. Jordan-Bychkov.)*

linear form to take full advantage of the turbines that were connected to waterwheels. What's more, given this locational requirement, these mill complexes were most often built in rural areas. Entire communities, most often with housing for the workers, were thus constructed along with the mills. These mill towns dotted the landscape of England, Scotland, and New England, the site of the first industrial developments in the United States (**Figure 9.19** and Mona's Notebook, page 342). Later—as water power was supplanted by steam, coal, and then electric power—factories were located near urban areas, taking advantage of the housing supply already there and the proximity to a large consumer base. These industrial landscapes were often located at the edge of downtowns, lining the railroad routes into the city, and were surrounded by working-class housing (see the section on class, race, and gender in the industrial city in Chapter 11). In the second half of the twentieth century—with the development of the interstate highway system and the trucking industry, as well as the switch to high-tech industries such as electronics—industrial landscapes took on a different form. Factories began to move to industrial parks at interstate exchanges, and their architecture

FIGURE 9.19 **Mill buildings in central Massachusetts.** Most textile mills in New England were abandoned by the mid-twentieth century, as textile production moved to southern states and then to countries outside the United States. ***What accounts for the footloose nature of the textile industry?*** *(Courtesy of Mona Domosh.)*

Mona's Notebook

Imagining the New England Landscape

Mona Domosh

Some of the extant buildings that housed the Lebanon Woolen Mill located along the Mascoma River in Lebanon, New Hampshire. These buildings are now home to offices and some small shops. *(Courtesy of Mona Domosh.)*

When most Americans imagine the New England landscape, the picture that comes to mind often resembles the December page of an illustrated calendar—church steeples, white clapboard houses, village greens. Yet we know that America's industrial revolution started in New England, first in areas close to Boston, like Lowell and Lawrence, but even as far north as Hanover, New Hampshire, where I teach at Dartmouth College. The water power needed to run mills could be supplied by the considerable number of rivers large and small, tempting industrialists to set up shop. When industries moved to urban areas (and, later, to different parts of the country), many of these mills went out of business, but still it's difficult to drive more than a couple of miles from where I live without seeing some relic of the region's industrial past: waterwheels to harness the power of the river; long, red-brick mill buildings; and rows of identical small houses for the mill workers.

Through my geographical reading and travels in the region, I've learned to "see" and appreciate this industrial landscape, and I've realized that it characterizes the New England landscape as much as white church steeples do. I've also learned that it is common for images of places to not always correspond to their geographical realities. For example, think of the place you live in and of its different landscapes. How do people who haven't visited your town or region describe it? Do you think their image corresponds to your own? Often outsiders' images of places and regions are based on commercial representations like those on TV, films, postcards, and guidebooks (and calendars!). Part of what is so exciting about seeing with a geographer's eye is that we can often see beyond these images to what is sometimes hidden to other people.

Posted by Mona Domosh

began to resemble other types of mass-produced architecture, such as "big-box" stores (**Figure 9.20**).

Given the degree of deindustrialization in certain parts of the old core industrial regions, it is not surprising that many of these factory complexes, particularly those that date from the nineteenth and early twentieth centuries, are derelict or are being retrofitted for housing or commercial uses. Others now house historical museums depicting past industrial technologies and ways of life. Lowell, Massachusetts, an early (mid-nineteenth-century) planned industrial town that produced textiles, is now the site of a national park. A good percentage of its factories, canals, and housing complexes have been preserved and can be toured by visitors. In Great Britain, many sites of industrial history are now preserved as museums. New Lanark, located outside of Glasgow in Scotland, is now a World Heritage Site

FIGURE 9.20 **Footwear factory in Picardie, France.** This factory is located right along the highway, providing easy access for its employees and for the trucks that transport its products. ***Are there any visual clues here that this structure is a factory rather than a "big-box" store?*** *(ForestierYves/Corbis Sygma.)*

and provides a particularly interesting example of an industrial landscape, given that it was originally established by Robert Owen as a utopian community. He included in his planning good schools, housing, and even a cooperative food store in order to create a benevolent community of workers (**Figure 9.21**).

In other parts of the world, as we've learned in this chapter, industrial landscapes are far from derelict; they are, in fact, being built anew. In Southeast Asia, China, and

Mexico, large industrial landscapes are under construction. Some, like the maquiladoras, are similar to the early mill towns in that they, too, are being constructed in nonurban sites without housing or other infrastructure (**Figure 9.22**). In the maquiladoras, as well as in many industrial regions of China, housing comes in the form of either dormitories or informal squatter settlements (**Figure 9.23**, page 344).

Service industries, too, produce a cultural landscape. Its visual content includes elements as diverse as high-rise bank

FIGURE 9.21 **New Lanark, Scotland.** Robert Owen's planned industrial town is now a World Heritage Site. ***Compare these buildings to the workers' dormitory in Figure 9.23 (page 344) and consider the changes that have occurred (or haven't occurred) in the provision of housing for industrial workers.*** *(Courtesy of Mona Domosh.)*

FIGURE 9.22 **Exterior of maquiladora factory in Mexico, just across the border from Brownsville, Texas.** ***What purpose(s) do you think the heavy-duty fence serves?*** *(© Bob Daemmrich/The Image Works.)*

FIGURE 9.23 Workers' dormitory in Dongguan, China. Notice the similarity in clothing hanging on each worker's balcony. Compare this image to Figure 9.21 (page 343). *(Courtesy of Charles Cowles Gallery, New York, and Robert Koch Gallery, San Francisco.)*

buildings, hamburger stands, gasoline stations, and the concrete and steel webs of highways and railroads. Some highway interchanges can best be described as a modern art form, but perhaps the aesthetic high point of the industrial landscape is found in bridges, which are often graceful and beautiful structures. The massive investment in these transportation systems has even changed the way we view the landscape. As geographer Yi-Fu Tuan commented, "In the early decades of the twentieth century vehicles began to displace walking as the prevalent form of locomotion, and street scenes were perceived increasingly from the interior of automobiles moving staccato-fashion through regularly spaced traffic lights." Los Angeles provides perhaps the best example of the new viewpoints provided by the industrial age. Its freeway system allows individual motorists to observe their surroundings at nonstop speeds. It also allows the driver to look down on the world. The view from the street, on the other hand, is not encouraged. In some areas of Los Angeles, streets actually have no sidewalks at all, so that the pedestrian viewpoint is functionally impractical. In addition, the shopping street is no longer scaled to the pedestrian—Los Angeles's Ventura Boulevard extends for 15 miles (24 kilometers).

Producer services related to financial activities, such as legal services, trade, insurance, and banking, traditionally were located in high-rise buildings in urban centers, but with suburbanization they have taken on a nonurban form. Many are now located in five- or six-story buildings, along the interstates surrounding cities, in what we've called high-tech corridors (**Figure 9.24**). Other producer service industries choose to maintain their downtown location for symbolic reasons. Some consumer services, particularly

retailing, have created distinctive and, within the American context at least, socially important landscapes. Shopping malls are now dominant features of the North American suburb and often serve as catalysts to suburban land development, in effect creating entirely new landscapes, all geared toward consumption. Chapter 11 provides more details about these new and emerging landscape elements.

FIGURE 9.24 Office buildings in suburban Florida. Most of the activities that take place in these offices are related to banking. ***Why are these buildings set back off the road?*** *(Courtesy of Mona Domosh.)*

CONCLUSION

As we have seen, two of the most significant events of our age—the diffusion of industrialization and the globalization of economic activities—have brought a host of far-reaching changes. Already these developments have modified the regions, habitats, cultures, and landscapes of some lands so greatly that people who lived there in the past would be bewildered by the modern setting. As it turns out, much of the process of industrialization and globalization has been carried out in cities. Indeed, industrialization is the principal cause of urbanization. In the following two chapters, we will turn our attention to the city as a geographical phenomenon.

DOING GEOGRAPHY

The Where and Why of What You Wear

As we've pointed out in this chapter, the textile and garment industries have historically been very footloose, able to move production facilities to locations that suit manufacturers, often because of the low cost of labor. In today's global world, that tendency is even more pronounced, with manufacturers of clothing and other garment-related materials outsourcing many aspects of production and often subcontracting with other companies to complete different tasks in different parts of the world. It's possible, then, that the clothes and shoes you are wearing right now were designed in one place, woven into fabric in another place, and assembled in yet another.

This exercise is about tracing the "origins" of the clothes and shoes you are wearing right now and about asking "Why?" The cost of labor, as we've learned, is important, but it certainly isn't the only factor in determining where garments and shoes are made. In manufacturing in general, other locational factors include the locations of markets as well as state and international policies, such as NAFTA. For clothing and shoes, another important factor is fashion. Styles change often, and there might be a need for manufacturers to be able to change their production quickly. This would lead companies to locate manufacturing facilities close to their main markets, which might mean in the United States or Canada.

Steps to Tracing the Origins of What You Wear

Step 1: Locate as many labels as you can from the shoes and all the items of clothing you are wearing right now.

Step 2: Read the labels carefully and create a list of the places mentioned in the "Made in" section of the labels.

Step 3: Locate those places on a map. This step alone should give you a good sense of the global nature of this industry.

Now that you have a better understanding of the global nature of the clothing industry, consider these questions:

- Why do you think certain items are manufactured in particular places?
- What are the similarities in these countries? What are the differences?

Consider each of the locational factors we've mentioned—labor, markets, state policies, consumer trends—and try to generalize the why of what you wear.

Workers in Shenzhen, Guangdong Province, China (*top*), produce clothing that more than likely will end up in the stores where this young couple (*bottom*) have been shopping. ***Think about all the other processes (packaging, shipping, advertising, etc.) and other forms of labor that enable clothing produced in China to make its way to your main street.*** (Top: *Rob Crandall/The Image Works*; Bottom: © *FogStock Collection/age fotostock.*)

SEEING GEOGRAPHY Factories in Guangdong Province, China

How is this Chinese landscape connected to the Walmart located in your neighborhood?

Buildings that house a set of shoe factories in Guangdong Province, China.

The buildings depicted in this image house a set of factories in Guangdong Province, China, that produce different types of shoes (Nike, New Balance, Reebok, etc.), some of which are contracted to be sold in the United States. Given that Walmart accounts for 12 percent of China's exports to the United States, it's not difficult to surmise that some of the shoes produced here, in southeastern China, will make their way to the shoe racks at your local store. China manufactures almost half of all the shoes in the world, and this particular manufacturing site, employing approximately 50,000 workers, is said to be the world's largest shoe production site.

The shoe industry is a particularly good example of the worldwide scale of contemporary economic globalization and of how this globalization is changing lives and landscapes in complex ways. The group of shoe factories shown here—called the Yu Yuan manufacturing plant—is actually owned by the Bao Cheng Group, a large corporation controlled by Taiwanese investors and entrepreneurs. With labor costs rising in Taiwan, these entrepreneurs began to invest heavily in China in the late 1980s and early 1990s. Most of the workers shown in this image have moved to this new urban center from surrounding rural areas, where it was difficult to make a living as farmers. This rural-to-urban migration promises to be the largest human migration in history. Almost 70 percent of the workers at the Yu Yuan factories are women, leading to transformations in family structure and domestic arrangements similar to what is happening in the border region in Mexico.

Even more changes are on the horizon. The Chinese economy is booming, bringing with it higher economic standards and therefore rising labor costs. As a result, the Taiwanese entrepreneurs who own the Bao Cheng Group have begun to look elsewhere for cheaper labor and have already moved a number of their factories to Vietnam. It may be only a matter of time before some of the factory buildings shown in this photograph are transformed to house other types of commercial activities or are torn down to make way for different land uses altogether.

Key Terms

Economic Geography on the Internet

You can learn more about economic geography on the Internet at the following web sites:

Global Policy Forum

http://www.globalpolicy.org/socecon/index.htm

This site provides access to resources about global economic development, poverty, and social justice issues that are relevant for policy makers around the globe.

Greenpeace

http://www.greenpeace.org

This is the site for information about an activist group that uses both orthodox and illegal methods in its attempts to bring attention to environmental crises.

Inter-American Development Bank

http://www.iadb.org/index.cfm

The Inter-American Development Bank (IDB) provides financing for social and economic development in Latin America and the Caribbean. This site provides information about its various development projects and allows you to download its publications, which include valuable data and sources for data concerning development in the region.

United Nations Industrial Development Organization

http://www.unido.org

This specialist agency of the United Nations is devoted to promoting sustainable industrial development in countries with developing and/or transition economies. The site contains industrial statistics and information on, for example, women in industrial development, and it allows you to access data and maps of the least developed countries in the world.

World Bank

http://www.worldbank.org

In addition to information about the workings of the World Bank, this site contains valuable data on measures of economic development for most countries in the world.

Worldwatch Institute, Washington, D.C.

http://www.worldwatch.org

This nongovernmental watchdog and research institute compiles and analyzes the latest information about such ecological problems as global warming, industrial pollution, and deforestation. It also seeks sustainable alternatives.

Sources

Airriess, Christopher A. 2001. "Regional Production, Information-Communication Technology, and the Developmental State: The Rise of Singapore as a Global Container Hub." *Geoforum* 32: 235–254.

Braudel, Fernand. 1972. *The Mediterranean and the Mediterranean World in the Age of Phillip II.* New York: Harper & Row.

Brown, Laurie, Martha Ronk, and Charles E. Little. 2000. *Recent Terrains: Terraforming the American West.* Baltimore: Johns Hopkins University Press.

Dean, Cornelia. "Study Sees 'Global Collapse' of Fish Species." *New York Times,* November 3, 2006.

Dicken, Peter. 2007. *Global Shift: Mapping the Changing Contours of the World Economy.* New York: Guilford.

Francaviglia, Richard V. 1991. *Hard Places: Reading the Landscape of America's Historic Mining Districts.* Iowa City: University of Iowa Press.

Holden, Andrew. 2000. *Environment and Tourism.* New York: Routledge.

Hudson, John C., and Edward B. Espenshade, Jr. (eds.). 2000. *Goode's World Atlas,* 20th ed. Chicago: Rand McNally.

Jakle, John A., and Keith A. Sculle. 1994. *The Gas Station in America.* Baltimore: Johns Hopkins University Press.

Jordan-Bychkov, Terry G., and Bella Bychkova Jordan. 2002. *The European Culture Area,* 4th ed. Lanham, Md.: Rowman & Littlefield.

National Atmospheric Deposition Program web site. http://nadp.sws.uiuc.edu

O'Hare, Greg. 2000. "Reviewing the Uncertainties in Climate Change Science." *Area* 32: 357–368.

Power, Thomas M. 1996. *Lost Landscapes and Failed Economies: The Search for the Value of Place*. Washington, D.C.: Island Press.

Rostow, W. W. 1962. *The Process of Economic Growth*. New York: Norton and Co.

Tuan, Yi-Fu. 1989. "Cultural Pluralism and Technology." *Geographical Review* 79: 269–279.

United Nations. 2006. *Statistical Yearbook*. New York: United Nations.

Warrick, Richard, and Graham Farmer. 1990. "The Greenhouse Effect, Climatic Change and Rising Sea Level: Implications for Development." *Transactions of the Institute of British Geographers* 15: 5–20.

Weber, Alfred. 1929. *Theory of the Location of Industries*. Carl J. Friedrich (trans. and ed.). Chicago: University of Chicago Press.

"Wind Power: Maybe This Time." (2001). *The Economist* (March 10): 30–31.

Yale Center for Environmental Law and Policy and the Center for International Earth Science Information Network. 2005. *2005 Environmental Sustainability Index*. New Haven, Conn.: Yale University.

Ten Recommended Books on Economic Geography

(For additional suggested readings, see *The Human Mosaic* web site: www.whfreeman.com/domosh12e)

Castree, Noel, Neil Coe, Kevin Ward, and Mike Samers. 2004. *Spaces of Work: Global Capitalism and Geographies of Labour*. London: Sage Publications. A very accessible introduction to how economic globalization is impacting different labor markets, and vice versa.

Cravey, Altha. 1998. *Women and Work in Mexico's Maquiladoras*. Lanham, Md.: Rowman & Littlefield. A detailed analysis of the relationships between gender and work in two different places along the U.S.-Mexican border.

Dicken, Peter. 2007. *Global Shift: Mapping the Changing Contours of the World Economy*. New York: Guilford. A comprehensive look at the causes and effects of the increasingly global and interlinked industries of the world.

Harrington, James W., and Barney Warf. 1995. *Industrial Location: Principles and Practice*. New York: Routledge. A useful basic primer on industrial location written for nonexperts.

Harvey, David. 2006. *Spaces of Global Capitalism: A Theory of Uneven Geographical Development*. New York: Verso. The latest book by a major theorist and critic of the effects of global capitalism throughout the world.

Lever-Tracy, Constance (ed.). 2010. *Routledge Handbook of Climate Change and Society*. New York: Routledge. The most up-to-date and comprehensive set of case studies concerning the diverse impacts of climate change on people, places, and cultures around the world.

McDowell, Linda. 1997. *Capital Culture: Gender at Work in the City*. 1997. Oxford: Blackwell. An interesting look at the importance of gender identity to the functioning of the financial sector.

Peck, Jamie. 1996. *Work-Place: The Social Regulation of Labor Markets*. New York: Guilford. A detailed analysis of the relationships among labor, place, and state policies.

Sheppard, Eric, and Trevor Barnes. 2002. *Companion to Economic Geography*. Oxford: Blackwell. A very accessible series of essays that serve to introduce the major concepts and issues in contemporary economic geography.

Williams, Stephen. 2009. *Tourism Geography: A New Synthesis*. New York: Routledge. A very good introduction to one of the fastest-growing global industries.

Journals in Economic Geography

Economic Geography. Published by Clark University. Volume 1 appeared in 1925.

Geoforum. Published by Elsevier Science. Volume 1 appeared in 1969.

Journal of Economic Geography. Published by Oxford University Press. Volume 1 appeared in 2000.

Journal of Transport Geography. Published by Elsevier Science. Volume 1 was published in 1993.

What are some of the major environmental and social impacts of an increasingly urbanized world?

Go to "Seeing Geography" on page 382 to learn more about this image.

10 URBANIZATION
The City in Time and Space

Imagine the 2 million years that humankind has spent on Earth as a 24-hour day. In this framework, settlements of more than a hundred people came about only in the last half hour. Towns and cities emerged only a few minutes ago. Yet it is during these "minutes" that we see the rise of civilization. *Civitas,* the Latin root word for *civilization,* was first applied to settled areas of the Roman Empire. Later it came to mean a specific town or city. *To civilize* meant literally "to citify."

Urbanization over the past 200 years has strengthened the links among culture, society, and the city. An urban explosion has gone hand in hand with the industrial revolution. According to United Nations' assessments, the year 2008 was a momentous one, as it marked the first time that the world's urban population exceeded its rural population. The world's urban population has more than quadrupled since 1950 (733 million in 1950 versus 3.5 billion in 2009) and will reach 4.94 billion by the year 2030. At that time, over 60 percent of the Earth's population will live in cities. The cultural geography of the world will change dramatically as we become a predominantly urban people and the ways of the countryside are increasingly replaced by urban lifestyles.

In this chapter, we consider overall patterns of urbanization, learn how urbanization began and developed, and discuss the differing forms of cities in the developing and developed worlds. In addition, we examine some of the external factors influencing city location. In Chapter 11, we look at the internal aspects of the city, as seen through the five themes of human geography.

Region

How are urban areas and urban populations spatially arranged? All of you know from your own travels locally or internationally that some regions contain lots of cities, while others contain relatively few, and that the size of those cities can vary greatly. How can we begin to understand the location, distribution, and size of cities? We start with a consideration of global patterns of urbanization, examining the general distribution of urban populations around the world. A quick look at **Figure 10.1** (pages 352–353) reveals differing patterns of **urbanized population**—the percentage of a nation's population living in towns and cities—around the world. For example, the countries of Europe, North America, Latin America, and the Caribbean have relatively high levels of urbanization, with approximately 75 percent of each country's population living in urban areas. The nations of Africa and Asia, on the other hand, are less urbanized, with approximately 38 percent of each country's population residing in urban areas. How do geographers explain these varying regional patterns of urbanization?

> **urbanized population**
> The proportion of a country's population living in cities.

Patterns and Processes of Urbanization

According to United Nations estimates, almost all the worldwide population growth in the next 30 years will be concentrated in urban areas, with the cities of the less

Urbanized Population

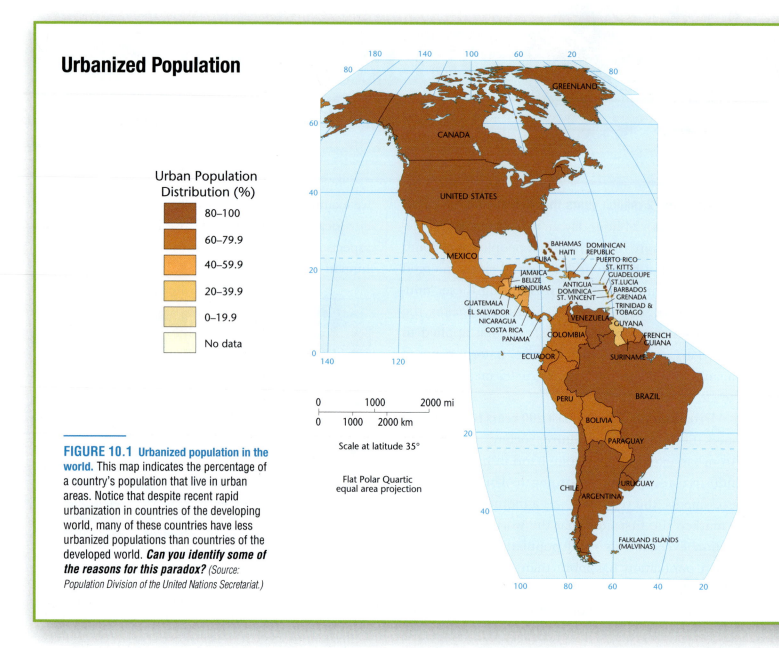

Urban Population Distribution (%)

- 80–100
- 60–79.9
- 40–59.9
- 20–39.9
- 0–19.9
- No data

FIGURE 10.1 Urbanized population in the world. This map indicates the percentage of a country's population that live in urban areas. Notice that despite recent rapid urbanization in countries of the developing world, many of these countries have less urbanized populations than countries of the developed world. ***Can you identify some of the reasons for this paradox?*** (*Source: Population Division of the United Nations Secretariat.*)

developed regions accounting for most of that increase. The reasons for this explosion in urban population growth and its uneven distribution around the world vary, as each country's unique history and society present a slightly different narrative of urban and economic development. Making matters more complex is the lack of a standard definition of what constitutes a city. Consequently, the criteria used to calculate a country's urbanized population differ from nation to nation. Using data based on these varying criteria would result in misleading conclusions. For example, the Indian government defines an urban center as an area having 5000 inhabitants, with an adult male population employed predominantly in nonagricultural work. In contrast, the U.S. Census Bureau defines a city as a densely populated area of 2500 people or more, and South Africa counts as a city any settlement of 500 or more people. Furthermore, some countries revise their definitions of urban settlements to suit specific purposes. China-watchers were baffled in 1983 when that country's urban population swelled by 13 percent in one year, only to learn that China had simply revised its census definitions for urban settlements, with criteria that vary from province to province. It is important to remember, then, that an international comparison of urbanized population data can be made only by taking into account the varying definitions of a city.

Nonetheless, several generalizations can be made about the differences in the world's urbanized population. First, there is a close link between urbanized population

and the more developed world. Put differently, highly industrialized countries have higher rates of urbanized population than do less developed countries. The second generalization, closely tied to the first, is that developing countries are urbanizing rapidly and that their ratio of urban to rural population is increasing dramatically (**Figure 10.2**, page 354).

Urban growth in these countries comes from two sources: the migration of people to the cities (see also the section on mobility) and the higher natural population growth rates of these recent migrants. People move to the cities for a variety of reasons, most of which relate to the effects of uneven economic development in their country. Cities are often the centers of economic growth, whereas

opportunities for land ownership and/or farming-based jobs are, in many countries, rare. Because urban employment is unreliable, many migrants continue to have large numbers of children to construct a more extensive family support system. Having a larger family increases the chances of someone's getting work. The demographic transition to smaller families comes later, when a certain degree of security is ensured. Often, this transition occurs as women enter the workforce (see Chapter 3).

Impacts of Urbanization

Although rural-to-urban migration affects nearly all cities in the developing world, the most visible cases are the

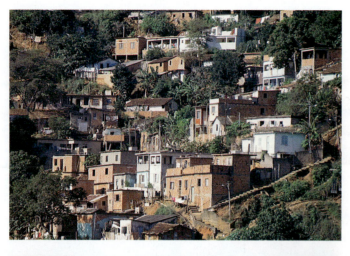

FIGURE 10.2 Urbanization. Shown here are scenes from a squatter settlement in Rio de Janeiro (*top*), the business district in Kolkata (Calcutta) (*middle*), and a residential district of Manila (*bottom*). For more information about the growth of such cities, see the section on globalizing cities in the developing world. As is evident from this scene in Kolkata, *the downtowns of cities in the developing world are often more vibrant than those of the developed world. Can you think of some reasons for this?* (Top: Najilah Feaney/Saba; Middle: Earl Young/Tony Stone Images; Bottom: Bruno Zehnder/Peter Arnold, Inc.)

extraordinarily large settlements we call **megacities,** those having populations of over 10 million. **Table 10.1** shows the world's 20 largest cities, a majority of which are in the developing world. This is a major change from 30 years ago, when the list would have been dominated by Western, industrialized cities, a trend that most expect to continue. Projections for future growth, however, must be qualified by two considerations. First, cities of the developing world will continue to explode in size only if economic development expands. If it stagnates because of political or resource problems, city growth will probably slow (although urban migration might increase if rural economies deteriorate). For example, Mexico City's growth is linked to that country's economic growth and, more specifically, to Mexico's oil industry, which fluctuates according to the world market for oil. Second, because these megacities are plagued by transportation, housing, employment, and ecological problems—such as an inadequate water supply, in the case of Mexico City—some countries are trying to control urban migration. The success or failure of these policies will influence city size in the next 10 to 20 years.

Nevertheless, the urban population in the developing world is growing at astounding rates. Even though the developed regions of the world are more urbanized overall than the less developed regions, the sheer scale and rate of growth in absolute numbers reveal a reversal in this pattern. According to geographer David Drakakis-Smith, there are now twice as many urban dwellers in the developing world as there are in developed countries. For example, the population of urbanites in the countries of Europe, North America, Latin America, and the Caribbean (according to the United Nations, 1.3 billion) is smaller than the population of urbanites in Asia (1.7 billion). And with this incredible increase in sheer numbers of urban dwellers in the less developed regions of the world comes a large list of problems. Unemployment rates in cities of the developing world are often over 50 percent for newcomers to the city; housing and infrastructure often cannot be built fast enough to keep pace with growth rates; water and sewage systems can rarely handle the influx of new people. Consequently, one of the world's ongoing challenges will be this radical restructuring of population and culture as people in developing countries move into cities.

The target for much urban migration is the **primate city.** This is a settlement that dominates the economic, political, and cultural life of a country and, as a result of rapid growth, expands its primacy or dominance. Buenos Aires is an excellent example of a primate city because it far exceeds Rosario, the second-largest city in Argentina, in size and importance. Although many

megacities
A term that refers to particularly large urban centers.

primate city
A city of large size and dominant power within a country.

TABLE 10.1 The World's 20 Largest Metropolitan Areas

Rank	Metropolitan Area	Country	Population (thousands)
1	Tokyo/Yokohama	Japan	34,000
2	Guangzhou	China	24,200
2	Seoul	South Korea	24,200
4	Mexico City	Mexico	23,400
5	Delhi	India	23,200
6	Mumbai	India	22,800
7	New York	USA	22,200
8	São Paulo	Brazil	20,900
9	Manila	Philippines	19,600
10	Shanghai	China	18,400
11	Los Angeles	USA	17,900
12	Osaka/Kobe/Kyoto	Japan	16,800
13	Kolkata	India	16,300
14	Karachi	Pakistan	16,200
15	Jakarta	Indonesia	15,400
16	Cairo	Egypt	15,200
17	Beijing	China	13,600
17	Dacca	Bangladesh	13,600
17	Moscow	Russia	13,600
20	Buenos Aires	Argentina	13,300

(Source: Th. Brinkhoff: The Principal Agglomerations of the World, http://www.citypopulation.de, 2010-07-18.)

developing countries are dominated by a primate city, often a former center of colonial power, urban primacy is not unique to these countries: think of the way London and Paris dominate their respective countries.

Reflecting on Geography

What types of historical, political, and economic factors account for nations that are dominated by a primate city? Why didn't Boston or New York City become the primate city for the United States?

Although these primate cities are often the ones that garner scholars' and policy makers' attention because of their dominance and large size, recent demographic trends suggest that the majority of urban growth today is taking place in smaller, less dominant cities—places such as Gabarone, the capital of Botswana, whose population of 186,000 people is expected to more than double by 2020. Urban planners and policy makers are beginning to focus their attention on these midsize cities that are scattered throughout the world, par-

ticularly in developing countries, in order to examine the impacts of increased urbanization.

Central-Place Theory

So far our analysis of regional patterns of urbanization has been focused on a global scale as we've examined the varying degrees of urbanization around the world. In this section we shift the focus to a much finer scale of analysis and consider the arrangement of cities within a particular region. Urban geographers have studied the spatial distribution of towns and cities to determine some of the economic and political factors that influence the pattern of cities. In doing so, they have created a number of models that collectively make up **central-place theory.**

central-place theory
A set of models designed to explain the spatial distribution of urban service centers.

Most urban centers are engaged mainly in the service industries. The service activities of urban centers include transportation, communication, and utilities—services that facilitate the movement of goods and that provide the networks for the exchange of ideas about those goods (see Chapter 9 for a more detailed examination of these different industrial activities). Towns and cities that support such activities are called **central places.**

central places
Towns or cities engaged primarily in the service stages of production; a regional center.

In the early 1930s, the German geographer Walter Christaller first formulated central-place theory as a series of models designed to explain the spatial distribution of urban centers. Crucial to his theory is the fact that different goods and services vary both in **threshold,** the size of the population required to make provision of the good or service eco-

threshold
In central-place theory, the size of the population required to make provision of goods and services economically feasible.

range
In central-place theory, the average maximum distance people will travel to purchase a good or service.

nomically feasible, and in **range,** the average maximum distance people will travel to purchase a good or service. For example, a larger number of people are required to support a hospital, university, or department store than to support a gasoline station, post office, or grocery store. Similarly, consumers are willing to travel a greater distance to consult a heart specialist, record a land title, or purchase an automobile than to buy a loaf of bread, mail a letter, or visit a movie theater. Because the range of central goods and services varies, urban centers are arranged in an orderly hierarchy. Some central places are small and offer a limited variety of services and goods; others are large and offer an abundance. At the top of this hierarchy are regional metropolises—cities such as New York, Beijing, or Mumbai—that offer all services associated with central places and that have very large tributary trade areas, or hinterlands. At the opposite extreme are small market villages and roadside hamlets, which may contain nothing more than a post office, service station, or café. Between these two extremes are central places of various degrees of importance. Each higher rank of central place provides all the goods and services available at a lower-rank center, plus one or more additional goods and services. Central places of lower rank greatly outnumber the few at the higher levels of the hierarchy. One regional metropolis may contain thousands of smaller central places in its tributary market area (**Figure 10.3**). The size of the market area is determined by the distance range of the goods and services it offers.

With this hierarchy as a background, Christaller evaluated the individual influence of three forces (the market, transportation, and political borders) in determining the spacing and distribution of urban centers by creating models. His first model measured the influence of the market and range of goods on the spacing of cities. To simplify the model, he assumed that the terrain, soils, and other environmental factors were uniform; that transportation was universally available; and that all regions would be supplied with goods and services from the minimum number

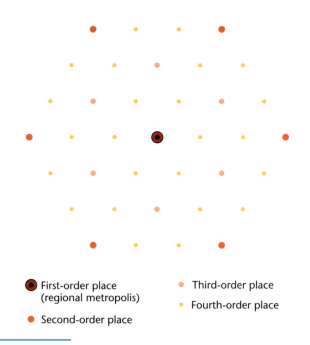

● First-order place (regional metropolis)   ● Third-order place

● Second-order place   ● Fourth-order place

FIGURE 10.3 Christaller's hierarchy of central places shows the orderly arrangement of towns of different sizes. This is an idealized presentation of places performing central functions. For each large central place, many smaller central places are located within the larger place's hinterland.

of central places. If market and range of goods were the only causal forces, the distribution of towns and cities would produce a pattern of nested hexagons, each with a central place at its center (**Figure 10.4a**).

In Christaller's second model, he measured the influence of transportation on the spacing of central places. He no longer assumed that transportation was universally and equally available in the **hinterland.** Instead, he assumed that as many demands

hinterland
The area surrounding a city and influenced by it.

for transport as possible would be met with the minimum expenditure for construction and maintenance of transportation facilities. Thus, as many high-ranking central places as possible would be on straight-line routes between the primary central places (**Figure 10.4b**). When transportation is considered the influential factor in shaping the spacing of central places, then the pattern is rather different from that created by the market factor. This transportation-driven pattern occurs because direct routes between adjacent regional metropolises do not pass through central places of the next-lowest rank. As a result, these second-rank central places are "pulled" from the "corners" of the hexagonal market area to the midpoints in order to be on the straight-line routes between adjacent regional metropolises.

Christaller hypothesized that the market factor would be the greater force in rural countries, where goods were seldom shipped throughout a region. In densely settled industrialized countries, however, he believed that the

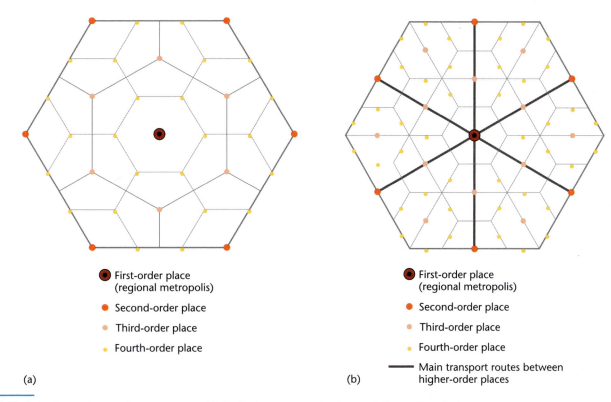

(a)

(b)

FIGURE 10.4 (a) **The influence of market area on Christaller's arrangement of central places.** If marketing was the only factor controlling the distribution of central places, this diagram would represent the arrangement of towns and cities. *Why, in this model, would hexagons be the shape to appear instead of a square, circle, or some other shape?* (b) **The distribution of central places according to Christaller's model.** If the availability of transportation was the determining factor in the location of central places, their distribution would be different from the distribution that would result if marketing was the determining factor. Note that the second-order central places are pulled away from the apexes of the hexagon and become located on the main transport routes between regional metropolises. *(After Christaller, 1966. By permission of the publisher.)*

transportation factor would be stronger because there were greater numbers of central places and more demand for long-distance transportation.

Christaller devised a third model to measure the effect of political borders on the distribution of central places. He recognized that political boundaries, especially within independent countries, would tend to follow the hexagonal market-area limits of each central place that was a political center. He also recognized that such borders tend to separate people and retard the movement of goods and services. Such borders necessarily cut through the market areas of many central places below the rank of regional metropolis. Central places in such border regions lose rank and size because their market areas are politically cut in two. Border towns are thus stunted, and important central places are pushed away from the border, which distorts the hexagonal pattern.

Market area, transportation, and political borders are but three of the many forces that influence the spatial distribution of central places. For example, in all three of these models, it is assumed that the physical environment is uniform and that people are evenly distributed. Of course, nei-

ther of these is true, leading to some distortions in the model. Nonetheless, central-place theory has proven to be a very useful tool for understanding the locations of cities and services in relationship to population. For example, with increasing concerns about environmental sustainability (see Chapter 9), regional planners have begun to use central-place theory to think about how to minimize transportation costs in a particular region by siting key services in central locations. Yet it is important to keep in mind that the model fails to take into account such cultural factors as people's historical attachment to places. For example, if people have shopped in a certain city for several generations, or even several years, and feel attached to that place, they often choose not to shop elsewhere, even if another city is located closer to them. In other words, the most economically rational (in terms of saving money by traveling less) path is not always the one most preferred by people.

Reflecting on Geography

If Christaller's assumptions did not hold, in what ways would central-place theory need to be altered?

Mobility

Where did cities begin and how did they diffuse to different regions in the world?

As noted in the beginning of this chapter, we are entering an era when more people live in cities than live in rural areas: quite a historical feat. When cities first began, only a very small portion of the population lived in them. Over time, cities began to diffuse throughout the globe, and established cities grew larger. In this section, we analyze two aspects of mobility in regard to urbanization. First, we take a historical look at the origins of city life and explore how, why, and where cities diffused around the world. Second, we examine the more contemporary processes of rural-to-urban migration.

Origin and Diffusion of the City

In seeking explanations for the origin of cities, we ultimately find a relationship among areas of early agricultural development, permanent village settlements, the emergence of new social forms, and urban life. The first cities resulted from a complicated transition that took thousands of years.

As early people, originally hunters and gatherers, became more successful at accumulating their resources and domesticating plants and animals, they began to settle, first semipermanently and then permanently. In the Middle East, where the first cities appeared, a network of permanent agricultural villages developed about 10,000 years ago. These farming villages were modest in size, rarely with more than 200 people, and were probably organized on a kinship basis. Jarmo, one of the earliest villages, located in present-day Iraq, had 25 permanent dwellings clustered together near grain storage facilities.

Although small farming villages like Jarmo predate cities, it is wrong to assume that a simple quantitative change took place whereby villages slowly grew, first into towns and then into cities. True cities, then and now, differ qualitatively from agricultural villages. All the inhabitants of agricultural villages were involved in some way in food procurement—tending the agricultural fields or harvesting and preparing the crops. Cities, however, were more removed, both physically and psychologically, from everyday agricultural activities. Food was supplied to the city, but not all city dwellers were involved in obtaining it. Instead, city dwellers supplied other services, such as technical skills or religious interpretations considered important in a particular society. Cities, unlike agricultural villages, contained a class of people who were not directly involved in agricultural activities.

agricultural surplus
The amount of food grown by a society that exceeds the demands of its population.

Two elements were necessary for this dramatic social change: the creation of an **agricultural surplus** and the development of a stratified social system. Surplus food, which is a food supply larger than the everyday needs of the agricultural labor force, is a prerequisite for supporting nonfarmers—people who work at administrative, military, or handicraft tasks. Social stratification, the existence of distinct socioeconomic classes, facilitates the collection, storage, and distribution of resources through well-defined channels of authority that can exercise control over goods and people. A society with these two elements—surplus food and a means of storing and distributing it—was set for urbanization.

Models for the Rise of Cities

One way to understand the transition from village to city life is to model the development of urban life assuming that a single factor is the trigger behind the transition. The question that scholars ask is: "What activity could be so important to an agricultural society that its people would be willing to give some of their surplus to support a social class that specializes in that activity?" Next we discuss answers to that question and clarify a multiple-factor explanation for the rise of cities.

Technical Factors The **hydraulic civilization** model, developed by Karl Wittfogel, assumes that the development of large-scale irrigation systems was the prime mover behind urbanization and that a class of technical specialists were the first urban dwellers. Irrigating agricultural crops yielded more food, and this surplus supported the development of a large nonfarming population. A strong, centralized government backed by an urban-based military could expand its power into the surrounding areas. Farmers who resisted the new authority were denied water. Continued reinforcement of the power elite came from the need for organizational coordination to ensure continued operation of the irrigation system. Labor specialization developed. Some people farmed; others worked on the irrigation system. Still others became artisans, creating the implements needed to maintain the system, or administrative workers in the direct employ of the power elite's court.

hydraulic civilization
A civilization based on large-scale irrigation.

Although the hydraulic civilization model fits several areas where cities first arose—China, Egypt, and Mesopotamia (present-day Iraq)—it cannot be applied to all urban hearths. In parts of Mesoamerica, for example, an urban civilization blossomed without widespread irrigated agriculture and therefore without a class of technical experts.

Religious Factors Geographer Paul Wheatley suggests that religion led to urbanization. In early agricultural societies, knowledge of such matters as meteorology and climate was considered an element of religion. Such societies depended on their religious leaders to interpret the heavenly bodies before deciding when and how to plant their crops. The propagation of this type of knowledge led to more successful harvests, which in turn allowed for the support of both a larger priestly class and a class of people engaged in ancillary activities. The priestly class exercised the political and social control that held the city together.

In this scenario, early cities were religious spaces. The first urban clusters and fortifications are seen as defenses not against human invaders but against spiritual ones: demons or the souls of the dead. This religious explanation is applicable in some ways to all the early centers of urbanization, although it seems especially successful in explaining Chinese urbanization.

Political Factors Other scholars suggest that the centralizing force in urbanization was political order. Urban historian Lewis Mumford described the agent of change in emerging urban centers as the institution of kingship, which involved the centralizing of religious, social, and economic aspects of a civilization around a powerful figure who became known as the king. This figure of authority, who in the pre-urban world was accorded respect for his or her human abilities, ascended to almost superhuman status in early urbanizing societies. By exercising power, the king was able to marshal the labor of others. The resulting social hierarchy enabled the society to diversify its endeavors, with different groups specializing in crafts, farming, trading, or religious activity. The institution of kingship provided

essential leadership and organization to this increasingly complex society, which became the city.

Multiple Factors At the onset of urbanization, and even much later in some places, sharp distinctions among economic, religious, and political functions were not always made. The king may also have functioned as priest, healer, astronomer, and scribe, thereby fusing secular and spiritual power. Critics of the kingship theory, therefore, point out that this explanation of urbanization may not be different from the religion-based model. Rather than attempting to isolate one trigger, a wiser course may be to accept the role of multiple factors behind the changes leading to urban life. Technical, religious, and political forces were often interlinked, with a change in one leading to changes in another. Instead of oversimplifying by focusing on one possible development schema, we must appreciate the complexities of the transition period from agricultural village to true city.

Urban Hearth Areas

The first cities appeared in distinct regions, such as Mesopotamia, the Nile River valley, Pakistan's Indus River valley, the Yellow River (or Huang Ho) valley of China, Mesoamerica, and the Andean highlands and coastal areas of Peru. These are called the **urban hearth areas** (**Figure 10.5**).

urban hearth areas Regions in which the world's first cities evolved.

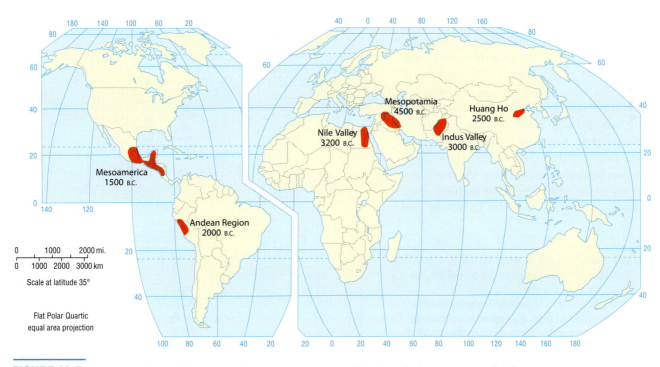

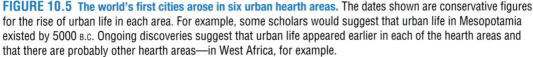

FIGURE 10.5 The world's first cities arose in six urban hearth areas. The dates shown are conservative figures for the rise of urban life in each area. For example, some scholars would suggest that urban life in Mesopotamia existed by 5000 B.C. Ongoing discoveries suggest that urban life appeared earlier in each of the hearth areas and that there are probably other hearth areas—in West Africa, for example.

Reflecting on Geography

Scholars are continually altering the dates for the emergence of urban life, as well as the location of the hearth areas. Why? Can you outline some of the reasons it is so difficult to pinpoint the places and dates for the emergence of urban life?

It is generally agreed that the first cities arose in Mesopotamia, the river valley of the Tigris and Euphrates in what is now Iraq. Mesopotamian cities, small by current standards, covered 0.5 to 2 square miles (1.3 to 5 square kilometers), with populations that rarely exceeded 30,000. Nevertheless, the densities within these cities could easily reach 10,000 people per square mile (4000 per square kilometer), which is comparable to the densities in many contemporary cities.

cosmomagical cities
Types of cities that are laid out in accordance with religious principles, characteristic of very early cities, particularly in China.

Scholars have referred to these first cities as **cosmomagical cities,** defined as cities that are spatially arranged according to religious principles. The spatial layouts of cosmomagical cities are similar in three important ways (**Figure 10.6**). First, great importance was accorded to the city's symbolic center, which was also thought to be the center of the known world. It was therefore the most sacred spot and was often identified by a vertical structure of monumental scale that represented the point on Earth closest to the heavens. This symbolic center, or **axis mundi,** took the form of the ziggurat in Mesopotamia, the palace or temple in China, and the pyramid in Mesoamerica.

axis mundi
The symbolic center of cosmomagical cities, often demarcated by a large vertical structure.

Often this elevated structure, which usually served a religious purpose, was close to the palace or seat of political power and to the granary. These three structures were often walled off from the rest of the city, forming a symbolic center that both reflected the significance of these societal functions and

dominated the city physically and spiritually (**Figure 10.7**). The Forbidden City in Beijing remains one of the best examples of this guarded, fortresslike "city within the city." The second spatial characteristic common to cosmomagical cities is that they were oriented toward the four cardinal directions. By aligning the city in the north-south and east-west directions, the geometric form of the city reflected the order of the universe. This alignment, it was thought, would ensure harmony and order over the known world, which was bounded by the city walls.

In all of these early cities, one sees evidence of a third spatial characteristic: an attempt to shape the form of the city according to the form of the universe. The ordering of the space of the city was thought to be essential to maintaining harmony between the human and spiritual worlds. In this way, the world of humans would symbolically repli-

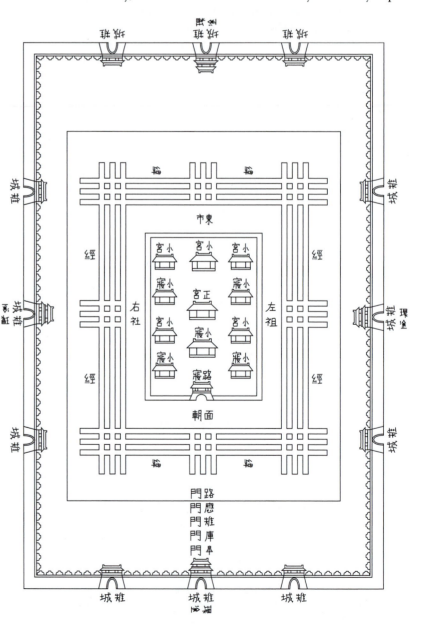

FIGURE 10.6 Plan of the city of Wang-Ch'eng in China.
The city as built did not follow this exact design, but the plan itself is of interest because it suggests the symbolic importance of the three spatial characteristics of the Chinese cosmomagical city. The four walls are aligned to the cardinal directions, and the axis mundi is represented by the walled-off center city containing ceremonial buildings. The physical space of the city (microcosmos) replicates the larger world of the heavens (macrocosmos). For example, each of the four walls represents one of the four seasons. ***What do you think the gates to the city represent?*** (Source: Wheatley, 1971.)

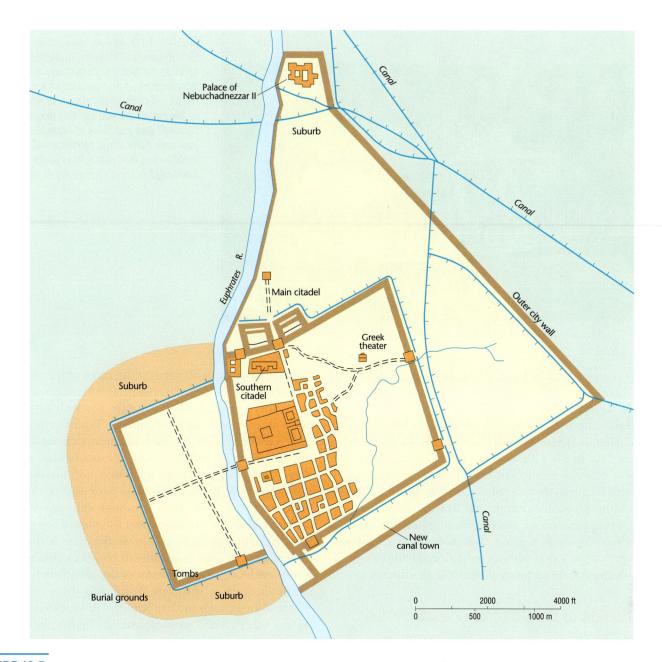

FIGURE 10.7 **Map of Babylon, illustrating the urban morphology of early Mesopotamian cities.** The citadel in the inner city is characterized by the ziggurat, main temple, palace, and granary. Beyond the citadel lay the residential areas; we can assume they extended out to the inner walls and occupied both sides of the river. Suburbs grew outside the major gates and were occupied by people not allowed to spend the night in the city, such as traders and noncitizens. *(After map in Encyclopaedia Brittanica, 1984: 555.)*

cate the world of the gods. This characteristic may have taken a literal form—a city laid out, for example, in a pattern of a major star constellation. Far more common, however, were cities that symbolically approximated mythical conceptions of the universe. Angkor Thom was an early city in Cambodia that presents one of the best examples of this parallelism. An urban cluster that spread over 6 square miles (15.5 square kilometers), Angkor Thom was a repre-

sentation in stone of a series of religious beliefs about the nature of the universe. Thus, the city was a microcosmos, a re-creation on Earth of an image of the larger universe.

Nevertheless, regional variations of this basic form certainly existed. For example, the early cities of the Nile were not walled, which suggests that a regional power structure kept individual cities from warring with one another. In the Indus Valley, the great city of Mohenjo-daro was laid out in

FIGURE 10.8 Mayan city of Chichén Itzá. Monumental and ceremonial architecture often dominated the morphology and landscape of urban hearth areas and reinforced ruling-class power. *In what part of the city would you say this monument is located? Can you think of examples from your own daily life of monumental architecture that symbolizes some ruling authority?* (Malcolm Kirk/Peter Arnold, Inc.)

a grid that consisted of 16 large blocks, and the citadel was located within the block that was central but situated toward the western edge.

The most important variations within the urban hearth areas occurred in Mesoamerican cities (**Figure 10.8**). Here, cities were less dense and covered large areas. Furthermore, these cities arose without benefit of the technological advances found in the other hearth areas, most notably the wheel, the plow, metallurgy, and draft animals. However, the domestication of maize compensated for these shortcomings. Maize is a grain that in tropical climates yields several crops a year without irrigation; in addition, it can be cultivated without heavy plows or pack animals.

The Diffusion of the City from Hearth Areas

Although urban life originated at several specific places in the world, cities are now found everywhere: North America, Southeast Asia, Latin America, and Australia. How did city life come to these regions? There are two possible explanations:

1. Cities evolved spontaneously as native peoples created new technologies and social institutions.

2. The preconditions for urban life are too specific for most cultures to have invented without contact with other urban areas; therefore, they must have learned these traits through contact with city dwellers. This scenario emphasizes the diffusion of ideas and techniques necessary for city life.

Diffusionists argue that the complicated array of ideas and techniques that gave rise to the first cities in Mesopotamia was shared with other people in both the Nile and the

Indus river valleys who were on the verge of the urban transformation. Indeed, archaeological evidence suggests that these three civilizations had trade ties with one another. Soapstone objects manufactured in Tepe Yahya, 500 miles (800 kilometers) to the east of Mesopotamia, have been uncovered in the ruins of both Mesopotamian and Indus River valley cities, which are separated by thousands of miles. Writings of the Indus civilization have also been found in Mesopotamian urban sites. Although diffusionists use this artifactual evidence to argue that the idea of the city spread from hearth to hearth, an alternative view is that trading took place only after these cities were well established. There is also evidence of contact across the oceans between early urban dwellers of the New World and those of Asia and Africa, although it is unclear whether this means that urbanization was diffused to Mesoamerica or simply that some trade routes existed between these peoples.

Nonetheless, there is little doubt that diffusion has been responsible for the dispersal of the city in historical times (**Figure 10.9**), because the city has commonly been used as the vehicle for imperial expansion. Typically, urban life is carried outward in waves of conquest as the borders of an empire expand. Initially, the military controls newly won lands and sets up collection points for local resources, which are then shipped back into the heart of the empire. As the surrounding countryside is increasingly pacified, the new collection points lose some of their military atmosphere and begin to show the social diversity of a city. Artisans, merchants, and bureaucrats increase in number; families appear; the native people are slowly assimilated into the settlement as workers and may eventually control the city. Finally, the process repeats itself as the empire pushes farther outward: first a military camp, then a collec-

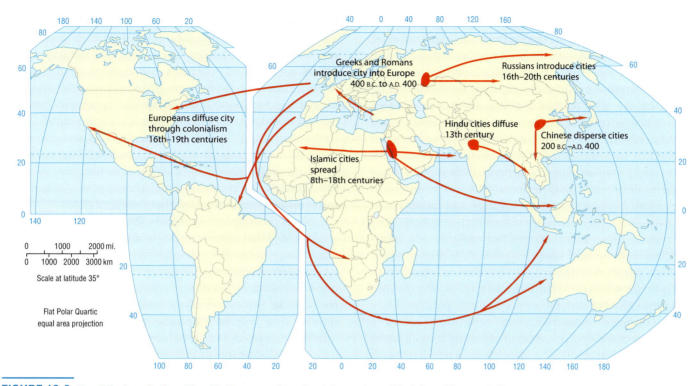

FIGURE 10.9 The diffusion of urban life with the expansion of certain empires. *What does this map tell us about the importance of urban life to military conquest?*

tion point for resources, then a full-fledged city expressing true division of labor and social diversity.

This process, however, did not always proceed without opposition. The imposition of a foreign civilization on native peoples was often met with resistance, both physical and symbolic. Expanding urban centers relied on the surrounding countryside for support. Their food was supplied by farmers living fairly close to the city walls, and tribute was exacted from the agricultural peoples living on the edges of the urban world. The increasing needs of the city required more and more land from which to draw resources. However, the peasants farming that land may not have wanted to change their way of life to accommodate the city. The fierce resistance of many Native American groups to the spread of Western urbanization is testimony to the potential power of folk society to defy urbanization, although the destructive long-term effects of such resistance suggest that the organized military efforts of urban society were difficult to overcome.

Rural-to-Urban Migration

This analysis of the diffusion of cities throughout the world helps provide a historical explanation for the increase in urban population, since it went hand-in-hand with the growing number of cities. In today's world, however, increasing urbanization is caused by two related phenomena: natural

population increase (see Chapter 3) and rural-to-urban migrations. Although the United Nations estimates that natural increase accounts for the majority of the recent growth in urban population, it was the rural-to-urban migrations of the past 20 to 30 years that brought millions of people originally into urban life. In countries throughout Africa and Asia, large numbers of people have left their rural villages and migrated to cities for better economic and social opportunities. And although people flock to cities in search of jobs, often there is not a perfect correlation between economic growth and urbanization. Cities increase in size not necessarily because there are jobs to lure workers, but rather because conditions in the countryside are much worse. In India, for example, rural poverty has exacerbated the rapid population increase in such cities as Mumbai and Kolkata. People leave in hope that urban life will offer a slight improvement, and often it does. Given the new global economy, however, many of the jobs that are available in these cities are low-skilled and low-paying manufacturing jobs with harsh working conditions. The result is that rural-to-urban migrants often find themselves either unemployed or with jobs that barely provide them a living.

Sociologist Alejandro Portes argues that the large internal migrations that bring impoverished agricultural people into the city are a phenomenon that can be traced back to colonial times. In colonial Latin America, for example, the city was essentially home to the Spanish elite.

When preconquest agricultural patterns were disrupted, peasants came to the city looking to improve their economic situation. These people usually lived on the margins of the city. Moreover, they were completely disenfranchised because only landowners had the right to hold office. The reaction by the elite to this ongoing movement of large masses of people into the city was a mixture of tolerance and indifference, with no one taking responsibility to integrate the migrants into the city. Rural-to-urban migrations continue to occur in many globalizing cities. Even though many of these modern migrants experience poor living conditions in the city and are torn from their village and family roots, the opportunities presented by urban life, such as better pay and exposure to new and more "Western" cultural forms (for example, music and film), are often difficult to resist.

China presents a particularly interesting example of rural-to-urban migration. Until the late 1970s, the country was predominantly rural, but state economic policies thereafter encouraged industrialization, and most of those industries were located in urban areas along its southeastern coast (see Chapter 9). The state lifted its restrictions on internal migration, which enabled its rural population to become more mobile. Today it is estimated that 18 million Chinese migrate to cities each year, a phenomenal rate of urban increase. These migrations are drastically altering the economy, society, and culture of China, as the country experiences this unprecedented shift from a rural, agricultural nation to one based on cities and industry.

Globalization

What is the relationship between cities and globalization? Many of the globalizing forces we have already discussed in this book—the integration of international economies, the merging of different cultures, and the reshaping of social organization—are centered in urban life. In other words, cities are the places where one can see both the multitude of benefits that come from globalization and the numerous pitfalls that it can bring. In this way, all cities can be considered global cities, places being shaped by the new global forces. However, scholars have identified two particular types of cities that are key to understanding globalization and that are themselves being critically reshaped by it.

Global Cities

global cities
Cities that are control centers of the global economy.

Global cities are defined as those cities that have become the control centers of the global economy—the places where major decisions about the world's commercial net-

works and financial markets are made. These cities house a concentration of multinational and transnational corporate headquarters, international financial services, media offices, and related economic and cultural services. While many industries have become global in the sense that their sites of production and consumption are spread throughout the world, the sites of decision making are now centralized. According to sociologist Saskia Sassen, there are only three such cities now operating at this level: New York City, London, and Tokyo. These cities have become, in many ways, the headquarters for a global economy and form the top level of a hierarchical global system of cities.

More recently, scholars have begun to broaden their analysis of global cities in order to also understand the next tier of cities within the urban hierarchy—those that contain a large percentage of producer service firms (law, accounting, advertising, financial services, consulting) that are international. In this new analysis, cities are categorized by the number and type of transnational firms they house—not only headquarters but also regional and national offices. This analysis has identified globally and regionally dominant cities and ones that are major participants in the new global economy. Geographers Yefang Huang, Yee Leung, and Jianfa Shen refer to these as international cities—places that are significant because they are centers of the new international economy. Their analysis allowed them to identify degrees of internationalization based on the number and locations of the international firms each city contained. The result is a very interesting list organized into classes of international cities, dominated by six in class A (London, New York, Hong Kong, Tokyo, Singapore, and Paris), followed by 10 in class B and 44 in class C. **Figure 10.10** maps those cities by class, revealing the dominance of certain regions within the global economy (the United States, Europe, parts of Asia) and also revealing another level within the global system of cities. In this way, we can begin to understand globalization as an uneven process, but one having impacts in many cities previously left unexamined.

Globalizing Cities

Globalizing cities are those that are being shaped by the new global economy and culture (see Doing Geography at the end of the chapter). This includes just about every city, present and past, to one degree or another. As geographer Brenda Yeoh indicates, cities have almost always been important hubs of activity beyond the national scale, and therefore it is not surprising that they figure prominently in contemporary discussions of globalization. But the degree of globalization in the last 30 years has sharpened and enhanced the ways in which global economies and cultures shape cities. As geographer Kris Olds points out (see Practic-

globalizing cities
Cities being shaped by the new global economy and culture.

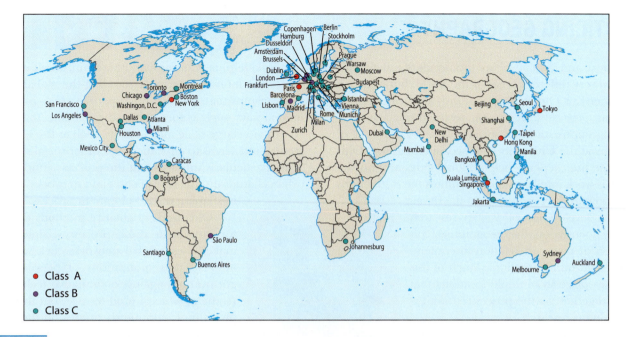

FIGURE 10.10 **The top 60 international cities in the world of class A, B, and C.** This map and index indicate the location of cities that contain international producer service firms. Class A cities contain over 88 percent of the 100 international firms that were surveyed, class B cities host over 70 percent of the same firms, and class C cities contain over 44 percent of those firms. *Look carefully at the distribution of these cities, and see if you can explain their locations by considering their national and regional contexts.*
(Source: Huang, Leung, and Shen, 2007.)

ing Geography, page 366), there are five interrelated dimensions of current globalizing processes that are shaping our cities: the development of an international financial system, the globalization of property markets, the prominence of transnational corporations, the stretching and intensifying of social and cultural networks, and the increased degree of international travel and networking. As he argues, it is impossible to understand developments in the urban landscapes of cities as diverse as Jakarta and London without referring to these global processes.

Yet not all cities are affected by these globalizing processes in the same way. Some cities like London, as we've just discussed, are the control centers of this economy, while others, like Jakarta, provide sites of production for the global economy. The differences between these two types of cities emerge out of cultural uniqueness and historical circumstance, particularly the colonial relationships that developed in the eighteenth and nineteenth centuries. With the end of colonialism and the movement toward political and economic independence, developing countries entered a period of rapid, sometimes tumultuous change. Cities have often been the focal point of this change, and as we have seen, millions of people have migrated to cities in search of a better life. Some of these newly globalizing cities are moving into the ranks of international cities, as you can see by looking at Figure 10.10. Other globalizing cities—like Gabarone in

Botswana, which we mentioned earlier—are just now beginning to feel the impacts of population growth and global economic integration, and their future is very much uncertain at this point. We will discuss the landscape forms and patterns of these globalizing cities in more detail in the section on cultural landscape, but for now we want to reiterate that globalization is an uneven process. Some cities are becoming globally and/or regionally dominant and are the primary beneficiaries of economic globalization, while others, many of which are located in the developing world, are experiencing more of the environmental and social problems that come with rapid and massive population growth, with less economic gain.

Nature-Culture

What is the relationship between cities and their physical settings? At first glance, cities seem totally divorced from the natural environment. What possible relationships could the shiny glass office buildings, paved streets, and high-rise apartment buildings that characterize most cities have with forests, fields, and rivers? In this section we examine several different ways to think about the relationships between urban life and its natural

PRACTICING GEOGRAPHY

Singapore and Vancouver are thousands of miles apart, in two very different areas of the world, but as urban geographer Kris Olds reminds us, they are being shaped by similar processes: economic restructuring, transnational migrations, and new social policies. A professor of geography at the University of Wisconsin, Madison, Olds studies the interdependence of global cities around the world. His work intelligently undermines the distinctions we used to make between cities in the developed and developing worlds. "My research primarily focuses on the geographical organization of power in relation to contemporary urban transformations. Much of this work takes place in multiple locations that are tied together via the processes I am examining in my research."

This type of research allows Olds to indulge two of his passions at the same time: traveling and solving real-world problems. "I've always had the travel bug. When I almost got kicked out of the University of British Columbia for atrocious grades (I thought that I would be better off becoming an engineer or a geologist instead of a geographer), I took some time off, traveled the world, and then returned to UBC . . . and stumbled into some great courses, all of them taught by urban geographers. Now, I get to travel while 'working.'" His work takes him to Bristol, Berlin, Singapore, and Hong Kong, but also to Vancou-

(Courtesy of Kris Olds.)

Kris Olds

ver, Philadelphia, Chicago, and Madison. In all of these places, his interest is in understanding the global power relationships that are disrupting people's lives in fundamental ways, such as forced housing evictions that often accompany large-scale mega-events, such as the Olympic games, and access to education and other social services. "I am particularly interested in the processes underlying urban change and the role of elites and networks of elites in shaping the development process."

Olds finds qualitative methods the best for getting at these processes, since what makes these global networks "tick" are the interests and motivations of people. "I've got nothing against quantitative techniques (some of which I used to use), but they simply cannot help me shed light on the issues that I find to be important. I now prefer to work with in-depth semistructured interviews, observation, and participant observation." Currently, he's observing and interviewing some of the key players involved in Singapore's push to become a global education hub, as well as working with an international nongovernmental organization to formulate concrete mechanisms that will prevent future forced housing evictions caused by mega-events. One of the most exciting things about being a practicing geographer, Olds says, is that "I am able to work on issues of real importance in the world."

---

setting. We start by exploring how the physical environment affects the locations of cities (site and situation) and then turn to an examination of the reciprocal relationship that exists between increasing urbanization and global environmental problems such as vulnerability to natural disasters.

Site and Situation

site
The local setting of a city.

situation
The regional setting of a city.

There are two components of urban location: site and situation. **Site** refers to the local setting of a city; the **situation** is the regional setting or location. As an example of site and situation, consider San Francisco. The original site of the Mexican settlement that became San Francisco was on a shallow cove on the eastern (inland) shore of a peninsula. The importance of its situation was that it drew on waterborne traffic coming across the bay from other, smaller settlements. Hence the town could act as a transshipment point.

Both site and situation are dynamic, changing over time. For example, in San Francisco during the gold rush period of the 1850s, the small cove was filled to create flatland for warehouses and to facilitate extending wharves into deeper bay waters. The filled-in cove is now occupied by the heart of the central business district (**Figure 10.11**). The geographical situation has also changed as patterns of trade and transportation technology have evolved. The original transbay situation was quickly replaced during the gold rush by a new role: supplying the mines and settlements of the gold country. Access to the two major rivers leading to the mines, plus continued ties to ocean trade routes, became the important components of the city's situation.

San Francisco's situation has changed dramatically in the last decade, for it is no longer the major port of the bay. The change in technology to containerized cargo was adopted more quickly by Oakland, the rival city on the opposite side of the bay, and San Francisco declined as a port city. Oakland was able to accommodate containerized

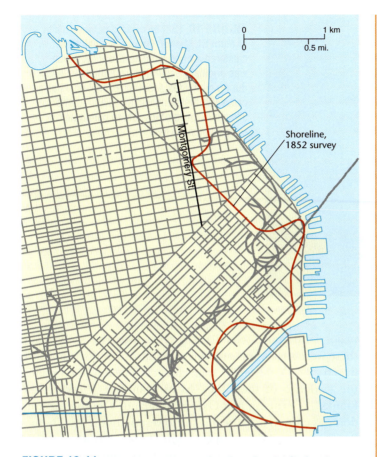

FIGURE 10.11 **This map shows how San Francisco's site has been changed by human activity.** During the late 1850s, shallow coves were filled, providing easier access to deeper bay waters as well as flatland near the waterfront for warehouses and industry.

cargo in part because it filled in huge tracts of shallow bay lands, creating a massive area for the loading, unloading, and storage of cargo containers.

Certain attributes of the physical environment have been important in the location of cities. Those cities with distinct functions, such as defense or trade, were located because of specific physical characteristics. The locations of many contemporary cities can be partially explained by decisions made in the past that capitalized on the advantages of certain sites. The following classifications detail some of the different location possibilities.

Defensive Sites There are many types of defensive sites for cities (some are diagrammed in **Figure 10.12**). A **defensive site** is a location that can be easily defended. The river-meander site, with the city located inside a loop where the stream turns back on itself, leaves only a narrow neck of land unprotected by water. Cities such as Bern, Switzerland, and New Orleans are situated inside river meanders. Indeed, the nickname for New Orleans, the Crescent City, refers to the curve of the Mississippi River.

> **defensive site**
> A location from which a city can be easily defended.

Even more advantageous was the river-island site, which, because the stream was split into two parts, often combined a natural moat with an easy river crossing. For example, Montreal is situated on a large island surrounded by the St. Lawrence River and other water channels. The offshore-island site—that is, islands lying off the seashore or in lakes—offered similar defensive advantages (**Figure 10.13**, page 368). Mexico City began as an Indian settlement on a lake island. Venice is the classic example of a city built on offshore islands in the sea, as is Hong Kong. Peninsula sites were almost as advantageous as island sites, because they offered natural water defenses on all but one side. Boston was founded on a peninsula for this reason, and a wooden palisade wall was built across the neck of the peninsula.

Danger of attack from the sea often prompted sheltered-harbor defensive sites, where a narrow entrance to the

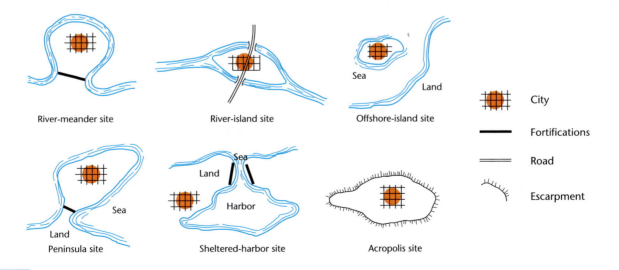

FIGURE 10.12 **Types of defensive city sites.** Natural protection is afforded by physical features.

FIGURE 10.13 **The classic defensive site of Mont St. Michel, France.** A small town clustered around a medieval abbey, which was originally separated from the mainland during high tides, Mont St. Michel now has a causeway that connects the island to shore, allowing armies of tourists to penetrate the town's defenses easily. *(Photo Researchers.)*

harbor could be defended easily. Examples of sheltered-harbor sites include cities such as Rio de Janeiro, Tokyo, and San Francisco.

High points were also sought out. These are often referred to as acropolis sites; the word *acropolis* means "high city." Originally the city developed around a fortification on the high ground and then spilled out over the surrounding lowland. Athens is the prototype of acropolis sites, but many other cities are similarly located.

trade-route sites
Locations for a city at a significant point on a transportation route.

Trade-Route Sites Defense was not always the primary consideration. Instead, urban centers were often built on **trade-route sites**— that is, at important points along already established trade routes. Here, too, the influence of the physical environment can be detected.

Especially common types of trade-route sites (**Figure 10.14**) are bridge-point sites and river-ford sites, places where

major land routes could easily cross over rivers. Typically, these were sites where streams were narrow and shallow with firm banks. Occasionally, such cities even bear in their names the evidence of their sites, as in Frankfurt ("ford of the Franks"), Germany, and Oxford, England. The site for London was chosen because it is the lowest point on the Thames River where a bridge—the famous London Bridge—could easily be built to serve a trade route running inland from Dover on the sea.

Confluence sites are also common. They allow cities to be situated at the point where two navigable streams flow together. Pittsburgh, at the confluence of the Allegheny and Monongahela rivers, is a fine example (**Figure 10.15**). Head-of-navigation sites, where navigable water routes begin, are even more common, because goods must be transshipped at such points. Iquitos, Peru, is located at the head-of-navigation site for the Amazon River, and Minneapolis–St. Paul, at the falls of the Mississippi River, also occupies a head-of-

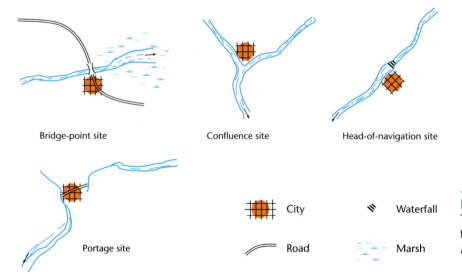

Bridge-point site   Confluence site   Head-of-navigation site

Portage site

City   Waterfall   Road   Marsh

FIGURE 10.14 **Trade-route city sites.** These sites are strategic positions along transportation arteries. *Is your city or one near you located on a trade-route site?*

FIGURE 10.15 **Pittsburgh's Golden Triangle.** At the confluence of the Allegheny and Monongahela rivers, Pittsburgh is a classic example of how an early trade-route site has evolved into a commercial center. *(Comstock Select/Corbis.)*

navigation site. Portage sites are very similar. Here, goods were portaged from one river to another. Chicago is near a short portage between the Great Lakes and the Mississippi River drainage basin. In these ways and others, an urban site can be influenced by the physical environment. But so, too, do cities impact that physical environment in a multitude of ways. In the next section we explore that general theme of urban sustainability.

Urbanization and Sustainability

As we examined in Chapter 9, industrialization goes hand-in-hand with urbanization, so the environmental impacts of industrialization are often found in cities. But urbanization generates its own set of environmental impacts: supplying enough energy, food, and water to large concentrations of people puts an array of stresses on the natural environment. Scholars now refer to these varied impacts of urban areas on the environment as the *urban footprint*. For example, Las Vegas is one of the fastest-growing urban areas in the United States, yet it is located in a desert (**Figure 10.16**). The city's primary water source is the Colorado River, but in more ways than one the costs of delivering that water are extremely high. It is expensive to construct the dams and infrastructure necessary to move the water into the city, and the environmental damage to the region is very costly. The dams alter the flows of water through the Colorado Valley, harming fish and disrupting aquatic life cycles, and the energy required to divert the water to the city leads to higher sulfur dioxide emissions, which have been shown to lead to global warming.

But the environmental effects of increased urbanization—the urban footprint—can spread beyond the immediate stresses caused by higher concentrations of population. Urban living often leads to rising levels of consumption, as new urbanites gain access to better jobs with more disposable income available to them. This growing demand for things like better food can impact areas far from urban centers. For example, as the United Nations report *Unleashing the Potential of Urban Growth* suggests, tropical forests in Tobasco, an area 400 miles away from Mexico City, have been turned into cattle-grazing areas in response to urbanites' demands for meat, while, in a second example, a major contributing factor to the deforestation of the Amazon is the increased

FIGURE 10.16 **Las Vegas, Nevada.** One of the fastest-growing cities in North America, Las Vegas is sited in the heart of the desert. As such, its growing demands for water are putting huge stresses on the natural environment, as water has to be diverted from the Colorado River. *(Robert Cameron/Getty Images.)*

demand for soybeans from the newly urbanizing regions of China as well as from the urbanites of the United States, Japan, and Europe. Urbanization, however, does not necessarily imply environmental degradation (see Subject to Debate on pages 372–373). For example, the higher densities of population that cities represent can be seen as a form of sustainable growth. Half the population of the world—the urban half—lives on approximately 3 percent of the landmass. The concentration of people in cities therefore opens up other areas that can be protected and left relatively free from human use. Many countries in the developing and developed world are working on local and regional planning projects that, on the one hand, maintain urban boundaries and prevent sprawl, and, on the other hand, protect natural environments outside urban areas from the sorts of environmental degradations we have just outlined. These sorts of urban environmental conservation projects will become increasingly important in future years.

Natural Disasters

Hurricane Katrina has become a critical reminder of the vulnerability of human settlements to natural disasters. The disaster, which struck the Gulf Coast of the United States in August 2005, killed 1836 people and left hundreds of thousands of people homeless. Urban areas, with their concentrations of people, are particularly vulnerable to such natural disasters, given that so many people's lives and livelihoods are at risk. In addition, since many cities like New Orleans are sited near water (see the section on site and situation), they are often located in the direct paths of disasters such as hurricanes and tsunamis. Making matters even worse, scientists have recently noted an increase in the number of natural disasters, attributable in part, they believe, to global climate change. This combination of increasing urbanization and a growing number of natural disasters means that more and more people's lives are being impacted adversely. Between 1980 and 2000, approximately 75 percent of the world's population lived in areas affected by natural disasters (Figure 10.17). And, as with Hurricane Katrina, it is most often poor people who are the most affected, since they tend to live in structures that are not well built and in sections of the city that are more vulnerable, such as low-lying areas. In cities in the developing world, squatter settlements (see the section on cultural landscape) are often sited on steep hillsides or poorly drained areas, making these areas far more vulnerable to landslides, flooding, and other natural disasters.

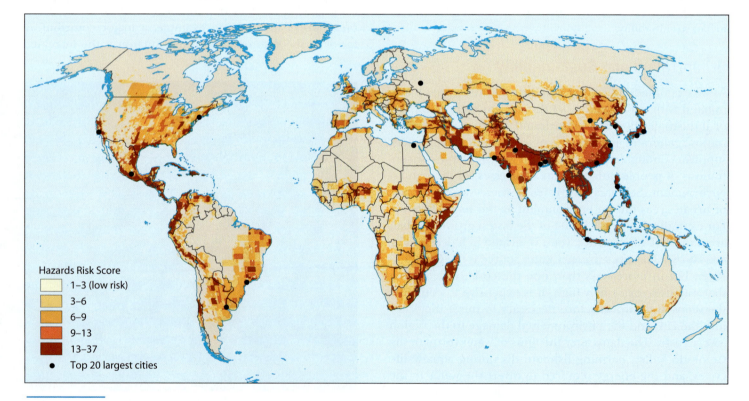

Hazards Risk Score
- 1–3 (low risk)
- 3–6
- 6–9
- 9–13
- 13–37
- ● Top 20 largest cities

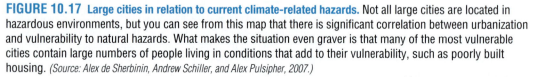

FIGURE 10.17 Large cities in relation to current climate-related hazards. Not all large cities are located in hazardous environments, but you can see from this map that there is significant correlation between urbanization and vulnerability to natural hazards. What makes the situation even graver is that many of the most vulnerable cities contain large numbers of people living in conditions that add to their vulnerability, such as poorly built housing. (*Source: Alex de Sherbinin, Andrew Schiller, and Alex Pulsipher, 2007.*)

CULTURAL LANDSCAPE

What do cities look like around the world? As you probably already know from your local, regional, or international travels, the look of cities varies from place to place. Some cities have bustling downtowns where most of the residents work, while others are characterized by nodes of activities clustered on the suburban fringe; some cities are built around a set of historical buildings with political or religious significance, while the centers of other cities are dominated by new skyscrapers housing financial offices. Finally, some cities have streets laid out in a gridlike pattern, while in other cities streets are a crisscrossed jumble without apparent order. How can we make sense of this range of urban landscapes? In this section, we explore several forms or models of urban landscapes, particularly those outside the developed world. In the cultural landscape section in Chapter 11, we focus on cities in North America and Europe, in order to explore in more detail how cities are arranged and to help you learn to "read" the landscape of your own city.

Globalizing Cities in the Developing World

In cities of the developing world, the combination of large numbers of migrants and widespread unemployment leads to overwhelming pressure for low-rent housing. Governments have rarely been able to meet these needs through housing projects, so one of the most characteristic landscape features of cities in the developing world has been construction of illegal housing: **squatter settlements.**

squatter settlements
Illegal housing settlements, usually made up of temporary shelters, that surround a large city.

Squatter settlements, or **barriadas** (**Figure 10.18**), are often referred to as slums—areas of degraded housing. The United Nations defines a slum household as a cohabiting group that lacks one or more of the following conditions: an adequate physical structure that protects people from extreme climatic conditions, sufficient living area such that no more than three people share a room, access to a sufficient amount of water, access to sanitation, and secure tenure or protection from forced eviction. Current estimates suggest that over 1 billion people in the world live in slums. In greater Manila alone, for example, scholars estimate that almost half of the city's 19 million inhabitants live in slum conditions within squatter settlements.

Squatter settlements usually begin as collections of crude shacks constructed from scrap materials; gradually, they become increasingly elaborate and permanent. Paths and walkways link houses, vegetable gardens spring up, and often water and electricity are bootlegged into the area so that a common tap or outlet serves a number of houses. At later stages, such economic pursuits as handicrafts and small-scale artisan activities take place in the squatter settlements. In many instances, these supposedly temporary settlements become permanent parts of the city and function as many neighborhoods do—that is, as social, economic, and cultural centers.

Squatter settlements are located not only in downtown areas but also close to places where people work. In many cities, this means places that just a few years ago were considered rural. Because of growing populations and pressures on space, many globalizing cities in the developing world are expanding outward. Some of these cities, such as Mexico City, Hanoi, and Manila, are expanding so rapidly out into the countryside that the city is swallowing up

barriadas
Illegal housing settlements, usually made up of temporary shelters, that surround large cities; often referred to as *squatter settlements*.

Text continues on page 374

FIGURE 10.18 Squatter settlements in Mexico City (*left*) and Kuala Lumpur (*right*). Migration to cities has been so rapid that often illegal squatter settlements have been the only solution to housing problems. *(Cameramann International, Ltd.)*

Subject to Debate

Can Urbanization Be Environmentally Sustainable?

The fact that 2008 marked the year when urban population in the world began to exceed rural population has refueled the debate over the environmental impacts of increasing urbanization. On the one hand, increasing urbanization concentrates population, potentially reducing the regional human *footprint* around cities, and creates economies of scale because more people can benefit from the urban infrastructure—transportation, power supply routes, water servicing, and so on. On the other hand, these same concentrations of population lead to greater demands on resources outside of the city (e.g., water and energy), and the wealth that is often correlated with urban life can lead to consumer demands that in turn impact large swaths of countryside outside the city. Are cities, then, sustainable?

First, we have to think about what *sustainable* actually means, and this issue itself is subject to debate! For our purposes here, sustainable urbanization is the creation of a situation whereby a society can meet the needs of contemporary urban dwellers for water, food, and shelter, while not damaging the ability of future urban dwellers to meet their needs. It was only in the late nineteenth and the twentieth centuries, with industrialization (see Chapter 9), that cities in Europe and North America grew rapidly in size, adding stress to the environment. That stress has been further exacerbated with the recent and rapid growth of cities in the developing world. And it has been this recent and rapid growth that has caused alarm among policy makers and scholars.

On one side of the debate are those who focus on the adverse environmental impacts of large concentrations of people. Not only do cities have a direct impact on the land, in terms of the amount of space they occupy and the pollution they create; they also have indirect impacts, since urban dwellers

typically have higher incomes and expectations in terms of material satisfaction than do rural dwellers, which places additional stress on the environment. More consumer demand for things like meat, wood, and metals from city dwellers in, say, Vancouver, can have adverse environmental impacts throughout Canada and other parts of the Asia-Pacific Rim. Those who are optimistic about creating a more sustainable future with increasing urbanization argue that higher-density human settlements are better for the environment in the long run than are less dense settlements. Even with the current growth rate of cities in terms of size and number, recent estimates based on satellite data show that urban settlements occupy only 2.8 percent of the total land on Earth. Urban spatial expansion per se does not appear to be the major stressor. Urban density actually helps maintain fragile ecosystems by keeping them free from human interference. Proponents of this side of the debate argue that policies geared toward dispersing the world's population are misguided. The real problem, they argue, rests with unsustainable forms of production and consumption, and it is on managing these issues that the world's attention should be focused.

Continuing the Debate

The debate over the environmental sustainability of urbanization is certain to grow in intensity as the world's population continues its urban course. After all, many, many people are drawn to urban life. Given what you have read here, consider the following questions:

- Can an increasingly urban world become sustainable? How?

- Does your city have a sustainability plan? If so, how might you learn more about this plan and its feasibility?

Aerial view of New York City. Because of the high density of its population and its mass transportation system, New York City is considered to be a relatively "green" city. Urban parks, like Central Park depicted here, often provide important sanctuaries for flora and fauna in the heart of the city. *(Panoramic Images/Getty Images.)*

FIGURE 10.19 Starbucks on Istiklal Street in Istanbul. An array of shops and cafés, including several Starbucks coffee shops, line Istiklal Street, one of the primary shopping streets in central Istanbul. *Can you see other signs of globalization in this landscape?* (Courtesy of Mona Domosh.)

extended metropolitan regions (EMRs)
New types of urban regions, complex in both landscape form and function, created by the rapid spatial expansion of cities in the developing world.

smaller villages and rural areas. In this way, these new urban forms are blurring the distinction between the rural and the urban, creating what some scholars have called **extended metropolitan regions (EMRs).** Multinational corporations often rely on the labor of new urban migrants, so the city grows rapidly at the margins as factories are built in export processing zones (EPZs) (see Chapter 9) and squatter settlements emerge around them. At the same time, the downtowns of these cities often experience not dispersal but concentration: high-rises accommodate the regional offices of corporations and the services associated with them (media, advertising, personnel management, and so on), and upper-class housing developments accommodate this new class of white-collar workers.

In addition, landscape forms associated with the global consumer, entertainment, and tourist economy are emerging: American-style shopping districts (with American stores) and new airports, hotels, restaurants, and entertainment facilities (**Figure 10.19**). Many of these landscape developments are funded by international investments. Entrepreneurs and governments in many parts of the world look to these new globalizing cities of the developing world as good places to make financial and real estate investments. Much of the extended metropolitan region of Hanoi, for example, was built with funds from a range of

countries (**Figure 10.20**): an export processing zone and a golf and entertainment facility were partly funded by Malaysia; South Korea invested in a hotel, another golf course, and a business center; Taiwan was involved in a tourism project and an industrial park. Hanoi's downtown (**Figure 10.21,** page 376) is being developed with more hotels and office construction funded by Japan and Singapore. Finland provided the infrastructure for the water supply to one of these new developments.

These new urban regions are complex in both their landscape form and function because they developed rapidly, without any planning, and have in many cases incorporated preexisting places. These EMRs, therefore, include multiple nodes of economic activity, with few of those nodes any older than a decade. Scholars who write about Mexico City, for example, find it impossible to speak of the city as one place; rather, they suggest that it can now only be represented as a pastiche of different places, each with its own center and outlying neighborhoods.

Latin American Urban Landscapes

To describe the landscapes of Latin American cities, urban geographers have developed a model, which is shown in Figure 1.3 (page 4), so you might want to refer back to it as you read this section. In contrast to many contemporary cities in the United States, the central business districts (CBDs)

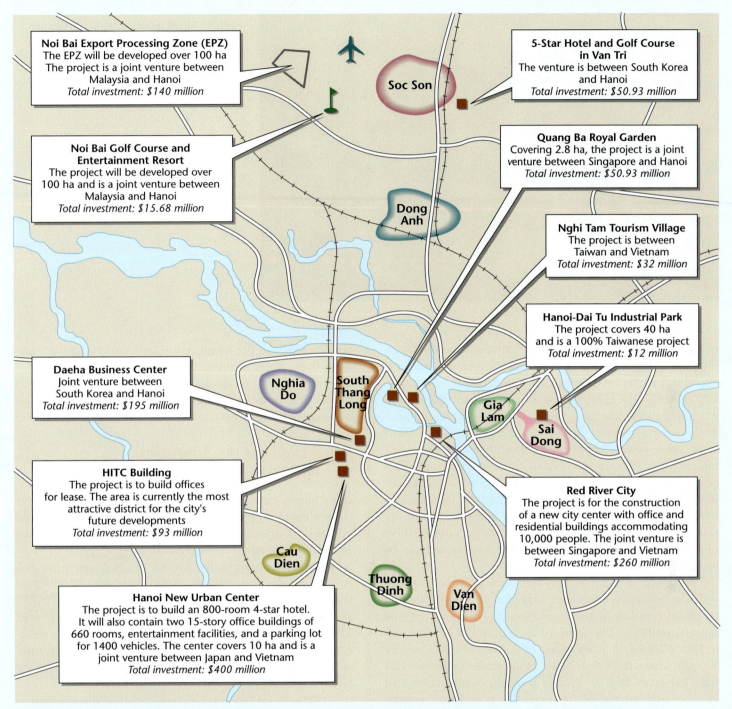

Noi Bai Export Processing Zone (EPZ)
The EPZ will be developed over 100 ha
The project is a joint venture between
Malaysia and Hanoi
Total investment: $140 million

5-Star Hotel and Golf Course in Van Tri
The venture is between South Korea and Hanoi
Total investment: $50.93 million

Noi Bai Golf Course and Entertainment Resort
The project will be developed over 100 ha and is a joint venture between Malaysia and Hanoi
Total investment: $15.68 million

Quang Ba Royal Garden
Covering 2.8 ha, the project is a joint venture between Singapore and Hanoi
Total investment: $50.93 million

Nghi Tam Tourism Village
The project is between Taiwan and Vietnam
Total investment: $32 million

Hanoi-Dai Tu Industrial Park
The project covers 40 ha and is a 100% Taiwanese project
Total investment: $12 million

Daeha Business Center
Joint venture between South Korea and Hanoi
Total investment: $195 million

HITC Building
The project is to build offices for lease. The area is currently the most attractive district for the city's future developments
Total investment: $93 million

Red River City
The project is for the construction of a new city center with office and residential buildings accommodating 10,000 people. The joint venture is between Singapore and Vietnam
Total investment: $260 million

Hanoi New Urban Center
The project is to build an 800-room 4-star hotel. It will also contain two 15-story office buildings of 660 rooms, entertainment facilities, and a parking lot for 1400 vehicles. The center covers 10 ha and is a joint venture between Japan and Vietnam
Total investment: $400 million

Soc Son · Dong Anh · Nghia Do · South Thang Long · Gia Lam · Sai Dong · Cau Dien · Thuong Dinh · Van Dien

Soc Son: Noi Bai International Airport is located in this area. Malaysia is investing in two projects there: a golf course and an EPZ.
Dong Anh: At the moment the area is one of the main vegetable-supplying sources for residents in Hanoi.
Nghia Do: Only five minutes from the West Lake, a property development area.
South Thang Long: Attracts a lot of foreign investment. The infrastructure is good, with the water supply system aided by the Finnish government.
Gia Lam Area: Called the "Daewoo area," it will soon become a satellite city of Hanoi.
Cau Dien: Recently approved project of traditional cultural tourism villages.
Thuong Dinh: Local industrial area.
Van Dien: An industrial park with infrastructure not yet developed. No foreign investment at present.

FIGURE 10.20 **Diagram of Hanoi's extended metropolitan region as it was expanding in the early 21st century.** Foreign investment has been crucial to the spatial expansion of Hanoi, as you can see from the various projects that were under construction. *Is this form of urban expansion different from what you know of cities in North America?* *(Source: Drakakis-Smith, 2000: 24.)*

of Latin American cities are vibrant, dynamic, and increasingly specialized. The dominance of the central district is explained partly by widespread reliance on public transit and partly by the existence of a large and relatively affluent population close to it. Outside the central district, the dominant component is a commercial spine surrounded by an elite residential sector. Because these two zones are interrelated, they are referred to as the spine/sector. This combination is an extension of the CBD down a major boulevard, along which the city's important amenities, such as parks, theaters, restaurants, and even golf courses, are located. Strict zoning and land controls ensure continuation of these activities and protect the elite from incursions by low-income squatters.

Somewhat less prestigious is the inner-city zone of maturity, a collection of homes occupied by people unable to afford housing in the spine/sector. This is an area of upward mobility. The zone of accretion is a diverse collection of housing types, sizes, and quality, which can be thought of as a transition between the zone of maturity and the next zone. It is an area of ongoing construction and change, emblematic of the explosive population growth that characterizes the Latin American city. Although some neighborhoods within this zone have city-provided utilities, other blocks must rely on water and butane delivery trucks for essential services.

The most recent migrants to the Latin American city are found in the zone of peripheral squatter settlements. This fringe of poor people and inadequate housing contrasts dramatically with the affluent and comfortable suburbs that ring North American cities. Streets are unpaved, open trenches carry waste, residents haul water from distant locations, and electricity is often pirated by attaching illegal wires to the closest utility pole. Although this zone's quality of life seems marginal, many residents transform these squatter settlements over time into permanent neighborhoods with minimal amenities (see the section on globalizing cities in the developing world for a discussion of squatter settlements).

Landscapes of the Apartheid and Postapartheid City

As we have seen in previous chapters, racism and the residential segregation that often results from it can have profound effects on landscapes. In South Africa, the state-sanctioned policy of segregating "races," known as **apartheid,** significantly altered urban patterns. Although racial segregation was not the only force shaping the apartheid city, it certainly was a dominant one. The intended effects of this policy on urban form are delineated in **Figure 10.22.** To understand this illustration fully, we need to outline some of the important components of the apartheid state.

> **apartheid**
> In South Africa, a policy of racial segregation and discrimination against non-European groups.

The policies of economic and political discrimination against non-European groups in South Africa were formalized and sharpened under National Party rule after 1948. To segregate the "races," the government passed two major pieces of legislation in 1950. The first was the Population Registration Act, which mandated the classification of the population into discrete racial groups. The three major groups were white, black, and colored, each of which was subdivided into smaller categories. The second piece of major legislation was called the Group Areas Act; its goal was, in the words of geographer A. J. Christopher, "to effect the total urban segregation of the various population groups defined under the Population Registration Act."

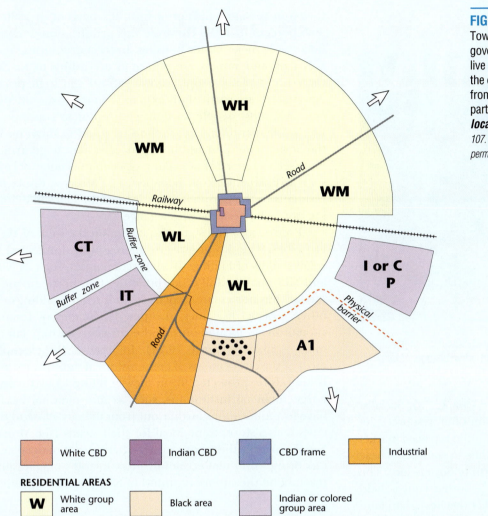

FIGURE 10.22 **The model apartheid city.**
Townships were areas set aside by the
government for members of nonwhite groups to
live in, often located close to industrial areas of
the city. Hostels were built to house black men
from the rural areas who were needed to work in
particular industries. *Why are nonwhite groups
located close to industry?* (From Christopher, 1994:
107. Copyright © 1994 by A. J. Christopher. Reprinted by
permission of Routledge.)

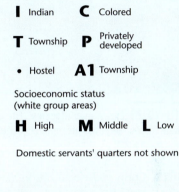

I Indian   **C** Colored

T Township   **P** Privately developed

• Hostel   **A1** Township

Socioeconomic status
(white group areas)

H High   **M** Middle   **L** Low

Domestic servants' quarters not shown

White CBD	Indian CBD	CBD frame	Industrial

RESIDENTIAL AREAS

W White group area	Black area	Indian or colored group area

Cities were thus divided into sections that were to be inhabited only by members of one population group.

Although the effects of these acts on the form of South African cities did not appear overnight, they were massive nonetheless. Members of nonwhite groups were by far the ones most adversely affected. Almost without exception, the downtowns of cities were restricted to whites, whereas those areas set aside for nonwhites were peripheral and restricted, often lacking any urban services, such as transportation and shopping. Large numbers of nonwhite families were displaced with little or no compensation (estimates suggest that only 2 percent of the displaced families were white). Buffer zones were established between residential areas to curtail contact between groups, further hindering access to the central city for those groups pushed into the periphery. In effect, this created a city similar to that described by the sector model (see Chapter 11), only here it was organized along racial lines.

After the change of government in 1994, apartheid came to an end as an official policy, and the laws that enforced racial segregation in South Africa were abolished.

Results from the 1996 census revealed a marked decline in degrees of racial segregation in South Africa's cities, a trend that characterizes the postapartheid city. However, historical patterns of residential segregation and racial prejudice are very difficult to change. A. J. Christopher found that despite a general overall decline in levels of segregation, particularly among the colored population, the majority of the white population in South Africa remained isolated from their nonwhite neighbors, while the black population remained relatively isolated due to the lack of mobility caused by poverty.

Landscapes of the Socialist and Postsocialist City

The effects of centralized state policies on urban land-use patterns are also evident in the cities of the countries that formed the Soviet Union and in other formerly socialist countries in eastern Europe. In the Soviet Union, for example, the Bolshevik Revolution of 1917 brought with it attempts by the state to confront and solve the urban problems

FIGURE 10.23 Downtown Moscow. The post-Soviet city, rekindled by market forces, demonstrates unprecedented urban change. *(Courtesy of Mona Domosh.)*

attendant upon industrialization (see Chapter 9). Socialist principles called for the nationalization of all resources, including land. Further, centralized planning replaced market forces as the means for allocating those resources. These ideas had profound effects on the form of Soviet cities. The principles that underlie capitalist cities—that land is held privately and that the economic market dictates urban land use—were no longer the prime influences on urban form. Instead, Soviet policies attempted to create a more equitable arrangement of land uses in the city. The results of those policies included (1) a relative absence of residential segregation according to socioeconomic status, (2) equitable housing facilities for most citizens, (3) relatively equal accessibility to sites for the distribution of consumer items, (4) cultural amenities (theater, opera, and so on) located and priced to be accessible to as many people as possible, and (5) adequate and accessible public transportation. Although these results created better living and working conditions for many people, the situation was far from ideal. By the 1970s and 1980s, many Soviet citizens realized that their standards of living were well below those in the West and that the centralized planning system was not successful.

National policies of economic restructuring introduced in the late 1980s, referred to as *perestroika,* led to mandates to privatize resources and the land market. In the postsocialist city, market forces are once again dominant in shaping urban land uses, and the pace and scale of urban change are unprecedented (**Figure 10.23**). One of the most significant of those changes is the privatization of the housing market. In Moscow, for example, the percentage of the housing stock that is in private hands grew from 9.3 percent in 1990 to 49.6 percent in 1994. However, this privatization does not necessarily mean better housing. In the new housing market, flats and homes are allocated according to market forces, and many people cannot afford the high prices. Apartments are particularly expensive in the center of Moscow, and most people have no choice but to continue to live in the communal apartments assigned to them under the Soviet system. In addition, processes akin to gentrification (see Chapter 11) are taking place in the center of cities such as Prague, Cracow, Budapest, and Moscow, displacing residents to the peripheral portions of the city while housing others in Western-style elegance.

Postsocialist cities are also taking on the look of Western cities. The downtowns are increasingly dominated by retail outlets of familiar Western companies, such as Nike and McDonald's. Tall office buildings that house financial service businesses are replacing industrial buildings. In other postsocialist cities, the new landscapes of finance capitalism are located outside the city center, where land is more available (**Figure 10.24**). In Prague, for example, most of the city center is of great historical significance and is therefore protected from new-building construction. Large financial institutions, such as international banks, have found locations farther out from the city, close to the city's metro stops, to be ideal for their needs (see Mona's Notebook and the photo in it). These new financial centers bring with them economic growth, and some are now resembling the suburban sprawl of North American cities.

FIGURE 10.24 A view across the Danube River from downtown Bratislava, Slovak Republic. The Soviet-era "UFO" bridge is rivaled on the skyline by glass skyscrapers housing new commercial and financial global companies. In the background is Petrzalka, a huge area of high-rises built in the 1970s that is home to many of Bratislava's middle-class citizens. *(Courtesy of Mona Domosh.)*

Mona's Notebook

Finding FedEx in a Most Unusual Place

Mona Domosh

Bank buildings clustered near a metro stop in Prague. Since most of downtown Prague is protected from new development because of its historical significance, the many financial institutions that want to locate in this burgeoning capitalist nation have constructed large skyscrapers to house their offices on the edge of the city, grouped around the metro stops. Contrast this to the location of bank office buildings in North American cities. *(Courtesy of Mona Domosh.)*

I found myself taking the Prague metro line C to the Budejŏvická station (near where I took the photo) for a simple reason, but what I found out by so doing was extremely (and unexpectedly) interesting! I was in Prague for an extended work visit and needed to ensure that some papers were delivered to New York City pronto. So I looked online for the closest FedEx office, got out my Prague transit map, and realized I needed to take the metro five stops out of the city center, and then take a short walk to the main office. Fine. But then I suddenly realized the strangeness of this: FedEx offices are almost always located close to the financial/commercial centers of cities, but this office was far outside the city center, in what I thought was a suburban residential area. I was puzzled, since all my geographical instincts and training told me this was the wrong location.

I boarded the train with trepidation, but sure enough the FedEx office was precisely where the online search had sent me—in a somewhat suburban residential area. But (and this is a big "but") it was also a neighborhood of tall glass buildings, most housing bank offices and all located within a block or so of the metro station. What was happening here? I *had* to investigate. I put together what I knew about the development of capitalist cities and socialist/postsocialist cities with a particular knowledge of the history of Prague, and I talked to Czech geographers. Here's what I found out: In most of the cities that I am familiar with (North American, western European), the heyday of the construction of large, high-rise financial office buildings was in the post–World War II era. During that time in socialist cities like Prague, investment in capitalist endeavors like banks just did not happen, for obvious reasons. It wasn't until the postsocialist era that large financial institutions emerged, and by then Prague's central city had been designated as a historical landmark, leaving little space for large-scale office construction. The solution for these companies was to build their offices in accessible locations outside the city center. Hence my explanation for Budějovická and my explanation for why FedEx was located there. FedEx was following the same principle of locating near commercial/financial centers, it was just that the center was not where I expected. It took me a while, but my little journey to Budějovická taught me a lot.

Posted by Mona Domosh

CONCLUSION

The first cities arose as new technologies—particularly the domestication of plants and animals—facilitated the concentration of people, wealth, and power in a few specific places. This transformation from village to city life was accompanied by new social organizations, a greater division of labor, and increased social stratification. These characteristics still distinguish rural from urban life. Although the first cities developed in specific hearths, urban life has now diffused worldwide, and all indications are that the majority of the Earth's population has become urban.

This increasing urbanization, however, leads to a host of problems, some environmental, some social, and some a bit of both. Many problems of globalizing cities in the developing world are the result of rapid and massive population growth and concentration. Such cities are bursting at the seams as thousands of new migrants crowd into them each day, seeking jobs, housing, and schooling. Because jobs are scarce, unemployment rates are often very high. Housing is also a problem. In some cities, over a third of the population lives in hastily constructed squatter settlements. But even many of these problems have historical roots, as an examination of the political and social history of colonial cities demonstrates. For example, massive rural-to-urban migration is not a new phenomenon. The disruptions caused by nineteenth-century colonial settlements deprived many people of their land and forced rural inhabitants into the city. This pattern (with modified causes) continues today.

The future of the world's cities is uncertain. Strong governmental planning measures might alleviate many present-day ills, but the long-range hope lies with decreased population growth and increased economic opportunities. Whether this is possible under contemporary conditions remains to be seen.

DOING GEOGRAPHY

Connecting Urban Population Growth with Globalization

As we've suggested in this chapter, most cities today are global in the sense that they are being shaped by global economic, cultural, and political forces. But sometimes it is difficult to actually see in what ways this is happening or in what ways this matters. In this exercise, you will focus on one simple measure of urbanization—population growth—in order to analyze how it is being affected by globalization.

Steps to Connecting Urban Population Growth with Globalization

Step 1: Choose one city from the list of megacities in Table 10.1 (on page 355).

Step 2: To provide context, track its population change over the past 50 years and determine its growth rate every decade (you will probably need to use online data, from the United Nations Statistics Division or other sources: http://unstats.un.org/unsd/citydata/).

Step 3: Concentrating on the past 20 years, see if you can determine the sources of the city's population growth. In other words, are people migrating to the city from rural areas? From other countries? Or is it natural population growth?

Step 4: Link the population growth of this city to global changes. Are people moving to this city to work for transnational companies? Are they moving because their rural homes are experiencing decline due to environmental degradation? Perhaps they must leave their homes in one country due to uneven global development?

As you work through this exercise, think about the ways in which population shifts are shaping the cultures of this city.

A view from Mumbai, India, looking south toward the downtown area. The squatter settlement in the foreground sits very close to expensive high-rises and new construction. *What could explain this apparent contradiction?* (Courtesy of Mona Domosh.)

Key Terms

The City on the Internet

You can learn more about the city in time and space on the Internet at the following web sites:

Globalization and World Cities

http://www.lboro.ac.uk/gawc/

A fantastic web site that details the work of the globalization and world cities research group and network, including data sets that document relationships among 100 world cities and recent publications of the group.

United Nations Statistics Division

http://www.un.org/Depts/unsd/

The source for world population statistics, this site contains a comprehensive section on urbanization as a social indicator.

U.S. Department of Housing and Urban Development

http://www.hud.gov

Here you can find information about housing issues and urban economic development and about how to get involved personally in your own local community.

Sources

Agnew, John A., and James S. Duncan (eds.). 1989. *The Power of Place: Bringing Together Geographical and Sociological Imaginations.* Boston: Unwin Hyman.

Bater, James H. 1980. *The Soviet City: Ideal and Reality.* Beverly Hills: Sage Publications.

Bater, James H. 1996. *Russia and the Post-Soviet Scene.* New York: Wiley.

Christaller, Walter. 1966. *The Central Places of Southern Germany.* C. W. Baskin (trans.). Englewood Cliffs, N.J.: Prentice-Hall.

Christopher, A. J. 1994. *The Atlas of Apartheid.* London: Routledge.

Christopher, A. J. 2001. "Urban Segregation in Post-Apartheid South Africa." *Urban Studies* 38: 449–466.

Cosgrove, Denis. 1984. *Social Formation and Symbolic Landscape.* London: Croom Helm.

Daniell, Jennifer, and Raymond Struyk. 1997. "The Evolving Housing Market in Moscow: Indicators of Housing Reform." *Urban Studies* 34: 235–254.

de Sherbinin, Alex, Andrew Schiller, and Alex Pulsipher. 2007. "The Vulnerability of Global Cities to Climate Hazards." *Environment and Urbanization* 19: 39–64.

Drakakis-Smith, David. 1987. *The Third World City.* London: Methuen.

Drakakis-Smith, David. 2000. *Third World Cities,* 2nd ed. London: Routledge.

Encyclopaedia Britannica. 1984. "Babylon." *New Encyclopaedia Britannica,* vol. 2. Chicago: Encyclopaedia Britannica.

Fowler, Melvin. 1975. "A Pre-Columbian Urban Center on the Mississippi." *Scientific American* (August): 93–102.

Griffin, Ernst, and Larry Ford. 1983. "Cities of Latin America," in Stanley Brunn and Jack Williams (eds.), *Cities of the World: World Regional Urban Development.* New York: Harper & Row, 199–240.

Huang, Yefang, Yee Leung, and Jianfa Shen. 2007. "Cities and Globalization: An International Perspective." *Urban Geography* 28: 209–231.

Mumford, Lewis. 1961. *The City in History.* New York: Harcourt Brace Jovanovich.

Population Division of the United Nations Secretariat. 2007. *World Urbanization Prospects: The 2007 Revision Population Database.*

Portes, Alejandro. 1977. "Urban Latin America: The Political Condition from Above and Below," in Janet Abu-Lughod and Richard Hag, Jr. (eds.), *Third World Urbanization.* New York: Methuen, 59–70.

Pounds, Norman J. G. 1969. "The Urbanization of the Classical World." *Annals of the Association of American Geographers* 59: 135–157.

Robinson, Jennifer. 1997. "The Geopolitics of South African Cities: States, Citizens, Territory." *Political Geography* 16: 365–386.

Samuels, Marwyn S., and Carmencita Samuels. 1989. "Beijing and the Power of Place in Modern China," in John A. Agnew and James S. Duncan (eds.), *The Power of Place.* Boston: Unwin Hyman, 202–227.

Sassen, Saskia. 1991. *The Global City: New York, London, Tokyo.* Princeton, N.J.: Princeton University Press.

Schneider, A., M.A. Friedl, and D. Potere. 2009. "A New Map of Global Urban Extent from MODIS Satellite Data." *Environmental Research Letters* 4 (October–December), doi:10.1088/1748-9326/4/4/044003.

Simon, D. 1984. "Third-World Colonial Cities in Context: Conceptual and Theoretical Approaches with Particular Interest to Africa." *Progress in Human Geography* 8: 493–514.

United Nations, Department of Economic and Social Affairs, Population Division. 2009. *World Urbanization Prospects, The 2009 Revision.*

United Nations Human Settlements Programme (UN-HABITAT). 2010. *State of the World's Cities 2010/11: Cities for All: Bridging the Urban Divide.*

SEEING GEOGRAPHY Rio de Janeiro

What are some of the major environmental and social impacts of an increasingly urbanized world?

A view of Rio de Janiero, Brazil, looking toward Sugarloaf Mountain.

Few cities boast such a spectacular site as Rio de Janeiro, located between the mountains and Guanabara Bay along the Atlantic Ocean. This view highlights the dramatic siting: the bustling city, the beautiful beaches, the rugged mountain peaks against the sky. One can barely discern in this image the "other" side of Rio: the favelas, or squatter settlements, located on the mountainsides (see Figure 10.2, page 354). Perched above the centers of economic activity and the middle- and upper-class residential areas located along the southern coastal areas of Rio, the favelas are an ever-present part of the urban landscape and are home to approximately one-fifth of the city's population.

The information in this chapter should help us to "read" this image of Rio de Janeiro. As a city of the developing world, Rio experienced rapid population growth in the twentieth century, and its metropolitan area now exceeds 11 million people. More striking is the dramatic increase in the urban population of Brazil. The percentage of the population living in metropolitan areas rose from approximately 31 percent in 1940 to about

84.2 percent in 2010. In Rio, that population growth is evident in the intensity of land use within the city, which is marked by the presence of skyscrapers, the metropolitan sprawl that extends well beyond the parameters of this image, and the presence of the favelas, home to many of the rural-to-urban migrants. Like other cities of the developing world, population growth has strained the city's infrastructure and its ability to provide services, which has led to traffic congestion, pollution, and crime. It has also exacerbated ecological problems. When vegetation covered the hillsides, the heavy summer rains Rio experiences were absorbed into the soil. Now the summer rains often flood the streets of the low-lying areas of the city and lead to landslides on the slopes that house the favelas. One such episode in 1996, for example, led to the death of 71 people, with approximately 2000 people left homeless. Yet Rio, like its larger neighbor 230 miles (370 kilometers) to the south, São Paulo, is now experiencing much slower population growth as a result of lower birthrates and less rural-to-urban migration.

Founded as a Portuguese colonial city in 1565, Rio grew quickly in the eighteenth century, when it became the primary port for exporting the gold and diamonds discovered in the interior of the country. Later, in the first decades of the twentieth century, Rio underwent industrialization. The southern and coastal portions of the city became home to the elite, while the factories and working classes moved north and west of the downtown.

Today, Rio is part colonial city and part global city. It is home to the regional headquarters of 10 multinational firms. Its beaches, particularly Ipanema and Copacabana, are icons for global jet-setters. Yet its favelas, some of which have now been recognized by the government as legal communities, continue to grow, with little infrastructure and few public services. Like other globalizing cities, it experiences both the bright lights and the grimmer realities of the twenty-first-century economic order.

United Nations Population Fund. 2007. *State of the World Population 2007: Unleashing the Potential of Urban Growth.* Online at: http://www.unfpa.org/swp/2007/english/introduction.html

Wheatley, Paul. 1971. *The Pivot of the Four Quarters.* Chicago: Aldine.

Wittfogel, Karl. 1957. *Oriental Despotism: A Comparative Study of Total Power.* New Haven, Conn.: Yale University Press.

Yeoh, Brenda S. A. 1999. "Global/Globalizing Cities." *Progress in Human Geography* 23: 607–616.

Ten Recommended Books on Urban Geography

(For additional suggested readings, see *The Human Mosaic* web site: www.whfreeman.com/domosh12e)

Brenner, Neil, and Roger Keil (eds.). 2006. T*he Global Cities Reader.* New York: Routledge. A rich collection of essays that explore the rise of the concept of the global city and examine a range of case studies of global cities throughout the world.

Davis, Mike. 2007. *Planet of Slums.* New York: Verso. A devastating overview of people's everyday struggles to maintain decent lives in the world's urban slums.

Harvey, David. 1985. *The Urban Experience.* Baltimore: Johns Hopkins University Press. A study of the relationship between capitalist economics and the cities it produces.

King, Anthony D. 2004. *Spaces of Global Culture: Architecture, Urbanism, Identity.* New York: Routledge. A fascinating examination of transnational urban forms.

Krueger, Rod, and David Gibbs (eds.). 2010. *The Sustainable Development Paradox: Urban Political Economy in the United States and Europe.* New York: Guilford. An incisive overview of the politics of sustainability within urban contexts, explored through a series of case studies.

Legates, Richard T., and Frederic Stout (eds.). 1999. *The City Reader,* 2nd ed. New York: Routledge. An extensive edited collection of readings covering the evolution of cities and the contemporary forces that are restructuring them.

Olds, Kris. 2001. *Globalization and Urban Change: Capital, Culture, and Pacific Rim Mega-Projects.* New York: Oxford University Press. An in-depth examination of how globalization actually operates in terms of large-scale urban developments in Vancouver and Shanghai.

Sassen, Saskia (ed.). 2003. *Global Networks, Linked Cities.* New York: Routledge. A collection of essays that examine the emerging networks of global commerce and communication that are reshaping the world's cities.

Smith, Michael. 2000. *Transnational Urbanism: Locating Globalization.* New York: Blackwell. An examination of how and why transnational linkages exist between a diverse array of cities.

Taylor, Peter. 2003. *World City Network: A Global Urban Analysis.* New York: Routledge. An empirically rich accounting of the myriad commercial and financial connections between cities that form a global network.

Journals in Urban Geography

Environment and Urbanization. Published by Sage, Beverly Hills. Volume 1 appeared in 1989.

International Journal of Urban and Regional Research. Published by Edward Arnold, London. Volume 1 appeared in 1987.

Urban Geography. Published by Bellwether Press, Lanham, Md. Volume 1 appeared in 1976.

Urban Studies. Published by Routledge, New York. Volume 1 appeared in 1964.

How has globalization affected urban ethnic neighborhoods?

Go to "Seeing Geography" on page 426 to learn more about this image.

INSIDE THE CITY
A Cultural Mosaic

Finding and understanding patterns in a city is a difficult matter. As you walk or drive through a city, its intricacy may dazzle you and its form can seem chaotic. It is often hard to imagine why city functions are where they are, why people cluster where they do. Why does one block have high-income housing and another, slum tenements? Why are ethnic neighborhoods next to the central business district? Why does the highway run through one neighborhood and around another? Just when you think you are beginning to understand some patterns in your city, you note that those patterns are swiftly changing. The house you grew up in is now part of the business district. The central city that you roamed as a child looks abandoned. A suburban shopping center thrives on what was once farmland.

Chapter 10 focused on cities as points in geographical space. In this chapter, we try to orient ourselves within cities to gain some perspective on their spatial patterns. In other words, the two chapters differ in scale. Chapter 10 presented cities from afar, as small dots diffusing across space and interacting with one another and with their environment. In this chapter, we use the five themes of human geography to study the city as if we were walking its streets.

Region

How are areas within a city spatially arranged? Much of the fascination with urban life comes from its diversity, from the excitement created by different groups of people and different types of activities packaged in a fairly small area. Yet within this diversity, it is possible to discern regional patterns, for cities are composed of a series of districts, each of which is defined by a particular set of land uses.

Downtowns

In the center of the typical city is the **central business district (CBD),** a dense cluster of offices and shops. The CBD is formed around the point within the city that is most accessible. As such, businesses and services located in the CBD experience the most "action" as measured by the volume of people, money, and ideas moving through this space. Competition for this space often leads to the construction of skyscrapers, creating a skyline that characterizes and symbolizes a city's CBD and the activities that are often located there: financial services, corporate headquarters, and related services such as advertising and public relations firms. The tallest buildings in the world are located in Asian cities such as Shanghai, Kuala Lumpur, and Taipei—cities that are vying to be major financial centers (**Figure 11.1**, page 386). Just beyond the skyscrapers are often four- and five-story buildings that comprise the

> **central business district (CBD)**
> The central portion of a city, characterized by high-density land uses.

FIGURE 11.1 **Skyline of Shanghai, China.** The skyline is a clear marker of the global importance of Shanghai as a new financial center. Compare these very recent skyscrapers to those of an American or Canadian city. *(Photographer's Choice/Getty.)*

city's main shopping district, traditionally centered around several department stores. Also within the CBD are concentrations of smaller retail establishments, transportation hubs such as railroad stations, and often civic centers such as a city hall and main library. Surrounding the CBD is a transitional zone, situated between the core commercial area and the outlying residential areas. It is a district of mixed land uses, characterized by older residential buildings, warehouses, small factories, and apartment buildings.

Residential Areas

Beyond the transitional zone are various types of residential communities, or regions. Geographers have studied these culture regions in depth, trying to discern patterns in the distribution of diverse peoples. Some have focused on the idea of a **social culture region:** a residential area characterized by socioeconomic traits, such as income, education, age, and family structure (**Figures 11.2** and **11.3**). Other researchers, who use the notion of **ethnic culture region,** highlight traits of ethnicity, such as language and migration history. Obviously the two concepts overlap, because social regions can exist within ethnic regions and vice versa. In addition, some researchers

social culture region
An area in a city where many of the residents share social traits such as income, education, and stage of life.

ethnic culture region
An area occupied by people of similar ethnic background who share traits of ethnicity, such as language and migration history.

FIGURE 11.2 **An inner-city neighborhood near the Capitol in Washington, D.C.** One of the most pressing problems facing the United States is reversing the continuing decay of inner cities. *What factors have led to this decay of the inner city?* (For help, look at the section on suburbanization and decentralization on pages 394–397.) *(Cameramann International, Ltd.)*

FIGURE 11.3 **Middle-income neighborhood in Reston, Virginia.** Social areas within the city can be delimited by certain traits taken from the census, such as income, education, or family size. *How would the social characteristics of this neighborhood differ from those in Figure 11.2?* (Cameramann International, Ltd.)

treat both social and ethnic culture regions as functions of the political and economic forces underlying and reinforcing residential segregation and discrimination. (More information on ethnic areas is found in Chapter 5.)

One way to define social culture regions is to isolate one social trait, such as income, and plot its distribution within the city. The U.S. Census is a common source of such information because the districts used to count population, called **census tracts,** are small enough to allow the subtle texture of social regions to show. These maps of social traits form the basis of many urban land-use models, which we discuss later in this section.

census tracts
Small districts used by the U.S. Census Bureau to survey the population.

Ethnicity is another characteristic that often distinguishes one urban residential area from another (see Chapter 5 for a fuller discussion of ethnicity). This is not surprising, given the history of immigration to North America and the propensity of immigrants to move to cities. During the middle to late nineteenth century, when waves of immigrants left eastern and southern Europe, North American port cities were a common destination, and employment as laborers in factories was the norm. With limited affordable housing available, most immigrants settled close to one another, forming ethnic communities or pockets within the mosaic of the city. These ethnic urban regions—with names such as Little Italy and Chinatown—often allowed the immigrants to maintain their native languages, holidays, foodways, and religions. Although the names of these urban enclaves still exist today, most of them have been transformed by new and different waves of immigrants arriving in North America in the past 20 years (see the section on globalization). New York's Little Italy, for example, is now home to people from East and Southeast Asia, with Italian restaurants vying for space with noodle houses.

Reflecting on Geography

As we look at these urban regions, various questions concern the geographer: How do ethnic and social regions differ? Why do people of similar social traits cluster together? What subtle patterns might be found within these regions?

Social culture regions are not merely statistical definitions. They are also areas of shared values and attitudes, of interaction and communication. **Neighborhood** is a concept often used to describe small social culture regions where people with shared values and concerns interact daily. For example, if we consider only census figures, we might find that parents between 30 and 45 years of age, with two or three children, and earning between $65,000 and $90,000 a year cover a fairly wide area in any given city. Yet, from our own observations, we know intuitively that this broad social area is probably composed of numerous neighborhoods where people associate a sense of community with a specific locale.

neighborhood
A small social area within a city where residents share values and concerns and interact with one another on a daily basis.

A conventional sociological explanation for neighborhoods is that people of similar values cluster together to reduce social conflict. Where a social consensus exists about such mundane issues as home maintenance, child rearing, acceptable behavior, and public order, there is little daily worry about these matters. People who deviate from this consensus will face social coercion that could force them to seek residence elsewhere, thus preserving the values of the neighborhood.

However, many neighborhoods have more heterogeneity than this traditional definition would allow. Consequently, the current understanding of neighborhoods is more flexible. While it embraces traditional components of locality, such as political outlook and shared economic

FIGURE 11.4 Woodside, Queens, New York City. Once home to Irish immigrants, Woodside is now the destination of immigrants from a wide range of countries. *In what ways are these new ethnic neighborhoods different from those of the Irish-Americans? In what ways do you think they are similar?* (James Estrin/NYT Pictures.)

characteristics, it also emphasizes the consensus that comes from the perception of both insiders and outsiders that a certain area is a neighborhood. So although a neighborhood might be ethnically and socially diverse, its residents may think of themselves as a community that shares similar political concerns, holds neighborhood meetings to address these problems, and achieves recognition at city hall as a legitimate group with political standing. For example, the neighborhood of Woodside in Queens has created a sense of community from a very diverse population base (**Figure 11.4**). For much of the late nineteenth century and into the late twentieth century, Woodside was predominantly an Irish-American community, but recent immigration has created a diverse neighborhood. In 2006, 46 percent of its residents were foreign-born, most of them from China and Latin America. In struggling for decent schools, housing, and jobs, many residents have united and forged a coalition of interests and a sense of communal identity—keystones of a neighborhood.

Homelessness

Neighborhoods are usually composed of people who have access to a permanent or semipermanent place of residence. In the cities of the United States, however, homelessness, and with it the loss of neighborhood, is increasingly common. It is nearly impossible to determine the exact number of homeless people in the United States because definitions of **homelessness** vary. For example, does living in a friend's house for

homelessness
A temporary or permanent condition of not having a legal home address.

more than a month constitute a homeless condition? How permanent does a shelter have to be before it is considered a "home"? To some people, home connotes a suburban middle-class house; to others, it simply refers to a room in a city-owned shelter.

Reflecting on Geography

Why is home such a difficult concept to define? How does it differ from the concept of a house?

In addition, homeless people are often not counted in the census or other population surveys. Estimates of the number of homeless, therefore, are only rough approximations. Recent studies suggest that there are up to 3 million homeless people in the United States, concentrated in the downtown areas of large cities, often in what we call the transitional zone (see the next section on models of the internal structure of American and Canadian cities).

The causes of homelessness are varied and complex. Many homeless people suffer from some type of disorder or handicap that contributes to their inability to maintain a job and obtain adequate housing. Most have been marginalized in some way by the economic problems that have plagued the United States since the early 1980s, and therefore they have been left out of the housing market. Deprived of the social networks that a permanent neighborhood provides, the homeless are left to fend for themselves. Most cities have tried to provide temporary shelter, but many homeless people prefer to rely on their own social ties for support in order to maintain some sense of personal pride and privacy. In a study of the Los Angeles

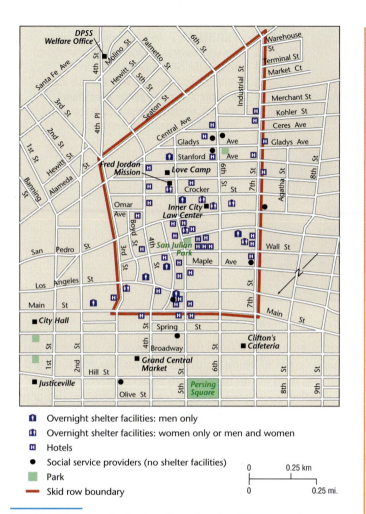

Overnight shelter facilities: men only

🏠 Overnight shelter facilities: men only

🏠 Overnight shelter facilities: women only or men and women

🅷 Hotels

● Social service providers (no shelter facilities)

🟩 Park

━━ Skid row boundary

0   0.25 km

0   0.25 mi.

FIGURE 11.5 The distribution of services for the homeless in the skid row district of Los Angeles. The population of the area is difficult to estimate, ranging from 6000 to 30,000. There are approximately 2000 shelter beds in the area, half of which are available to women. Single-room-occupancy hotels provide about 6700 units of longer-term housing. More than 50 social service programs are run through agencies, missions, and shelters. Love Camp and Justiceville are the sites of informal street encampments of homeless people. *(After Rowe and Wolch, 1990.)*

skid row district, Stacy Rowe and Jennifer Wolch explored how homeless women formed new types of social networks and established a sense of community to cope with the day-to-day needs of physical security and food (**Figure 11.5**). This study points to the importance of social ties in maintaining personal identity and helps us understand the magnitude of a problem that deprives people of their home and neighborhood.

Models of the Internal Structure of American and Canadian Cities

Beginning in the 1920s, urban geographers in the United States began to recognize and create models to describe

the ways that central business districts and residential areas are located in relation to each other. They examined these relationships in a range of North American cities and ended up devising three different spatial models of urban land use. Below we describe each of these models and discuss what each has to offer in terms of describing land use within cities. We then discuss some of the important critiques of these models that highlight their limitations.

Concentric-Zone Model The **concentric-zone model** was developed in 1925 by Ernest W. Burgess, a sociologist at the University of Chicago. **Figure 11.6** shows the concentric-zone model with its five zones. At first glance, you can see the effects of residential decentralization. There is a distinct pattern of income levels from zone 1, the CBD, out to the commuter residential zone. This pattern shows that even at the beginning of the automobile age, American cities reflected a clear separation of social groups. The extension of trolley lines into the surrounding countryside had a lot to do with this pattern.

> **concentric-zone model** A social model that depicts a city as five areas bounded by concentric rings.

Zone 2, a transitional area between the CBD and residential zone 3, is characterized by a mixed pattern of industrial and residential land use. Rooming houses, small apartments, and tenements attract the lowest-income segment of the urban population. Often this zone includes slums and skid rows. In the past, many ethnic ghettos took root here as well. Landowners, while waiting for the CBD to reach their land, erected shoddy tenements to house a massive influx of foreign workers. An aura of uncertainty was characteristic of life in zone 2, because commercial activities rapidly displaced residents as the CBD expanded. Today, this area is often characterized by physical deterioration (**Figure 11.7**, page 390).

Zone 3, the "workingmen's quarters," is a solid blue-collar arc, located close to the factories of zones 1 and 2.

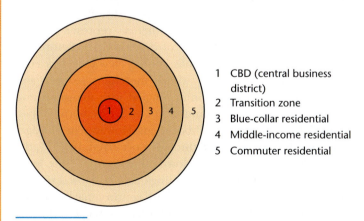

1 CBD (central business district)
2 Transition zone
3 Blue-collar residential
4 Middle-income residential
5 Commuter residential

FIGURE 11.6 The concentric-zone model. Each zone represents a different type of land use in the city. ***Can you identify examples of each zone in your community?***

FIGURE 11.7 **An abandoned building in the uptown area of Chicago.** The transitional zone in the city contains vacant and deteriorated buildings. ***Why has this area become a likely target for gentrifiers?*** (See the section on gentrification on pages 398–401.) *(Cameramann International, Ltd.)*

Yet zone 3 is more stable than the zone of transition around the CBD. It is often characterized by ethnic neighborhoods: blocks of immigrants who broke free from the ghettos in zone 2 and moved outward into flats or single-family dwellings. Burgess suggested that this working-class area, like the CBD, was spreading outward because of pressure from the zone of transition and because blue-collar workers demanded better housing.

Zone 4 is a middle-class area of better housing. From here, established city dwellers—many of whom moved out of the central city with the construction of the first streetcar network—commute to work in the CBD.

Zone 5, the commuters' zone, consists of higher-income families clustered together in suburbs, either on the farthest extension of the trolley or on commuter railroad lines. This zone of spacious lots and large houses is the growing edge of the city. From here, the rich press outward to avoid the increasing congestion and social heterogeneity brought to their area by an expansion of zone 4.

Burgess's concentric-zone theory represented the American city in a new stage of development. Before the 1870s, an American metropolis, such as New York, was a city of mixed neighborhoods where merchants' stores and sweatshop factories were intermingled with mansions and hovels. Rich and poor, immigrant and native-born rubbed shoulders in the same neighborhoods. However, in Chicago, Burgess's hometown, something else occurred. In 1871, the Great Chicago Fire burned down the core of the city, leveling almost one-third of its buildings. As the city was rebuilt, it was influenced by late-nineteenth-century market forces: real estate speculation in the suburbs, inner-city industrial development, new streetcar systems, and the need for low-cost working-class housing. The result was more clearly demarcated social patterns than existed in other large cities. Chicago became a segregated city with a concentric pattern working its way out from the downtown in what one scholar called "rings of rising affluence." It was this rebuilt city that Burgess used as the basis for his concentric-zone model.

However, as you can see from **Figure 11.8,** the actual residential map of Chicago does not exactly match the simplicity of Burgess's concentric zones. For instance, it is evident that the wealthy continue to monopolize certain high-value sites within the other rings, especially Chicago's

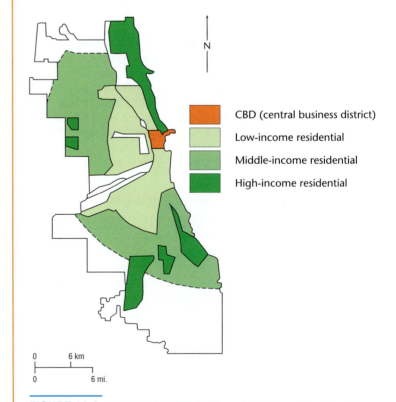

CBD (central business district)

Low-income residential

Middle-income residential

High-income residential

0   6 km

0   6 mi.

FIGURE 11.8 **Residential areas of Chicago in 1920** were used as the basis for many studies and models of the city. Compare this pattern with the concentric-zone and sector models.

"Gold Coast" along Lake Michigan on the Near North Side. According to the concentric-zone theory, this area should have been part of the zone of transition. Burgess accounted for some of these exceptions by noting how the rich tended to monopolize hills, lakes, and shorelines, whether they were close to or far from the CBD. Critics of Burgess's model also were quick to point out that even though portions of each zone did exist in most cities, rarely were they linked in such a way as to totally surround the city. Burgess countered that there were distinct barriers, such as old industrial centers, that prevented the completion of the arc. Still other critics felt that Burgess, as a sociologist, overemphasized residential patterns and did not give proper credit to other land uses—such as industry, manufacturing, and warehouses—in describing urban patterns.

sector model
An economic model that depicts a city as a series of pie-shaped wedges.

Sector Model Homer Hoyt, an economist who studied housing data for 142 American cities, presented his **sector model** of urban land use in 1939. He maintained that high-rent residential districts (*rent* meaning capital outlay for the occupancy of space, including purchase, lease, or rent in the popular sense) were instrumental in shaping the land-use structure of the city. Because these areas were reinforced by transportation routes, the pattern of their development was one of sectors or wedges (**Figure 11.9**), not concentric zones.

Hoyt suggested that the high-rent sector would expand according to four factors. First, a high-rent sector moves from its point of origin near the CBD, along established routes of travel, toward another nucleus of high-rent buildings. That is, a high-rent area directly next to the CBD will naturally head in the direction of a high-rent suburb, eventually linking the two in a wedge-shaped sector. Second, this sector will progress toward high ground or along waterfronts when these areas are not used for industry. The rich have always preferred such environments for their residences. Third, a high-rent sector will move along the route of fastest transportation. Fourth, it will move toward open space. A high-income community rarely moves into an occupied lower-income neighborhood. Instead, the wealthy prefer to build new structures on vacant land where they can control the social environment.

As high-rent sectors develop, the areas between them are filled in. Middle-rent areas move directly next to them, drawing on their prestige. Low-rent areas fill in the remaining areas. Thus, moving away from major routes of travel, rents go from high to low.

There are distinct patterns in today's cities that echo Hoyt's model. He had the advantage over Burgess in that he wrote later in the automobile age and could see the tremendous impact that major thoroughfares were having on cities. However, when we look at today's major transportation arteries, which are generally freeways, we see that the areas surrounding them are often low-rent districts. According to Hoyt's theory, they should be high-rent districts. Freeways are rather recent additions to the city, coming only after World War II, that were imposed on an existing urban pattern. To minimize the economic and political costs of construction, they were often built through low-rent areas, where the costs of land purchase for the rights-of-way were less and where political opposition was kept to a minimum because most people living in these low-rent areas had little political clout. This is why so many freeways rip through ethnic ghettos and low-income areas. Economically speaking, this is the least expensive route.

Multiple-Nuclei Model Both Burgess and Hoyt assumed that a strong central city affected patterns throughout the urban area. However, as cities increasingly decentralized, districts developed that were not directly linked to the CBD. In 1945, two geographers, Chauncey Harris and Edward Ullman, suggested a new model: the **multiple-nuclei model.** They maintained that a city developed with equal intensity around various points, or multiple nuclei (**Figure 11.10**, page 392). In their eyes, the CBD was not the only focus of activity. Equal weight had to be given to an old community on the city outskirts around which new suburban developments clustered; to an industrial district that grew from an original waterfront location; or to a low-income area that developed because of some social stigma attached to the site.

multiple-nuclei model
A model that depicts a city growing from several separate focal points.

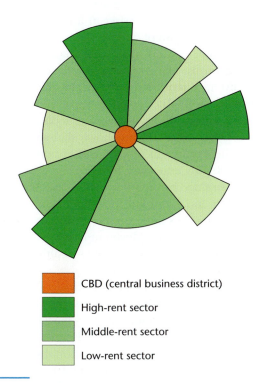

CBD (central business district)

High-rent sector

Middle-rent sector

Low-rent sector

FIGURE 11.9 The sector model. In this model, zones are pie-shaped wedges radiating along main transportation routes.

CBD (central business district)

Light industry and warehouses

Heavy industry

Low-rent residential

Middle-rent residential

High-rent residential

FIGURE 11.10 **The multiple-nuclei model.** This model was devised to show that the CBD is not the sole force in creating land-use patterns within the city. Rather, land-use districts may evolve for specific reasons at specific points elsewhere in the city—hence the name *multiple nuclei*. **Compare this model to Figure 11.9 (page 391). How and why are they different?**

Harris and Ullman rooted their model in four geographical principles. First, certain activities require highly specialized facilities, such as accessible transportation for a factory or large areas of open land for a housing tract. Second, certain activities cluster together because they profit from mutual association. Car dealers, for example, are commonly located near one another because automobiles are very expensive and so people will engage in comparative shopping—moving from one dealer to another until their decisions are made. Third, certain activities repel each other and will not be found in the same area. Examples would be high-rent residences and industrial areas, or slums and expensive retail stores. Fourth, certain activities could not make a profit if they paid the high rent of the most desirable locations and would therefore seek lower-rent areas. For example, furniture stores may like to locate where pedestrian traffic is greatest to lure the most people into their showrooms. However, they need large amounts of space for showrooms and storage. Thus, they cannot afford the high rents that the most accessible locations demand. They compromise by finding an area of lower rent that is still relatively accessible.

The multiple-nuclei model, more than the other models, seems to take into account the varied factors of decentralization in the structure of the North American city. Many geographers criticize the concentric-zone and sector models as being rather simplistic, for they emphasize a single factor (residential differentiation in the concentric-zone model and rent in the sector model) to explain the pattern of the city. But the multiple-nuclei model encompasses a larger spectrum of economic and social factors. Harris and Ullman could probably accommodate the variety of forces working on the city because they did not confine themselves to seeking simply a social or economic explanation. As geographers, they tried to integrate the disparate elements of culture into a workable model. Most urban scholars agree that they succeeded.

Critiques of the Models Most of the criticisms of the models just discussed focus on their simplification of reality or their inability to account for all the complexities of actual urban forms. More recently, feminist geographers have noticed some flaws in the models and in how they were constructed that call into question their descriptive power.

All three models assume that urban patterns are shaped by an economic trade-off between the desire to live in a suburban neighborhood appropriate to one's economic status and the need to live relatively close to the central city for employment opportunities. These models assume that only one person in the family is a wage worker—the male head of the family. They ignore dual-income families and households headed by single women, who contend with a larger array of factors in making locational decisions, including distances to child-care and school facilities and other services important for other members of a family. For many of these households, the traditional urban models that assume a spatial separation of workplace and home are no longer appropriate.

For example, a study of the activity patterns of working parents shows that women living in a city have access to a wider array of employment opportunities and are better able to combine domestic and wage labor than are women who live in the suburbs. Many of these middle-class women will choose to live in a gentrified inner-city location, hoping that this type of area will offer the amenities of the suburbs (good schools and safety) while also accommodating their work schedules. Other research has shown that some businesses will locate their offices in the suburbs because they rely on the labor of highly educated, middle-class women who are spatially constrained by their domestic work. As geographers Susan Hanson and Geraldine Pratt found in their study of employment practices and gender in Worcester, Massachusetts, most women seek employment locations closer to their homes than do men, and this applies to almost all women, not just those with small children.

The traditional models are also criticized for being created by men who all shared certain assumptions about how cities operate and thus presented a very partial view of urban life. Geographers David Sibley and Emily Gilbert, for instance, have both brought to our attention the development of other theories about urban form and structure during the same time. These theories incorporate the alternative perspectives of female scholars. Drawing on the urban reform work done by Jane Addams at Hull House in Chicago, scholars in the first decades of the twentieth century examined the causes of and possible solutions to urban problems. For example, Edith Abbott, Sophonisba Breckinridge, and Helen Rankin Jeter, faculty at the School of Social Service Administration at the University of Chicago, worked with their mostly female students to produce a number of studies about "race," ethnicity, class, and housing in Chicago. These studies differed in several ways from those of such theorists as Burgess and Hoyt. For instance, they emphasized the role of landlords in shaping the housing market and included an awareness of how racism is related to the allocation of housing and a sensitivity to the different urban experiences of ethnic groups.

Much of what these researchers at the School of Social Service uncovered in the 1930s is applicable to urban areas today. For example, a study by urban historian Raymond Mohl chronicles the making of black ghettos in Miami between 1940 and 1960. His research reveals the role of public policy decisions, landlordism, and discrimination in that process—forces identified by Abbott and others that continue to operate today.

Reflecting on Geography

Consider why the insights gained from the studies done by these women have been ignored until recently. How do you think our knowledge of urban life would have been different if these studies had become part of our accepted urban curriculum?

Mobility

How can we understand the spatial movement of people and activities in the city? The patterns of activities we see in the city are the result of thousands of individual decisions about location: Where should we locate our store—in the central city or the suburbs? Where should we live—downtown or outside the city? And so on. The result of such decisions might be expansion at the city's edge or the relocation of activities from one part of the city to another. The cultural geographer looks at such decisions in terms of expansion and relocation diffusion (see Chapter 1).

To understand the role of diffusion, let us divide the city into two major areas—the inner city and the outer city.

centralizing forces
Diffusion forces that encourage people or businesses to locate in the central city.

decentralizing forces
Diffusion forces that encourage people or businesses to locate outside the central city.

Those diffusion forces that result in residences, stores, and factories locating in the inner or central city are **centralizing forces.** Those that result in activities locating outside the central city are called **decentralizing forces,** or suburbanizing forces. The pattern of homes, neighborhoods, offices, shops, and factories in the city results from the constant interplay of these two forces.

Centralization

Centralization has two primary advantages: economic and social.

Economic Advantages An important economic advantage of central-city location has always been accessibility. For example, imagine that a department store seeks a new location. If its potential market area is viewed as a full circle, then naturally the best location is in the center. There, customers from all parts of the city can gain access with equal ease. Before the automobile, a central-city location was particularly necessary because public transportation—such as the streetcar—was usually focused there. A central location is also important to those who must deliver their goods to customers. Bakeries and dairies were usually located as close to the center of the city as possible to maximize the efficiency of their delivery routes.

Location near regional transportation facilities is another aspect of accessibility and is thus an economic advantage. Many a North American city developed with the railroad at its center. Hence, any activity that needed access to the railroad had to locate in the central city. In many urban areas, giant wholesale and retail manufacturing districts grew up around railroad districts, producing "freight-yard and terminal cities" that supplied the produce of the nation. Although many of these areas have been abandoned by their original occupants, a walk by the railroad tracks today will give the most casual pedestrian a view of the modern "ruins" of the railroad city.

Another major economic advantage of the inner city is **agglomeration,** or clustering, which results in mutual benefits for businesses. For example, retail stores locate near one another to take advantage of the pedestrian traffic each generates. Because a large department store generates a good deal of foot traffic, any nearby store will also benefit.

agglomeration
A snowballing geographical process by which secondary and service industrial activities become clustered in cities and compact industrial regions in order to share infrastructure and markets.

Historically, offices clustered together in the central city because of their need for communication. Remember, the telephone was invented only in 1875. Before that, messengers hand-carried the work of banks, insurance firms, lawyers, and many other services. Clustering was essential for rapid communication. Even today, office buildings tend to cluster because face-to-face communication is still important for businesspeople. In addition, central offices take advantage of the complicated support system that grows up in a central city and aids everyday efficiency. Printers, bars, restaurants, travel agents, and office suppliers are within easy reach.

Social Advantages Three social factors have traditionally reinforced central-city location: historical momentum, prestige, and the need to locate near work. The strength of historical momentum should not be underestimated. Many activities remain in the central city simply because they began there long ago. For example, the financial district in San Francisco is located mainly on Montgomery Street. This street originally lay along the waterfront, and San Francisco's first financial institutions were established there in the mid-nineteenth century because it was the center of commercial action. In later years, however, landfill extended the shoreline (see Figure 10.11, page 367). Today, the financial district is several blocks from the bay; consequently, the district that began at the wharf head remained at its original location, even though other activity moved with the changing shoreline.

The prestige associated with the downtown area is also a strong centralizing force. Some activities still necessitate a central-city address. Think how important it is for some advertising firms to be on New York's Madison Avenue or for a stockbroker to be on Wall Street. This factor extends to many activities in cities of all sizes. The "downtown lawyer" and the "uptown banker" are examples. Residences

have often been located in the central city because of the prestige associated with it.

Probably the strongest social force for centralization has been the desire to live near one's place of employment. Until the development of the electric trolley in the 1880s, most urban dwellers had little alternative but to walk to work, and as most employment was in the central city, people had no choice but to live nearby. Upper-income people had their carriages and cabs, but others had nothing. Even after the introduction of electric streetcar lines in the 1880s, which made possible the exodus of some middle-class residents, many people continued to walk to work, particularly those who could not afford the new housing being constructed in what Sam Bass Warner has called "streetcar suburbs."

Suburbanization and Decentralization

The past 50 years have witnessed massive changes in the form and function of most Western cities (**Figure 11.11**). In the United States in particular, the suburbanization of residences and the decentralization of workplaces have emptied many downtowns of economic vitality. How and why has this happened? Geographer Neil Smith argues

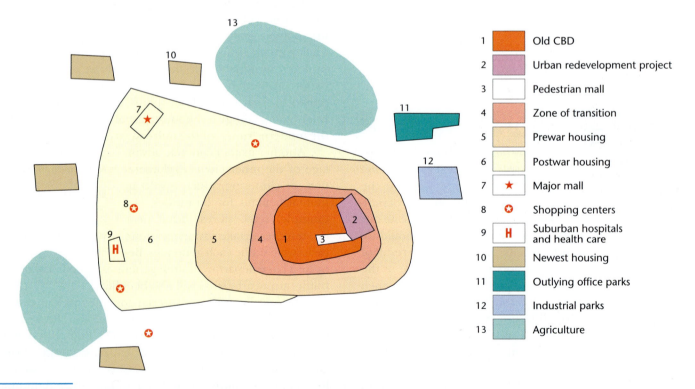

FIGURE 11.11   A hypothetical decentralized city. While the old central business district struggles (vacant stores and upper floors), newer activities locate either in the urban redevelopment project (offices, convention center, hotel) or in outlying office parks, malls, or shopping centers. However, some new specialty shops might be found around the new downtown pedestrian mall. New industry locates in suburban industrial parks that, along with outlying office areas, form major destinations for daily lateral commuting. *In what ways does your city follow this hypothetical pattern? In what ways does it differ?*

that the processes of suburbanization and the decline of the inner city are fundamentally linked: capital investment in the suburbs is often made possible by disinvestment, or the removal of money, from the central city. In post–World War II America, investors received greater returns on their money in the new suburbs than they did in the inner city, and therefore much of the economic boom of this period took place in the suburbs at the expense of the city. Smith refers to these processes as **uneven development** (Figure 11.12). This type of explanation gives us a broad picture of the economic reasons that many cities are now decentralized. We will now look more closely, examining the specific socioeconomic and public policy causes for the decentralization of our cities and for the problems that have resulted.

uneven development
The tendency for industry to develop in a core-periphery pattern, enriching the industrialized countries of the core and impoverishing the less industrialized periphery. This term is also used to describe urban patterns in which suburban areas are enriched while the inner city is impoverished.

Socioeconomic Factors Changes in accessibility have been a major reason for decentralization. The department store that was originally located in the central city may now find that its customers have moved to the suburbs and no longer shop downtown. As a result, the department store may move to a suburban shopping mall. The same process affects many other industries as well. The activities that were located downtown because of its proximity to the railroad may now find trucking more cost-effective. They relocate closer to a freeway system that skirts the downtown area. Finally, many offices now locate near airports so that their executives and salespeople can fly in and out more easily.

Although agglomeration once served as a centralizing force, its former benefits have now become liabilities in many downtown areas. These disadvantages include rising rents as a result of the high demand for space; congestion in the support system, which causes delays in getting supplies or means standing in endless lines for lunch; and traffic congestion, which makes delivery to market time-consuming and costly. Some downtown areas are so congested that traffic moves more slowly today than it did at the turn of the twentieth century. Often dissatisfied with the inconveniences of central-city living, employees may demand higher wages as compensation. This adds to the cost of doing business in the central city, and many firms choose to leave instead. For example, many firms have left New York City for the suburbs. They claim that it costs less to locate there and that their employees are happier and more productive because they do not have to put up with the turmoil of city life.

Clustering in new suburban locations can also have benefits. In industrial parks, for example, all the occupants share the costs of utilities and transportation links. Similar benefits can come from residential agglomeration. Suburban real estate developments take advantage of clustering by sharing the costs of schools, parks, road improvements, and utilities. Potential residents much prefer moving into a new development when they know that a full range of services is available nearby. Then they will not have to drive miles to find, say, the nearest hardware store. It is to the developer's advantage to encourage construction of nearby shopping centers.

The need to be near one's workplace has historically been a great centralizing force, but it can also be a very strong decentralizing force. At first the suburbs were "bedroom communities" from which people commuted to their jobs in the downtown area. This is no longer the case. In many metropolitan areas, most jobs are not in the central city but in outlying districts. Now people work in suburban industrial parks, manufacturing plants, office buildings, and shopping centers. Thus, a typical journey to work involves **lateral commuting:** travel from one suburb to another. As a result, most people who live away from the city center actually live closer to their workplaces (see the discussion of edge cities on pages 419–420).

lateral commuting
Traveling from one suburb to another in going from home to work.

FIGURE 11.12 Abandoned row houses in North Philadelphia. Often the economic neglect of these areas is directly linked to economic investment in the suburbs. *(Associated Press.)*

The prestige of the downtown area might once have lured people and businesses into the central city. But once it begins to decay, once shops close and offices are empty, a certain stigma develops that may drive away residents and commercial activities. Investors will not sink money into a downtown area that they think has no chance of recovery, and shoppers will not venture downtown when streets are filled with vacant stores, transients, pawnshops, and secondhand stores. One of the persistent problems faced by cities is how to reverse this image of the downtown area so that people once again consider it the focus of the city.

Reflecting on Geography

Identify some of the efforts that your city has undertaken to create a better image of itself. Have these efforts been successful? Does this reimagining of the city actually help residents of the inner city?

Public Policy   Many public policy decisions, particularly at the national level, have contributed greatly to the decentralization and abandonment of our cities. Both the Federal Road Act of 1916 and the Interstate Highway Act of 1956 directed government spending on transportation to the advantage of the automobile and the truck. Urban expressways, in combination with the emerging trucking industry, led to massive decentralization of industry and housing. In addition, the ability to deduct mortgage interest from income for tax purposes favors individual homeownership, which has tended to support a move to the suburbs.

The federal government in the United States has also intervened more directly in the housing market. In *Crabgrass Frontier,* Kenneth Jackson outlines the implications of two federal housing policies for the spatial patterning of our metropolitan areas. The first was the establishment of the Federal Housing Administration (FHA) in 1934 and its supplement, known as the GI Bill, enacted in 1944. These federal acts, which insured long-term mortgage loans for home construction, were meant to provide employment in the building trades and to help house the returning soldiers from World War II. Although the FHA legislation contained no explicit antiurban bias, most of the houses it insured were located in new residential developments in the suburbs, thereby neglecting the inner city.

Jackson identifies three reasons that this happened. First, by setting particular terms for its insurance, the FHA favored the development of single-family over multifamily projects. Second, FHA-insured loans for repairs were of short duration and were generally small. Most families, therefore, were better off buying a new house that was probably in the suburbs than updating an older home in the city.

Jackson regards the third factor as the most important. To receive an FHA-insured loan, the applicant and the neighborhood of the property had to be assessed by an "unbiased professional." This requirement was intended to guarantee that the property value of the house would be greater than the debt. This policy, however, encouraged bias against any neighborhood that was considered a potential risk in terms of property values. The FHA explicitly warned against neighborhoods with a racial mix, assuming that such a social climate would bring property values down, and encouraged the inclusion of **restrictive covenants** in property deeds, which prohibited certain "undesirable" groups from buying property. The agency also prepared extensive maps of metropolitan areas depicting the locations of African-American families and predicting the spread of that population. These maps often served as the basis for **redlining,** a practice whereby banks and mortgage companies demarcated areas (often by drawing a red line around them on these maps) considered to be at high risk for loans.

restrictive covenant
A statement written into a property deed that restricts the use of the land in some way; often used to prohibit certain groups of people from buying property.

redlining
A practice by banks and mortgage companies of demarcating areas considered to be high risk for housing loans.

These policies had two primary effects. First, they encouraged construction of single-family homes in suburban areas while discouraging central-city locations. Second, they intensified the segregation of residential areas and actively promoted homogeneity in the new suburbs.

The second federal housing policy that had a major impact on the patterning of metropolitan areas, the United States Housing Act, was intended to provide public housing for those who could not afford private housing. Originally implemented in 1937, the legislation did encourage the construction of many low-income housing units. Yet most of those units were built in the inner city, thereby contributing to the view of the suburbs as the refuge of the white middle class. This growing pattern of racial and economic segregation arose in part because public housing decisions were left up to local municipalities. Many municipalities did not need federal dollars and therefore did not want public housing. In addition, the legislation required that for every unit of public housing erected, one inferior housing unit had to be eliminated. Thus, only areas with inadequate housing units could receive federal dollars, again ensuring that public housing projects would be constructed in the older, downtown areas, not the newer suburbs. As Jackson claims, "The result, if not the intent, of the public housing program of the United States was to segregate the races, to concentrate the disadvantaged in inner cities, and to reinforce the image of suburbia as a place of refuge for the problems of race, crime, and poverty."

The Costs of Decentralization   Decentralization has taken its toll on North American cities. Many of the urban problems they now face are the direct result of the rapid

decentralization that has taken place in the last 50 years. Those people who cannot afford to live in the suburbs are forced to live in inadequate and run-down housing in the inner city, areas that currently do not provide good jobs. Vacant storefronts, empty offices, and deserted factories testify to the movement of commercial functions from central cities to suburbs. Retail stores in North American central cities have steadily lost sales to suburban shopping centers. Even offices are finding advantages to suburban location when they can capitalize on lower costs and easier access to new transportation networks.

Decentralization has also cost society millions of dollars in problems brought to the suburbs. Where rapid suburbanization has occurred, *sprawl* has usually followed. A common pattern is leapfrog or **checkerboard development,** where housing tracts jump over parcels of farmland, resulting in a mixture of open lands with built-up areas. This pattern occurs because developers buy cheaper land farther away from built-up areas, thereby cutting their costs. Furthermore, home buyers are often willing to pay premium prices for homes in subdivisions surrounded by farmland (**Figure 11.13**). The developer's gain is the area's loss, for it is more expensive to provide city services—such as police, fire protection, sewers, and electrical lines—to those areas that lie beyond open parcels that are not built up. Obviously, the most cost-efficient form of development is the addition of new housing directly adjacent to built-up areas so that the costs of providing new services are minimal.

checkerboard development
A mixture of farmlands and housing tracts.

Sprawl also extracts high costs because of the increased use of cars. Public transportation is extremely costly and inefficient when it must serve a low-density checkerboard development pattern—so costly that many cities and transit firms cannot extend lines into these areas. This means that the automobile is the only form of transportation. More energy is consumed for fuel, more air pollution is created by exhaust, and more time is devoted to commuting and everyday activities in a sprawling urban area than in a centralized city.

Moreover, we should not overlook the costs of losing valuable agricultural land to urban development. Farmers cultivating the remaining checkerboard parcels have a hard time earning a living. They are usually taxed at extremely high rates because their land has high potential for development, and few can make a profit when taxes eat up all their resources. Often the only recourse is to sell out to subdividers. So the cycle of leapfrog development goes on.

Many cities are now taking strong measures to curb this kind of sprawling growth. San Jose, California, for example, one of the fastest-growing cities of the 1960s, is now focusing new development on empty parcels of the checkerboard pattern. This is called **in-filling.** Other cities are tying the number of building permits granted each year to the availability of urban services. If schools are already crowded, water supplies inadequate, and sewage plants overburdened, the number of new dwelling units approved for an area will reflect this lower carrying capacity.

in-filling
New building on empty parcels of land within a checkerboard pattern of development.

Gentrification

Beginning in the 1970s, urban scholars began to observe what seemed to be a trend opposite to suburbanization. This trend, called **gentrification,** is the movement of middle-class people into deteriorated areas of city centers. Gentrification often begins in an inner-city residential district, with gentrifiers moving into an area that had been run down and is therefore more affordable than suburban housing (**Figure 11.14**). The infusion of new capital into the housing market usually results in higher property values, and this, in turn, often displaces residents who cannot afford the higher prices. Displacement opens up more housing for gentrification, and the gentrified district continues its spatial expansion.

gentrification
The displacement of lower-income residents by higher-income residents as buildings in deteriorated areas of city centers are restored.

Commercial gentrification usually follows residential, as new patterns of consumption are introduced into the inner city by the middle-class gentrifiers. Urban shopping malls and pedestrian shopping corridors bring the conveniences of the suburbs into the city, and bars and restaurants catering to this new urban middle class provide entertainment and nightlife for the gentrifiers.

The speed with which gentrification has proceeded in many of our downtowns and the scale of landscape changes that accompany it are causing dramatic shifts in the urban mosaic. What factors have led to this reshaping of our cities?

Economic Factors Some urban scholars look to broad economic trends in the United States to explain gentrification. We already know from our discussion of suburbanization that throughout the post–World War II era most investments in metropolitan land were made in the suburbs; as a result, land in the inner city was devalued. By the 1970s, many home buyers and commercial investors found land in the city much more affordable, and a better economic investment, than in the higher-priced suburbs. This situation brought capital into areas that had been undervalued and accelerated the gentrifying process.

In addition, most Western countries have been experiencing **deindustrialization,** a process whereby the economy is shifting from one based on secondary industry to one based on the service sector. This shift has led to the abandonment of older industrial districts in the inner city, including waterfront areas. Many of these areas are prime targets of gentrifiers, who convert the waterfront from a noisy, commercial port area into an aesthetic asset. In Buenos Aires, for example, one of the new gentrified neighborhoods is Puerto Madero, an area that was once home to docking facilities and a wholesale market (**Figure 11.15** and Mona's Notebook, page 400). The shift to an economy based on the service sector also means that the new productive areas of the city will be dedicated to white-collar activities. These activities often take place in relatively clean and quiet office buildings, contributing to a view of the city as a more livable environment.

deindustrialization
The decline of primary and secondary industry, accompanied by a rise in the service sectors of the industrial economy.

Social Factors Other scholars look to changes in social structure to explain gentrification. The maturing of the baby-boom generation has led to significant modifications of our

FIGURE 11.14 Gentrification in Vancouver, British Columbia, Canada. Gentrification often occurs in older neighborhoods with historic buildings. These Victorian houses have been carefully restored to reflect their new owners' interest in preserving the past, although this might have come at the cost of displacing the previous tenants. As you can see from this image, *this neighborhood is also very close to downtown. Why?* (Canada Stock Photographs.)

traditional family structure and lifestyle. With a majority of women in the paid labor force and many young couples choosing not to have children or to delay that decision, a suburban residential location looks less appealing. A gentrified location in the inner city attracts this new class because it is close to their managerial or professional jobs downtown, is usually easier to maintain, and is considered more interesting than the bland suburban areas where they grew up. Living in a newly gentrified area is also a way to display social status. Many suburbs have become less exclusive, while older neighborhoods in the inner city frequently exploit their historical associations as a status symbol.

Political Factors Many metropolitan governments in the United States, faced with the abandonment of the central city by the middle class and therefore with the erosion of their tax base, have enacted policies to encourage commercial and residential development in downtown areas. Some policies provide tax breaks for companies willing to locate downtown; others furnish local and state funding to redevelop central-city residential and commercial buildings.

At a more comprehensive level, some larger metropolitan areas have devised long-term planning agendas that target certain neighborhoods for revitalization. Often this is accomplished by first condemning the targeted area, thereby transferring control of the land to an urban-development authority or other planning agency. Such areas are often older residential neighborhoods that were originally built to house people who worked in nearby factories, which are usually torn down or transformed into lofts or office space. The redevelopment authority might locate a new civic or arts center in the neighborhood. Public-sector initiatives often lead to private investment, thereby increasing property values. These higher property values, in turn, lead to further investment and the eventual transforma-

tion of the neighborhood into a middle- to upper-class gentrified district.

Sexuality and Gentrification Gentrified residential districts are often correlated with the presence of a significant gay and lesbian population. It is fairly easy to understand this correlation. First, the typical suburban life tends not to appeal to people whose lifestyle is often regarded as different and whose community needs are often different from those of people living in the suburbs. Second, gentrified inner-city neighborhoods provide access to the diversity of city life and amenities that often include gay cultural institutions. In fact, the association of urban neighborhoods with gays and lesbians has a long history. For example, urban historian George Chauncey has documented gay culture in New York City between 1890 and World War II, showing that a gay world occupied and shaped distinctive spaces in the city, such as neighborhood enclaves, gay commercial areas, and public parks and streets.

Yet, unlike this earlier period, when gay cultures were often forced to remain hidden, the gentrification of the postwar period has provided gay and lesbian populations with the opportunity to reshape entire neighborhoods actively and openly. Urban scholar Manuel Castells argues that in cities such as San Francisco, the presence of gay men in institutions directly linked to gentrification, such as the real estate industry, significantly influenced that city's gentrification processes in the 1970s.

Geographers Mickey Lauria and Lawrence Knopp emphasize the community-building aspect of gay men's involvement in gentrification, recognizing that gays have seized an opportunity to combat oppression by creating neighborhoods over which they have maximum control and that meet long-neglected needs. Similarly, geographer Gill Valentine argues that the limited numbers and types of

Mona's Notebook

"Seeing" New Places

Mona Domosh

The landscape of Boston's gentrified waterfront displays similarities to Puerto Madero but is also unique. *(Courtesy of Mona Domosh.)*

As a cultural geographer, I've thought a bit about how people often see new places through the "eyes" of familiar ones. You, too, have most likely noticed this. For example, when you travel somewhere new, you often compare it to what you know, saying things like "This street reminds me of the one I grew up on" or "Don't we have almost an identical building in our town?" I shouldn't have been surprised then when on a trip to Argentina I noticed that parts of Buenos Aires looked like other cities I was familiar with, particularly the recently gentrified neighborhoods in the city (see Figure 11.15, page 399). The brick, four-story buildings that had once lodged warehouses and now were home to fancy apartments and upscale restaurants and bars looked to my North American eye to be almost identical to the gentrified waterfronts of Boston and New York City. But they weren't. There are similarities, of course, since these waterfronts served similar purposes in the past, but each city and its history has put a particular stamp on its urban landscape. Boston's gentrified waterfront, for example, was built much earlier and in a style different from the Buenos Aires waterfront, and granite was used as a building material as much as brick.

I *was* surprised, however, when I visited other parts of Buenos Aires and realized that some neighborhoods were making explicit their similarity to and connection with North American cities through their place-names. Fairly recently, Palermo, a neighborhood in the process of gentrification in northeast Buenos Aires, has been subdivided into smaller districts, including Palermo Soho and Palermo Hollywood. The name Palermo itself is already a place-name taken from somewhere else (a relatively common feature of immigrant nations), but the twinning of that now well-established name with Soho and Hollywood was something new to me. As it turns out, I learned that other places in the world have taken the Soho name and used it to lend a certain trendy and hip connotation to their neighborhoods. Most likely, developers and entrepreneurs thought that using Soho, a term derived from *So*uth of *Ho*uston Street (a very chic neighborhood of lofts and trendy restaurants and bars in New York City), would boost the area's commercial potential. And the global symbol of Hollywood, I realized, needed no explanation: it connoted a West Coast version of glamour. The local and the global often meet in very interesting ways!

Posted by Mona Domosh

---

lesbian spaces in cities also serve as community-building centers for lesbian social networks.

According to geographer Tamar Rothenberg, the gentrified neighborhood of Park Slope in Brooklyn is home to the heaviest concentration of lesbians in the United States. Its extensive social networks are marking the neighborhood as both a center of lesbian identity and a visible lesbian social space.

The Costs of Gentrification Gentrification often results in the displacement of lower-income people, who are forced to leave their homes because of rising property values. This displacement can have serious consequences for the city's social fabric. Because many of the displaced people come from disadvantaged groups, gentrification frequently contributes to racial and ethnic tensions. Displaced people are often forced into neighborhoods more peripheral to the city, a trend that only adds to their disadvantages. In addition, gentrified neighborhoods usually stand in stark contrast to surrounding areas where investment has not taken place, thus creating a very visible reminder of the uneven distribution of wealth within our cities.

The success of a gentrification project is usually measured by its appeal to an upper-middle-class clientele. This suggests that gentrified neighborhoods are completely homogeneous in their use of land. Residential areas are consciously planned to be separate from commercial districts and are themselves sorted by cost and tenure type (homeownership versus rental). Thus, gentrification often draws on the suburban notion of residential homogeneity and eliminates what many people consider to be a great asset of urban life—its diversity and heterogeneity (see Subject to Debate, page 402).

Globalization

In what ways has globalization shaped cities? The photograph of Chinatown in New York City that opened this chapter suggests some of the diverse impacts that globalization is having on the urban mosaic. Transnational firms, for example, create a workforce that is increasingly mobile, moving between countries with relative ease and impacting cities in a multitude of ways (as we saw in Chapter 10), while global workplaces bring people into cities and often across national borders to provide labor, creating new ethnic neighborhoods within cities. Globalization is also apparent in creating similar urban forms around the world: global cities such as Shanghai and New York, Delhi and London, for example, are beginning to share common landscape characteristics, as we will see in the section on global urban form.

New Ethnic Neighborhoods

According to the United States Department of Homeland Security, approximately 3.3 million people immigrated to the United States between 2007 and 2009: the highest percentage of immigrants came from Mexico (approximately 16 percent), followed by China (approximately 7 percent), the Philippines (approximately 6 percent), and India (approximately 5 percent). The vast majority of these immigrants found job opportunities and cultural connections that drew them to major metropolitan regions in six states—California, New York, Texas, Florida, New Jersey, and Illinois (**Figure 11.16**). This spatial concentration of America's new immigrants has created diverse communities with distinctive landscapes, both within the downtown areas of these cities and in the suburban regions. Miami's Little Havana, for example, is easily recognized by the commercial signs in Spanish, Spanish street names, and the colors and styles of buildings. Parts of what were once run-down neighborhoods

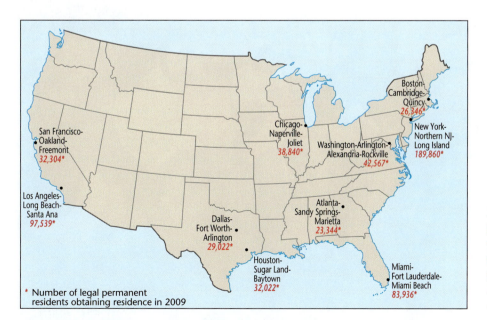

* Number of legal permanent residents obtaining residence in 2009

FIGURE 11.16 Map indicating the location of 10 metropolitan areas in the United States with the largest number of persons obtaining legal permanent resident status, 2009. *Why is immigration focused on these particular cities?*
(*Source: U.S. Department of Homeland Security, 2009.*)

Subject to Debate

Can Gentrification Be Socially Just?

As we have seen, many scholars and policy makers criticize gentrification because it can lead to homelessness and the displacement of disadvantaged groups. Yet gentrification also helps downtowns become more lively social centers and can lead to an economic rejuvenation that benefits many groups of urban dwellers. In other words, gentrification has benefits as well as costs. But is there some way of diminishing the costs so that all residents can take advantage of the social and economic benefits of gentrification? Can gentrification be socially just?

Critics of the gentrification process point to a range of studies by geographers, planners, and sociologists in cities around the world that show similar outcomes: poor or disadvantaged groups of people are often left homeless or are displaced from their homes and forced to move into marginal areas of the city as wealthier and more privileged groups of people move into the neighborhood, often encouraged to do so by urban governments through a variety of public policy decisions (e.g., tax breaks and zoning changes). The neighborhood is then subject to inflows of capital that create urban amenities such as restaurants, clubs, and services. The rich and advantaged are able to benefit from these new services; the poor and disadvantaged are not. In addition, critics point out that gentrification often exacerbates racism, since commonly gentrifiers are white, while those being displaced are people of color.

On the other side of the debate are those who believe that gentrification does not necessarily lead to displacement and that its benefits can be shared by all. Recent studies have shown that urban gentrification brings benefits to the entire city: lower crime rates, better schools, higher property taxes (monies that can be used to improve urban services). And these advantages can be shared by all urban residents. In addition, they point to recent studies of several cities that have been successful in "managing" gentrification so that is does not lead to displacement. Some of these management techniques include enacting rent control, providing eviction controls, strengthening tenants' rights, and maintaining some control over new housing developments through governmental oversight. Many cities have grassroots organizations that fight displacement by bringing different interests together from diverse parts of the community: schools, churches, businesses, and residents. One particularly successful group in Chicago is called ONE (Organization of the Northeast), whose goal is to build a "successful multi-ethnic, mixed-economic community on the northeast side of Chicago" (for more information, see its web site: http://www.onechicago.org/index.html).

Continuing the Debate

Some scholars and policy makers believe carefully managed gentrification can be socially just in that its benefits can be enjoyed by a diverse range of residents; others say that the lure of more money will eventually lead to the displacement of the poor in favor of the rich. Keeping this in mind, consider these questions:

- How has gentrification affected your city?

- Have you or a family member been affected by gentrification? How so?

- What have been the positive and negative effects of gentrification?

- Do you think tenants' rights groups and other similar organizations can control gentrification?

Organizers from Chicago ONE protest the lack of affordable housing in the city. (Courtesy of Chicago ONE.)

of the city have been remade into vibrant commercial and residential communities. Another example is Los Angeles: 80 years ago, a Saturday morning stroll through the CBD of Los Angeles would have led through streets lined with department stores, movie theaters, and offices. Now it is filled with the sounds of Latin music and vendors selling everything from electronics to mango ice cream.

But these new ethnic landscapes are not limited to the central city. Portions of America's suburbs have also become diverse. The decentralization of the downtown has created new economic centers in suburban regions, and immigrants are drawn to these centers. According to the 2000 census, approximately 40 percent of new immigrants to the United States are settling in the suburbs. Geographer Wei Li refers to these new immigrant communities located in the suburbs as ethnoburbs. In Montgomery County, Maryland, a suburban community outside of Washington, D.C., almost a quarter of all households are headed by a person who is foreign-born. Instead of the dense residential and commercial districts that characterize downtown ethnic enclaves, the new suburban ethnicity is proclaimed within shopping centers, suburban cemeteries, and dispersed churches and temples. According to geographer Joseph Wood, their presence in the landscape is often not visible to observers precisely because it is suburban. For example, most of the 50,000 Vietnamese-Americans who migrated to the Washington, D.C., area from the 1970s through the 1990s settled in suburbs in northern Virginia. The focal point of the Vietnamese community there is the Eden Center, a typical L-shaped shopping center that has been transformed into a Vietnamese-American economic and social center (**Figure 11.17**). This pattern of shopping plazas serving as ethnic community markers for a dispersed

FIGURE 11.18 **A Muslim family in the neighborhood of Belleville in Paris.** Many cities have ethnic enclaves where immigrants from other countries live, creating a different scale of community within the larger scale of the city. *How are these ethnic enclaves related to globalization?* (Peter Turnley/Corbis.)

FIGURE 11.17 **Eden Shopping Center in Fairfax, Virginia.** This shopping plaza serves as a social and cultural center for the Vietnamese community of northern Virginia. (Courtesy of Joseph Wood.)

immigrant population is not peculiar to northern Virginia; it is commonplace in the metropolitan areas of most major North American cities.

Urban ethnic enclaves are not limited to U.S. or Canadian cities. Many cities around the world are experiencing the effects of immigration as globalization makes it easier for people from one country to seek employment in thriving metropolitan areas of a different country. As we've learned from previous chapters, globalization has created conditions that support the movement of peoples across national and ethnic boundaries (**Figure 11.18**). There are North African communities in Paris, Turkish neighborhoods in Berlin, and Kurdish neighborhoods in Istanbul, and each of these is impacting the culture, economy, and landscape of these cities in profound ways.

A Global Urban Form?

As we have seen, the increasing mobility brought about by globalization can bring diversity (new immigrants living in what once were all-white suburbs, bringing their own cultural forms into new contexts), but it can also lead to homogeneity. The relative ease of communication and transportation today helps erode cultural barriers, and global capital allows international and transnational development companies to finance, design, and construct similar buildings throughout the world. In terms of the urban mosaic, this means that cities are beginning to resemble one another, as particular urban elements that used to be specific to one place or one culture are now being exported to other places outside their original context. For example, as we discovered in our section on downtowns, skyscrapers—once thought to be a uniquely American urban feature—are now reaching their greatest heights in Asia. The Burj Dubai building in Dubai, United Arab Emirates, is currently the world's tallest structure. The building was developed by a team of companies that are incredibly international: the land development corporation is from Dubai, the architecture firm is from the United States, and the main contractor is the Samsung Corporation from South Korea. Taipei 101 in Taiwan (**Figure 11.19**) is the second-tallest building in the world, followed by the Shanghai World Financial Center and the twin Petronas Towers in Kuala Lampur.

Increasingly, much of the new construction in the world's cities is being impacted by these global development corporations and teams, including commercial structures, such as skyscrapers and shopping malls (see the section on the new urban landscape), and residential structures. An interesting example in this regard is the global spread of gated communities. Gated communities are residential areas that are inaccessible to the public. These private communities can take several forms, either using literal gates to enclose streets and yards or in some way introducing forms of surveillance such as guards posted at all street entrances. Like skyscrapers, gated communities are often considered "American," since they became common in the United States in the late twentieth century. But they now can be found in many places throughout the world, often with architectural styles similar to those in the United States. Gated communities have become particularly prevalent in large cities in Brazil, Argentina, South Africa, Indonesia, India, and China, places with an emerging middle and upper class who desire residences that are secure while also expressing their new economic status. Gated communities serve this purpose for them, since they are privately controlled and guarded, and are considered symbols of wealth and prestige. Many of these communities resemble one another in architectural style as well as form (i.e., gated), contributing to what some believe is the homogenization of the urban landscape, a sort of global residential form. However, as urban geographers are now show-

FIGURE 11.19 **Taipei 101, in downtown Taipei, Taiwan.** Like other growing cities, Taipei is betting that having the tallest building in the world will bring more and more attention to its bid for regional and perhaps global economic dominance. *Notice how the building dwarfs the surrounding, older skyscrapers and is built in a very different architectural style. What accounts for these differences?* (Digital Vision Ltd./SuperStock.)

ing, even though many of these gated communities are financed and developed by large transnational corporations or teams, the notion that they are uniform or somehow dominated by American landscape tastes is simply not true. Local cultural contexts, either at the regional or national level, are important shapers both of the styles and meanings of these communities.

For example, geographers Choon-Piew Pow and Lily Kong interviewed residents of Vanke Garden City in Shanghai, a well-established gated community, and examined advertisements for many of the gated communities in the city (**Figure 11.20**). What they found was that these communities fulfill the aspiration of the new Chinese middle class for exclusivity and prestige based on their association to the West, but they do so partly by drawing on meanings that are rooted in Chinese history and culture. In fact, gated com-

FIGURE 11.20 A gated community in the Pudong area of Shanghai. Pudong, an area of Shanghai that was predominantly agricultural until the 1990s, is now home to many of the city's nouveaux riches. Gated communities here are common, as signs of status and for security. *From the outside, this community looks little different from what you would see outside Atlanta. Why is that?* (Kevin Hulsey/KHI Inc.)

munities are nothing new in China but have deep roots in Chinese history (see Chapter 10, on the cosmomagical city) and were part of the collectivist vision of China's socialist government. In Shanghai, Pow and Kong discovered, for example, that houses with private gardens, a feature of these new gated communities, allure buyers not necessarily because of their association to prestige and the West, but because gardens in Chinese history have long been associated with the highest ideals of civilization—the contemplation of nature from a secluded location. They also discovered that Western housing plans were altered in these communities to accommodate Chinese-style kitchens (more open) and Chinese notions of feng shui—the arrangement of spaces to create harmonious energy flows. As an example of what is happening in today's global cities, gated residential communities show us that on the one hand they are being produced by similar forces (global capital flows, transnational development firms), but on the other they are also shaped by local, regional, and national cultural contexts.

Nature-Culture

How can we understand the relationships between the urban mosaic and the physical environment? The physical environment affects cities, just as urbanization profoundly alters natural environmental processes (**Figure 11.21**). The theme of nature-culture helps us to organize information about these city-nature relationships. Although we discuss these topics in general terms in the next pages, one should not lose sight of how the differing cultural fabric within and between cities affects the relationship between city and nature. Urban ecology differs greatly from place to place because of different physical environments and varying cultural patterns.

Urban Weather and Climate

Cities alter virtually all aspects of local weather and climate. Temperatures are higher in cities, rainfall increases, the incidence of fog and cloudiness is greater, and levels of atmospheric pollution are much higher.

The causes of these changes are no mystery. Because cities cover large areas of land with streets, buildings, parking lots, and rooftops, about 50 percent of the urban area is a hard surface. Rainfall is quickly carried into gutters and

FIGURE 11.21 Suburban homes built on landfills, Treasure Island, Florida. When land values are high and pressure for housing intense, terrain rarely stands in the way of the developer. In fact, particular characteristics of physical sites can actually increase land values. *What site characteristics evident in this photo tell you that Treasure Island is a very expensive place to live?* (Cameramann International, Ltd.)

sewers, so that little standing water is available for evaporation. Because evaporation removes heat from the air, when moisture is reduced, evaporation is lessened and air temperatures are higher.

Moreover, cities generate enormous amounts of heat. This heat comes not just from the heating systems of buildings but also from automobiles, industry, and even human bodies. One study showed that on a winter day in Manhattan, the amount of heat produced in the city is two and a half times the amount that reaches the ground from the sun. This results in a large mass of warmer air sitting over the city, called the urban **heat island** (**Figure 11.22**). The heat island causes yearly temperature averages in cities to be 3.5°F (2°C) higher than in the countryside; during the winter, when there is more city-produced heat, the average difference can easily reach 7°F to 10°F (4°C to 5.6°C).

Urbanization also affects precipitation (rainfall and snowfall). Because of higher temperatures in the urban area, snowfall will be about 5 percent less than in the surrounding countryside. However, rainfall can be 5 to 10 percent higher. The increased rainfall results from two factors: the large number of dust particles in urban air and the

heat island
An area of warmer temperatures at the center of a city, caused by the urban concentration of heat-retaining concrete, brick, and asphalt.

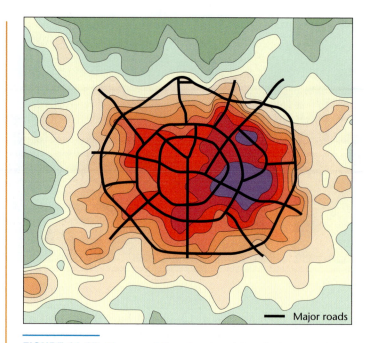

FIGURE 11.22 Diagram of the urban heat island in Chengdu, China. The deep colors (purple and red) indicate higher temperatures, while the shades of yellow and green indicate lower temperatures. Notice the marked contrast in temperature between the built-up part of the city and the surrounding rural areas. *(Source: Shangming and Bo, 2001.)*

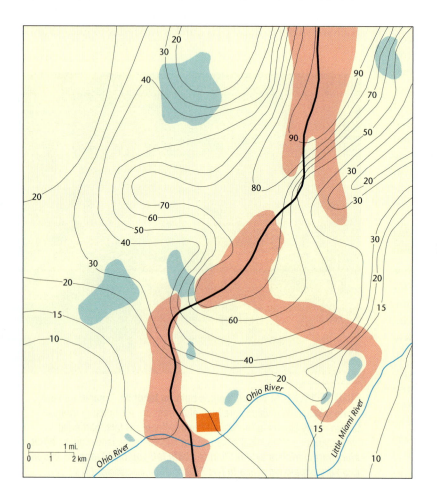

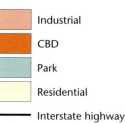

FIGURE 11.23 The dust dome over Cincinnati, Ohio. Numbers show the concentration of particulate matter in the air at an elevation of 3000 feet (914 meters). The higher the value, the greater the amount of particulate matter. ***Does land use (industrial, central business district, and so on) have an effect on the concentration of particulate matter in the air?*** *(After Bach and Hagedorn, 1971.)*

higher city temperatures. Dust particles are a necessary precondition for condensation, offering a nucleus around which moisture can adhere. An abundance of dust particles, then, facilitates condensation. That is why fog and clouds (**dust domes**) are usually more frequent around cities (**Figure 11.23**).

dust dome
A pollution layer over a city that is thickest at the center of the city.

Urban Hydrology

Not only is the city a great consumer of water, but it also alters runoff patterns in a way that increases the frequency and magnitude of flooding. Within the city, residential areas are usually the greatest consumers of water. Water consumption can vary, but generally each person in the United States uses about 60 gallons (264 liters) per day in a residence. Of course, residential demand varies. It is greater in drier climates as well as in middle- and high-income neighborhoods. Higher-income groups usually have a larger number of water-using appliances, such as washing machines, dishwashers, and swimming pools.

Urbanization can increase both the frequency and the magnitude of flooding because cities create large impervious areas where water cannot soak into the earth. Instead, precipitation is converted into immediate runoff. It is forced into gutters, sewers, and stream channels that have been straightened and stripped of vegetation, which results in more frequent high water levels than are found in a comparable area of rural land. Furthermore, the time between rainfall and peak runoff is reduced in cities; there is more lag in the countryside, where water runs across soil and vegetation into stream channels and then into rivers.

So, because of hard surfaces and artificial collection channels, runoff in cities is concentrated and immediate.

Urban Vegetation

Until a decade ago, it was commonly thought that cities were made up mostly of artificial materials: asphalt, concrete, glass, and steel. Studies, however, show that about two-thirds of a typical North American city is composed of trees and herbaceous plants (mostly weeds in vacant lots and cultivated grass in lawns). This urban vegetation, usually a mix of natural and introduced species, is a critical component of the urban ecosystem because it affects the city's topography, hydrology, and meteorology.

More specifically, urban vegetation influences the quantity and quality of surface water and groundwater; reduces wind velocity as well as turbulence and temperature extremes; affects the pattern of snow accumulation and melting; absorbs thousands of tons of airborne particulates and atmospheric gases; and offers a habitat for mammals, birds, reptiles, and insects, all of which play some useful role in the urban ecosystem. Furthermore, urban vegetation influences the propagation of sound waves by muffling much of the city's noise; affects the distribution of natural and artificial light; and, finally, is an extremely important component in the development of soil profiles that, in turn, control hillside stability.

Our urban settlements are still closely tied to the physical environment. Cities change these natural processes in profound ways, and we must understand these disturbances in order to make better decisions about adjustments and control.

CULTURAL LANDSCAPE

What do the urban patterns we have been discussing look like? How can we recognize different types of cities from their three-dimensional forms, and how are these forms changing? Cities, like all places humans inhabit, demonstrate an intriguing array of cultural landscapes, the reading of which gives varied insights into the complicated interactions between people and their surroundings. In this section, we offer some thoughts about how to view and "read" urban landscapes. We begin by discussing some geographical reference points—one might say helpful hints—for investigating and reading **cityscapes.** We then turn to a discussion of one of the key

cityscape
An urban landscape.

factors in reading a cityscape—understanding a city's landscape history. A brief discussion of the new components in urban landscapes then follows. Much of what we say will strike a familiar chord, because our urban scene is the basis of so much of our life. You will find that you have a great deal of intuitive knowledge about cityscapes.

Ways of Reading Cityscapes

Understanding urban landscapes requires an appreciation of large-scale urbanizing processes and local urban environments both past and present. The patterns we see today in the city, such as building forms, architecture, street plans, and land use, are a composite of past and present cultures. They reflect the needs, ideas, technology, and institutions of human occupancy. Two concepts underlie our examination of urban landscapes. The first is **urban morphology,** or the physical form of the city, which consists of street patterns, building sizes and shapes, architecture, and density. The second concept is **functional zonation,** which refers to the pattern of land uses within a city or, put another way, the existence of areas with differing functions, such as residential, commercial, and governmental. Functional zonation also includes social patterns—whether, for example, an area is occupied by the power elite or by people of low status, by Jews or by Christians, by the wealthy or by the poor. Both concepts are central to understanding the cultural landscape of cities, because both make statements about how cultures occupy and shape space.

Human geographers look to cityscapes for many different kinds of information (see Doing Geography at the end of the chapter). Here we discuss four interconnected themes that are commonly used as organizational frameworks for landscape research (**Figure 11.24**).

Landscape Dynamics Think of some familiar features of the cityscape: downtown activities creeping into residential areas, deteriorated farmland on a city's outskirts, older buildings demolished for the new. These are all signs of specific processes that create urban change; the landscape faithfully reflects these dynamics.

When these visual clues are systematically mapped and analyzed, they offer evidence for the currents of change expressed in our cities. Of equal interest is where change is *not* occurring—those parts of the city that, for various reasons, remain relatively static. An unchanging landscape also conveys an important message. Perhaps that part of the city is stagnant because it is removed from the forces that are producing change in other parts. Or perhaps there is a conscious attempt by local residents to inhibit change—to preserve open space by resisting suburban development, for example, or to preserve a historic landmark. Documenting landscape changes over time gives valuable insight into the paths of settlement development.

The City as Palimpsest Because cityscapes change, they offer a rich field for uncovering remnants of the past. A **palimpsest** is an old parchment used repeatedly for written messages. Before a new missive was written, the old was erased, yet rarely were all the previous characters and words completely obliterated—so remnants of

FIGURE 11.24 Boston's central city. There are various ways of looking at cityscapes: as indicators of change, as palimpsests, as expressions of visual biases, and as manifestations of symbolic traditions. This photo offers evidence of all approaches. *Which clues would you select to illustrate each cityscape theme?* (Steve Dunwell/The Image Bank.)

earlier messages were still visible. This record of old and new is called a *palimpsest,* a word geographers use to describe the visual mixture of past and present in cultural landscapes.

Cities are full of palimpsestic offerings, scattered across the contemporary landscape. How often have you noticed an old Victorian farmhouse surrounded by new tract homes, or a historic street pattern obscured or highlighted by a recent urban redevelopment project, or a brick factory shadowed by new high-rise office buildings? All of these give clues to past settlement patterns, and all are mute testimony to the processes of change in the city.

Our interest in this historical accumulation is more than romantic nostalgia. A systematic collection of these urban remnants provides us with glimpses of the past that might otherwise be hidden. All societies pick and choose, consciously or not, what they wish to preserve for future generations, and, in this process, a filtering takes place that often excludes and distorts information. But the landscape does not lie.

The urban palimpsest, then, offers a way to find the past in the contemporary landscape. We can evaluate these remnants to glean a better understanding of historical settlements.

Symbolic Cityscapes Landscapes contain much more than literal messages. They are also loaded with figurative or metaphorical meaning and can elicit emotions and memories. To some people, skyscrapers are more than high-rise office buildings: they are symbols of progress, economic vitality, downtown renewal, or corporate identities. Similarly, historical landscapes—those parts of the city where the past has been preserved—help people to define themselves in time; establish social continuity with the past; and codify a largely forgotten, yet sometimes idealized, past.

D. W. Meinig, a geographer who has given much thought to urban landscapes, maintains that there are three highly symbolic townscapes in the United States: the New England village, with its white church, commons, and tree-lined neighborhoods; Main Street of Middle America, a string street of a small midwestern town, with storefronts, bandstand, and park (**Figure 11.25**); and what Meinig calls California Suburbia, suburbs of quarter-acre lots, effusive garden landscaping, swimming pools, and ranch-style houses. As Meinig explains: "Each is based upon an actual landscape of a particular region. Each is an image derived from our national experience . . . simplified . . . and widely advertised so as to become a commonly understood symbol. Each has . . . influenced the shaping of the American scene over broader areas."

More politically and problematically, the cultural landscape is an important vehicle for constructing and maintaining, subtly and implicitly, certain social and ethnic distinctions. For example, geographers James and Nancy Duncan have found that because conspicuous consumption is a major way of conveying social identity in our culture, elite landscapes are created through large-lot zoning, imitation country estates, and the preservation of undeveloped land. They see the residential landscapes in upper-income areas as controlled and managed in order to reinforce class and status categories. Their study of elite suburbs near Vancouver and New York sensitizes us to how the cultural landscape can be thought of as a repository of symbols used by our society to differentiate itself and protect vested interests.

Perception of the City During the last 25 years, social scientists have been concerned with measuring people's perceptions of the urban landscape. They assume that if we really know what people see and react to in the city, we can ask architects and urban planners to design and create a more humane urban environment.

Kevin Lynch, an urban designer, pioneered a method for recording people's images of the city. On the basis of interviews conducted in Boston, Jersey City (New Jersey), and Los Angeles, Lynch suggested five important elements in mental maps (images) of cities:

1. *Pathways* are the routes of frequent travel, such as streets, freeways, and transit corridors. We experience the city from the pathways, and they become the threads that hold our maps together.

FIGURE 11.25 Main Street, Ferndale, California. The symbolism of Main Street, USA, is a powerful force in shaping communities today, particularly because an ersatz Main Street is the central element of Disney World. Think of the ways this symbol is used in art, literature, film, and television and of the messages and emotions conveyed by this landscape. *(ChromoSohm/Sohm/Photo Researchers, Inc.)*

2. *Edges* are boundaries between areas or the outer limits of our image. Mountains, rivers, shorelines, and even major streets and freeways are commonly used as edges. They tend to define the extremes of our urban vision. Then we fill in the details.

3. *Nodes* are strategic junction points, such as breaks in transportation, traffic circles, or any place where important pathways come together.

4. *Districts* are small areas with a common identity, such as ethnic areas and functional zones (for instance, the CBD or a row of car dealers).

5. *Landmarks* are reference points that stand out because of shape, height, color, or historical importance. The city hall in Los Angeles, the Washington Monument, and the golden arches of a McDonald's are all landmarks.

legible city
A city that is easy to decipher, with clear pathways, edges, nodes, districts, and landmarks.

Using these concepts, Lynch saw that some parts of the cities were more **legible,** or easier to decipher, than others. Lynch discovered that, in general, legibility increases when the urban landscape offers clear pathways, nodes, districts, edges, and landmarks. Further, some cities are more legible than others. For example, Lynch found that Jersey City is not very legible. Wedged between New York City and Newark, Jersey City is fragmented by railroads and highways. Residents' mental maps of Jersey City have large blank areas in them. When questioned, they can think of few local landmarks. Instead, they tend to point to the New York City skyline just across the river.

Landscape Histories of American, Canadian, and European Cities

When you walk around any city, you are looking at buildings and roads and parks that were built in different times for diverse purposes. Understanding and "reading" urban landscapes, therefore, requires in-depth knowledge of the people who have created and inhabited them, in both the past and the present. In this next section, we provide a guide to the history of urban forms that have characterized Western cities—cites that trace their origins to Greece and Rome. We provide this guide not only to help you read the landscape of your own city, but also to enrich your understanding of the diverse array of urban landscapes located throughout North America and much of Europe.

The Greek City Western civilization and the Western city both trace their roots back to ancient Greece. City life diffused to Greece from Mesopotamia. By 600 B.C., there were more than 500 towns and cities on the Greek mainland and surrounding islands. As Greek civilization expanded, cities spread with it throughout the Mediterranean, reaching as far as the north shore of Africa, Spain, southern France, and Italy. These cities were of modest size, rarely containing more than 5000 inhabitants. Athens, however, may have reached a population of 300,000 in the fifth century B.C.

Greek cities had two distinctive functional zones: the acropolis and the agora. In many ways, the acropolis was similar to the citadel of Mesopotamian cities. Here were the temples of worship, the storehouse of valuables, and the seat of power. The acropolis also served as a place of retreat in time of siege (**Figure 11.26**). If the acropolis was the domain

FIGURE 11.26 The Acropolis in Athens. The Acropolis dominates the contemporary city and reminds us that many cities throughout the world have been centered on fortified places that eventually became more symbolic than functional. ***How does this landscape compare with other defensive acropolis sites?*** *(James Hanley/ Photo Researchers.)*

of power, the agora was the province of the citizens. As originally conceived, the agora was a place for public meetings, education, social interaction, and judicial matters. In other words, it was the civic center, the hub of democratic life for Greek men (women were excluded from political life).

The early Greek cities were probably not planned but rather grew spontaneously, without benefit of formal guidelines. However, some scholars think that many ceremonial areas within these cities were designed to be seen according to prescribed lines of vision and that those lines of vision included not only the buildings but also the natural landscape that surrounded it. The human aesthetic sense was given a degree of authority that it did not have in the cosmomagical city.

More formalized city design and plan are apparent in later Greek cities that were built in areas of colonial expansion. One of the best examples of such planned cities

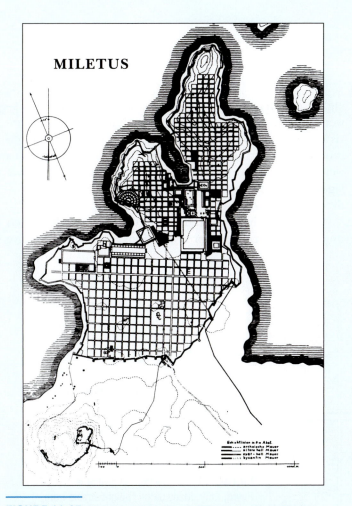

FIGURE 11.27 The plan of the city of Miletus by Hippodamus, circa 450 B.C. Notice how the strict grid is imposed on the irregular coastline. The central agora is also regularized, characteristic of this colonial phase of the Greek city. *(Source: Vance, 1990.)*

is Miletus, on the eastern shore of the Mediterranean, in Ionia (present-day Turkey). The city was laid out in a rigid grid pattern, imposing its geometry onto the physical conditions of the site (**Figure 11.27**). Although the source of such a plan is debatable, clearly this orderly and coherent layout indicates an abstract and highly rational notion of urban life and seems to fit well with the functional needs of a colonial city.

Roman Cities By 200 B.C., Rome had replaced Greece as the chief urbanizing force in the West. The Romans adopted many urban traits from the Greeks as well as from the Etruscans, a civilization of central Italy that Rome had conquered. As the Roman Empire expanded, city life diffused farther into France, while also reaching Germany, England, interior Spain, the Alpine countries, and parts of eastern Europe—areas that had not previously experienced urbanization. Most of these cities were military and trading outposts of the Roman Empire. The military camp, or *castrum*, was the basis for many of these new settlements. **Figure 11.28** (page 412) shows the diffusion of urban life into Europe as the Greek and Roman frontiers advanced.

The landscape of these Roman cities shared several traits with that of their Greek predecessors. The gridiron street pattern, used in later Greek cities, was fundamental to Roman cities. This pattern can still be seen in the heart of such Italian cities as Pavia (**Figure 11.29**, page 412). The straight streets and right-angle intersections make a striking contrast to the curved, wandering lanes of the later medieval quarters or the streets of Rome itself. At the intersection of a city's two major thoroughfares was the *forum*, a zone combining elements of the Greek acropolis and agora. Here were not only the temples of worship, administrative buildings, and warehouses, but also the libraries, schools, and marketplaces that served the common people.

Rome's most important legacy probably was not its architectural and engineering feats, although they remain landmarks in European cities to this day, but rather the Roman method for choosing the site of a city, which remains applicable today. The Romans consistently chose sites with transportation in mind. The Roman Empire was held together by a complicated system of roads and highways linking towns and cities. In choosing a site for a new settlement, the Romans made access to transportation a major consideration. The significance of Roman location was such that even though urban life declined dramatically with the collapse of the empire, many cities—such as Paris, London, and Vienna—were established centuries later on the same old Roman sites because they offered the advantage of access to the surrounding countryside.

With the decline of the Roman Empire by A.D. 400, urban life also declined. Historians attribute the fall of Rome

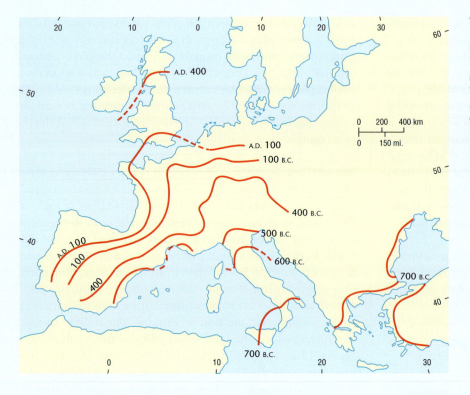

FIGURE 11.28 The diffusion of urbanization in Europe. The early spread of urban development moved in waves across Europe. The nucleus of city life was well established in the Greek lands by 700 B.C. In the following centuries, urbanization diffused westward and northward until it reached the British Isles. ***What do you think were some of the effects of this imposition of urban life on agricultural peoples?*** *(Reproduced by permission from Pounds, 1969.)*

to internal decay, the invasion of the Germanic peoples, and other factors. Cities were sapped of their vitality. The highway system that linked them fell into disrepair, so that cities could no longer exchange goods and ideas. As symbols of a conquering empire, Roman outposts were either actively destroyed or, devoid of purpose, simply left to decay.

Yet there were exceptions. Some cities of the Mediterranean survived because they established trade with the Eastern Roman Empire, centered in Constantinople. After the

eighth century, some cities—particularly those in Spain—were infused with new vigor by the Almohad and Almoravis empires of Morocco, which spread across the Mediterranean from northern Africa. The cities of northern regions were unable to survive, however. Cities became small villages. Where thousands had formerly thrived, a few hundred eked out a living.

Urban decline occurred only in the areas that had been under Roman rule. Other civilizations continued to thrive

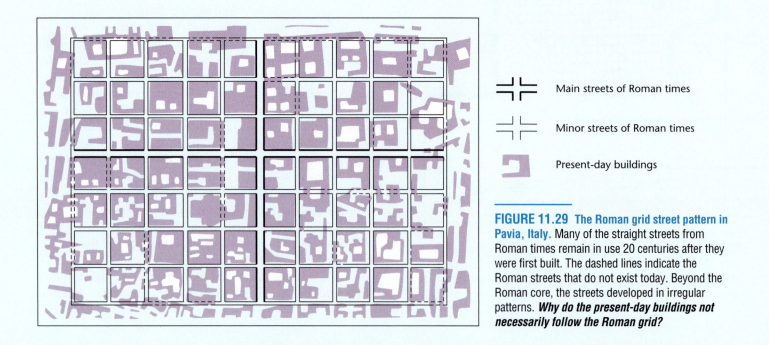

Main streets of Roman times

Minor streets of Roman times

Present-day buildings

FIGURE 11.29 The Roman grid street pattern in Pavia, Italy. Many of the straight streets from Roman times remain in use 20 centuries after they were first built. The dashed lines indicate the Roman streets that do not exist today. Beyond the Roman core, the streets developed in irregular patterns. ***Why do the present-day buildings not necessarily follow the Roman grid?***

throughout this period. The achievements of Chinese civilization and the great cities of the Mayan Empire remind us that this collapse was limited to a particular area of Europe.

The Medieval City The medieval period, lasting roughly from A.D. 1000 to 1500, was a time of renewed urban expansion in Europe that also deeply influenced the future of urban life. As the Germanic and Slavic peoples expanded their empires, urban life spread beyond the borders of the former Roman Empire, into the north and east of Europe. In only four centuries, 2500 new German cities were founded. Most cities of present-day Europe were established during this period, including many built on old Roman sites.

Scholars have debated why urban life began to regain vigor in the eleventh century. In essence, it was the result of the revival of both local and long-distance trade, which was itself the consequence of a combination of factors, including population increase, political stability and unification, and agricultural expansion through new land reclamations along with the development of new agricultural technologies. Sustained trading networks required protected markets and supply centers, functions that renewed life in cities. In addition, trading—particularly over long distances—led to the development of a new social class: the merchant class. Members of the merchant class breathed new life into early medieval cities, providing the impetus and the wealth for sustained city building.

The medieval city can be characterized by the presence of four features: the charter, the wall, the marketplace, and the cathedral. The charter was a governmental decree from a regional power, usually a feudal lord, granting political autonomy to the town. This act had important implications, as it freed the population from feudal restrictions, made the city responsible for its own defense and government, and

often allowed it to coin money. The wall served a defensive purpose, but it was also a symbol of the sharp distinction between country and city (**Figure 11.30**). Within the wall, most inhabitants were, by charter, free; outside, most were serfs.

Another central feature was the marketplace. It symbolized the important role of economic activities in the medieval city. The city depended on the countryside for its food and produce, which were traded in the market. The market was also a center for long-distance trade. Textiles, salt, ore, and other raw materials were bought and sold in the marketplace. At one end of the marketplace stood the town hall, a fairly tall structure that provided meeting space for the city's political leaders. The town hall often served as a market hall as well, with many of its rooms used to store and display the finer goods that could not be exposed to the natural elements outside on the market square. Yet, in many of the larger commercial cities, civic and economic functions were located in separate buildings. Brugge, Belgium, an important trading center for northern Europe, had two distinct complexes of buildings at its center (**Figure 11.31**, page 414).

The crowning glory of a medieval town was usually the cathedral, a dominating architectural symbol of the importance of the church. Often the cathedral, the marketplace, and the town hall were close together, indicating close ties among religion, commerce, and politics. However, the church was frequently the prevailing political force in medieval towns.

The functional zonation of the medieval city differs markedly from that of our modern cities. The city was divided into small quarters, or districts, each containing its own center that served as its focal point. Within each of these districts lived people who were engaged in similar occupations. Coopers (people who made and repaired wooden barrels), for example, lived in one particular district,

FIGURE 11.30 **The medieval hill town of Carcassonne in southern France.** Notice the double set of fortified walls that surround this medieval town. *(Jonathan Blair/Corbis.)*

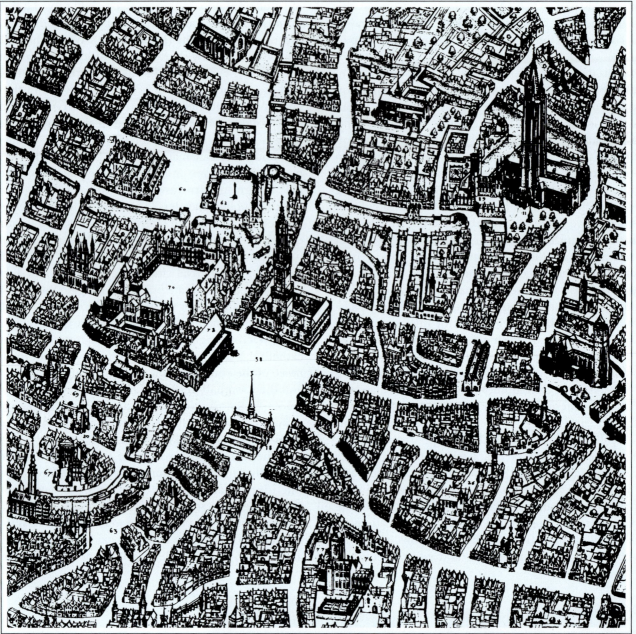

11. Cathedral of Notre Dame.
12. Church of St Sauveur.
18. Chapel of St Christopher.
20. Chapel of St John.
21. Chapel of St Amanda.
22. Chapel of St Peter.
26. Chapel of the Painters.

58. Fish market.
60. Grain market.
62. Leather market.
63. Bourse.
70. Castle, with the Town Hall and Chancellery.
71. Halle.

72. Waterhalle.
75. Prison.
76. Prince's Hall.
77. Mint.
88. So-called 'Castle of the Seven Turrets'.

FIGURE 11.31 **Part of a 1562 panoramic map of the city of Brugge, Belgium, showing the central area.** Directly in the center of the image is the great Halle building, the economic heart of the city. Just in front of it and to the left is the Waterhalle, so named because it straddled the canal, allowing goods to be delivered directly into the building. To the left is the old castle surrounded by guildhalls and the town hall. To the extreme right is the cathedral building. ***How do the reasons for this organic urban plan compare and contrast with the reasons for the grid plan of Roman cities?*** (See Figure 11.29, page 412.) *(Source: Benevolo, 1980.)*

attended the same local church, and belonged to the same guild. Their church and guildhall were located in the small center area of their district. Along the narrow, winding streets surrounding this center area were the houses and workplaces of the coopers. Many worked in the first story of their houses and lived above the shop, with their apprentices living above them.

Some of these districts, however, were defined not by occupation but by ethnicity—these areas have been referred to as ghettos. The origin of the term **ghetto** is somewhat unclear, although one plausible explanation suggests that the word dates from the early sixteenth century, when Venetians decided to restrict Jewish settlement in the city to an area already known as Ghetto Nuovo, or the new foundry. This area was physically separated from the rest of the city and had a single entrance that could be guarded. The practice of spatially segregating the Jewish population was not limited to Venice. In most medieval cities, Jews were forced to live in their own districts. In Frankfurt am Main, Jews lived on the *Judengasse,* a street that was formed from the dried-up moat that had run along the old wall to the city. The *Judengasse* was enclosed by walls with only one guarded gate for entrance and exit. Because the area was not allowed to expand beyond those walls, a growing population led to denser living conditions. In 1462, the population of the *Judengasse* was only 110 inhabitants; by 1610, 3000 people lived in the Jewish ghetto, creating one of the densest districts in the city.

In summary, there are three fundamental points to be made about the role of the medieval period in the evolution of the Western city: (1) most European cities were founded during this period; (2) many of the traditions of Western urban life began then; and (3) the medieval landscape is still with us, providing a visible history of the city and a distinctive form into which twenty-first-century activities are placed.

ghetto
Traditionally, an area within a city where an ethnic group lives, either by choice or by force. Today in the United States, the term typically indicates an impoverished African-American neighborhood.

The Renaissance and Baroque Periods During the Renaissance (approximately 1500–1600) and the Baroque period (1600–1800), the form and function of the European city changed significantly. Absolute monarchs arose to preside over unified countries. The burghers, or rising middle class, of the cities slowly gave up their freedoms to join with the king in pursuit of economic gain. City size increased rapidly because the bureaucracies of regional power structures came to dominate cities and because trade patterns expanded with the beginnings of European imperial conquest. One city, the national capital, rose to prominence in most countries.

A new concern with city planning went hand-in-hand with these developments. Rulers considered the city a stage on which to act out their destinies, and as a stage, the city could be rearranged at will. Most planning measures, then, were meant to benefit the privileged classes. Typical of the time was the infatuation with wide, grandiose boulevards (**Figure 11.32**). The rich could ride along them in carriages, and the army could march along them in an impressive display of power. Other features of the Baroque city were large, open squares; palaces; and public buildings.

The Capitalist City Underlying many of the innovations in Renaissance and Baroque city planning was a sweeping socioeconomic transformation that reshaped western

FIGURE 11.32 A view of Paris, showing boulevards designed by Baron Haussmann. The boulevard was a favorite of Baroque planners. It was a ceremonial street that often led to public buildings and monuments, was lined with trees and upper-income housing, and offered public space for the wealthy. Boulevards were often created at the expense of thousands of poorer citizens, who were displaced as older housing was destroyed by the boulevard builders. *Has similar displacement occurred in your city or one near you as freeways have been built?* (*Jeff Greenberg/Photo Researchers.*)

Europe. The transition from a feudal order to a capitalist one, which stretched from the mid-sixteenth century to the mid-eighteenth century, involved drastic changes in class structure, economic systems, political allegiances, cultural patterns, and human geographies. The countryside was reordered with the introduction of commercialized and specialized agriculture and with the enclosure of individual land units. The city was also reshaped, as the value of two-dimensional location and three-dimensional form in the city acquired economic significance.

Perhaps of greatest significance was how the capitalist mind-set introduced a notion of urban land as a source of income. Proximity to the center of the city, and therefore to the most pedestrian traffic, added economic value to land. Other specialized locations, such as areas close to the river or harbor, or along the major thoroughfares into and out of the city, also increased land value. This fundamental change in the value accorded to urban land led to the gradual disintegration of the medieval urban pattern.

In the emerging capitalist city, the ability to pay determined where one would live. The city's residential areas thus became segregated by economic class. The wealthy lived in the desirable neighborhoods; those without much money were forced to live in the more disagreeable parts of the city. In addition, places of work were separated from home, so that a merchant, for example, lived in one part of the city and traveled to another to conduct his business. This spatial separation of work from home, of public space from private space, both reflected and helped to shape the changing social worlds of men and women. In general, men generated economic income from work outside the home

and therefore came to be associated with the public space of the city. Women, who were primarily engaged in domestic work, were considered the keepers of the private world of the home. This association of women with private domestic space and men with public work space deepened and became more complex throughout the next few hundred years.

Reflecting on Geography

How did (and does) the association of private domestic space with women and of public work space with men affect the daily lives of men and women? Can you think of counterexamples, that is, instances of the merging of private and public spaces?

The center of the capitalist city was not the cathedral or town hall, but instead the buildings devoted to business enterprises. A downtown defined by economic activity emerged that, with the coming of industrialization, eventually expanded and subdivided into specialized districts. The new upper classes of the city, whose status was based on their accumulation of wealth, not only made money from buying and selling urban land but also used urban land as a basis for expressing their wealth. With the downtown devoted to mercantile and emerging industrial uses, the upper classes sought newer land on the edge of the city for their residential enclaves. These new areas often acted as three-dimensional symbols of relatively recent wealth, conferring on their residents the legitimacy of upper-class membership.

One of the first and finest of these new enclaves for the wealthy was London's Covent Garden Piazza, a residential

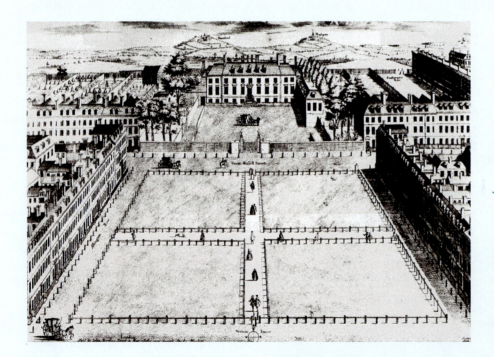

FIGURE 11.33 A 1730 view of Bloomsbury Square, laid out in 1661 by the Earl of Southampton. Southampton House occupies the far end of the square, lending an aristocratic air to this speculative, mercantile development. Notice the men and women parading in their finery, suggesting the wealthy and leisurely life of the inhabitants. *What other signs of wealth are evident in this image?* (Source: Hayes, 1969.)

square designed by Inigo Jones in the early 1630s. The inhabitants of Covent Garden included some of London's nobility and wealthier bourgeoisie. The presence of nobility lent an aristocratic aura to the area and provided social legitimacy to the newly affluent who lived there. The economic success of this speculative real estate venture led to many imitations, and similar residential squares cropped up throughout the West End of London (**Figure 11.33**). These upper-class squares were transplanted to America throughout the seventeenth and eighteenth centuries, arising in such cities as Boston, New York, Philadelphia, and Savannah.

Class, "Race," and Gender in the Industrial City Until the industrial period, the rate of urbanization in Western countries was relatively low. For example, in 1600, urban dwellers made up only 2 percent of the German, French, and English populations; in the Netherlands and Italy, 13 percent of the population were urban dwellers. However, as millions of people migrated to the cities over the next 200 years, the rate of urbanization skyrocketed. By 1800, England was 20 percent urbanized, and around 1870 it became the world's first urban society. By the 1890 census, 60 percent of its people lived in cities. The United States was 3 percent urbanized in 1800, 40 percent in 1900, and 51 percent in 1920 (when it became an urban country); today about 75 percent of its population live in towns and cities.

The industrial revolution greatly accelerated the growth of cities and brought large numbers of people into the city (see Chapter 9). With the development of steam power, factories began to cluster together in cities to share the benefits of agglomeration—that is, to share labor, transportation costs, and utility costs—and to take advantage of financial institutions found in the city. Land use intensified drastically. With the increased competition for land in the industrial period, land transactions and speculation became an everyday part of city life. Land parcels became the property of the owner, who had no obligations to society in deciding how to use them. The historical urban core was often destroyed, the older city replaced. The result was a mosaic of mixed land uses: factories directly next to housing, slum tenements next to public buildings, open spaces and parks violated by railroad tracks. A planned attempt to bring order to the city came only in the twentieth century with the concept of zoning.

Class Laissez-faire industrialism did little for the working classes that labored in shops and plants. In their slum dwellings, direct sunlight was seldom available and open spaces were nonexistent. In Liverpool, England, for instance, one-sixth of the population reportedly lived in "underground cellars." A study from the middle of the nineteenth century showed that in Manchester, England, there was but one toilet for every 212 people. In 1893, the life expectancy of a male worker in Manchester was 28; his country cousin might live to 52. The death rate in New York City (**Figure 11.34**) in 1880 was 25 per 1000; it was half that in the rural counties of the state. Legislation correcting such ills came only in the latter part of the nineteenth century.

"Race" With all its faults, industrialization created some of the most vibrant centers of urban activity in modern times. American industrial cities relied on a diverse labor force, and each social group fought for its place in the

FIGURE 11.34 New York City, looking southwest from the Bronx. The rows and rows of working-class housing in the foreground and the factories visible to the left indicate the horizontal spatial expansion of industrial urban form; the massive skyline indicates the degree of vertical expansion. *(Courtesy of Mona Domosh.)*

urban land market. Despite the harsh living conditions, various groups of laborers carved out identities in the urban landscape.

Industrialization in the United States drew its workforce not only from European immigrants but also from African-Americans. After the Civil War, many former slaves in the South either migrated to northern cities to work in a diverse array of skilled and semiskilled jobs or moved from the countryside to industrializing cities. In both northern and southern cities, the African-American population lived in segregated neighborhoods, forced by discrimination and often by law to keep its distance from Anglo-American residential districts.

Although the services provided to these neighborhoods were usually minimal, many people did find opportunities for cultural expression in the new urban mosaic. A study of African-Americans in Richmond, Virginia, after the Civil War found that residents effectively used public rituals in the streets and buildings of the city to carve out their own civic representations as well as to challenge the dominant Anglo-American order. For example, African-American militias were formed that marched through the streets of Richmond on holidays certified by the African-American community as their own political calendar: January 1, George Washington's birthday, April 3 (Emancipation Day), and July 4. As urban historians Elsa Barkley Brown and Gregg Kimball state: "Richmonders watched in horror as former slaves claimed civic holidays white residents believed to be their own historic possession, and as black residents occupied spaces, like Capitol Square, that formerly had been reserved for white citizens." Other spaces, such as churches, schools, and beauty shops, served as both community centers and public statements of an African-American identity. In this way, the urban landscape acted as one arena for the struggle to control the meanings and uses of an environment often thought to be totally dominated by Anglo-American culture (**Figure 11.35**).

Reflecting on Geography

Consider how important the appropriation of space is to forming and maintaining cultural identity in the city. Why might such appropriation be more important for groups that are not dominant in a society? What do examples from your city tell you about struggles to control the meaning of its built environment?

Gender Industrialization, as we have seen, not only destroyed sections of cities to make way for railroads and factories but also made possible the creation of new urban identities and neighborhoods. Throughout the nineteenth century, the industrial city became increasingly segregated by function; large areas of the city were dedicated to the production of goods and services, surrounded by working-class neighborhoods. At the same time, the center of the city was remade into an area of consumption and leisure, with large department stores, theaters, clubs, restaurants, and nightclubs. In New York City during the last half of the nineteenth century, one of the foremost displays of such a culture of consumption was located along Broadway and Sixth Avenue between Union and Madison squares. This area was called Ladies' Mile because, as the new class of consumers, middle-class women were the major patrons of the large department stores that architecturally dominated the streets.

Although industrialization led to a particular form of separate spheres—the female sphere centered on the home and domestic duties, the male sphere dominating the public spaces and duties—it also created the need for mass consumption to keep the factories running profitably.

FIGURE 11.35 Sketch of an African-American congregation in Washington, D.C. African-American churches often served as centers of community organizations and as public statements of identity in the industrializing cities of the North.

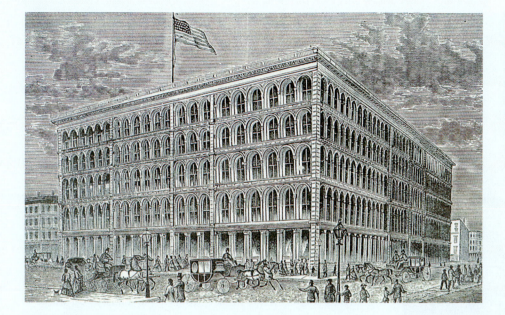

FIGURE 11.36 Stewart's department store. One of the most ornate department stores along Ladies' Mile in New York City was Stewart's, located on Broadway at 10th Street. Because the store catered to the needs of Victorian women, it can be considered an example of "feminine" space. *Can you think of other examples of such "feminine" spaces in your city or in a city with which you are familiar?*

With men as the class of producers, the duties of consumption fell to the women. Moreover, the locational logic of the urban land market meant that retailers were located in the most central parts of the city. This established what some scholars have referred to as a feminized downtown, meaning not only that the downtown was characterized by the presence of middle- and upper-class women but also that the retailers themselves created spaces considered appropriately "feminine." Interiors were orderly and well arranged, the facades of buildings were heavily ornamented, and streets were paved and well lit (**Figure 11.36**).

Although this type of ornate and "feminine" downtown retailing area is still evident in large cities such as New York and San Francisco, the decentralizing forces in the twentieth-century city led to the abandonment of many of these areas, which have been replaced by the suburban shopping mall. It would be interesting to speculate in what ways the shopping mall is also a "feminine" space.

Megalopolis   In the nineteenth century, cities grew at unprecedented rates because of the concentration of people and commerce. The inner city became increasingly dominated by commerce and the working class. In the twentieth century, particularly after World War II, new forms of transportation and communication led to the decentralization of many urban functions. As a result, one metropolitan area blends into another, until megacities are created that stretch for hundreds of miles. The geographer Jean Gottmann coined the term **megalopolis** to describe these megacities.

This term is now used worldwide in reference to giant metropolitan regions such as Tokyo-Yokohama in Japan and the New York City–northern

> **megalopolis**
> A large urban region formed as several urban areas spread and merge, such as Boswash, the region including Boston, New York, and Washington, D.C.

New Jersey areas. These urban regions are characterized by high population densities extending over hundreds of square miles or kilometers; concentrations of numerous older cities; transportation links formed by freeways, railroads, air routes, and rapid transit; and an extremely high proportion of the nation's wealth, commerce, and political power.

Edge Cities   The past 25 years have witnessed an explosion in metropolitan growth in areas that had once been peripheral to the central city. Many of the so-called bedroom communities of the post–World War II era have been transformed into urban centers, with their own retail, financial, and entertainment districts (**Figure 11.37**, page 420). Author Joel Garreau refers to these new centers of urban activity as **edge cities**, although many other terms have been used in the past, including *suburban downtowns*, *galactic cities*, and *urban villages*. As Garreau mentions, most Americans now live, work, play, worship, and study in this type of settlement. What differentiates an edge city from the suburbs is that it is a place of work, of productive economic activity, and therefore is the destination of many commuters. In fact, the conventional work commute from the suburbs to the inner city has been replaced by commuting patterns that completely encircle the inner city. People live in one part of an edge city and commute to their workplace in another part of that city. Garreau argues that an edge city is a place defined by five characteristics:

> **edge city**
> A new urban cluster of economic activity that surrounds nineteenth-century downtowns.

1. Has five million square feet or more of leasable office space.

2. *Has 600,000 square feet or more of leasable retail space.*

3. *Has more jobs than bedrooms.* The idea here is that these are places that people go to for their work—in

FIGURE 11.37 Edge city. These new centers of economic activity are located on the "edges" of traditional downtowns. *(Walter Jimenez/TexStock Photo Inc.)*

other words, like all cities, the number of people here increases at 9 A.M.

4. *Is perceived by the population as one place.* In other words, this place is recognized locally as a defined place, with a name.

5. *Was nothing resembling a "city" as recently as 30 years ago.*

Many scholars are wary of referring to these new nodes of activity as cities because they do not resemble our nineteenth- or early-twentieth-century vision of a city. Edge cities contain all the functions of old downtowns, but they are spread out and less dense, with clusters along major freeways and off-ramps. This new form is attributable to changes in Americans' lifestyles and to the development of new transportation and communication technologies. The interstate highway system made possible an effective trucking system to transport consumer goods, thereby enabling new industries to locate outside the downtown. Breakthroughs in computer and communication technologies have allowed corporate executives to move company headquarters out of the downtowns and into sleek new glass buildings with parking

garages, jogging paths, and picnic tables under the trees. Real estate speculation in emerging edge cities has fueled their development, resulting in an environment that many people feel is ugly and chaotic. As Garreau points out, however, even a place as revered by designers as Venice, Italy, began as an ad hoc mercantilist adventure; the Piazza San Marco was the result not of great urban planning but of the coincidence of centuries of building and rebuilding by people concerned with making money.

The New Urban Landscape

Within the past 30 to 35 years, our cityscapes have undergone massive transformations. The impact of suburbanization and decentralization has led to edge cities, while in older downtown areas, we have seen that gentrification, redevelopment, and immigration have also created novel urban forms. What we see now in our cityscapes, then, is a new urban landscape, composed of distinctive elements that we now will discuss.

Shopping Malls Many people consider the image of the shopping mall, surrounded by mass-produced suburbs, to be one of the most distinctive landscape symbols of modern urbanity. Yet, oddly, most malls are not designed to be seen from the outside. As a matter of fact, without appropriate signs, a passerby could proceed past a mall without noticing any visual display. Unlike the retail districts of nineteenth- and early-twentieth-century cities, where grand architectural displays along major boulevards were the norm, shopping malls are enclosed, private worlds that are meant to be seen from the inside. Often located near an off-ramp of a major freeway or beltway of a metropolitan area, and close to the middle- and upper-class residential neighborhoods, shopping malls can be distinguished more by their extensive parking lots than by their architectural design.

For all that, shopping malls do have a characteristic form. The early malls of the 1960s tended to have a simple, linear form, with two department stores at each end that functioned as "anchors" and 20 to 30 smaller shops connecting the two ends. In the 1970s and 1980s, much larger malls were built, and their form became more complex.

Malls today are often several stories high and may contain five or six anchor stores and up to 400 smaller shops. In addition, many malls now serve more than a retail function—they often contain food courts and restaurants, professional offices, movie complexes, hotels, chapels, and amusement arcades and centers (**Figure 11.38**). A shopping center in Kuala Lumpur, known simply as The Mall, contains Malaysia's largest indoor amusement park, a replica of the historic city of Malacca, a cineplex, and a food court, in addition to hundreds of stores (**Figure 11.39**). Many scholars consider these megamalls to be the new centers of urban life because they seem to be the major

FIGURE 11.38 **Food court at the Fashion Mall in Plantation, Florida.** A large percentage of the third story of this mall is devoted to fast-food outlets. The fountain and natural lighting are meant to create a gardenlike setting for mall dining. *Why do you think these fast-food outlets are all clustered together within the mall?* (Courtesy of Mona Domosh.)

sites for social interaction. For example, after workplace, home, and school, the shopping mall is where most Americans spend their time.

However, unlike the open-air marketplaces of an earlier era, shopping malls are private, not public, spaces. The use of the shopping mall as a place for social interaction, therefore, is always of secondary importance to its private, commercial function. If a certain group of people were considered a nuisance to shoppers, the mall owners could prevent them from what Jeffrey Hopkins calls "'mallingering'—the act of lingering about a mall for economic and noneconomic social purposes."

Office Parks With the connection of metropolitan areas by major interstate highways and with the development of new communication technologies, office buildings no longer need to be located in the central city. Cheaper rent in suburban locations, combined with the convenience of easy-access parking and the privacy of a separate location, has led to the construction of **office parks** throughout suburban America. Many of these office parks are occupied by the regional or national headquarters of large corporations or by local sales and professional offices. To take advantage of economies of scale, many of these offices will locate together and rent or buy space from a land development company.

> **office park**
> A cluster of office buildings, usually located along an interstate, often forming the nucleus of an edge city.

The use of the word *park* to identify this new landscape element points to the consciously antiurban imagery of these complexes. Many of these developments are surrounded by a well-landscaped outdoor space, often incorporating artificial lakes and waterfalls (**Figure 11.40**, page 422). Jogging paths, fitness trails, and picnic tables all cater to the new lifestyle of professional and managerial employees.

Many office parks are located along what have been called **high-tech corridors:** areas along limited-access highways that contain offices and other services associated with new high-tech industries. As our discussion of edge cities suggests, this new type of commercial landscape is gradually replacing our downtowns as the workplace for most Americans.

> **high-tech corridor**
> An area along a limited-access highway that houses offices and other services associated with high-tech industries.

Master-Planned Communities Many newer residential developments on the suburban fringe are planned and built as complete neighborhoods by private development

FIGURE 11.39 **Interior of a shopping mall in Kuala Lumpur, Malaysia.** *In what ways does this mall look similar to one near you? In what ways is it different?* (© Mark Henley/Impact/HIP/The Image Works.)

master-planned communities

Large-scale residential developments that include, in addition to architecturally compatible housing units, planned recreational facilities, schools, and security measures.

companies. These **master-planned communities** include not only architecturally compatible housing units but also recreational facilities (such as tennis courts, fitness centers, bike paths, and swimming pools) and security measures (gated or guarded entrances). In Weston, a master-planned community that covers approximately 10,000 acres (4000 hectares) in southern Florida, land use is completely regulated not only within the gated residential complexes but also along the road system that connects Weston to the interstate (**Figure 11.41**). Shrubbery is planted strategically to shield residents from views of the roadway, and the road signs are uniform in style and encased in stylish, weathered-gray wood frames. This massive community—Weston is now home to 50,000 people—contains various complexes catering to particular lifestyles, ranging from smaller patio homes to equestrian estates. For example, in the mid-1990s, homes in one such complex—Tequesta Point—cost $250,000–$300,000 and came with gated entranceways, split-level floor plans, and Roman bathtubs. Those in Bermuda Springs, on the other hand, which originally cost $115,000–$120,000, were significantly smaller and did not offer the same interior features. Typical of master-planned communities, the name of the development itself was chosen to convey a hometown feeling. In this instance, developers carried out an extensive marketing survey before they settled on the name Weston.

Festival Settings In many cities, gentrification efforts focus on a multiuse redevelopment scheme that is built around a particular setting, often one with a historical association. Waterfronts are commonly chosen as focal points for these

large-scale projects, which Paul Knox has referred to as **festival settings.** These complexes integrate retailing, office, and entertainment facilities and incorporate trendy shops, restaurants, bars and nightclubs, and hotels. Knox suggests that these developments are "distinctive as new landscape elements merely because of their scale and their consequent ability to stage—or merely to be—the spectacular." Such festival settings as Faneuil Hall in Boston, Bayside

festival setting

A multiuse redevelopment project that is built around a particular setting, often one with a historical association.

FIGURE 11.41 **A typical streetscape in Weston, Florida.** Notice the manicured lawns and evenly spaced palm trees that line the streets. (Larry Luxner.)

FIGURE 11.42 **An outdoor café near Faneuil Hall in Boston.** As you can see from this image, many people enjoy the spaces of festival settings for the social interactions they produce. Not everyone in the city, though, enjoys the benefits of these "staged" landscapes. *Can you think of some people who have suffered as a result of this multiuse development?* (Andre Jenny/Newscom.)

in Miami, and Riverwalk in San Antonio serve as sites for concerts, ethnic festivals, and street performances; they also serve as focal points for the more informal human interactions that we usually associate with urban life (**Figure 11.42**). In this sense, festival settings do perform a vital function in the attempt to revitalize our downtowns. Yet, like many other gentrification efforts, these massive displays of wealth and consumption often stand in direct contrast to neighboring areas of the inner city that have received little, if any, monetary or other social benefit from these projects.

"Militarized" Space Considered together, these new elements in the urban landscape suggest some trends that many scholars find disturbing. Urbanist Mike Davis has called one such trend the "militarization" of urban space, meaning that increasingly space is used to set up defenses against people the city considers undesirable. This includes landscape developments that range from the lack of street furniture to guard against the homeless living on the streets, to gated and guarded residential communities, to the complete segregation of classes and "races" within the city. Particularly in downtown redevelopment schemes, the goal of city planners and others is usually to provide safe and homogeneous environments, segregated from the diversity of cultures and lifestyles that often characterizes the central area of most cities. As Davis says, "Cities of all sizes are rushing to apply and profit from a formula that links together clustered development, social homogeneity, and a perception of security."

Although this "militarization" is not completely new, it has taken on epic proportions as whole sections of such cities as Los Angeles, Atlanta, Dallas, Houston, and Miami have become "militarized" spaces (**Figure 11.43**).

Reflecting on Geography

Some scholars might argue that the increasing "militarization" of our urban spaces will lead to situations little different from what happened to South African cities under apartheid (see Figure 10.22, page 377). Identify similarities and differences between these two urban situations.

The Decline of Public Space Related to the increase in "militarized" space is the decline of public spaces in most of our cities. For example, the shift in shopping patterns from the downtown retailing area to the suburbs indicates a change of emphasis from the public space of city streets to the privately controlled and operated shopping malls. Similarly, many city governments, often joined by private developers, have built enclosed walkways either above or below the city streets. These walkways serve partly to provide a climate-controlled means of passing from one building to another and partly to provide pedestrians with a "safe" environment that avoids possible confrontations on the street. Again, the public space of the street is being replaced by controlled spaces that do not provide the same access to all members of the urban community. It remains to be seen whether this trend will continue or whether new public spaces will be formed that accommodate all groups of people within the urban landscape.

FIGURE 11.43 **The Metro Dade Cultural Center in downtown Miami.** Inside this literal fortress is Dade County's public library and Center for the Fine Arts. The building was designed to be completely removed from the street, with the few entrances heavily monitored by cameras. *Does this seem like a welcoming space? Should it be?* (Courtesy of Mona Domosh.)

CONCLUSION

We have examined various components of the intricate urban mosaic. Culture regions are found at a smaller scale than others that we have previously explored; in the city, neighborhoods and census units can be thought of as social regions and culture regions. Models of the internal structure of cities describe how social and economic activities sort themselves out in space. The simplification of these models calls into question their usefulness for describing the contemporary American city as well as other cities around the world.

We also see two major forces at work in the city that are examples of locational mobility. One works to centralize activities within the city, the other to decentralize activities into the suburbs. The costs of decentralization run high, not just in the suburbs, where unplanned growth takes its toll, but also in the inner cities, which are left with decayed and stagnant cores.

Globalization is a major force in reshaping cities today. On the one hand, the creation of transnational development firms means that urban landscapes around the world are beginning to share some common elements, such as skyscrapers, shopping malls, and gated communities. On the other hand, local cultural contexts are actively shaping these similar forms, leading to more diversity within the urban mosaic.

As dense collections of human artifacts, cities offer a fascinating array of cultural landscapes that tell us much about ourselves and our interaction with the environment. These cityscapes reveal contemporary change, the past, prevalent scenic values, and our storehouse of symbols. New elements in the urban landscape relay information about our current social, economic, and political reality, and they indicate trends for the future that many scholars find disturbing.

DOING GEOGRAPHY

Reading "Your" Urban Landscape

Even without extensive training in geography, most of us are avid "readers" of our landscapes. We drive through residential neighborhoods and make judgments about the types of people who live there; we walk downtown and, finding that the streets get narrower, surmise that they were built a long time ago, before automobile traffic was a factor; we marvel at how high buildings rise into the sky, knowing that those with offices at the top make "top" dollar. But reading landscapes can be tricky business; what we see in front of us may not always be as it appears, nor built when and for whom we think. This exercise, then, is about interpreting your own urban landscape and about understanding the limitations of that interpretation.

Your goal in this exercise is to create an interpretative walking tour of part of your city that could be used by tourists (or anyone, really) and that is both informative and engaging. In other words, the idea is to get people to look, think, and make connections—but also to understand the limitations of "just looking." To do this exercise, follow the steps below.

Steps to Reading Your City

Step 1: Pick your place. Decide on a route that takes people through a particular part of your city—downtown, an ethnic landscape, a commercial area, a local neighborhood, and so on. Almost anywhere will work. From reading this chapter, you should have a pretty good idea of the types of urban landscapes and the ones that are of interest to you.

Step 2: Walk the route yourself, marking places where you want people to stop, look, and think.

Step 3: Investigate the places you marked. You should be able to use the information in this chapter and elsewhere in the book to do a good bit of the investigative work, but the specifics of your city will require more information: archival research, perhaps some informal interviews, or something else altogether. At each point, you should be asking yourself: What can I learn from looking at this landscape and what could I never have guessed?

Step 4: Based on what you have learned, write your interpretative tour, providing information and questions for each particular stop on the route.

Finally, try it out—give your route map and written descriptions and questions to a friend, and see what happens.

A young woman observes the urban landscape. *(© Exactostock/ SuperStock.)*

Key Terms

The Urban Mosaic on the Internet

You can learn more about cities and suburbs on the Internet at the following web sites:

The Centre for Urban History at the University of Leicester

http://www.le.ac.uk/ur/index.html

This site is a wonderful portal for researching the history of urban form online.

Levittown: Documents of an Ideal American Suburb

http://www.uic.edu/~pbhales/Levittown

Here you can find a history of Levittown, New York, the first mass-produced suburb, with interesting historical and contemporary photographs.

The Skyscraper Museum

http://www.skyscraper.org/home.htm

The Skyscraper Museum in New York City runs this site and provides extremely useful information and virtual tours of tall buildings around the world.

The State of the Nation's Cities: A Comprehensive Database on American Cities and Suburbs

http://www.policy.rutgers.edu/cupr/sonc/sonc.htm

This site is a comprehensive database on 77 American cities and suburbs that you can download onto your computer and display in graphic form.

Sources

Abbott, Edith, and Mary Zahrobsky. 1936. "The Tenement Areas and the People of the Tenements," in E. Abbott (ed.), *The Tenements of Chicago.* Chicago: University of Chicago Press, 72–169.

Agnew, John, John Mercer, and David Sopher (eds.). 1984. *The City in Cultural Context.* Boston: Allen & Unwin.

Anderson, Kay. 1987. "The Idea of Chinatown: The Power of Place and Institutional Practice in the Making of a Racial Category." *Annals of the Association of American Geographers* 77: 580–598.

Bach, Wilfred, and Thomas Hagedorn. 1971. "Atmospheric Pollution: Its Spatial Distribution over an Urban Area." *Proceedings of the Association of American Geographers* 3: 22.

Bell, David, and Gill Valentine (eds.). 1995. *Mapping Desire: Geographies of Sexualities.* London: Routledge.

Bell, Morag. 1995. "A Woman's Place in 'a White Man's Country.' Rights, Duties and Citizenship for the 'New' South Africa, c. 1902." *Ecumene* 2: 129–148.

Benevolo, Leonardo. 1980. *The History of the City.* Cambridge, Mass.: MIT Press.

Brown, Elsa Barkley, and Gregg D. Kimball. 1995. "Mapping the Terrain of Black Richmond." *Journal of Urban History* 21: 296–346.

Burgess, Ernest W. 1925. "The Growth of the City: An Introduction to a Research Project," in Robert E. Park, Ernest W. Burgess, and Roderick D. McKenzie (eds.), *The City.* Chicago: University of Chicago Press, 47–62.

Castells, Manuel. 1983. *The City and Grassroots: A Cross-Cultural Theory of Urban Social Movements.* Berkeley: University of California Press.

Chauncey, George. 1994. *Gay New York: Gender, Urban Culture, and the Making of the Gay Male World, 1890–1940.* New York: Basic Books.

Clay, Grady. 1973. *Close-Up: How to Read the American City.* New York: Praeger.

Davis, Mike. 1992. "Fortress Los Angeles: The Militarization of Urban Space," in Michael Sorkin (ed.), *Variations on a Theme Park: The New American City and the End of Public Space.* New York: Hill & Wang, 154–180.

Domosh, Mona. 1995. "The Feminized Retail Landscape: Gender Ideology and Consumer Culture in Nineteenth-Century New York City," in Neil Wrigley and Michelle Lowe (eds.), *Retailing, Consumption and Capital.* Essex, U.K.: Longman, 257–270.

Doxiades, Constantine A. 1972. *Architectural Space in Ancient Greece.* Cambridge, Mass.: MIT Press.

SEEING GEOGRAPHY Chinatown, New York City

How has globalization affected urban ethnic neighborhoods?

A street in Chinatown, New York City.

"Reading" this urban landscape is relatively simple: the signs of the stores and restaurants leave no doubt that this place is marked as a "Chinatown." Interpreting this scene is easy for most Americans because Chinatown has become a symbolic landscape: a place that connotes America's history and geography of immigration, the idea of the melting pot, and now a commitment to cultural pluralism. But there are other ways to understand this landscape as well, ways that rely on a keen eye and an urban geographical analysis. That some of the writing is in English and some in Chinese tells us that this streetscape is located in a place where both languages are spoken; the style of architecture, the types of cars, the street signs, and the American flag indicate a location in the United States. This Chinatown happens to be in New York City, but one could have guessed San Francisco, Boston, or Philadelphia.

Based on our understanding of the models of urban land use, we could surmise that Chinatown is located in a transitional zone, just outside the CBD. This certainly was the case in the late nineteenth and early twentieth centuries, when most of the brick buildings in this photograph were built. In this part of New York City, land values were relatively low. The financial center of Wall Street was to the south, and the retail areas along Broadway and Fifth Avenue were farther north, as were the middle- and upper-class residences. With few choices available, Chinese immigrants (and many others, including Italians and eastern European Jews) settled into this area and found jobs in the small industries and businesses nearby. The result was a series of neighborhoods of mixed land uses, on the edge of both the financial CBD and the retail CBD, in the zone of transition. Each of these neighborhoods came to be defined by the dominant Anglo culture as an ethnic region—in this case, Chinatown.

Today, Chinatown is still home to new migrants, but many of them come from other East Asian countries such as Korea and the Philippines. In the United States between 1997 and 2006, the Korean population grew 71 percent and the Filipino population, 52 percent. Although we know that an increasing proportion of new immigrants to the United States move to suburban areas, Chinatown in New York City still attracts many immigrants because of its accessibility to a fairly diverse job market and to the services available for immigrants who may not speak English and do not drive automobiles.

The Häagen-Dazs sign reminds us of the presence of globalizing forces and also indicates that Chinatown has become a tourist destination for visitors to the city. After a trip to the top of the Empire State Building or to the Metropolitan Museum of Art, tourists might go downtown for "authentic" Chinese food, stopping off for ice cream afterward. If they really wanted to sample the avant-garde club scene of the city, they might find themselves in a trendy bar on the edge of Chinatown. Much of the Lower East Side is already gentrified, and Chinatown is a likely next target. If that happens, the Mandarin Court sign might soon read Starbucks Coffee.

Drakakis-Smith, David. 2000. *Third World Cities,* 2nd ed. London: Routledge.

Duncan, James, and Nancy Duncan. 1984. "A Cultural Analysis of Urban Residential Landscapes in North America: The Case of the Anglophile Elite," in John Agnew, John Mercer, and David Sopher (eds.), *The City in Cultural Context.* Boston: Allen & Unwin, 255–276.

Duncan, James, and Nancy Duncan. 2003. *Landscapes of Privilege: The Politics of the Aesthetic in an American Suburb.* New York: Routledge.

Garreau, Joel. 1991. *Edge City: Life on the New Frontier.* New York: Doubleday.

Gilbert, Emily. 1994. "Naturalist Metaphors in the Literatures of Chicago, 1893–1925." *Journal of Historical Geography* 20: 283–304.

Gordon, Margaret T., et al. 1981. "Crime, Women, and the Quality of Urban Life," in Catherine R. Stimpson, Elsa Dixler, Martha J. Nelson, and Kathryn B. Yatrakis (eds.), *Women and the American City.* Chicago: University of Chicago Press, 141–157.

Gottmann, Jean. 1961. *Megalopolis.* Cambridge, Mass.: MIT Press.

Hanson, Susan. 1992. "Geography and Feminism: Worlds in Collision?" *Annals of the Association of American Geographers* 82: 569–586.

Hanson, Susan, and Geraldine Pratt. 1995. *Gender, Work, and Space.* New York: Routledge.

Harris, Chauncey D., and Edward L. Ullman. 1945. "The Nature of Cities." *Annals of the Association of the American Academy of Political and Social Science* 242: 7–17.

Hartshorn, Truman A., and Peter O. Muller. 1989. "Suburban Downtowns and the Transformation of Metropolitan Atlanta's Business Landscape." *Urban Geography* 10: 375–395.

Hayes, John. 1969. *London: A Pictorial History.* New York: Arco.

Hopkins, Jeffrey S. P. 1991. "West Edmonton Mall as a Centre for Social Interaction." *The Canadian Geographer* 35: 268–279.

Hoyt, Homer (ed.). 1939. *Structure and Growth of Residential Neighborhoods in American Cities.* Washington, D.C.: Federal Housing Administration.

Jackson, Kenneth T. 1985. *Crabgrass Frontier.* New York: Oxford University Press.

Kenny, Judith T. 1995. "Climate, Race and Imperial Authority: The Symbolic Landscape of the British Hill Station in India." *Annals of the Association of American Geographers* 85: 694–714.

Knopp, Lawrence. 1995. "Sexuality and Urban Space: A Framework for Analysis," in David Bell and Gill Valentine (eds.), *Mapping Desire.* London: Routledge, 149–161.

Knox, Paul L. 1991. "The Restless Urban Landscape: Economic and Sociocultural Change and the Transformation of Metropolitan Washington, D.C." *Annals of the Association of American Geographers* 81: 181–209.

Lauria, Mickey, and Lawrence Knopp. 1985. "Toward an Analysis of the Role of Gay Communities in the Urban Renaissance." *Urban Geography* 6: 152–169.

Li, Wei. 1998. "Anatomy of a New Ethnic Settlement: The Chinese Ethnoburb in Los Angeles." *Urban Studies* 35: 479–501.

Lynch, Kevin. 1960. *The Image of the City.* Cambridge, Mass.: MIT Press.

McDowell, Linda. 1983. "Towards an Understanding of the Gender Division of Urban Space." *Environment and Planning D: Society and Space* 1: 59–72.

Meinig, D. W. (ed.). 1979. *The Interpretation of Ordinary Landscapes: Geographical Essays.* New York: Oxford University Press.

Mills, Sara. 1996. "Gender and Colonial Space." *Gender, Place and Culture* 3: 125–148.

Mohl, Raymond A. 1995. "Making the Second Ghetto in Metropolitan Miami, 1940–1960." *Journal of Urban History* 21: 395–427.

Mumford, Lewis. 1961. *The City in History.* New York: Harcourt Brace Jovanovich.

Pirenne, Henri. 1996. *Medieval Cities.* Garden City, N.Y.: Doubleday (Anchor Books).

Portes, Alejandro, and Ruben G. Rumbaut. 1996. *Immigrant America: A Portrait,* 2nd ed. Berkeley: University of California Press.

Pounds, Norman J. G. 1969. "The Urbanization of the Classical World." *Annals of the Association of American Geographers* 59: 135–157.

Pow, Choon-Piew, and Lily Kong. 2007. "Marketing the Chinese Dream House: Gated Communities and Representations of the Good Life in (Post-)Socialist Shanghai." *Urban Geography* 28: 129–169.

Pratt, Geraldine. 1990. "Feminist Analyses of the Restructuring of Urban Life." *Urban Geography* 11: 594–605.

Rothenberg, Tamar. 1995. "'And She Told Two Friends': Lesbians Creating Urban Social Space," in David Bell and Gill Valentine (eds.), *Mapping Desire.* London: Routledge, 165–181.

Rowe, Stacy, and Jennifer Wolch. 1990. "Social Networks in Time and Space: Homeless Women in Skid Row, Los Angeles." *Annals of the Association of American Geographers* 80: 184–204.

Secor, Anna. 2004. "'There Is an Istanbul That Belongs to Me': Citizenship, Space and Identity in the City." *Annals of the Association of American Geographers* 94: 352–368.

Shangming, Dan, and Dan Bo. 2001. "Analysis of the Effects of Urban Heat Island by Satellite Remote Sensing." Paper presented at the 22nd Asian Conference on Remote Sensing, November 2001, Singapore, 1–6.

Sibley, David. 1995. "Gender, Science, Politics and Geographies of the City." *Gender, Place and Culture: A Journal of Feminist Geography* 2: 37–49.

Sjoberg, Gideon. 1960. *The Preindustrial City.* New York: Free Press.

Smith, Neil. 1984. *Uneven Development.* New York: Blackwell.

Smith, Neil. 1996. *The New Urban Frontier: Gentrification and the Revanchist City.* New York: Routledge.

Sorkin, Michael (ed.). 1992. *Variations on a Theme Park: The New American City and the End of Public Space.* New York: Hill & Wang.

Spain, Daphne. 1992. *Gendered Spaces*. Chapel Hill: University of North Carolina Press.

Summerson, John. 1946. *Georgian London*. New York: Charles Scribner's Sons.

U.S. Department of Homeland Security. 2009. *Yearbook of Immigration*. Washington, D.C.: U.S. Government Printing Office.

Valentine, Gill. 1993. "Desperately Seeking Susan: A Geography of Lesbian Friendships." *Area* 25: 109–116.

Vance, James E., Jr. 1971. "Land Assignment in the Pre-Capitalist, Capitalist and Post-Capitalist City." *Economic Geography* 47: 101–120.

Vance, James E., Jr. 1990. *The Continuing City: Urban Morphology in Western Civilization*. Baltimore: Johns Hopkins University Press.

Ward, David. 1971. *Cities and Immigrants*. New York: Oxford University Press.

Warner, Sam Bass. 1962. *Streetcar Suburbs: The Process of Growth in Boston, 1870–1900*. Cambridge, Mass.: Harvard University Press.

Warr, Mark. 1985. "Fear of Rape Among Urban Women." *Social Problems* 32: 238–250.

Watson, Sophie, with Helen Austerberry. 1986. *Housing and Homelessness: A Feminist Perspective*. London: Routledge & Kegan Paul.

Women and Geography Study Group of the Institute of British Geographers. 1997. *Feminist Geographies: Explorations in Diversity and Difference*. London: Prentice Hall.

Wood, Joseph. 1997. "Vietnamese-American Place Making in Northern Virginia." *The Geographical Review* 87: 58–72.

Ten Recommended Books on the Urban Mosaic

(For additional suggested readings, see *The Human Mosaic* web site: www.whfreeman.com/domosh12e)

Allen, James P., and Eugene Turner. 1997. *The Ethnic Quilt: Population Diversity in Southern California*. Northridge, Calif.: Center for Geographical Studies. A fascinating visual exploration of Southern California's diverse population.

Blunt, Alison, and Robyn Dowling. 2006. *Home.* New York: Routledge. An insightful and concise examination of the concept and everyday experiences of home.

Crutcher, Michael E., Jr. 2010. *Tremé: Race and Place in a New Orleans Neighborhood.* Athens: University of Georgia Press. A case study of a neighborhood in New Orleans where issues of race and gentrification come to the fore.

Fincher, Ruth, and Jane M. Jacobs. 1998. *Cities of Difference.* New York: Guilford. A series of case studies that examine the relationships between urban space and social identities and differences.

Ford, Larry. 2003. *America's New Downtowns: Revitalization or Reinvention?* Baltimore: Johns Hopkins Press. An interesting assessment of 16 contemporary American downtowns.

Harvey, David. 2003. *Paris, Capital of Modernity.* New York: Routledge, 2003. A revealing look at Paris during its mid-nineteenth-century political turmoil.

Hayden, Dolores. 2003. *Building Suburbia: Green Fields and Urban Growth, 1820–2000.* New York: Pantheon. A critical historical treatment of American suburban development.

Lees, Loretta, Tom Slater, and Elvin Wylie. 2007. *Gentrification.* New York: Routledge. An overview of the major theoretical and conceptual frameworks for understanding gentrification.

Marshall, Richard. 2002. *Global Urban Projects in the Asia Pacific Rim.* New York: Routledge. A fascinating and well-illustrated look at how global cities in the Asia Pacific Rim are creating and embodying a new urban vision.

Pattilo, Mary. 2007. *Black on the Block: The Politics of Race and Class in the City.* Chicago: University of Chicago Press. An in-depth examination of the many conflicts faced by residents in an African-American Chicago community experiencing gentrification.

What do these images convey about empire and globalization and their similarities and differences?

Go to "Seeing Geography" on page 456 to learn more about this image.

ONE WORLD OR MANY?
The Cultural Geography of the Future

This final chapter is about the future. Unless one is precognizant, writing about the future is "a fool's game." The best we can do is collect evidence about past and present conditions, examine ongoing trends, and then let our imaginations go to work. Even when we do this well, it is difficult to predict the outcome of anything with certainty. Who, for example, would have thought that we would still be combating slavery and the slave trade in the twenty-first century? It is even more difficult to predict wholly new phenomena. Who would have predicted the dramatic social changes that the revolution in information technology is bringing about, especially the ascendance of the World Wide Web in daily life? Despite the lack of certainty, looking ahead is a necessary activity if we want to be at all prepared for the world in which we will build our careers, raise our families, and work toward the fulfillment of ourselves and others. Globalization, whose effects we have traced throughout the pages of this text, will have a profound influence on the cultural geography of the future. As noted in Chapter 1, the phenomenon of globalization has created an interlinked world of instantaneous communication where people, goods, and money move increasingly as if international borders did not exist. While there's no denying the general trends produced through globalization, we will see in this chapter that the future is far from predetermined by them.

We are posing perhaps the most essential and difficult cultural geographical questions of all in this chapter. Will our future contain one world or many? Will we face a world where more people will have more opportunities and choices, or where deprivation and powerlessness will spread? Will we witness a global-scale process of cultural diversification or of homogenization? Are we headed toward a human mosaic or a human monochrome?

Throughout *The Human Mosaic*, we have witnessed the transformation of folk and indigenous cultures, the spread of a few world religions at the expense of many local ones, the rise of a handful of languages and the demise of countless others, the erosion of ethnicity through acculturation and assimilation, the decline of the power of independent states, the urbanization of the world, the rise of corporate agribusiness, the spread of the green revolution, and other homogenizing trends that reflect globalization and undermine diversity.

Pick up a newspaper almost any day and you can read about globalization, though the word may not always be used. New suburbs in India are built to resemble those in Southern California; the citizens of the United Kingdom lament membership in the European Union, worrying that the forces of globalization will erode their distinct English identity; commercial loggers deforest Rendova Island in the Solomon

FIGURE 12.1 Australian aboriginal children at play with a laptop. The blending of modern technology and traditional culture is increasingly common under globalization. *Will such interactions enhance or reduce the world's cultural heterogeneity?* (Robert Essel NYC/Corbis.)

Islands and in the process destroy the culture and lifestyle of the native Haporai people; and so on—endlessly, it seems.

But does globalization truly have the ability to render cultural differences irrelevant? Can't groups of people selectively choose from what globalization has to offer and still retain their cultural identity and attachment to place (**Figure 12.1**)? Does globalization act to homogenize the world or, instead, to widen the differences between haves and have-nots? In fact, much of the resistance to globalization is based on the belief that it enriches and empowers the few at the expense of the many, heightening class differences.

Many—perhaps most—geographers believe that the future will continue to contain many worlds. They speak of profound and irreducible cultural contrasts and of the capacity of indigenous peoples to weave Western elements into their own cultures, creating new and distinct hybrids. Globalization, they argue, produces different results in different lands. The global

need not and cannot abolish the regional or the local. In fact, throughout the chapter we will explore many examples of groups using the technologies and networks of globalization in campaigns that defend local cultural geographies and reassert the relevance of the local. Often local cultures are not merely defending ancient tradition but rather are interacting with outside influences to produce new and distinctive forms of cultural expression. So which view is correct or most likely? Will it be one world or many? No one can say for certain, but we can gain additional insight into the question by looking at the future of regions.

Region

Are culture regions weakening and fading? Do the diverse hues of the human mosaic as revealed in maps shine less brightly than before? Can we detect any such trends, using the theme of culture region? Some geographers look at the maps and do indeed see fading colors.

David Nemeth, for example, suggests that regions are "being crushed and recycled into a bland, ambiguous amalgam," producing "something akin to a vast parking lot of global scale." And without question, people in all parts of the world must seek a new form of self-identity, torn as they are between their traditional attachment to place and local culture on the one hand and their inevitable attraction to globalization-driven cosmopolitanism on the other. They feel the opposite tugs of hearth (region) and cosmos (world).

The Uneven Geography of Development

Most cultural geographers believe that culture regions will persist. In particular, the familiar core-periphery concept is just as relevant as ever, especially where development is concerned. Some regions—forming the core—are moving ahead and prospering, while many others—the periphery—fall further and further behind. Global interdependence has not evened out the differences between the haves and have-nots; in many ways it increases them. What do we mean by falling behind or moving ahead? One clear measure is global wealth distribution. There is now a transnational class of the super-rich, the 0.25 percent of the world's population that owns as much wealth as the other 99.75 percent. Most of them live in a First World core consisting of Anglo America, the European Union (EU), the coastal zone of East Asia, and Australia–New Zealand. As the greatest beneficiaries of globalizing processes, these regions are moving ahead of the remaining Third World peripheral lands.

Not all observers, however, agree that the global income gap is widening. In *The New Geography of Global Inequality*, Glenn Firebaugh argues that world core-periphery differences are not actually increasing. He points out that if one takes into account only the fact that the incomes of the richest nations are increasing at a greater rate than those of the poorest, it does look like the rich are becoming richer faster, while the poor remain poor. Yet this overlooks another fact: the many poorer countries contain only 10 percent of the world's population, while those that are developing very rapidly (mostly Asian countries) contain more than 40 percent. Taking national population numbers into account, global income gaps are in fact decreasing.

In the meantime, income gaps *within* nation-state boundaries are growing rapidly. Instead of core-periphery, another way to think about the unevenness of development is in terms of "fast" and "slow" worlds. In many ways economic development speeds up daily life. The fast world is at the forefront of instantaneous global communication, where one can download the latest music and films from around the world or play the Tokyo stock market from one's home computer in Seattle. The fast world is fully wired for the Internet; located in the world's megacities; adaptable to rapid shifts in global trends of investment, trade, production, and consumption; and home to the transnational super-rich. The slow world consists of the hollowed-out rural landscapes, declining or abandoned manufacturing zones, and slums and shantytowns. Computers and Internet broadband service, even if available, are priced out of reach of most of the inhabitants of the slow world. Pieces of the periphery that lie adjacent to the core, such as Mexico's northern border, scramble to join the privileged part of the world, to be fast rather than slow. As Mexican border cities' recent experience with job loss to China demonstrates, for such regions, membership in the fast world can be fleeting.

British geographer Rob Shields, in *Places on the Margin,* concentrates on an array of places and regions of varying size that have been "left behind in the modern race for progress." He finds peripheries even within the core. Often these places and regions become sites of illicit or stigmatized activities, such as the international trades in sex and illegal drugs. Says Shields, such "margins become signifiers of everything centers deny or repress."

Sociologist Elijah Anderson's classic study of Philadelphia neighborhoods provides a good urban-scale illustration of how peripheries are created and become homes to drug trafficking. The neighborhoods he studied were historically home to stable, racially and ethically diverse middle-class families rooted in a manufacturing-based economy. When larger-scale economic changes produced by globalization led to the movement of manufacturing jobs "offshore," city factories closed their doors, poverty rates increased, and the future prospects of youth dimmed. The old manufacturing economy was replaced by an underground economy of drug trafficking, prostitution, and associated violent crime.

Tragically, places on the margin have also become key sites in the international trafficking of slave labor. Slavery, it turns out, is not a shameful practice consigned to the distant past, but an increasingly common phenomenon under globalization. Some believe that as many as 27 million people live in slavery worldwide, on every continent in the world save Antarctica. Traffickers prey on the most desperate and powerless of society's castoffs. In the United States, California, Texas, and Florida lead the way in the increase in slave labor cases. Some of Florida's famous orange juice, for example, was recently found to be harvested by illegal Mexican immigrants held against their wills in remote rural camps (**Figure 12.2**, page 434). The insatiable international demand for Brazil's timber and beef has been met through the labor of captive workers held deep in the Amazon. Labor recruiters in Brazilian cities lure jobless workers to cut remote forestlands through false promises of good wages and housing. Once there, workers' wages are withheld and they are prevented from leaving. Some are forced to

FIGURE 12.2 **The mobile Florida Modern Slavery Museum in front of the United States Capitol building.** The museum has been driven to cities across the United States to bring the labor abuses of Florida's agricultural sector to public attention and promote reform. *What reasons can you think of to explain the existence of modern slavery in agriculture?* (Fritz Myer.)

work for years clearing forests to make way for cattle ranching. According to one Brazilian labor official, "Slave labor in Brazil is directly linked to deforestation." The example of slavery shows how illicit activities can thrive in peripheries, such as rural Florida and the Brazilian Amazon.

One Europe or Many?

Looking at just one region of the world, geographer Ray Hudson echoes our question: "One Europe or many?" Local identities are asserting themselves within states at the same time that supranationalism is touted as the path to "one Europe." He opts decidedly for "many," saying that one main role of the European Union should be to promote "complex geographies of identities." Hudson also concludes that power and wealth will not be evenly distributed geographically in Europe, contributing to the maintenance of many cultural regions.

A similar outlook leads geographer Michael Keating to speak of a "new regionalism" in Europe, and David Hooson went so far as to suggest that globalization actually *strengthens* people's bond between place and identity. This strengthening is suggested by the rise of ethnic separatism in countries as diverse as Spain, the United Kingdom, and Serbia and Montenegro. As people find that the forces of globalization

are eroding cultural norms, customs, and practices, local political movements that seek to reinvigorate ethnic identities rooted in place may arise in reaction.

Major political-geographic events such as the disintegration of the Soviet Union or the creation and expansion of the EU can have complex effects on cultural identities. The recent expansion of the EU is likely to bring a host of unintended and surprising outcomes for culture regions. In some former Soviet territories, for example, ethnic Russian immigrants have become a disadvantaged minority group. New post-Soviet states have used their membership in the EU to reassert national ethnic heritage and redefine themselves culturally as "Western."

Glocalization

The theme of culture region, then, seems to suggest that a new human mosaic is forming in the age of globalization. Rapid change is pervasive, but its direction differs from one location to the next. The future will not be like the past, but it will also not be monochromatic.

The interaction between global and local prompted geographer Erik Swyngedouw to promote the term **glocalization** to describe the consequences. In brief, he argues that the outcome of this interaction involves change both

glocalization
The process by which global forces of change interact with local cultures, altering both in the process.

in the regional way of life *and* in the globalizing force. For example, transnational corporations often have to adapt their product lines to local norms and preferences or adjust their production practices to local labor and environmental laws. Put differently, glocalization is a process that ensures the survival of culture regions and places in the future. These considerations, prompted by the culture region theme, lead us to conclude that a planetary culture is almost certainly illusory and that potent forces are at work to prevent homogenization.

The Geography of the Internet

Has the Internet reduced the importance of geography or the viability of culture regions? In 2001, *The Economist* described the Internet as being "everywhere, yet nowhere in particular." The Internet can cross borders, both politi-

cal and cultural; it breaks down barriers; it eliminates the effects of distance. In 2003, however, *The Economist* drew the opposite conclusion. Perhaps it was naive of people to think that the Internet would make geography irrelevant. Geography is "far from dead," *The Economist* concluded.

The reality of the Internet's influence is very complex and ultimately leaves culture regions intact. In truth, the Internet is much "constrained by the realities of geography." Rather than making national borders irrelevant, the Internet can sometimes highlight differences between different countries. For example, the Internet can now be screened, censored, regulated, and blocked by governments (**Figure 12.3**; see Subject to Debate, page 440).

Although the constraints of geography are increasingly recognized, there is no question that the Internet has been very effective in creating new forms of social interaction, including the creation of virtual communities. We might ask, as John Barlow does, "Is there a *there* in cyberspace?" Does the Internet contain a geography at all? Certainly,

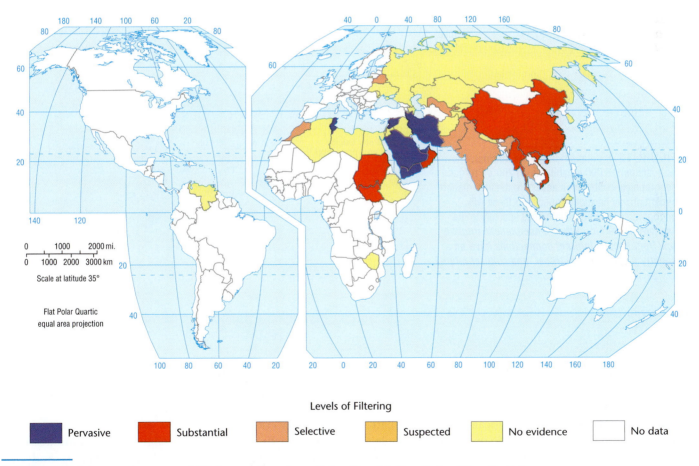

Levels of Filtering

Pervasive   Substantial   Selective   Suspected   No evidence   No data

FIGURE 12.3 This map of government filtering practices shows that nation-state boundaries still matter in the global flow of information on the Internet. *Based on your knowledge of world affairs, do you find the various government filtering activities depicted here predictable or surprising?* (Source: Open Net Initiative web site.)

Internet Connections

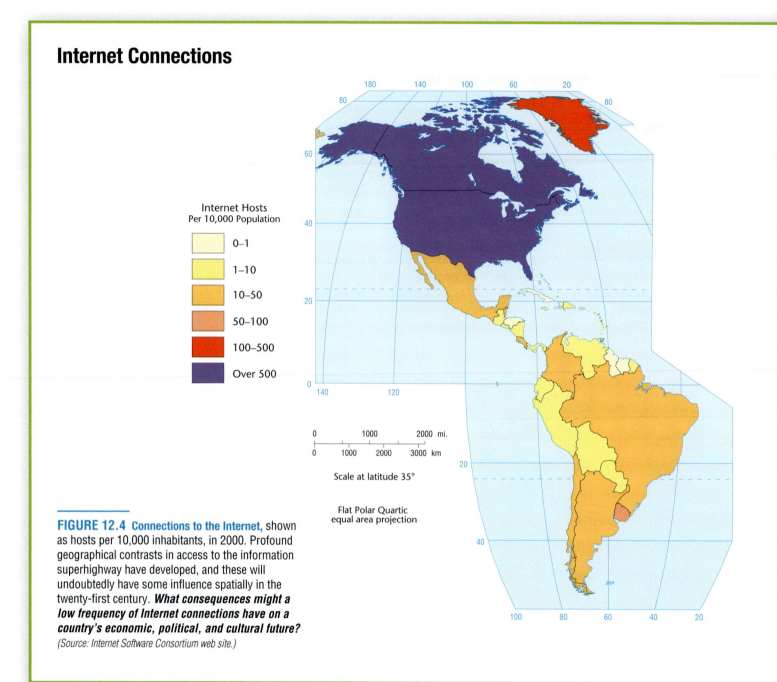

Internet Hosts
Per 10,000 Population

- 0–1
- 1–10
- 10–50
- 50–100
- 100–500
- Over 500

Scale at latitude 35°

Flat Polar Quartic
equal area projection

FIGURE 12.4 Connections to the Internet, shown as hosts per 10,000 inhabitants, in 2000. Profound geographical contrasts in access to the information superhighway have developed, and these will undoubtedly have some influence spatially in the twenty-first century. *What consequences might a low frequency of Internet connections have on a country's economic, political, and cultural future?* (Source: Internet Software Consortium web site.)

---

places—at least as understood by cultural geographers—cannot be created on the Internet. For starters, these "virtual places" lack a cultural landscape and a cultural ecology. In the broader context, on a worldwide scale, human diversity is poorly portrayed in cyberspace. "Old people, poor people, the illiterate, and the continent of Africa" seem not to be "there," as Barlow notes (**Figure 12.4**). Users usually end up "meeting" others pretty much like themselves on the Internet. More important, the breath and spirit of place cannot

exist in cyberspace. Barlow, resorting to a Hindu term, calls this missing essence *prana*. These are not real places, nor can they ever be.

Other critics point out that virtual communities do not have the defining qualities of geographic communities: communion among citizens, shared responsibilities, and civic duty. Communication across the Internet does not a community make. Nevertheless, people can use virtual communities to establish bonds that carry over into the real world. To

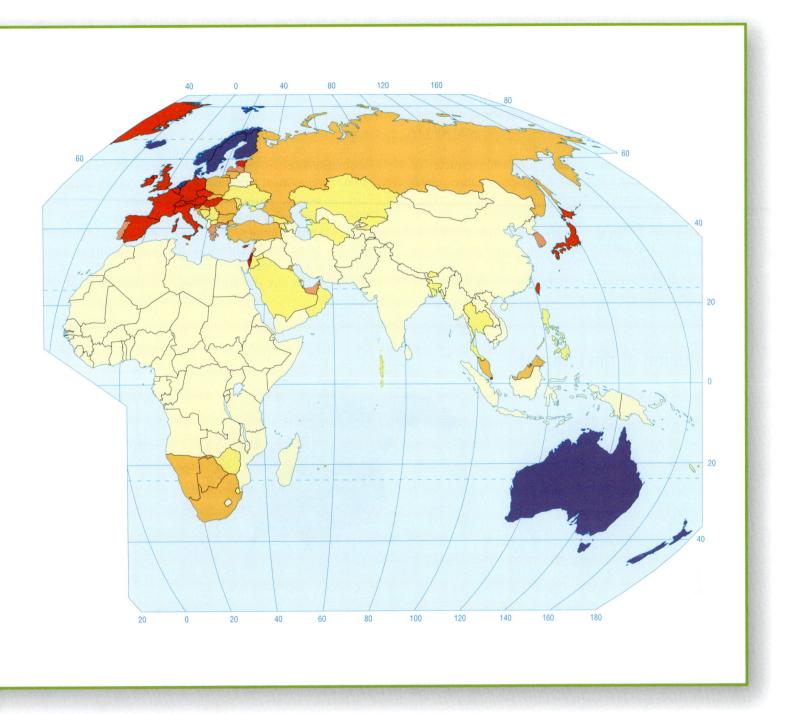

keep things in historical perspective, we should remember that throughout the modern period families and individuals have left their geographic communities and dispersed widely. However, these people have constructed social networks to maintain cultural and emotional ties with their geographic communities of origin. From this perspective we might view the Internet as just another vehicle for maintaining those ties and preserving cultural identities (see Subject to Debate, page 440).

Clearly the virtual regions and places created in cyberspace lack the materiality of actual regions and places that geographers study. However, cyberspace is itself earthbound and rooted in actual places. The fiber-optic cables of the Internet have a location, and in fact the whole enterprise remains largely city-bound. A high-speed digital subscriber line requires proximity to a telephone exchange. As the 2003 report in *The Economist* sums up, "So much for the borderless internet."

Mobility

How is the theme of mobility relevant to the debate over "one world or many"? How will changing patterns of movement help configure the cultural geography of the future? Movement on a global scale—of ideas, people, money, and commodities—in many ways will characterize the future. The computer, the Internet, satellite television, compact discs, and other globally available technologies greatly facilitate access to and the diffusion of ideas and information, while also accelerating their spread. At the same time, the spread of new ideas and innovations often produces disruptions that change the world in unexpected and unintended ways. At the very least, the theme of mobility cautions us against counting on predetermined outcomes.

The Information Superhighway

Terms such as *information superhighway* and *infobahn* imply the enhanced ability to achieve cultural diffusion. Moreover, the use of more efficient transportation systems, such as containerized cargo units, has greatly increased "the spatial dispersion of the production and consumption of economic goods," in the words of geographer Christopher Airriess, whose work we introduced in Chapter 9.

The Internet allows ideas, music, artwork, and indeed any form of cultural expression that can be turned into digital data to be transmitted around the world in a matter of seconds. The circulation of people and things, although

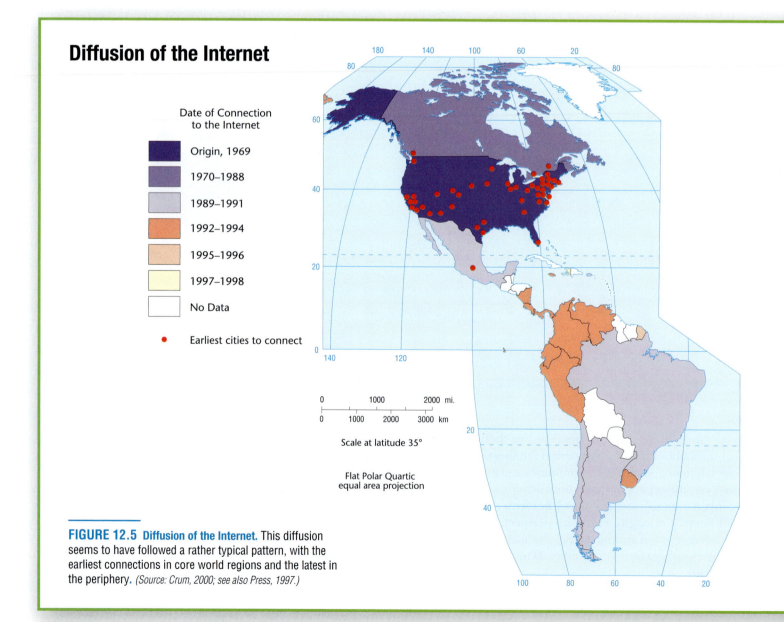

Diffusion of the Internet

Date of Connection to the Internet

- Origin, 1969
- 1970–1988
- 1989–1991
- 1992–1994
- 1995–1996
- 1997–1998
- No Data

● Earliest cities to connect

Scale at latitude 35°

Flat Polar Quartic equal area projection

FIGURE 12.5 Diffusion of the Internet. This diffusion seems to have followed a rather typical pattern, with the earliest connections in core world regions and the latest in the periphery. *(Source: Crum, 2000; see also Press, 1997.)*

it takes considerably longer, can nonetheless reach virtually any part of the globe in only a matter of hours or days. With the development of the Internet and rapid global transportation systems, transnational corporations are able to advertise, market, and deliver virtually anywhere. Every *Fortune* 500 company now has its own web site, allowing them to market cultural products directly to consumers in every corner of the globe. What's more, nongovernmental organizations, art museums, indigenous peoples, individual artists and musicians, and just about all imaginable culture groups or producers of culture have their own web sites.

The Internet clearly has the potential to profoundly alter the speed and character of cultural interaction, but to what effect? The spread of more democratic forms of governance, for example, has accompanied the more rapid movement of information over the Internet. Dictatorships thrive by controlling and manipulating information—an increasingly difficult task in the age of the Internet. And the Internet is certainly not the only means by which information spreads rapidly. Cellular telephones, satellite-beamed cable television, and news channels such as CNN contribute to the diffusion of information, as does the increased volume of international travelers.

The theme of mobility also applies to the spread of the Internet itself. If the future is to bring the universal diffusion of cultural elements, then surely this new homogenizing tendency ought to be revealed in the spread of this most essential element of globalization (**Figure 12.5**). The diffusion of the Internet has now spread across much of the world, following the models set down in Chapter 1.

But although the Internet now reaches into almost every land, its use varies profoundly (see Figures 12.4 and 12.5).

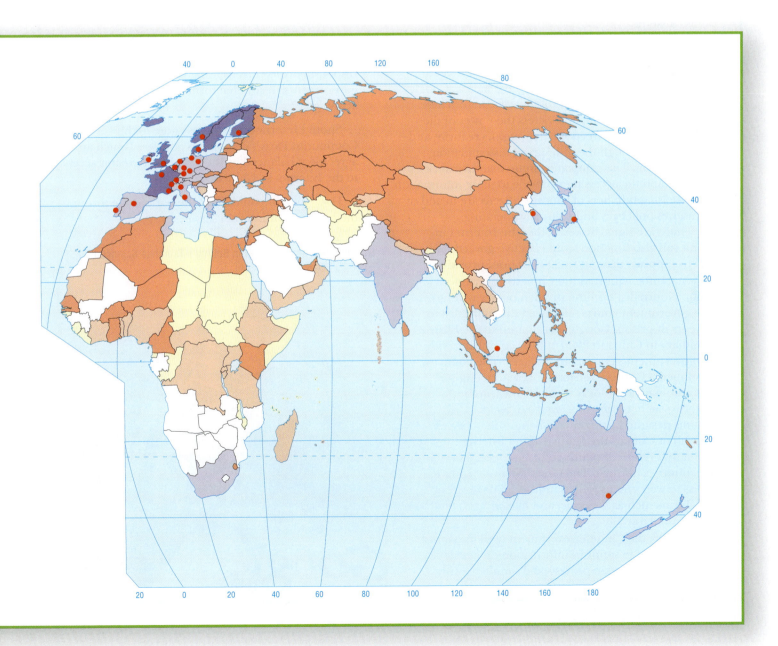

Subject to Debate

The Internet: Global Tool for Democracy or Repression?

In a few short years the Internet has become the most important medium for the global exchange of information and ideas. Many argue that the expansion of this global network will spread democratic ideals of citizen participation in government, freedom of expression, and individual liberty. Others suggest that the Internet, like other technologies of mass communication, can be used for any number of political and cultural ends, including the repression of public dissent.

The case of China illustrates the debate over the global expansion of the Internet. China represents the largest and fastest-growing market for Western-based Internet companies, and industry giants such as Google and Yahoo! have established subsidiaries there. These and other Western companies stress the power of the Internet to promote freedom of thought, creativity, and social equity through the widespread dissemination of information. The companies argue that their very presence and the access they provide to more information for increasing numbers of Chinese citizens are in and of themselves politically and culturally liberating.

There are signs, however, that the Internet in China can also function as an effective government tool to suppress dissent. In 2004, Yahoo! provided personal information to the Chinese government about a journalist's online pro-democracy writings. The government courts sentenced the journalist to 10 years in prison. Google was allowed to establish a subsidiary in China in 2006 only by complying with the government's censorship policies. The company developed a Chinese version of its search engine that automatically blocks results from searches on terms such as *Tibet* or *Tiananmen Square*. Cisco Systems designed the software behind the "Great Firewall of China," the feature of the Internet in China that filters all online information as it crosses the country's borders. The Great Firewall can block access to web sites as well as individual pages from a site. Terms such as *democracy* and *human rights* tend to trigger the Firewall.

In 2008, the Chinese government used its control of the Internet to suppress information on its military crackdown on Tibetan protests and simultaneously foment nationalistic outrage among Han Chinese against the Tibetans. Then, in 2010, Google announced that it would no longer cooperate with China's Internet censorship program. Google had discovered that a cyberattack on its computers originating from China was an attempt to access the accounts of Gmail users who were advocates of human rights in China. For the time being, the Chinese government has allowed Google to continue operating in China.

Continuing the Debate

What does the case of Google in China tell us about the power of the Internet? Consider these questions:

- Is China a unique case or just an extreme example of repressive uses of the Internet?

- Do private corporations have an ethical responsibility to citizens living under authoritarian governments? If so, what form should corporate action take?

- Should Internet companies agree to the Communist Party's censorship policies? Is such an agreement balanced by the benefits of greater access to the Internet by Chinese citizens?

- How can companies limit the harm done through the provision of Internet services?

Google China headquarters in Beijing's Tsinghua Science Park.
(Frederic J. Brown/AFP/Getty Images.)

Just because an innovation is available does not mean that most or even many people will accept it. In much of the world, use of the Internet remains arrested, and we should not necessarily expect it to diffuse much further. Barriers of wealth, education, and governmental opposition prove formidable. It may be correct, in the words of Frances Cairncross, to refer to "the death of distance" caused by the revolution in communications, but it clearly does not follow that everything will diffuse everywhere.

New (Auto)Mobilities

Back in the middle of the twentieth century, there was much popular debate about the future of transportation in the United States. People speculated that anything from a national network of high-speed railways to nuclear-powered flying machines would replace automobiles (**Figure 12.6**). Hardly anyone predicted that the twenty-first century would be more of the same old thing: the internal combustion engine–driven family car. No one predicted that the American fascination with the automobile would diffuse to East and South Asia. At that time, these regions were largely agri-

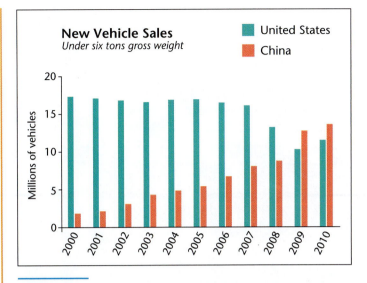

FIGURE 12.7 China is surpassing the United States as the world's leader in car sales. This is a phenomenal turn of events, considering China's level of economic development only a generation ago. *(Source: J. D. Power and Associates.)*

cultural, facing recurring famine, and among the poorest regions of the world.

In 2010, however, China overtook the United States to become the largest automobile market in the world in terms of both sales and production (**Figure 12.7**). Between 2007 and 2012, China is expected to account for an astounding 18 percent of the growth in worldwide auto sales. India, while expected to account for only 9 percent of the growth, will not be far behind the expected 11 percent growth of the United States.

The transformation in China's urban culture is as profound as it is sudden, with cars now clogging city streets built for pedestrians and bicycles (**Figure 12.8**, page 442). Barely a decade ago, only state bureaucrats bought cars and had only six models from which to choose. Now there are over 90 models on the market and businessmen, flush with profits from the economic boom, are paying cash. China's new millionaires have adopted the "Beemer," the Yuppie status symbol of the go-go 1980s in the United States, as their own. They have made China the world's number one market for BMW's 760Li model.

Crowding and associated pollution will only intensify in the future as cities such as Beijing register 1000 new drivers *every day*. The environmental consequences of an annual sales growth rate of 48 percent (2009 data) are numerous and of global importance. Among the biggest concerns is the increase in carbon dioxide emissions, caused by increased car exhausts, auto-manufacturing activities, and power plant growth to meet manufacturing demands for electricity. Carbon dioxide is a greenhouse gas that is the main contributor to the problem of global warming. China

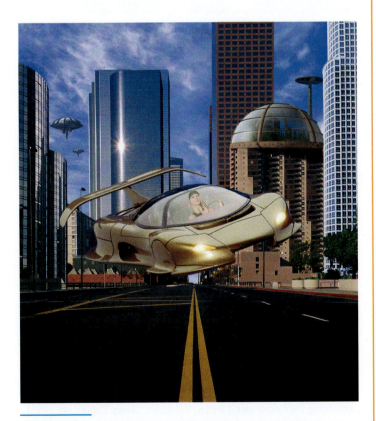

FIGURE 12.6 A futuristic vision of urban transportation. Such imaginative portrayals of the near future were common in the mid-twentieth century, but relatively little has changed. Our twenty-first century transportation mode is more similar to Henry Ford's Model-T than to this flying machine. *(Glen Wexler/Masterfile.)*

FIGURE 12.8 Traffic congestion in Shanghai. China is now the second largest automobile market in the world. Here automobiles and bicycles compete for space in the increasingly congested streets of Shanghai, China. *(Thomas Imo/photothek.net/drr.net.)*

has passed the United States as the world's number-one source of carbon dioxide and is projected to double its 2000 emissions by 2020 (**Figure 12.9**).

China's embrace of the family car is not simply imitation. Perhaps because of the suddenness and profundity of the changes in mobility in urban China, Chinese drivers have developed driving habits that would seem peculiar if not downright dangerous to most Americans. For example, the use of turn signals, windshield wipers, and nighttime headlights is often considered a distraction—Beijing had outlawed the use of headlights prior to the mid-1980s—both for the driver and for other cars on the road. Chinese buying habits are different as well. Since the automobile is so new to Chinese culture, buyers rely heavily on dealers' advice. Dealer discounts and customer bargaining, common features of American auto culture, are absent in China, where cars—despite the trends outlined above—are still seen as scarce, highly desirable commodities.

American auto manufacturers have scrambled to get a foothold in emerging markets such as China. General Motors Corporation, for example, began establishing manufacturing plants in China in the late 1990s (**Figure 12.10**). It now has nine joint ventures and two wholly owned foreign enter-

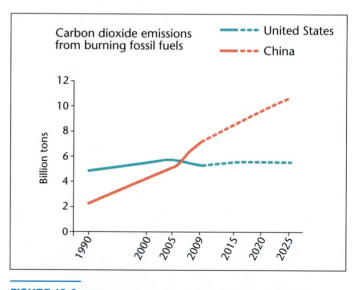

FIGURE 12.9 China's greenhouse gas emissions are skyrocketing, surpassing the former world leader, the United States. China's per capita output, however, remains far below that of the United States. This chart reflects projections that assume the success of a national efficiency campaign; experts in both China and the West, however, are increasingly concerned that emissions may rise even faster than projected here. *(Source: International Energy Association, www.iea.org.)*

The Place(s) of the Global Tourist

Tourism will be one of the key types of mobility of the twenty-first century. By some estimates, tourism is or soon will be the largest industry in the world, second only to oil as the most important source of revenue for the Third World. Cultural geographers are increasingly interested in understanding this phenomenon. What sort of cultural interactions will take place under global tourism? Will the end result be the preservation, conversion, or elimination of distinct culture regions?

As with many global-scale patterns of mobility, there are distinct differences in power that structure the tourist experience. The most glaring inequality is that between the visiting and visited cultures. The vast majority of global tourists are from the First World. The cultures visited are often located in the marginalized sites of the global economy, such as on the famed African safaris or spring-break excursions to Cancún, Mexico.

The global tourist industry presumes the existence of unique places that appear "exotic" to the Westerner—or at least serve to emphasize the difference between the experience of home and the experience of the travel destination. Tourism can thus be a double-edged sword. On the one hand, there is an imperative to preserve and nurture folk and indigenous cultures, and, on the other, an imperative to merely create the illusion of cultural difference. The former imperative can gain support from tourism, such as in the case of agricultural tourism, where traditional but uneconomical food production persists with the aid of profits gained from tourist visits. For example, the European Union provides funds to member states to support the production of a range of traditional agricultural products such as local cheeses and rare livestock breeds. The EU believes that these investments keep rural economies intact and spark the tourist trade because urbanites and foreign tourists are attracted to the countryside to sample unique and historically significant agricultural products. The latter imperative is often driven by commercial enterprises that make their profits from entertaining their customers with sights and sounds of the "exotic." In such cases, "traditional" dress and performance is for the benefit of tourists, not a part of everyday life (**Figure 12.12**, page 444). Some cultural geographers have concluded from these observations that the tourism business packages its offerings to conform to the tastes of the affluent global tourist and so results in an "inauthentic" experience of place.

It would be wrong, however, to simply view local cultures as "made-up" or "invented" by the global tourism industry or to suggest that only inauthentic cultural interactions are possible (see Rod's Notebook, page 445). Geographers Peggy Teo and Lim Hiong Li's study of global and

FIGURE 12.10 GM plant in Shanghai. Factory workers check a car body on the assembly line at the Shanghai General Motors factory in Pudong, China. Other American, Japanese, and European carmakers are building factories throughout China. *(Natalie Behring/drr.net.)*

prises in China that are involved in all aspects of the auto industry, including vehicle manufacturing, sales, financing, and parts distribution. Shanghai went from five GM dealerships in 2000 to twenty-seven in 2010! Carmakers are betting that the symbolic importance of the luxury automobile will diffuse not only to China but also to other regions of East and South Asia. For example, Buick, GM's top model in China, is perceived as plush, elegant, and a marker of high social standing. It appears as though a prime status symbol in American middle-class culture is poised to become the same for a new global middle class (**Figure 12.11**).

FIGURE 12.11 General Motors Corporation launches its Cadillac line at the Beijing Auto Show. *What brand name better symbolizes America's car culture than Cadillac?* *(Ng Han Gua/AP Photo.)*

FIGURE 12.12 Global tourism and local culture. Tourists line up to photograph a colorfully dressed "native." *What do such encounters suggest about the effects of global tourism on local cultures?* (AP World Wide Photos.)

local interactions at a tourist site in Singapore—a former private mansion and fantasy garden—provides an excellent case in point. They favor the view that local groups can use the forces of global tourism to strengthen their cultural identities and traditions. Their study focused on a government-sponsored effort to create from the mansion and garden an "Oriental Disneyland," a Western-style theme park loosely based on Disney's technologically enhanced tourist attractions. After the venture failed, local people successfully lobbied the government to redesign the site to more closely replicate its original condition. The motivation behind these efforts was to defend local cultural and historical meanings associated with the site. They conclude that "the global does not annihilate the local." Rather, tourism can result in "unique outcomes in different locations."

Globalization

How does globalization operate to shape the future? How does it relate to the phenomena of the past, such as European colonialism? These are some of the questions geographers and other scholars are investigating in their research. We can start by looking at some of this groundbreaking work.

A Deeper Look at Globalization

As we have seen, the one-world-or-many debate hinges on globalization, a theme that has helped structure every chapter. It is vital, therefore, that we organize our thoughts about the future by examining what people mean by the term. Chap-

ter 1 introduced and defined globalization, and we have seen in subsequent chapters just how complex and encompassing the phenomenon is. Perhaps we are now prepared to analyze the concept at an even deeper level by looking at how practicing geographers define it in their research.

Carolyn Cartier began her book on the economic growth and sociocultural transformation of South China, *Globalizing South China*, with a description of globalization. For Cartier, it includes many different cultural, economic, and political phenomena, both material and symbolic. Globalization is most closely associated with the transnational activities of large corporations, which operate sometimes in alliance with and sometimes against states.

In *Global Shift*, Peter Dicken views globalization as interrelated processes working on a global scale to produce effects that can vary greatly from place to place. He distinguishes "internationalization"—a decades- if not centuries-old practice of simply extending economic activities beyond state borders—from globalization. He explains that globalization is qualitatively different from internationalization because it involves the full integration of human activities on a global scale. Increasingly, globalization processes are linking far-flung places around the world in a complex, integrated system for the production, transport, and marketing of many everyday commodities

Our last view of globalization comes from R. J. Johnston, Peter Taylor, and Michael Watts's *Geographies of Global Change*. They think the most important element of globalization is the global scale of social activities. Similarly to Dicken, they differentiate between the international aspects of social interaction and globalization, arguing that the latter is constituted of "trans-state" processes. That is, under globalization, activities and outcomes "do not merely cross borders, these processes operate as if borders were not there." They suggest that we carefully examine the processes of globalization from a "geohistorical perspective" to find out what, exactly, is different now and what the historical precedents are. This involves tracing the flow of things, people, and information through time and across spaces. It also involves identifying and explaining how specific places respond to and are transformed by globalization in different and unique ways.

Globalization is an extremely complex and variable phenomenon involving many cultural, economic, and political processes. The best way for us to reach a better understanding is to look at concrete examples.

History, Geography, and the Globalization of Everything

From these ideas we can identify several defining features of globalization. First, globalization involves global-scale interactions among cultural, political, economic, and envi-

Rod's Notebook

Rethinking Global Tourism on the Quetzal Quest

Rod Neumann

The resplendent quetzal, which frequents the cloud forests of Central America. *(Danita Delimont/Alamy.)*

As dusk became night on the narrow two-lane highway climbing Costa Rica's Cerro de la Muerte (Mountain of Death), we leaned toward the windshield, scanning for some sign of our destination. We knew from the web site we had visited back home that we could expect a crude wooden sign signaling the turnoff. We had come for a glimpse of the resplendent quetzal (*Pharomachrus mocinno*), a bird revered by ancient Aztecs and contemporary nature tourists alike, though no doubt based on very different sets of cultural values. We found our way down a muddy track to a random collection of rough-hewn wooden buildings. The owner of the *finca* (farm), a solidly built farmer with a crooked smile, served us rice and beans and cold beer. It was a delightful and memorable meal.

Our experience in the quetzal's cloud forest home led me to ponder the meaning of the term *global tourism* and Western tourists' endless quest for the authentic encounter. Global tourism seems to imply huge institutions or corporations shuttling masses of camera-toting vacationers to every corner of the world. Clearly there were many multinational corporations involved in this tourist encounter, beginning with the airline that had transported us from Miami to San Jose. On the other hand, the key global entity in this story was the worldwide Internet, which put us in direct contact with a peasant farmer high in the mountains 1995 kilometers away. And although we were "global tourists," we were also guests of the farm owner. Our host had been born and raised on this steeply tilted piece of forest that happened to be key quetzal habitat. Our conversations at mealtimes and predawn bird-watching adventures had a feeling of intimacy absent in large, packaged tours. There was no sense of alienation or performance. We talked about his childhood in the mountains, farming, and forest conservation. I came away from our experience thinking that the degree of authenticity in the tourist encounter has a lot to do with who owns the means of producing the tourist experience.

Posted by Rod Neumann

---

ronmental phenomena. Second, globalization is characterized by relentless movement—a constant flow of money, people, information, ideas, goods, and services through all parts of the world. Third, the effects of globalization are felt unevenly in different places at different times. Fourth, transnational corporations are the primary institutions driving globalization. While states remain important, the relevance of international borders has diminished; several transnational corporations now have total sales greater than entire national economies (**Figure 12.13**, page 446). Fifth, local- and global-scale movements have formed to oppose globalization, hoping to limit or mitigate its effects. Sixth, globalization, although historically rooted in processes of internationalization, is qualitatively different.

Many scholars trace the origins of globalization to the age of European exploration and imperialism or to the

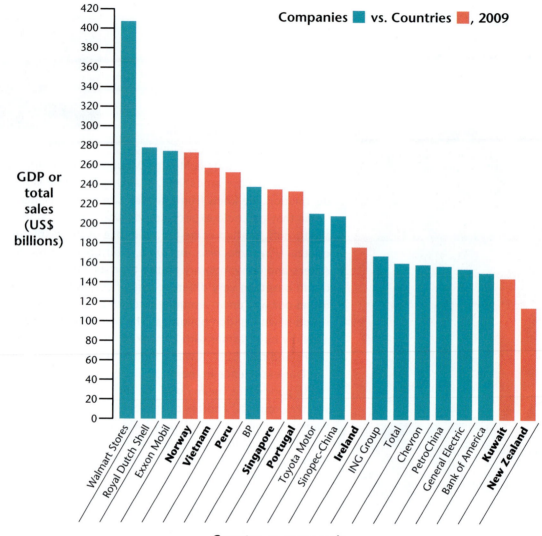

Country or corporation

FIGURE 12.13 The economic power of global corporations. The annual sales of the world's largest privately owned corporations are greater than many countries' GDPs. *How might this affect these countries' abilities to regulate the activities of foreign companies?* (*Source: Elwood, 2001. 2009 data from CIA Factsheet and Forbes.com.*)

industrial revolution. While its historical roots may run deep, globalization is a recent phenomenon. The processes of globalization first became prominent in the 1960s, when the capitalist world economy, increasingly dominated by huge multinational corporations and high technology, produced a new, efficient, and integrated system of production, marketing, transportation, communications, and information processing. Globalization functions as a single system—dynamic, yet highly organized. Globalization—unlike, say, the British Empire—has no single seat of power, and no single country can control it. In that sense, globalization represents something radically new and quite different from older forms of empire.

Geographers are interested in globalization because it bears so profoundly on the central geographical issue of human diversity. It produces a multitude of historically unique, contradictory, and paradoxical cultural conditions. The most evident paradox is that globalization—which would seem to erase cultural difference—has been accompanied by a reassertion of distinct religious, national, and ethnic identities. The start of the twenty-first century has been marked by heightened ethnic divisions and conflicts, the resurgence of nationalisms, and numerous highly visible identity movements—gay, feminist, green, and born-again. Contradictions and paradoxes such as these lead geographers to ask whether we are headed for one world or many.

Globalization and Its Discontents

Many people view the consolidation of transnational corporate power, global environmental decline, and cultural homogenization as threatening to local livelihoods and cultural practices. We see on television and in newspapers the accounts of people from many countries and cultures participating, usually nonviolently, in mass demonstrations against "globalization." In popular news accounts, for example, we learn of antiglobalization protests, such as those trying to disrupt the meetings of the World Trade Organization (WTO) in Seattle in 1999 and in Cancún, Mexico, in 2003. The WTO regulates exchange of goods and services between nation-states with the goal of liberalizing trade (i.e., reducing tariffs and other barriers).

Who are the opponents of globalization, what are they protesting, and what do they want? To find an answer, we need to move beyond the six o'clock news footage and front-page headlines of big, noisy demonstrations. Recognizing that there is not just one kind of antiglobalization activist, we need to carefully examine how specific groups in specific locations organize to maintain local cultural practices, identities, and landscapes in the face of globalizing processes. Let us take a look at local opposition to globalizing processes in two different locations, among very different social groups: indigenous peoples in Mexico and unionized workers in New York City.

On January 1, 1994—the day that the North American Free Trade Agreement (NAFTA) among Canada, the United States, and Mexico took effect—hundreds of indigenous Mayan peasants forcibly occupied government offices in the southern Mexican state of Chiapas. Calling themselves *Zapatistas*, they declared that they had no choice but to take up arms (**Figure 12.14**). The impoverished Mayan farmers saw globalization in the form of NAFTA as sounding the death knell for their culture and way of life. Globalization meant that cheap corn from the midwestern region of the United States would flood Mexico, making it impossible for Mayas to continue farming for a living and to maintain their communities. Interestingly, the initial armed rebellion soon gave way to a skillful campaign that used key instruments of globalization—satellite TV and the Internet—to gain worldwide support for their campaign. The *Zapatistas* have won some concessions from the Mexican government, but their struggle to protect their culture and homeland continues. Their web site details the latest developments.

The response of the International Ladies' Garment Workers' Union (ILGWU) to globalization provides an example of protest and resistance very different from that of the *Zapatistas*. In this case, manufacturing jobs in the garment district in New York City began to disappear in the 1980s under the demands of the new global economy. Garment-manufacturing jobs were being moved out of the United States to regions with cheaper labor costs. Labor unions in the United States have consistently argued that this is a significant drawback of globalization and have fought to slow the loss of jobs with limited success. In this case, however, a labor union found an indirect way to keep manufacturing jobs in the United States. At the same time manufacturing jobs were leaving New York, the

FIGURE 12.14 Local resistance to globalization. *Zapatista* commanders sit at the negotiating table in San Andres Larrainzar, Mexico, in 1996. The Mayan-based *Zapatista* leaders have been periodically involved in talks with the Mexican government over cultural and economic rights since NAFTA took effect in 1994. *(AP Photo/Scott Sady.)*

new economy raised demand for office space and upscale urban housing, encouraging landlords to convert garment factories to more profitable spaces. Labor geographer Andrew Herod showed how the union fought job losses by lobbying for changes in the city zoning laws that would prevent the conversion of factories to high-rent offices and condos. A special garment center district was created that prohibited conversions, thereby giving the workers increased job security. The ILGWU managed to slow New York City manufacturing job losses resulting from globalization by defending the local historical urban landscape.

Many more examples of local efforts to resist globalization or at least mitigate its worst effects could be recounted. Some movements mobilize globally; others, locally. Some movements take up arms; others work through existing democratic institutions. In most cases, however, these movements are rooted in place and are expressed through local cultural values and beliefs. They all ultimately seek to gain a voice in controlling the speed and extent of globalization's transformative forces. As a consequence, the effects of globalization are not predetermined; cultural homogeneity is not the only possible outcome. Rather, the actions of local culture groups—be they made up of indigenous peoples, urban workers, or rural farmers—also shape its effects. The local-global link, it seems, operates in two directions.

Blending Sounds on a Global Scale

On the popular music scene, debates about the effects of globalization abound. These debates follow the general terms of the broader globalization debates. Is globalization a force for homogenization or new forms of diversification, hybridity, and synthesis? If we think about music historically, synthesis and hybridization have deep roots. It is actually fairly difficult to find a truly "authentic" or "pure" locally bound musical genre that has not been influenced by extra-local musical forms. Distinctive regional sounds, such as Memphis soul or Tibetan throat singing, can definitely be identified. In many cases these musical genres are associated with culture regions. At the same time, musical genres rarely develop in geographic isolation. Prior to globalization, migration and movement produced all manner of blending of ideas, styles, and genres—probably since the time humans began imitating the rhythms and sounds of nature. Modern technological innovations, from the invention of the gramophone to the release of online file-sharing software, have accelerated, intensified, and added to the complex processes of musical hybridization.

Take the case of soukous, a musical genre centered in Africa's Congo River basin. Soukous originated in the folk music and dance traditions of various Congolese ethnic groups. Some of these performers began to adopt Western instruments and jazz arrangements during the Belgian colo-

nial period of the early twentieth century. As the population urbanized in the 1950s, record companies found a burgeoning market and began importing Cuban "rumba" 78-rpm vinyl recordings. These "Latin" rhythms were incorporated and helped launch soukous as the first pan-African sound. It is interesting to note that this transfer was part of a historical process of multidirectional diffusion, since the rhythms imported from Cuba originated among African slave laborers who had carried them from West Africa a century earlier. Congolese musicians were given record contracts and brought to Paris studios, where they continue to blend new sounds and techniques and pump out the pulsing soukous beat for the world music scene. One can get dizzy just trying to keep track of the many multidirectional pathways of cultural interaction across time and space.

Cultural interaction in music has been ongoing for centuries but has been greatly speeded up and geographically expanded under globalization. From one perspective, globalization has sparked a creative cultural interaction by mixing musical traditions from around the world to produce new hybrid forms, many of which are highly localized. As a result, new and innovative regionally based soundscapes emerge continuously, reinforcing old or helping to construct new culture regions. From another perspective, globalization has enriched First World transnational entertainment corporations without providing due compensation for the creative labor of local cultures. The debate over which perspective best reflects the actual effects of globalization on music is complex and will undoubtedly continue.

Nature-Culture

How is the theme of nature-culture related to the question of "one world or many"? The central issues include the impact of globalization processes on ecosystems and local communities and the promise and peril of new technologies for the natural world. Globalization represents for some an ever-expanding world economy. Where will the natural resources come from to fuel this expansion? Oil prices are rising in response to the growing demands of China's and India's economies. Will the entire world be able to replicate the post–World War II boom of the United States, or are gas prices a sign that we are reaching the planet's limits? What about alternative technologies that don't rely so completely on oil?

Sustainable Futures

Whether fearful or hopeful, most forecasters agree that recent trends indicate rising levels of consumption worldwide. Specifically, the cultures of mass consumption that developed first in the United States and Europe are spreading to every corner of the globe. China is a case in point. As we enter the twenty-first century, China is or soon will be the world's largest consumer of coal and automobiles. The cultures of mass consumption at the heart of globalization require enormous amounts of natural resources and produce prodigious quantities of pollutants. Given the ecological problems associated with the mass consumption of commodities such as cars, refrigerators, and so forth, we are compelled to ask ourselves whether current trends in consumption are sustainable for much longer.

The question of sustainable development on a global scale has been around since the 1970s. Cambridge geographer William Adams has produced the most comprehensive and carefully researched history of sustainable development. He suggests that the first attempt to create a plan for globally sustainable development came in 1980 with the publication of the World Conservation Strategy. This idea was refined a few years later by the UN-sponsored World Commission on Environment and Development. The commission brought sustainable development into the mainstream through its 1987 book *Our Common Future*. It identified poverty as a fundamental cause of the world's ecological problems and concluded that to reduce global ecological problems, we need to reduce global poverty through the promotion of economic development. Today, just about everyone from the barefoot "tree hugger" to the well-heeled international bank executive advocates sustainable development. This is what

Adams labeled the "mainstream" version of sustainable development, by which he meant that the concept had been redefined in a way that posed no serious challenge to the status quo of continual global expansion of consumption and economic growth.

But let us take a closer look at the mainstream approach to sustainable development. In essence it says that the familiar industrial model of continual economic expansion is the cure for both global poverty and ecological problems. However, while globalization has brought new prosperity and higher levels of consumption to some parts of the world, ecological problems only seem to be increasing. China's rising coal consumption has led to increased emissions of carbon dioxide, a leading greenhouse gas. It is now the world's largest producer of carbon dioxide and is projected to far exceed the United States, now in second place. Furthermore, as we noted previously, the rewards of globalization are distributed unevenly, with the majority of poor people remaining poor and only a few people becoming richer.

Some observers conclude that the term *sustainable development* is an oxymoron. Given the ecological record of modern industrialization, the faith in economic growth as a cure for environmental ills seems misplaced (**Figure 12.15**). Cultural and political ecologists have been strong critics of the mainstream approach to sustainable development. They claim that it does not take into account the larger-scale historical and structural causes of poverty, such as the lasting effects of European colonialism. Unless these are addressed,

FIGURE 12.15 Mountaintop mining in West Virginia. Landscapes and ecosystems are permanently altered to meet the industrial demand for increasing amounts of fossil fuels. A popular technique in coal mining involves the removal of entire mountaintops to access the deposit. *(Vivian Stockman.)*

it is unlikely that the mainstream approach to sustainable development will substantially decrease poverty levels. A further criticism is that the focus on the link between poverty and ecological degradation downplays the environmental impact from high levels of consumption in affluent countries. For instance, Americans alone consume one-fourth of the world's petroleum output and generate one-fourth of the carbon dioxide pollution.

One suggested alternative is to formulate sustainable development "from below" rather than through a top-down global program. The idea is to assist local initiatives and employ local knowledge to craft different economic paths for developing countries and communities that will not degrade land and resources along the way to higher living standards. Such alternatives, because they are informed by the communities they most immediately concern, would also be designed to maintain cultural identities, landscapes, and regions. Across the globe there are now hundreds of such efforts, which go by a variety of titles, such as "community conservation," "joint forest management," and "indigenous peoples' reserves." Approaching sustainable development from below means that the future will be defined by cultural heterogeneity, not homogeneity.

Think Globally, Act Locally

Perhaps sustainable development from below can mitigate widespread poverty while minimizing the ecological degradation that accompanies many top-down development initiatives. But what about the environmental effects of modern affluence—how can those be addressed? Throughout the 1990s a series of international conferences were held that sought to identify the world's most pressing environmental problems and propose global-scale initiatives to address them. The most prominent of these was the United Nations Conference on Environment and Development, held in Rio de Janeiro in 1992. Now known simply as the Rio Conference, it produced international conventions that sought to reduce global ecological problems such as species extinction and global warming. The most prominent of these was the Convention on Biological Diversity. This convention committed signatory states to protecting wildlife habitats and pursuing economic development policies that minimize species loss.

The approach taken at the Rio Conference is best captured in the slogan "Think Globally, Act Locally." If we plan carefully, actions taken at the local scale in places around the world will collectively result in an improved global environment. For example, the establishment of local parks and reserves will provide a global network of protected habitats that will help to maintain the Earth's biodiversity.

The recent introduction of hybrid cars—vehicles that combine traditional fossil-fuel engines with electric motors to greatly reduce gas consumption and pollution—is another case in point. When the state of California passed a law requiring 10 percent of all car sales to be hybrids, it forced carmakers to produce more fuel-efficient cars. Since California is the largest car market in the United States, its action has had ripple effects nationwide and, ultimately, worldwide. China is now promoting hybrid cars for its burgeoning market, and several of the new auto factories will produce hybrids. These local- and national-level initiates should help considerably in reducing global levels of carbon dioxide and other pollutants.

CULTURAL LANDSCAPE

Is the debate over whether the future will reflect one world or many visible? Can we get some clue about the outcome by observing the cultural landscape under globalization? Of course we can. Philip Kelly even speaks of "landscapes of globalization." He is referring mostly to the urban, corporate architecture that has sprung up on every continent, replicating itself in city after city. Such homogenization is often resisted by local communities that hope to preserve existing cultural landscapes.

Globalized Landscapes

Geographer David Keeling sought visible evidence of globalization in the landscape of Buenos Aires, Argentina—the capital of a country that desperately wants to become enmeshed in the world economy but struggles, perhaps in vain, against its peripheral location. He found abundant evidence of "a homogenized landscape" of glass-and-steel corporate office towers, of luxury hotels and conference centers for the corporate power brokers of the world economy (**Figure 12.16**).

Clearly, urban landscapes—cityscapes—can serve as an index to the level and type of engagement with the globalization process. We have now begun witnessing the diffusion of

FIGURE 12.16 Modern office buildings in Buenos Aires, Argentina. The forces of globalization have accelerated the construction of landscapes such as this one near the docklands area of Puerto Madero Harbor basin. *(wim wiskerke/Alamy.)*

California-style suburban housing tracts to affluent regions of India. Still, Keeling notes, these homogenizing processes are neither omnipresent nor omnipotent. Certain other Latin American capitals—such as Quito, Ecuador; La Paz, Bolivia; and Havana, Cuba—reveal minimal global influences in their cityscapes. Given the improbability of a global culture, visible differences among cities seem likely to persist. Moreover, people all over the world value their cultural landscapes, whether as visual reminders of their heritage or as lucrative attractions in the global tourism industry, and will thus want to preserve them. Landscape and place still matter in a globalizing world.

Striving for the Unique

Urban landscapes in the age of globalization reveal another element: the enduring spirit of place. One city after another has preserved or erected some building or monument so unique as to be a symbol or icon of that particular city. When you see a photograph of this structure, you know at once where it is located (**Figure 12.17**). Television journalists often stand in front of such visible icons to prove to viewers that they are actually reporting on location. Examples would include the Petronas Towers in Kuala Lumpur, the arch linking east and west in St. Louis, the Space Needle in Seattle, and the Eiffel Tower in Paris. True, many or most of these icons predate the era of globalization, but that is not the issue. Rather, their retention and protection offer the relevant message. In the 1960s, the city of Copenhagen restored Nyhavn (New Harbor), then a derelict seventeenth-century canal, into one of the most visually striking and instantly recognizable historical

ports in the world (**Figure 12.18**, page 452). The Kremlin walls in Moscow may retain few if any of their original red bricks, as they fall victim to weathering and are replaced, but the structure is renewed and endures as a symbol of the city.

Neolocalism is the desire evident in many local communities to embrace the uniqueness and authenticity of place. Governments and electorates at all levels—from local to national—have a far greater say about globalization than one might imagine. A backlash against chain stores and conformity can find strength in local ordinances. A community can actually prevent McDonald's or Walmart from establishing outlets. Neolocalism, then, pits the cultural power of place against the economic power of globalization.

> **neolocalism**
> The desire to reembrace the uniqueness and authenticity of place, in response to globalization.

Wal-Martians Invade Treasured Landscape!

Local communities and their governments attempt to preserve cultural landscapes in numerous ways. Zoning laws,

FIGURE 12.17 The Petronas Towers in Kuala Lumpur, have become a symbol for the city. Their unique design helps establish Kuala Lumpur's identity as a city different from others. Uniqueness of design, a feature of much modern architecture, stands in opposition to cultural homogenization. *(age fotostock/SuperStock.)*

FIGURE 12.18 Copenhagen's Nyhavn district. A key port from the 1600s to the early 1900s, Nyhavn was abandoned by the shipping industry as ship size increased after World War II. In the mid-1960s the city slapped on fresh coats of bright paint and hauled in antique sailing ships to create one of the most recognizable historic ports in the world. *Can you think of other historic waterfronts that have been similarly restored to create iconic urban landscapes?* (Courtesy of Roderick Neumann.)

architectural guidelines, minimum lot-size requirements, building codes, and conservation easements can all be used to maintain the distinct character of landscapes. At the global scale we have World Heritage Sites, places that the United Nations Educational, Scientific, and Cultural Organization (UNESCO) has deemed to be of such cultural significance that they should be given international protection. The temples at Angkor Wat in Cambodia (**Figure 12.19**) and the "Old Stone Town" on Zanzibar Island, Tanzania, are two examples of the dozens of UNESCO World Heritage Sites worldwide.

Globalizing processes are putting new pressures on cultural landscapes. Often small community groups and town governments are pitted against powerful transnational corporations whose investment choices can profoundly transform a landscape. The phenomenal expansion of Walmart Stores, Inc., is an often-cited example of how small-town landscapes are transformed by corporate investment. The biggest impact is the "hollowing out" of small-town main streets. Owner-run small businesses cannot compete with the giant retailer and soon have to lock their doors. Since "big box stores" such as Walmart are typically located outside of city centers, the old downtowns are turned into empty shells.

Some communities have welcomed Walmart and others have tried hard to keep it out. Perhaps the most novel opposition campaign has been conducted in Vermont. In a confrontation that author Barbara Ehrenreich labeled "Earth People vs. Wal-Martians," the National Trust for Historic Preservation declared the entire state of Vermont to be

FIGURE 12.19 Angkor Wat, UNESCO World Heritage Site. Located 192 miles from Phnom Penh, Cambodia, and built between A.D. 1113 and 1150, this temple is regarded as the pinnacle of the Khmer Empire's architecture. UNESCO named it and surrounding structures a World Heritage Site in 1992. Such a designation helps efforts to safeguard and restore cultural landscapes of global significance. (Courtesy of Ari Dorfsman.)

FIGURE 12.20 Residents and tourists believe the historic town squares of New England are threatened by the construction of Walmart stores. Defenders of Vermont's cultural landscape argue that large retail ventures like Walmart are both visually disruptive and economically destructive of local businesses. ***What do you think might happen to small town centers when large retail stores open nearby?*** (Left: David Frazier/Index Stock/Corbis; Right: © Chuck Franklin/Alamy.)

an endangered landscape. Vermont is the only state ever to make the list of endangered historic places, which generally comprise individual buildings and historic urban districts. According to the National Trust, building more supersized Walmart stores in Vermont would degrade the state's "sense of place." The National Trust had employed a similar strategy in 1993, which forced Walmart to build stores more appropriate to Vermont's landscape (**Figure 12.20**).

Protecting Europe's Rural Landscape

Another aspect of globalization, the drive to eliminate territorial barriers to the free trade of commodities, threatens rural landscapes in Europe. The fear is that cheap food imports will put small farmers out of business, which in turn will lead to the demise of treasured rural landscapes (**Figure 12.21**). European farmers and their national and EU representatives have argued that agriculture is not solely about food production—that it performs multiple functions, such as maintaining cultural landscapes and providing environmental services. Geographer Gail Hollander has observed that farmers and their advocates are using what they call agriculture's "multifunctionality" to gain exemptions from the WTO's strict rules on free trade. The exemptions are warranted, they argue, in order to preserve the character of cultural landscapes that agriculture supports. Hollander concludes that multifunctionality could be used to preserve the landscapes of a few communities in Europe or, in stronger form, to challenge the very logic of

the WTO's rules on the global trade of agricultural products. Given their symbolic value, the preservation of cultural landscapes may be a key tool used to slow or mitigate globalization's homogenizing effects.

FIGURE 12.21 Rural town in Andalusia, Spain. The ancient olive terraces, gardens, woodlots, and pastures of this town are typical of much of Mediterranean Europe. Globalization and other economic forces are making the traditional, extensive agricultural systems that produced and support such landscapes increasingly obsolete. *(Courtesy of Roderick Neumann.)*

CONCLUSION

As powerfully transformative as the processes of globalization are, they are unlikely to result in cultural homogenization any time soon. As we learned by examining globalization's interaction with each of cultural geography's five themes, we do not yet live in a placeless world. In fact, we have observed many trends suggesting that globalization will produce more geographic diversity and many unintended and unforeseen outcomes in our future.

In closing, it is our hope that we have excited your interest in the world's cultural diversity. To paraphrase the words of Aldous Huxley, we hope that our vicarious world travels have left you "poorer by exploded convictions" and "perished certainties," but richer by what you have seen. Perhaps we, like Huxley, set out on this journey with preconceptions of how people should best "live, be governed, and believe." When one travels—even if just through the pages of a geography book as we have—such convictions often get mislaid. The main message of *The Human Mosaic* is that we will best be prepared to thrive in this new millennium if we maintain a willingness to question even our most closely held convictions and remain open to the boundless capacity of human cultural expression to surprise and amaze.

DOING GEOGRAPHY

Interpreting the Imagery of Globalization

The mandate of large corporations today, nearly regardless of the type of service or good they produce, is to go global or go bankrupt. In addition to staying competitive, going global gives a company a certain cachet and consumer appeal, the way that being "modern" did in previous decades. Transnational corporations also stress other popular notions related to globalization, such as respect for the world's cultural diversity and concern for the global environment.

For this activity, you will use your interpretive skills to look at the way the processes of globalization are represented in corporate-produced visual imagery and text. Using the steps below, try to find representations of globalization in more than one medium, including product packaging, magazine advertisements, and corporate web sites.

Steps to Interpreting the Imagery of Globalization

Step 1: Concentrate on one type of industry, such as pharmaceuticals or automobiles, or several.

Step 2: Look for materials that include visual and textual representations of the globe, the Earth, or the world, and remember to make use of the five themes of cultural geography.

Step 3: Analyze and interpret each of the samples that you select by setting up a series of questions such as: What popular notions about globalization are emphasized? How much validity do these representations carry? This is not the same as asking about the truth or falseness of an ad. Rather it is to ask, for example: What does going to the Hard Rock Café and buying a T-shirt have to do with "saving the planet"?

In the cases of visual imagery, look carefully at the way objects are arranged and scaled in relation to one another:

- How is power represented in the imagery?
- Does power lie with the individual consumer or the corporation?
- How are local-global linkages represented?
- Are the activities of global corporations given a moral authority, and if so, how?

Step 4: Provide a written summary of your findings and analysis. Include a copy of the image or images that you use and make sure to reference your sources.

As you do this exercise, bear in mind the power of transnational corporations in an era of globalization. Think about the importance of understanding how the images and texts they produce give meaning to our world, our places, and our landscapes.

Spanning the world. Corporations increasingly use such global imagery in advertising and public relations to signify global reach and engagement with the forces of globalization. *(Corbis/SuperStock.)*

Key Terms

glocalization	p. 434
neolocalism	p. 451

The Geography of the Future on the Internet

You can learn more about the geography of the future on the Internet at the following web sites:

An Atlas of Cyberspaces, Martin Dodge
http://www.cybergeography.org/atlas/atlas.html
Here cyberspaces are made visible: graphic representations of the geography of the electronic territories of the Internet, the World Wide Web, and other new cyberspaces help you to visualize and comprehend the digital "landscapes" beyond your computer screen.

The Hawaii Research Center for Futures Studies
http://www.futures.hawaii.edu/
One of the best-known institutions for future studies. The Hawaii State Legislature created it in 1971 to train students in future thinking for work in government and business.

Millennium Project: World Federation of UN Associations
http://www.millennium-project.org/
A global, participatory, futures-research think tank of futurists, scholars, business planners, and policy makers. It produces numerous future scenarios and such documents as the annual "State of the Future" report.

World Future Society
http://www.wfs.org/
A scientific and educational association exploring how social and technological developments are shaping the future. The society serves as a clearinghouse for forecasts, recommendations, and alternative scenarios. It publishes, among other periodicals, the bimonthly journal *The Futurist*.

World Futures Studies Federation
http://www.wfsf.org/
Founded in 1967 to further research and education in future studies. It is comprised of hundreds of individuals and institutions worldwide that together create a global network of researchers, teachers, policy analysts, and activists.

Sources

Adams, William. 2001. *Green Development: Environment and Sustainability in the Third World,* 2nd ed. London: Routledge.

Airriess, Christopher A. 2001. "Regional Production, Information-Communication Technology, and the Developmental State: The Rise of Singapore as a Global Container Hub." *Geoforum* 32: 235–254.

Anderson, Elijah. 1990. *Streetwise: Race, Class, and Change in an Urban Community.* Chicago: Chicago University Press.

Anderson, Janna Quitney, and Lee Rainie. 2006. "The Future of the Internet II." Washington, D.C.: The Pew Internet and American Life Project.

Barber, Benjamin. 1995. *Jihad vs. McWorld: How Globalism and Tribalism Are Reshaping the World.* New York: Ballantine.

Barlow, John P. 1995. "Cyberhood Versus Neighborhood." Special issue of *Utne Reader* 68(3): 52–64.

Beaverstock, Jonathan, Phillip Hubbard, and John Short. 2004. "Getting Away with It? Exposing the Geographies of the Super-Rich." *Geoforum* 35(4): 401–407.

Brewer's Digest Buyers' Guide and Directory. 1992. Chicago: Siebel.

Bruntland, H. 1987. *Our Common Future.* Oxford: Oxford University Press (for the World Commission on Environment and Development).

Bunge, William. 1973. "The Geography of Human Survival." *Annals of the Association of American Geographers* 63: 275–295.

Cairncross, Frances. 1997. *The Death of Distance: How the Communications Revolution Will Change Our Lives.* Cambridge, Mass.: Harvard Business School Press.

Cartier, Carolyn. 2001. *Globalizing South China.* Oxford: Blackwell.

Cosgrove, Denis. 2001. *Apollo's Eye: A Cartographic Genealogy of the Earth in the Western Imagination.* Baltimore: Johns Hopkins University Press.

Crouch, David (ed.). 1999. *Leisure/Tourism Geographies: Practices and Geographical Knowledge.* London: Routledge.

Crum, Shannon L. 2000. "The Spatial Diffusion of the Internet." Ph.D. dissertation, University of Texas at Austin.

Dicken, Peter. 2003. *Global Shift: Reshaping the Global Economic Map in the 21st Century,* 4th ed. New York: Guilford.

Drummond, Ian, and Terry Marsden. 1999. *The Condition of Stability: Global Environment Change.* New York: Routledge.

Ehrenreich, Barbara. 2004. "Earth People vs. Wal-Martians." *New York Times,* 25 July, p. 11.

Elwood, Wayne. 2001. The No-Nonsense Guide to Globalization. London: Verso.

Feld, Stephen. 2001. "A Sweet Lullaby for World Music," in Arjun Appadurai (ed.), *Globalization.* Durham, N.C.: Duke University Press, 189–216.

Ferguson, Andrew. 2004. "Wal-Mart Opponents Launch Two-Front Attack." *Pittsburgh Post-Gazette,* 20 June, p. C2.

Ferry, Luc. 1995. *The New Ecological Order.* Carol Volk (trans.). Chicago: University of Chicago Press.

Firebaugh, Glenn. 2003. *The New Geography of Global Inequality.* Cambridge, Mass.: Harvard University Press.

Flack, Wes. 1997. "American Microbreweries and Neolocalism." *Journal of Cultural Geography* 16(2): 37–53.

Fryer, Donald. 1974. "A Geographer's Inhumanity to Man." *Annals of the Association of American Geographers* 64: 479–482.

"Geography and the Internet." 2001. Special report. *The Economist* (August 11): 18–20.

Gilbert, Anne, and Paul Villeneuve. 1999. "Social Space, Regional Development, and the Infobahn." *Canadian Geographer* 43: 114–117.

Greider, William. 1997. *One World, Ready or Not: The Manic Logic of Global Capitalism.* New York: Touchstone.

Hardt, Michael, and Antonio Negri. 2000. *Empire.* Cambridge, Mass.: Harvard University Press.

Herod, Andrew. 2001. *Labor Geographies: Workers and the Landscapes of Capitalism.* New York: Guilford.

SEEING GEOGRAPHY Global Reach

What do these images convey about empire and globalization and their similarities and differences?

Two images of global reach, more than a century apart (Queen Victoria with world map; a JVC advertisement).

Before we explore an answer, let us pause to recall our Practicing Geography profile of Denis Cosgrove in Chapter 1. Remember that Professor Cosgrove's primary interest in cultural geography was in "interpreting" rather than "explaining." His book *Apollo's Eye* is a good example. In it, Cosgrove attempted to interpret the power represented in images of the globe and to show how the practices of globalization are historically rooted in a Western cultural history of imagining, seeing, and representing the globe. We will try a little of this interpretive method on these two images

The image on the left shows Queen Victoria circa 1850 in front of a world map oriented so that the majority of Britain's territorial empire is displayed. We know that during this period of European history the queen was sovereign, meaning that she personified, even embodied, Britain and its colonial empire. The image is scaled such that the queen's arm span matches the span of the British Empire. She is positioned in front of the map, emphasizing her authority and power over it. With power and authority comes responsibility; the viewer is meant to read in the queen's pose and dress a moral role as protector and civilizing force. Foregrounded as she is, then, all lines of power, authority, and moral right and responsibility in empire run through her.

The image on the right is an advertisement for JVC, a transnational consumer electronics company that began as the Victor Talking Machine Company of Japan, Limited, in 1927. In the advertisement, photographed in 2004, the company's logo, "JVC," is scaled to continental size. The message conveyed is one of global dominance and global reach. It also conveys the message of one company bringing the world together, a goal that JVC's web site describes as "contributing to the global community through cultural activities" with corporate underwriting. Finally, the advertisement is meant to express through the image of the globe the fact that JVC now has a network of manufacturing sites throughout Asia, the Americas, and Europe as well as sales subsidiaries in many more regions.

So what do these images tell us about continuity and change from empire to globalization? We see a common theme in the claim of global reach. In the JVC ad, it is expressed as the company's ability to span the globe with its products and services. In the image with Queen Victoria, it is expressed in the cartographic representation of Britain's global empire. We can also identify significant differences between the emotional and affective meaning of British Empire and globalization conveyed by these images. Under empire, the queen personified British imperial rule, and allegiance to the queen was required of all imperial subjects. Under globalization, on the other hand, allegiance is constructed between the consumer and transnational corporations. Corporations are faceless rather than personified, represented by abstract logos rather than by living, breathing sovereigns. In summary, we can see in these images both the roots of globalization in empire as well as the significant differences between the two kinds of global power.

Hessler, Peter. 2007. "Wheels of Fortune: The People's Republic Learns to Drive." *The New Yorker,* 26 November, p. 104.

Hollander, Gail. 2004. "Agricultural Trade Liberalization, Multi-functionality, and Sugar in the South Florida Landscape." *Geoforum* 35: 299–312.

Hooson, David (ed.). 1994. *Geography and National Identity.* Oxford: Blackwell.

Hudson, Ray. 2000. "One Europe or Many? Reflections on Becoming European." *Transactions of the Institute of British Geographers* 25: 409–426.

Huxley, Aldous. 1926. *Jesting Pilate: An Intellectual Holiday.* New York: George H. Doran.

Iyer, Pico. 2000. *The Global Soul: Jet Lag, Shopping Malls, and the Search for Home.* New York: Knopf.

James, Harold. 2001. *The End of Globalization.* Cambridge, Mass.: Harvard University Press.

Johnston, R. J., Peter Taylor, and Michael Watts (eds.). 2002. *Geographies of Global Change: Remapping the World,* 2nd ed. Oxford: Blackwell.

Keating, Michael. 1998. *The New Regionalism in Western Europe.* Northampton, Mass.: Edward Elgar.

Keeling, David J. 1999. "Neoliberal Reform and Landscape Change in Buenos Aires." *Yearbook, Conference of Latin Americanist Geographers* 25: 15–32.

Kelly, Philip F. 2000. *Landscapes of Globalization.* London: Routledge.

Kimble, George H. T. 1951. "The Inadequacy of the Regional Concept," in L. Dudley Stamp and Sidney W. Wooldridge (eds.), *London Essays in Geography: Rodwell Jones Memorial Volume.* Cambridge, Mass.: Harvard University Press, 151–174.

Kitchen, Rob, and Martin Dodge. 2002. "Emerging Geographies of Cyberspace," in R. J. Johnston, Peter Taylor, and Michael Watts (eds.), *Geographies of Global Change: Remapping the World,* 2nd ed. Oxford: Blackwell, 340–354.

Knox, Paul. 2002. "World Cities and the Organization of Global Space," in R. J. Johnston, Peter Taylor, and Michael Watts (eds.), *Geographies of Global Change: Remapping the World,* 2nd ed. Oxford: Blackwell, 328–339.

Kraus, Clifford. 2002. "Returning Tundra's Rhythm to the Inuit, in Film." *The New York Times,* 30 March, p. A4.

Kunstler, James H. 1993. *The Geography of Nowhere: The Rise and Decline of America's Man-Made Landscape.* New York: Simon & Schuster.

Leyshon, Andrew, David Matless, and George Revill (eds.). 1998. *The Place of Music.* New York: Guilford.

McDowell, Linda (ed.). 1997. *Undoing Place? A Geographical Reader.* London: Arnold.

Naughton, Keith. 2004. "China Hits the Road." *Newsweek,* 28 June, p. E22.

Nemeth, David J. 2000. "The End of the Re(li)gion?" *North American Geographer* 2: 1–8.

O'Loughlin, John, et al. 1998. "The Diffusion of Democracy, 1946–1994." *Annals of the Association of American Geographers* 88: 545–574.

Poon, Jessie P. H., Edmund R. Thompson, and Philip F. Kelly. 2000. "Myth of the Triad? The Geography of Trade and Investment Blocs." *Transactions of the Institute of British Geographers* 25: 427–444.

Press, Larry. 1997. "Tracking the Global Diffusion of the Internet." *Communications of the Association of Computing Machinery* 40(11): 11–17.

Relph, Edward. 1976. *Place and Placelessness.* London: Pion.

"The Revenge of Geography." 2003. *The Economist Technology Quarterly* (March 15): 19–22.

Rohter, Larry. 2002. "Brazil's Prized Exports Rely on Slaves and Scorched Land." *The New York Times,* 25 March, pp. A1, A6.

Sessions, George (ed.). 1995. *Deep Ecology for the 21st Century.* Boulder, Colo.: Shambala.

Shields, Rob. 1991. *Places on the Margin: Alternative Geographies of Modernity.* London: Routledge.

Suvantola, Jaakko. 2002. *Tourist's Experience of Place.* Aldershot, U.K.: Ashgate.

Swerdlow, Joel L. (ed.). 1999. "Global Culture." Special issue of *National Geographic* 196(2): 2–132.

Swyngedouw, Erik. 1997. "Neither Global nor Local," in Kevin R. Cox (ed.), *Spaces of Globalization: Reasserting the Power of the Local.* New York: Guilford, 137–166.

Teo, Peggy, and Lim Hiong Li. 2003. "Global and Local Interactions in Tourism." *Annals of Tourism Research* 30(2): 287–306.

"Used and Abused: Five Recent Cases with Slavery Convictions." 2003. *Palm Beach Post,* 7 December, p. 2.

Wood, William B. 2001. "Geographic Aspects of Genocide: A Comparison of Bosnia and Rwanda." *Transactions of the Institute of British Geographers* 26: 57–75

Ten Recommended Books on the Geography of the Future

(For additional suggested readings, see *The Human Mosaic* web site: www.whfreeman.com/domosh12e)

Bagchi-Sen, Sharmistha, and Helen Smith (eds). 2006. *Economic Geography: Past, Present and Future.* London: Routledge. Economic geographers point the way for the future development of the subfield in 20 wide-ranging chapters.

Crang, Mike, Phil Crang, and Jon May (eds.). *Virtual Geographies: Bodies, Space and Relations.* London: Routledge. Explores how new communications technologies produce new types of space and even new geographies.

Firebaugh, Glenn. 2003. *The New Geography of Global Inequality.* Cambridge, Mass.: Harvard University Press. Makes the argument that income inequalities among and within world regions are misunderstood. There are a lot of economic data to wade through, but they make the case stronger. The author raises important questions about the effects of globalization.

Gabel, Medard, and Henry Bruner. 2003. *Global Inc.: An Atlas of the Multinational Corporation.* New York: New Press. A wonderful atlas mapping everything from the historical rise of multinational companies to the latest geographic expansions of Walmart. It's full of facts on every important global industry, including food, cars, and pharmaceuticals. It also maps the impacts of multinational corporations, including cultural and environmental ones.

Johnston, R. J., Peter Taylor, and Michael Watts (eds). 2002. *Geographies of Global Change: Remapping the World,* 2nd ed. Oxford: Blackwell. Considers such issues as post–cold war geopolitics, global environmental governance, and cultural changes related to mass consumption, the Internet, and ethnic identity.

Jussila, Heikki, Roser Majoral, and Fernanda Delgado-Cravidao. 2001. *Globalization and Marginality in Geographical Space.* Aldershot, U.K.: Ashgate. Case studies from Europe, the Americas, Africa, and Australia illustrate how geographical research aids our understanding of the way in which the policies and politics of globalization affect the more marginalized areas of the world.

Kotkin, Joel. 2000. *The New Geography: How the Digital Revolution Is Reshaping the American Landscape.* New York: Random House. Argues that computer and telecommunication technology has freed people and businesses to locate wherever they wish, thereby weakening venerable core-periphery patterns but strengthening place distinctiveness.

Miles, Malcolm, and Tim Hall (eds). 2003. *Urban Futures: Critical Commentaries on Shaping Cities.* This volume brings together experts from a range of disciplines to debate the cultures and forms of tomorrow's cities.

Norwine, Jim, and Jonathan M. Smith (eds). 2000. *Worldview Flux: Perplexed Values Among Postmodern Peoples.* Lanham, Md.: Lexington Books. An irreverent, occasionally funny look at how groups as diverse as Cajuns, South African whites, and Pacific coast Native Americans are coping with the new age of globalization and the need to restructure their self-identities and place attachments.

Skelton, Tracey, and Tim Allen (eds.). 1999. *Culture and Global Change.* London: Routledge. No fewer than 27 authors consider the interaction of culture and globalization, in the process rejecting both cultural and economic determinism.

GLOSSARY

absorbing barrier A barrier that completely halts diffusion of innovations and blocks the spread of cultural elements. (page 12)

acculturation The adoption by an ethnic group of enough of the ways of the host society to be able to function economically and socially. (page 155)

acid rain Rainfall with much higher acidity than normal, caused by sulfur and nitrogen oxides derived from the burning of fossil fuels being flushed from the atmosphere by precipitation, with lethal effects for many plants and animals. (page 336)

adaptive strategy The unique way in which each culture uses its particular physical environment; those aspects of culture that serve to provide the necessities of life—food, clothing, shelter, and defense. (page 255)

agglomeration A snowballing geographical process by which secondary and service industrial activities become clustered in cities and compact industrial regions in order to share infrastructure and markets. (page 393)

agribusiness Highly mechanized, large-scale farming, usually under corporate ownership. (page 295)

agricultural landscape The cultural landscape of agricultural areas. (page 307)

agricultural region A geographic region defined by a distinctive combination of physical and environmental conditions; crop type; settlement patterns; and labor, cultivation, and harvesting practices. (page 275)

agricultural surplus The amount of food grown by a society that exceeds the demands of its population. (page 358)

agriculture The cultivation of domesticated crops and the raising of domesticated animals. (page 275)

agroforestry A cultivation system that features the interplanting of trees with field crops. (page 54)

amenity landscapes Landscapes that are prized for their natural and cultural aesthetic qualities by the tourism and real estate industries and their customers. (page 60)

Anatolian hypothesis A theory of language diffusion holding that the movement of Indo-European languages from the area in contemporary Turkey known as Anatolia followed the spread of plant domestication technologies. (page 122)

animists Adherents of animism, the idea that souls or spirits exist not only in humans but also in animals, plants, rocks, natural phenomena such as thunder, geographic features such as mountains or rivers, or other entities of the natural environment. (page 244)

apartheid In South Africa, a policy of racial segregation and discrimination against non-European groups. (page 376)

aquaculture The cultivation, under controlled conditions, of aquatic organisms, primarily for food but also for scientific and aquarium uses. (page 284)

assimilation The complete blending of an ethnic group into the host society, resulting in the loss of all distinctive ethnic traits. (page 155)

axis mundi The symbolic center of cosmomagical cities, often demarcated by a large vertical structure. (page 360)

barriadas Illegal housing settlements, usually made up of temporary shelters, that surround large cities; often referred to as *squatter settlements*. (page 371)

bilingualism The ability to speak two languages fluently. (page 117)

biofuel Broadly, this term refers to any form of energy derived from biological matter, increasingly used in reference to replacements for fossil fuels in internal combustion engines, industrial processes, and the heating and cooling of buildings. (page 305)

border zones The areas where different regions meet and sometimes overlap. (page 7)

buffer state An independent but small and weak country lying between two powerful countries. (page 193)

cadastral pattern The shapes formed by property borders; the pattern of land ownership. (page 307)

carrying capacity The maximum number of people that can be supported in a given area. (page 74)

census tracts Small districts used by the U.S. Census Bureau to survey the population. (page 387)

central business district (CBD)   The central portion of a city, characterized by high-density land uses. (page 385)

central places   Towns or cities engaged primarily in the service stages of production; a regional center. (page 356)

centralizing forces   Diffusion forces that encourage people or businesses to locate in the central city. (page 393)

central-place theory   A set of models designed to explain the spatial distribution of urban service centers. (page 355)

centrifugal force   Any factor that disrupts the internal unity of a country. (page 195)

centripetal force   Any factor that supports the internal unity of a country. (page 195)

chain migration   The tendency of people to migrate along channels, over a period of time, from specific source areas to specific destinations. (page 165)

checkerboard development   A mixture of farmlands and housing tracts. (page 397)

circulation   A term that implies an ongoing set of movements of people, ideas, or things that have no particular center or periphery. (page 12)

cityscape   An urban landscape. (page 407)

cleavage model   A political-geographic model suggesting that persistent regional patterns in voting behavior, sometimes leading to separatism, can usually be explained in terms of tensions pitting urban against rural, core against periphery, capitalists against workers, and power group against minority culture. (page 212)

colonialism   The forceful appropriation of a territory by a distant state, often involving the displacement of indigenous populations to make way for colonial settlers (page 51); the building and maintaining of colonies in one territory by people based elsewhere. (page 190)

concentric-zone model   A social model that depicts a city as five areas bounded by concentric rings. (page 389)

consumer nationalism   A situation in which local consumers favor nationally produced goods over imported goods as part of a nationalist political agenda. (page 48)

consumer services   The range of economic activities that facilitate the consumption of goods. (page 328)

contact conversion   Spread of religious beliefs by personal contact. (page 246)

contagious diffusion   A type of expansion diffusion in which cultural innovation spreads by person-to-person contact, moving wavelike through an area and population without regard to social status. (page 11)

conventional agriculture   The widely adopted commercial, industrialized form of farming that uses a range of synthetic fertilizers, insecticides, and herbicides to control pests and maximize productivity; a term that emerged following the creation of alternative forms, such as organic farming. (page 304)

convergence hypothesis   A hypothesis holding that cultural differences among places are being reduced by improved transportation and communications systems, leading to a homogenization of popular culture. (page 47)

cool chain   The refrigeration and transport technologies that allow for the distribution of perishables. (page 294)

core area   The territorial nucleus from which a country grows in area and over time, often containing the national capital and the main center of commerce, culture, and industry. (page 203)

core periphery   A concept based on the tendency of both formal and functional culture regions to consist of a core or node, in which defining traits are purest or functions are headquartered, and a periphery that is tributary and displays fewer of the defining traits. (pages 7, 293)

cornucopians   Those who believe that science and technology can solve resource shortages. In this view, human beings are our greatest resource rather than a burden to be limited. (page 100)

cosmomagical cities   Types of cities that are laid out in accordance with religious principles, characteristic of very early cities, particularly in China. (page 360)

cottage industry   A traditional type of manufacturing in the preindustrial revolution era, practiced on a small scale in individual rural households as a part-time occupation and designed to produce handmade goods for local consumption. (page 322)

creole   A language derived from a pidgin language that has acquired a fuller vocabulary and become the native language of its speakers. (page 117)

cultural diffusion   The spread of elements of culture from the point of origin over an area. (page 285)

cultural ecology   Broadly defined, the study of the relationships between the physical environment and culture; narrowly (and more commonly) defined, the study of culture as an adaptive system that facilitates human adaptation to nature and environmental change. (page 16)

cultural landscape   The visible human imprint on the land. (page 24)

cultural maladaptation   Poor or inadequate adaptation that occurs when a group pursues an adaptive strategy that, in the short run, fails to provide the necessities of life or, in the long run, destroys the environment that nourishes it. (page 172)

cultural preadaptation   A complex of adaptive traits and skills possessed in advance of migration by a group, giving it survival ability and competitive advantage in occupying the new environment. (page 171)

cultural simplification   The process by which immigrant ethnic groups lose certain aspects of their traditional culture in the process of settling overseas, creating a new culture that is less complex than the old. (page 167)

culturally preadapted   A complex of adaptive traits and skills possessed in advance of migration by a group, giving it survival ability and competitive advantage in occupying the new environment. (page 304)

culture   A total way of life held in common by a group of people, including such learned features as speech, ideology, behavior, livelihood, technology, and government; or the local, customary way of doing things—a way of life; a never-changing process in which a group is actively engaged; a dynamic mix of symbols, beliefs, speech, and practices. (page 2)

culture hearth   A focused geographic area where important innovations are born and from which they spread. (page 246)

decentralizing forces   Diffusion forces that encourage people or businesses to locate outside the central city. (page 393)

defensive site   A location from which a city can be easily defended. (page 367)

deindustrialization   The decline of primary and secondary industry, accompanied by a rise in the service sectors of the industrial economy. (pages 325, 398)

demographic transition   A term used to describe the movement from high birth and death rates to low birth and death rates. (page 77)

depopulation   A decrease in population that sometimes occurs as the result of sudden catastrophic events, such as natural disasters, disease epidemics, and warfare. (page 106)

desertification   A process whereby human actions unintentionally turn productive lands into deserts through agricultural and pastoral misuse, destroying vegetation and soil to the point where they cannot regenerate. (page 303)

dialect   A distinctive local or regional variant of a language that remains mutually intelligible to speakers of other dialects of that language; a subtype of a language. (page 117)

diffusion   The movement of people, ideas, or things from one location outward toward other locations. (page 10)

dispersed   A type of settlement form in which people live relatively distant from each other. (page 24)

domesticated animal   An animal kept for some utilitarian purpose whose breeding is controlled by humans and whose survival is dependent on humans; domesticated animals differ genetically and behaviorally from wild animals. (page 287)

domesticated plant   A plant deliberately planted and tended by humans that is genetically distinct from its wild ancestors as a result of selective breeding. (page 286)

double-cropping   Harvesting twice a year from the same parcel of land. (page 277)

dust dome   A pollution layer over a city that is thickest at the center of the city. (page 407)

ecofeminism   A doctrine proposing that women are inherently better environmental preservationists than men because the traditional roles of women involved creating and nurturing life, whereas the traditional roles of men too often necessitated death and destruction. (page 21)

ecotheology   The study of the influence of religious belief on habitat modification. (page 248)

ecotourism   Responsible travel that does not harm ecosystems or the well-being of local people. (page 339)

edge city   A new urban cluster of economic activity that surrounds nineteenth-century downtowns. (page 419)

electoral geography   The study of the interactions among space, place, and region and the conduct and results of elections. (page 199)

enclave   A piece of territory surrounded by, but not part of, a country. (page 191)

environmental determinism   The belief that cultures are directly or indirectly shaped by the physical environment. (page 18)

environmental perception   The belief that culture depends more on what people perceive the environment to be than on the actual character of the environment; perception, in turn, is colored by the teachings of culture. (page 20)

environmental racism   The targeting of areas where ethnic or racial minorities live with respect to environmental contamination or failure to enforce environmental regulations. (page 174)

ethnic cleansing   The removal of unwanted ethnic minority populations from a nation state through mass killing, deportation, or imprisonment. (page 166)

ethnic culture region   An area occupied by people of similar ethnic background who share traits of ethnicity, such as language and migration history. (page 386)

ethnic flag   A readily visible marker of ethnicity on the landscape. (page 175)

ethnic group   A group of people who share a common ancestry and cultural tradition, often living as a minority group in a larger society. (page 151)

ethnic homelands   Sizable areas inhabited by an ethnic minority that exhibits a strong sense of attachment to the region and often exercises some measure of political and social control over it. (page 156)

ethnic islands   Small ethnic areas in the rural countryside; sometimes called folk islands. (page 157)

ethnic neighborhood   A voluntary community where people of like origin reside by choice. (page 157)

ethnic religion   A religion identified with a particular ethnic or tribal group; does not seek converts. (page 232)

ethnic substrate   Regional cultural distinctiveness that remains following the assimilation of an ethnic homeland. (page 157)

ethnoburbs   A suburban ethnic neighborhood, sometimes home to relatively affluent immigration populations. (page 159)

ethnographic boundary   A political boundary that follows some cultural border, such as a linguistic or religious border. (page 194)

ethnolect   A dialect spoken by a particular ethnic group. (page 127)

Eurocentric   Using the historical experience of Europe as the benchmark for all cases. (page 79)

exclave   A piece of national territory separated from the main body of a country by the territory of another country. (page 192)

expansion diffusion   The spread of innovations within an area in a snowballing process, so that the total number of knowers or users becomes greater and the area of occurrence grows. (page 10)

export processing zones (EPZs)   Designated areas of countries where governments create conditions conducive to export-oriented production. (page 331)

extended metropolitan regions (EMRs)   New types of urban regions, complex in both landscape form and function, created by the rapid spatial expansion of cities in the developing world. (page 374)

farm villages   Clustered rural settlements of moderate size, inhabited by people who are engaged in farming. (page 105)

farmstead   The center of farm operations, containing the house, barn, sheds, and livestock pens. (page 105)

federal state   An independent country that gives considerable powers and even autonomy to its constituent parts. (page 194)

feedlot   A factorylike farm devoted to either livestock fattening or dairying; all feed is imported and no crops are grown on the farm. (page 280)

festival setting   A multiuse redevelopment project that is built around a particular setting, often one with a historical association. (page 422)

folk   Traditional, rural; the opposite of "popular." (page 32)

folk architecture   Structures built by members of a folk society or culture in a traditional manner and style, without the assistance of professional architects or blueprints, using locally available raw materials. (page 56)

folk culture   A small, cohesive, stable, isolated, nearly self-sufficient group that is homogeneous in custom and race; characterized by a strong family or clan structure, order maintained through sanctions based in the religion or family, little division of labor other than that between the sexes, frequent and strong interpersonal relationships, and a material culture consisting mainly of handmade goods. (pages 32, 277)

folk fortress   A stronghold area with natural defensive qualities, useful in the defense of a country against invaders. (page 216)

folk geography   The study of the spatial patterns and ecology of traditional groups; a branch of cultural geography. (page 33)

foodways   Customary behaviors associated with food preparation and consumption. (page 180)

formal region   A cultural region inhabited by people who have one or more cultural traits in common. (page 6)

functional region   A cultural area that functions as a unit politically, socially, or economically. (page 8)

functional zonation   The pattern of land uses within a city; the existence of areas with differing functions, such as residential, commercial, and governmental. (page 408)

fundamentalism   A movement to return to the founding principles of a religion, which can include literal interpretation of sacred texts, or the attempt to follow the ways of a religious founder as closely as possible. (page 234)

Gaia hypothesis   The theory that there is one interacting planetary ecosystem, Gaia, that includes all living things and the land, waters, and atmosphere in which they live; further, that Gaia functions almost as a living organism, acting to control deviations in climate and to correct chemical imbalances, so as to preserve Earth as a living planet. (page 260)

gender roles   What it means to be a man or a woman in different cultural and historical contexts. (page 86)

generic toponym   The descriptive part of many place-names, often repeated throughout a culture area. (page 139)

genetically modified (GM) crops   Plants whose genetic characteristics have been altered through recombinant DNA technology. (page 298)

genocide   The systematic killing of a racial, ethnic, religious, or linguistic group. (page 170)

gentrification   The displacement of lower income residents by higher income residents as buildings in deteriorated areas of city centers are restored. (page 398)

geography   The study of spatial patterns and of differences and similarities from one place to another in environment and culture. (page 1)

geometric boundary   A political border drawn in a regular, geometric manner, often a straight line, without regard for environmental or cultural patterns. (page 194)

geopolitics   The influence of the habitat on political entities. (page 216)

gerrymandering   The drawing of electoral district boundaries in an awkward pattern to enhance the voting impact of one constituency at the expense of another. (page 199)

ghetto   Traditionally, an area within a city where an ethnic group lives, either by choice or by force. Today in the United States, the term typically indicates an impoverished African-American neighborhood. (pages 158, 415)

global cities   Cities that are control centers of the global economy. (page 364)

global warming   The pronounced climatic warming of the Earth that has occurred since about 1920 and particularly since the 1970s. (page 336)

globalization   The binding together of all the lands and peoples of the world into an integrated system driven by capitalistic free markets, in which cultural diffusion is rapid, independent states are weakened, and cultural homogenization is encouraged. (pages 13, 295)

globalizing cities   Cities being shaped by the new global economy and culture. (page 364)

glocalization   Forces of change interact with local cultures, altering both in the process. (page 434)

green revolution   The recent introduction of high-yield hybrid crops and chemical fertilizers and pesticides into traditional Asian agricultural systems, most notably paddy rice farming, with attendant increases in production and ecological damage. (page 288)

greenhouse effect   A process in which the increased release of carbon dioxide and other gases into the atmosphere, caused by industrial activity and deforestation, permits solar short-wave heat radiation to reach the Earth's surface but blocks long-wave outgoing radiation, causing a thermal imbalance and global heating. (page 337)

Greens   Activists and organizations, including political parties, whose central concern is addressing environmental deterioration. (page 339)

guild industry   A traditional type of manufacturing in the pre-industrial revolution era, involving handmade goods of high quality manufactured by highly skilled artisans who resided in towns and cities. (page 322)

hamlet   A small rural settlement, smaller than a village. (page 307)

heartland   The interior of a sizable landmass, removed from maritime connections; in particular, the interior of the Eurasian continent. (page 217)

heartland theory   A 1904 proposal by Halford Mackinder that the key to world conquest lay in control of the interior of Eurasia. (page 217)

heat island   An area of warmer temperatures at the center of a city, caused by the urban concentration of heat-retaining concrete, brick, and asphalt. (page 406)

hierarchical diffusion   a type of expansion diffusion in which innovations spread from one important person to another or from one urban center to another, temporarily bypassing other persons or rural areas. (page 11)

high-tech corridor   An area along a limited-access highway that houses offices and other services associated with high-tech industries. (pages 328, 421)

hinterland   The area surrounding a city and influenced by it. (page 356)

homelessness   A temporary or permanent condition of not having a legal home address. (page 388)

human geography   The study of the relationships between people and the places and spaces in which they live. (page 2)

hunting and gathering   The killing of wild game and the harvesting of wild plants to provide food in traditional cultures. (page 285)

hydraulic civilization   A civilization based on large-scale irrigation. (page 358)

independent invention   A cultural innovation that is developed in two or more locations by individuals or groups working independently. (page 10)

indigenous culture   A culture group that constitutes the original inhabitants of a territory, distinct from the dominant national culture, which is often derived from colonial occupation. (page 33)

indigenous technical knowledge (ITK)   Highly localized knowledge about environmental conditions and sustainable land-use practices. (pages 52, 287)

infant mortality rate   The number of infants per 1000 live births who die before reaching one year of age. (page 88)

in-filling   New building on empty parcels of land within a checkerboard pattern of development. (page 397)

intensive agriculture   The expenditure of much labor and capital on a piece of land to increase its productivity. In contrast, *extensive* agriculture involves less labor and capital. (page 302)

intercropping   The practice of growing two or more different types of crops in the same field at the same time. (page 275)

involuntary migration   Also called forced migration, refers to the forced displacement of a population, whether by government policy (such as a resettlement program), warfare or other violence, ethnic cleansing, disease, natural disaster, or enslavement. (page 166)

isogloss   The border of usage of an individual word or pronunciation. (page 125)

Kurgan hypothesis   A theory of language diffusion holding that the spread of Indo-European languages originated with animal domestication in the central Asian steppes and grew more aggressively and swiftly than proponents of the Anatolian hypothesis maintain. (page 122)

labor-intensive industry   An industry for which labor costs represent a large proportion of total production costs. (page 329)

land-division patterns   A term that refers to the spatial patterns of different land uses. (page 24)

language   A mutually agreed-on system of symbolic communication that has a spoken and usually a written expression. (page 115)

language family   A group of related languages derived from a common ancestor. (page 117)

language hotspots   Those places on Earth that are home to the most unique, misunderstood, or endangered languages. (page 133)

lateral commuting   Traveling from one suburb to another in going from home to work. (page 395)

legible city   A city that is easy to decipher, with clear pathways, edges, nodes, districts, and landmarks. (page 410)

leisure landscapes   Landscapes that are planned and designed primarily for entertainment purposes, such as ski and beach resorts. (page 60)

lingua franca   An existing, well-established language of communication and commerce used widely where it is not a mother tongue. (page 117)

linguistic refuge area   An area protected by isolation or inhospitable environmental conditions in which a language or dialect has survived. (page 136)

livestock fattening   A commercial type of agriculture that produces fattened cattle and hogs for meat. (page 275)

local consumption cultures   Distinct consumption practices and preferences in food, clothing, music, and so forth formed in specific places and historical moments. (page 48)

Malthusian   Those who hold the views of Thomas Malthus, who believed that overpopulation is the root cause of poverty, illness, and warfare. (page 100)

marchland   A strip of territory, traditionally one day's march for infantry, that served as a boundary zone for independent countries in premodern times. (page 193)

mariculture   A branch of aquaculture specific to the cultivation of marine organisms, often involving the transformation of coastal environments and the production of distinctive new landscapes. (page 284)

market gardening   Farming devoted to specialized fruit, vegetable, or vine crops for sale rather than consumption. (page 279)

market   The geographical area in which a product may be sold in a volume and at a price profitable to the manufacturer. (page 330)

master-planned communities   Large-scale residential developments that include, in addition to architecturally compatible housing units, planned recreational facilities, schools, and security measures. (page 422)

material culture   All physical, tangible objects made and used by members of a cultural group, such as clothing, buildings, tools and utensils, instruments, furniture, and artwork; the visible aspect of culture. (page 32)

mechanistic view of nature   The view that humans are separate from nature and hold dominion over it and that the habitat is an integrated mechanism governed by external forces that the human mind can understand and manipulate. (page 21)

megacities   A term that refers to particularly large urban centers. (page 354)

megalopolis   A large urban region formed as several urban areas spread and merge, such as Boswash, the region including Boston, New York, and Washington, D.C. (page 419)

migrant workers   Most broadly, this term refers to people working outside of their home country. Migrant workers are particularly critical to large-scale commercial agriculture. (page 289)

migrations   The large-scale movements of people between different regions of the world. (page 13)

mobility   The relative ability of people, ideas, or things to move freely through space. (page 10)

model   An abstraction, an imaginary situation, proposed by geographers to simulate laboratory conditions so that they can isolate certain causal forces for detailed study. (page 4)

monoculture   The raising of only one crop on a huge tract of land in agribusiness. (page 296)

monotheistic religion   The worship of only one god. (page 233)

multiple-nuclei model   A model that depicts a city growing from several separate focal points. (page 391)

nationalism   The sense of belonging to and self-identifying with a national culture. (page 195)

nation-state   An independent country dominated by a relatively homogeneous culture group. (page 206)

natural boundary   A political border that follows some feature of the natural environment, such as a river or mountain ridge. (page 194)

natural hazard   An inherent danger present in a given habitat, such as floods, hurricanes, volcanic eruptions, or earthquakes; often perceived differently by different peoples. (page 20)

nature-culture   A term that refers to the complex relationships between people and the physical environment, including how culture, politics, and economies affect people's ecological situation and resource use. (page 16)

neighborhood   A small social area within a city where residents share values and concerns and interact with one another on a daily basis. (page 387)

neolocalism   The desire to reembrace the uniqueness and authenticity of place, in response to globalization. (page 451)

neo-Malthusians   Modern-day followers of Thomas Malthus. (page 101)

node   A central point in a functional culture region where functions are coordinated and directed. (page 8)

nomadic livestock herder   A member of a group that continually moves with its livestock in search of forage for its animals. (page 282)

nonmaterial culture   The wide range of tales, songs, lore, beliefs, values, and customs that pass from generation to generation as part of an oral or written tradition. (page 32)

nucleation   A relatively dense settlement form. (page 24)

office park   A cluster of office buildings, usually located along an interstate, often forming the nucleus of an edge city. (page 421)

organic agriculture   A form of farming that relies on manuring, mulching, and biological pest control and rejects the use of synthetic fertilizers, insecticides, herbicides, and genetically modified crops. (page 304)

organic view of nature   The view that humans are part of, not separate from, nature and that the habitat possesses a soul and is filled with nature-spirits. (page 21)

orthodox religions   Strands within most major religions that emphasize purity of faith and are not open to blending with other religions. (page 234)

outsource   The physical separation of some economic activities from the main production facility, usually for the purpose of employing cheaper labor. (page 329)

paddy rice farming   The cultivation of rice on a paddy, or small flooded field enclosed by mud dikes, practiced in the humid areas of the Far East. (page 276)

palimpsest   A term used to describe cultural landscapes with various layers and historical "messages." Geographers use this term to reinforce the notion of the landscape as a text that can be read; a landscape palimpsest has elements of both modern and past periods. (page 408)

peasant   A farmer belonging to a folk culture and practicing a traditional system of agriculture. (page 277)

permeable barrier A barrier that permits some aspects of an innovation to diffuse through it but weakens and retards continued spread; an innovation can be modified in passing through a permeable barrier. (page 12)

physical environment All aspects of the natural physical surroundings, such as climate, terrain, soils, vegetation, and wildlife. (page 2)

pidgin A composite language consisting of a small vocabulary borrowed from the linguistic groups involved in commerce. (page 117)

pilgrimages Journeys to places of religious importance. (page 249)

place A term used to connote the subjective, idiographic, humanistic, culturally oriented type of geography that seeks to understand the unique character of individual regions and places, rejecting the principles of science as flawed and unknowingly biased. (page 5)

placelessness A spatial standardization that diminishes regional variety; may result from the spread of popular culture, which can diminish or destroy the uniqueness of place through cultural standardization on a national or even worldwide scale. (page 35)

plantation A large landholding devoted to specialized production of a tropical cash crop. (page 278)

plantation agriculture A system of monoculture for producing export crops requiring relatively large amounts of land and capital; originally dependent on slave labor. (page 278)

political geography The geographic study of politics and political matters. (page 189)

polyglot A mixture of different languages. (page 121)

polytheistic religion The worship of many gods. (page 233)

popular culture A dynamic culture based in large, heterogeneous societies permitting considerable individualism, innovation, and change; having a money-based economy, division of labor into professions, secular institutions of control, and weak interpersonal ties; producing and consuming machine-made goods. (page 33)

population density A measurement of population per unit area (for example, per square mile). (page 72)

population explosion The rapid, accelerating increase in world population since about 1650 and especially since about 1900. (page 99)

population geography The study of the spatial and ecological aspects of population, including distribution, density per unit of land area, fertility, gender, health, age, mortality, and migration. (page 71)

population pyramid A graph used to show the age and sex composition of a population. (page 84)

possibilism A school of thought based on the belief that humans, rather than the physical environment, are the primary active force; that any environment offers a number of different possible ways for a culture to develop; and that the choices among these possibilities are guided by cultural heritage. (page 8)

postindustrial phase The phase of a society characterized by the dominance of the service sector of economic activity. (page 325)

primary industry An industry engaged in the extraction of natural resources, such as agriculture, lumbering, and mining. (page 321)

primate city A city of large size and dominant power within a country. (page 354)

producer services The range of economic activities required by producers of goods. (page 327)

proselytic religions Religions that actively seek new members and aim to convert all humankind. (page 232)

proxemics The study of the size and shape of people's envelopes of personal space. (page 112)

push-and-pull factors Unfavorable, repelling conditions (push factors) and favorable, attractive conditions (pull factors) that interact to affect migration and other elements of diffusion. (page 96)

race A classification system that is sometimes understood as arising from genetically significant differences among human populations, or from visible differences in human physiognomy, or as a social construction that varies across time and space. (page 150)

ranching The commercial raising of herd livestock on a large landholding. (page 283)

range In central-place theory, the average maximum distance people will travel to purchase a good or service. (page 356)

redlining A practice by banks and mortgage companies of demarcating areas considered to be high risk for housing loans. (page 396)

refugees Those fleeing from persecution in their country of nationality. The persecution can be religious, political, racial, or ethnic. (page 96)

region A grouping of like places or the functional union of places to form a spatial unit. (page 1)

regional trading blocs Agreements made among geographically proximate countries that reduce trade barriers in order to better compete with other regional markets. (page 197)

relic boundary   A former political border that no longer functions as a boundary. (page 194)

religion   A social system involving a set of beliefs and practices through which people seek harmony with the universe and attempt to influence the forces of nature, life, and death. (page 231)

relocation diffusion   The spread of an innovation or other element of culture that occurs with the bodily relocation (migration) of the individual or group responsible for the innovation. (pages 10, 287)

restrictive covenant   A statement written into a property deed that restricts the use of the land in some way; often used to prohibit certain groups of people from buying property. (page 396)

return migration   A type of ethnic diffusion that involves the voluntary movement of a group of migrants back to their ancestral or native country or homeland. (page 166)

rimland   The maritime fringe of a country or continent; in particular, the western, southern, and eastern edges of the Eurasian continent. (page 217)

sacred spaces   Areas recognized by a religious group as worthy of devotion, loyalty, esteem, or fear to the extent that they become sought out, avoided, inaccessible to nonbelievers, and/or removed from economic use. (page 266)

satellite state   A small, weak country dominated by one powerful neighbor to the extent that some or much of its independence is lost. (page 194)

secondary industry   An industry engaged in processing raw materials into finished products; manufacturing. (page 321)

sector model   An economic model that depicts a city as a series of pie-shaped wedges. (page 391)

sedentary cultivation   Farming in fixed and permanent fields. (page 282)

services   The range of economic activities that provide services to industry. (page 321)

settlement forms   The spatial arrangement of buildings, roads, towns, and other features that people construct while inhabiting an area. (page 24)

sex ratio   The numerical ratio of males to females in a population. (page 85)

site   The local setting of a city. (page 366)

situation   The regional setting of a city. (page 366)

slang   Words and phrases that are not part of a standard, recognized vocabulary for a given language but that are nonetheless used and understood by some of its speakers. (page 126)

social culture region   An area in a city where many of the residents share social traits such as income, education, and stage of life. (page 386)

sovereignty   The right of individual states to control political and economic affairs within their territorial boundaries without external interference. (page 189)

space   A term used to connote the objective, quantitative, theoretical, model-based, economics-oriented type of geography that seeks to understand spatial systems and networks through application of the principles of social science. (page 4)

squatter settlements   Illegal housing settlements, usually made up of temporary shelters, that surround a large city. (page 371)

state   A centralized authority that enforces a single political, economic, and legal system within its territorial boundaries. Often used synonymously with "country." (page 189)

stimulus diffusion   A type of expansion diffusion in which a specific trait fails to spread but the underlying idea or concept is accepted. (page 12)

subcultures   Groups of people with norms, values, and material practices that differentiate them from the dominant culture to which they belong. (page 31)

subsistence agriculture   Farming to supply the minimum food and materials necessary to survive. (page 276)

subsistence economies   Economies in which people seek to consume only what they produce and to produce only for local consumption rather than for exchange or export. (page 52)

suitcase farm   In American commercial grain agriculture, a farm on which no one lives; planting and harvesting are done by hired migratory crews. (page 281)

supranational organization   A group of independent countries joined together for purposes of mutual interest. (page 197)

supranationalism   A situation that occurs when states willingly relinquish some degree of sovereignty in order to gain the benefits of belonging to a larger political-economic entity. (page 197)

survey pattern   A pattern of original land survey in an area. (page 307)

sustainability   The survival of a land-use system for centuries or millennia without destruction of the environmental base, allowing generation after generation to continue to live there. (page 301)

swidden cultivation   A type of agriculture characterized by land rotation in which temporary clearings are used for several years and then abandoned to be replaced by new clearings; also known as slash-and-burn agriculture. (page 275)

symbolic landscapes   Landscapes that express the values, beliefs, and meanings of a particular culture. (page 24)

syncretic religions   Religions, or strands within religions, that combine elements of two or more belief systems. (page 233)

technopole   A center of high-tech manufacturing and information-based industry. (page 325)

territoriality   A learned cultural response, rooted in European history, that produced the external bounding and internal territorial organization characteristic of modern states. (page 189)

threshold   In central-place theory, the size of the population required to make provision of goods and services economically feasible. (page 356)

time-distance decay   The decrease in acceptance of a cultural innovation with increasing time and distance from its origin. (page 12)

toponym   A place-name, usually consisting of two parts, the generic and the specific. (page 139)

total fertility rate (TFR)   The number of children the average woman will bear during her reproductive lifetime (15–44 or 15–49 years of age). A TFR of less than 2.1, if maintained, will cause a natural decline of population. (page 75)

trade-route sites   Locations for a city at a significant point on a transportation route. (page 368)

transnational corporations   Companies that have international production, marketing, and management facilities. (page 325)

transnational migrations   The movements of groups of people who maintain ties to their homelands after they have migrated. (page 13)

transportation/communication services   The range of economic activities that provide transport and communication to businesses. (page 327)

uneven development   The tendency for industry to develop in a core-periphery pattern, enriching the industrialized countries of the core and impoverishing the less industrialized periphery. This term is also used to describe urban patterns in which suburban areas are enriched while the inner city is impoverished. (pages 15, 324, 395)

unitary state   An independent state that concentrates power in the central government and grants little authority to the provinces. (page 194)

universalizing religions   Also called proselytic religions, they expand through active conversion of new members and aim to encompass all of humankind. (page 232)

urban agriculture   The raising of food, including fruit, vegetables, meat, and milk, inside cities, especially common in the Third World. (page 283)

urban hearth areas   Regions in which the world's first cities evolved. (page 359)

urban morphology   The form and structure of cities, including street patterns and the size and shape of buildings. (page 408)

urbanized population   The proportion of a country's population living in cities. (page 351)

vernacular (culture) region   A (culture) region perceived to exist by its inhabitants, based in the collective spatial perception of the population at large, and bearing a generally accepted name or nickname (such as "Dixie"). (pages 8, 41)

zero population growth   A stabilized population created when an average of only two children per couple survive to adulthood, so that, eventually, the number of deaths equals the number of births. (page 75)

INDEX

Note: Page numbers followed by f indicate figures; those followed by t indicate tables.